Nanomaterials: Green Synthesis, Characterization and Applications

Nanomaterials: Green Synthesis, Characterization and Applications

Editor: Ollie Moore

NY RESEARCH
P R E S S

New York

Published by NY Research Press
118-35 Queens Blvd., Suite 400,
Forest Hills, NY 11375, USA
www.nyresearchpress.com

Nanomaterials: Green Synthesis, Characterization and Applications
Edited by Ollie Moore

International Standard Book Number: 978-1-64725-437-7 (Hardback)

Cataloging-in-Publication Data

Nanomaterials : green synthesis, characterization and applications / edited by Ollie Moore.
    p. cm.
Includes bibliographical references and index.
ISBN 978-1-64725-437-7
1. Nanostructured materials. 2. Nanotechnology. 3. Microstructure. I. Moore, Ollie.
TA418.9.N35 N36 2023
620.115--dc23

Contents

Preface...VII

Chapter 1 **Green and Traditional Synthesis of Copper Oxide
Nanoparticles**...1
Obakeng P. Keabadile, Adeyemi O. Aremu,
Saheed E. Elugoke and Omolola E. Fayemi

Chapter 2 **Flower-Based Green Synthesis of Metallic Nanoparticles: Applications
beyond Fragrance**...20
Harsh Kumar, Kanchan Bhardwaj, Kamil Kuča, Anu Kalia,
Eugenie Nepovimova, Rachna Verma and Dinesh Kumar

Chapter 3 **Green Synthesis of Gold and Silver Nanoparticles using Leaf Extract of
Clerodendrum inerme; Characterization, Antimicrobial and
Antioxidant Activities**...36
Shakeel Ahmad Khan, Sammia Shahid and Chun-Sing Lee

Chapter 4 **Facile and Robust Solvothermal Synthesis of Nanocrystalline
CuInS$_2$ Thin Films**...61
Anna Frank, Jan Grunwald, Benjamin Breitbach and Christina Scheu

Chapter 5 **Green Synthesis of High Temperature Stable Anatase Titanium Dioxide
Nanoparticles using Gum Kondagogu: Characterization and
Solar Driven Photocatalytic Degradation of Organic Dye**.......................79
Kothaplamoottil Sivan Saranya, Vinod Vellora Thekkae Padil, Chandra Senan,
Rajendra Pilankatta, Kunjumon Saranya, Bini George,
Stanisław Wacławek and Miroslav Černík

Chapter 6 **Development of Effective Lipase-Hybrid Nanoflowers Enriched with
Carbon and Magnetic Nanomaterials for Biocatalytic Transformations**.......98
Renia Fotiadou, Michaela Patila, Mohamed Amen Hammami,
Apostolos Enotiadis, Dimitrios Moschovas, Kyriaki Tsirka,
Konstantinos Spyrou, Emmanuel P. Giannelis, Apostolos Avgeropoulos,
Alkiviadis Paipetis, Dimitrios Gournis and Haralambos Stamatis

Chapter 7 **Ionic Nanocomplexes of Hyaluronic Acid and Polyarginine to
form Solid Materials**...115
María Gabriela Villamizar-Sarmiento, Ignacio Moreno-Villoslada,
Samuel Martínez, Annesi Giacaman, Victor Miranda, Alejandra Vidal,
Sandra L. Orellana, Miguel Concha, Francisca Pavicic, Judit G. Lisoni,
Lisette Leyton and Felipe A. Oyarzun-Ampuero

Chapter 8 **"Chocolate" Gold Nanoparticles — One Pot Synthesis and Biocompatibility**.......129
Neelika Roy Chowdhury, Allison J. Cowin, Peter Zilm and Krasimir Vasilev

Chapter 9   **Effects of Sample Preparation on Particle Size Distributions of
               Different Types of Silica in Suspensions**...**139**
               Rodrigo R. Retamal Marín, Frank Babick, Gottlieb-Georg Lindner,
               Martin Wiemann and Michael Stintz

Chapter 10   **Green Micro- and Nanoemulsions for Managing Parasites,
               Vectors and Pests**..**157**
               Lucia Pavoni, Roman Pavela, Marco Cespi, Giulia Bonacucina,
               Filippo Maggi, Valeria Zeni, Angelo Canale, Andrea Lucchi,
               Fabrizio Bruschi and Giovanni Benelli

Chapter 11   **Environmentally-Friendly Green Approach for the Production of
               Zinc Oxide Nanoparticles and their Anti-Fungal, Ovicidal and
               Larvicidal Properties**...**188**
               Naif Abdullah Al-Dhabi and Mariadhas Valan Arasu

Chapter 12   **Elucidating the Chemistry behind the Reduction of Graphene Oxide
               using a Green Approach with Polydopamine**...**201**
               Cláudia Silva, Frank Simon, Peter Friedel,
               Petra Pötschke and Cordelia Zimmerer

Chapter 13   **A Polyol-Mediated Fluoride Ions Slow-Releasing Strategy for
               the Phase-Controlled Synthesis of Photofunctional Mesocrystals**.......................**219**
               Xianghong He, Yaheng Zhang, Yu Fu, Ning Lian and Zhongchun Li

Chapter 14   **Eco-Friendly Method for Tailoring Biocompatible and Antimicrobial
               Surfaces of Poly-L-Lactic Acid**...**229**
               Magdalena Aflori, Maria Butnaru,
               Bianca-Maria Tihauan and Florica Doroftei

               Permissions

               List of Contributors

               Index

Preface

Every book is a source of knowledge and this one is no exception. The idea that led to the conceptualization of this book was the fact that the world is advancing rapidly; which makes it crucial to document the progress in every field. I am aware that a lot of data is already available, yet, there is a lot more to learn. Hence, I accepted the responsibility of editing this book and contributing my knowledge to the community.

The synthesis of nanoparticles can be done using biological, physical and chemical pathways. Green synthesis of nanomaterials refers to the synthesis of diverse metal nanoparticles utilizing bioactive agents. These include microorganisms, plant materials, and several biowastes such as eggshells, vegetable waste, agricultural waste, and fruit peel trash. Green synthesis of nanoparticles is aimed to reduce waste and develop sustainable techniques. It is a bottom-up strategy in which nanoparticles are synthesized by oxidation/reduction of metallic ions through organic moieties derived from biological resources. Microorganisms such as actinomycetes, fungi, algae, bacteria and yeast may be found in these biological systems. This book elucidates the concepts and innovative models around prospective developments with respect to the green synthesis of nanomaterials, their characterization and applications. It covers in detail some existent theories and innovative concepts revolving around this domain. This book is a vital tool for all researching and studying the green synthesis of nanomaterials.

While editing this book, I had multiple visions for it. Then I finally narrowed down to make every chapter a sole standing text explaining a particular topic, so that they can be used independently. However, the umbrella subject sinews them into a common theme. This makes the book a unique platform of knowledge.

I would like to give the major credit of this book to the experts from every corner of the world, who took the time to share their expertise with us. Also, I owe the completion of this book to the never-ending support of my family, who supported me throughout the project.

Editor

Green and Traditional Synthesis of Copper Oxide Nanoparticles

Obakeng P. Keabadile [1,2], Adeyemi O. Aremu [3], Saheed E. Elugoke [1,2] and Omolola E. Fayemi [1,2,*]

1   Department of Chemistry, Faculty of Natural and Agricultural Sciences, North-West University (Mafikeng Campus), Private Bag X2046, Mmabatho 2735, South Africa; 29582172@student.g.nwu.ac.za (O.P.K.); elugokesaheed@gmail.com (S.E.E.)
2   Material Science Innovation and Modelling (MaSIM) Research Focus Area, Faculty of Natural and Agricultural Sciences, North-West University (Mafikeng Campus), Private Bag X2046, Mmabatho 2735, South Africa
3   Indigenous Knowledge Systems Centre, Faculty of Natural and Agricultural Sciences, North-West University (Mafikeng Campus), Private Bag X2046, Mmabatho 2735, South Africa; Oladapo.Aremu@nwu.ac.za
*   Correspondence: Omolola.Fayemi@nwu.ac.za.

Abstract: The current study compared the synthesis, characterization and properties of copper oxide nanoparticles (CuO) based on green and traditional chemical methods. The synthesized CuO were confirmed by spectroscopic and morphological characterization such as ultraviolet-visible (UV-vis) spectroscopy, fourier transform infrared (FTIR) spectroscopy, zeta potential, scanning electron microscopy (SEM) and energy dispersed X-ray (EDX). Electrochemical behavior of the modified electrodes was done using cyclic voltammetry (CV) in ferricyanide/ferrocyanide ($[Fe(CN)_6]^{4-}$/$[Fe(CN)_6]^{3-}$) redox probe. As revealed by UV spectrophotometer, the absorption peaks ranged from 290–293 nm for all synthesized nanoparticles. Based on SEM images, CuO were spherical in shape with agglomerated particles. Zeta potential revealed that the green CuO have more negative surface charge than the chemically synthesized CuO. The potential of the green synthesized nanoparticles was higher relative to the chemically synthesized one. Cyclic voltammetry studies indicated that the traditional chemically synthesized CuO and the green CuO have electrocatalytic activity towards the ferricyanide redox probe. This suggests that the green CuO can be modified with other nanomaterials for the preparation of electrochemical sensors towards analytes of interest.

Keywords: green copper oxide; nanoparticles; spectroscopy; cyclic voltammetry; terminalia phanerophlebia

1. Introduction

Nanotechnology generally involves the application of extremely small particles that are used across all fields of science including chemistry, biology, medicine and material science [1–4]. According to Sathiyavimal et al. [5], nanotechnology deals with the synthesis of metal and metal oxide nanoparticles of different sizes, shapes, disparity and chemical composition. Nanoparticles, which by definition are the clusters of atoms in the size range of 1–100 nm are the major building blocks of nanotechnology [6,7].

There are two methods of synthesizing nanoparticles, which are the chemical and physical methods [8]. The chemical method of synthesis includes chemical reduction, electrochemical techniques and photochemical reduction [9,10]. The classical chemical method in which a reducing agent such as sodium borohydride and hydrazine [6] as well as radiation chemical method generated by ionization radiation are also used for chemical synthesis of nanoparticles. Even though most chemical methods successfully produce pure and well-defined nanoparticles, they are expensive, inefficient and can release harmful wastes to the environment hence a better eco-friendly methods are preferred [10]. On the other hand, the chemical method of nanoparticles synthesis give room for the use of desired reducing agents with lower economic implication compared to the physical method of synthesis such as the laser ablation technique which attracts extra cost of the laser system procurement.

The physical methods of synthesis include condensation, evaporation and laser ablation [8]. Green synthesis method (plant-mediated synthesis of nanoparticles) is established as an alternative to physical and chemical methods. It is simple, rapid process that involve the use of less toxic and environmentally-friendly materials as compared to other methods of synthesis [7,11]. Green synthesis mitigates environmental problems such as solar conversation, catalysis and agricultural production [12]. According to Zhang et al. [13], the plant material in the green synthesis often have a vital role in the size and surface morphology of the nanoparticle synthesized. The plant extract can act as both the reducing and capping agent. In terms of size of the nanoparticles formed, the products of green synthesis are larger compared to the ones obtained by chemical methods [8]. In addition, nanomaterials produced from the green source have high biocompatibility which contributes to their biofunctionality [14]. It has also been reported that the green synthetic route produces higher yield of nanoparticles than the chemical means [15].

Nanoparticles of a wide range of elements such as copper (Cu) [16], gold (Au) [17], silver (Ag) [18], titanium (Ti) [19], zinc (Zn) [20], silicon (Si) [21] and palladium (Pd) [22] and their oxides are abound in literature. Specifically, copper is a good conductor of heat or electricity and it is cheaper than silver and gold [23]. Copper oxide nanoparticles (CuO) are used as water purifiers, antimicrobials and are also used in batteries, gas sensors, catalysts for different cross-coupling reactions and high temperature superconductors [7,24–27]. CuO have a high surface-to-volume ratio that makes them very reactive and interact easily with other materials [28]. Moreover, CuO have moderate band gap, good catalytic activity and high optical transparency which further extend their industrial applications [29,30].

Green synthesis of CuO with good physico-chemical properties have been accomplished with the use of microbial precursors such as *Penicillium aurantiogriseum*, *Penicillium citrinum*, *Penicillium waksmanii* [31] and *Seratia* sp. [32] as reductants. Likewise, leaf extracts from plants such as *Aloe vera* [33], *Psidium guajava* [34], *Abutilon indicum* [35], *Malva sylvetris* [36], *Carica papaya* [37], *Camellia sinensis* [38] and *Ixiro coccinea* [39] have been used for the preparation of CuO. Interestingly, most of the CuO from these green sources have demonstrated noteworthy antimicrobial activity, particle sizes below 100 nm and photocatalytic activity towards the degradation of contaminants were investigated. For instance, Ijaz et al. [33] prepared CuO from the leaf extract of *Abutilon indicum* and copper (II) nitrate trihydrate. The resultant nanoparticles had noteworthy antimicrobial activity against *Escherichia coli*, *Staphyloccus aureus* and *Bacillus subtilis*, excellent oxidation properties towards linolenic acid as well as good photocatalytic effect on the degradation of Acid Black 210 organic dye.

These precedents informed our interest in the preparation of CuO using *Terminalia phanerophlebia* leaf extracts. Green CuO were prepared from the leaf extract of *Terminalia phanerophlebia* obtained using ethanol (eCuO), deionized water (wCuO) and acetone (aCuO) as solvent. We also synthesized CuO through the traditional chemical means (cCuO) for a comparative study with the green CuO. Precisely, comparative analysis of the spectroscopic, morphological and electrochemical properties of the two nanoparticles were investigated. To the best of our knowledge, this is the first time a CuO prepared from this green precursor will be used in a comparative analysis with the chemical counterpart.

2. Materials and Methods

2.1. Reagents and Materials

Copper sulphate pentahydrate ($CuSO_4.5H_2O$) (99.5%), sodium borohydride ($NaBH_4$) (95%), ethanol and acetone were purchased from Sigma-Aldrich, Gauteng, 1600, South-Africa. All reagents were of analytical grade. Leaves of Terminalia phanerophlebia (Family: Combretaceae) were collected from the Botanical Garden, University of KwaZulu-Natal (Pietermaritzburg campus), South Africa and were positively identified by the curator of the Bews Herbarium, University of KwaZulu-Natal, South Africa.

2.2. Preparation of Plants Extracts

Extract from the oven-dried leaves of *Terminalia phanerophlebia* was prepared by weighing 2 g of the ground powder and dissolved in 150 mL deionized water, ethanol and acetone, respectively, as previously reported in literature. The three extracts were filtered and used for the synthesis of nanoparticles.

2.3. Synthesis of CuO

Traditional Chemical Synthesis of CuO

The traditional chemical synthesis was done according to a method reported by Khatoon et al. [6] Briefly, 60 mL of $NaBH_4$ was transferred into 250 mL Erlenmeyer flask which was subsequently placed on a hot plate with magnetic stirrer. The resultant solution was heated for 10 min while stirring at 80 °C. Thereafter, 50 mL of 0.02 M $CuSO_4.5H_2O$ was added dropwise to the flask and the final solution was heated for 7 min with continuous stirring. The flask was covered with aluminum foil after completion to avoid oxidation. The particles were allowed to settle leaving the solution colorless and copper nanoparticles with green color settled at the bottom of the flask. The solution was kept in a hot air oven for 24 h to convert the aqueous mixture to powder. The dried powder was collected, dissolved in ethanol and distilled water and centrifuged for purification. This purification was done thrice to remove all impurities and dried for further use. The XRD studies were done with a Rontgen PW3040/60 X'Pert Pro diffractometer by Bruker, Hamburg, Germany, from diffraction angles of 10–80 degrees at room temperature.

2.4. Green Synthesis of CuO

The green synthesis was carried out as reported by Sathiyavimal et al. [5] Briefly, 30 mL of 0.1 M $CuSO_4.5H_2O$ was added to 10 mL of the plant extract in a flask and then the solution was stirred and heated at 90 °C for 5 h. Color change of the solution (from yellowish green to brownish black) indicated the formation of CuO. The solution was kept overnight at room temperature. The CuO obtained were centrifuged and washed twice with distilled water. The nanoparticles were collected and dried in a hot air oven. Similar procedure was applied for the ethanol and acetone plant extract using temperatures of 68 °C and 46 °C, respectively. The purification of the nanoparticles was done with the approach used for the CuO nanoparticles through the traditional chemical synthetic route.

2.5. Characterization of CuO

CuO synthesized were characterized by using Uviline 9400UV-vis spectrophotometer (Sl Analytics, Mainz, Germany) at wavelengths ranging from 200 to 700 nm for UV-vis spectroscopy. Opus Alpha-P Fourier-transform infrared spectrophotometer (Brucker Corporation, Billerica, MA, USA) in the range of 4500–400 cm^{-1} was applied for the IR-spectroscopy. Quanta FEG 250 scanning electron microscope (ThermoFisher Scientific, USA) was used to investigate the morphology and the elemental composition via SEM and EDX analysis. In preparation for SEM analysis, clear suspension of the nanoparticles was made in ethanol and subsequently deposited on a mica solid substrate. The accelerating image, brightness, working distance and contrast of the microscope were adjusted to optimum level prior to imaging. Zeta analyzer was used to obtain the surface charge and the long-term stability of the synthesized nanoparticles. This was accomplished with the aid of a Malvern multi-purpose titrator instrument.

2.6. Electrochemical Studies

The cyclic voltammetry characterization was done using a potentiostat with a three electrode system. DropSense with inbuilt Dropview 200 software (Metrohm South Africa (Pty) Ltd., Johannesburg, South Africa) was fitted with screen printed carbon electrode (SPC) having inner carbon working electrode (diameter of 4 mm), a silver pseudo-reference (Ag/AgCl) electrode (RE) and carbon counter electrode (CE). SPC was modified with cCuO, wCuO, aCuO and eCuO to obtain the modified working electrode used for the electrochemical studies. SPC electrode was modified by dropping about 0.5 µL of the ultrasonicated CuO nanoparticles samples on the surface of the electrode and air dried for 20 min. Noteworthy, the ultrasonication of the nanoparticles was done at room temperature. Each of the modified working electrodes were characterized using the $Fe(CN)_6]^{4-}/[Fe(CN)_6]^{3-}$ (10 mM) redox probe in the presence of a phosphate buffer saline (PBS) (0.1 M, pH 7.0) supporting electrolyte. All cyclic voltammetry (CV) measurements were done at a scan rate of 50 mVs^{-1} over a potential window of 1.0–1.2 V.

3. Results and Discussion

3.1. Spectroscopic and Microscopic Characterization

3.1.1. UV-vis Spectroscopy Analysis

The UV absorption peak of CuO usually shows an absorption peak ranging from 280 nm to 360 nm [19]. As shown in Figure 1a, a UV absorption peak of 289 nm was recorded for CuO synthesized by a traditional chemical method. As an indication of the formation of different sized CuO, the UV absorption peaks observed for the CuO nanoparticles from green synthesis using water, acetone and ethanol as solvent were 291, 291 and 290 nm for wCuO, aCuO and eCuO, respectively (Figure 1b–d). These observed peaks are within the range reported in literature [25,37,39]. The extra broad peak at 410 nm is similar to the peak reported for green mediated CuO by Saif et al., 2016 [40]. The band gap calculated for wCuO, aCuO and eCuO from the UV-visible spectra were 1.25, 1.75 and 1.63 eV, respectively. These band gaps are higher than a value of 1.20 eV reported for CuO which suggest that the green CuO nanoparticles have higher conductivity than the traditional chemically synthesized CuO nanoparticles [41].

3.1.2. FT-IR Analysis

According to Prakash et al. [42], characteristic peaks of IR spectra of is in the range 400–650 cm^{-1}. In the FTIR spectra of CuO from the traditional chemical synthesis, Cu-O stretching was observed at 494 and 595 cm^{-1} (Figure 2a). The peaks observed at 3404 cm^{-1} is associated with OH stretching present as a result of the hydroxyl group on the surface of the CuO. The FTIR spectra of wCuO, aCuO and eCuO were presented in Figure 2b–d. In the spectra of wCuO, Cu-O stretching band was observed at 444 and 579 cm^{-1} and OH stretching at 3117 cm^{-1} (Figure 2b). Figure 2c shows a Cu-O stretching at 450 cm^{-1} and OH stretching at 3353 cm^{-1}. In Figure 2d, the absorption peak situated at 450 cm^{-1} corresponds to the Cu-O stretching and the one at 3353 cm^{-1} is due to the OH stretching vibration. Absorption peaks at 1696 cm^{-1} which appeared only in the spectra of wCuO (Figure 2b) is associated with the stretching vibrations of C=C of a water-soluble unsaturated component of the plant extract. As shown in Figure 2, the OH stretching band for a green method are broader when compared to that of the traditional chemical method. This could be due to the hydroxyl group present in the plant extract attached to the CuO.

3.1.3. Energy Dispersed X-ray and SEM Analysis

The dominance of copper and oxygen in all the plots confirms that the synthesized nanoparticles are indeed that of copper oxide (Figure 3. The presence or origin of sodium is a result of the reducing agent that was used for the chemical synthesis (Figure 3a). The percentage of Cu is lower in cCuO and also in eCuO. The standard deviation for the percentage composition of copper in the nanoparticles are 1.27, 4.87, 3.19 and 1.84 for cCuO, wCuO, eCuO and aCuO, respectively.

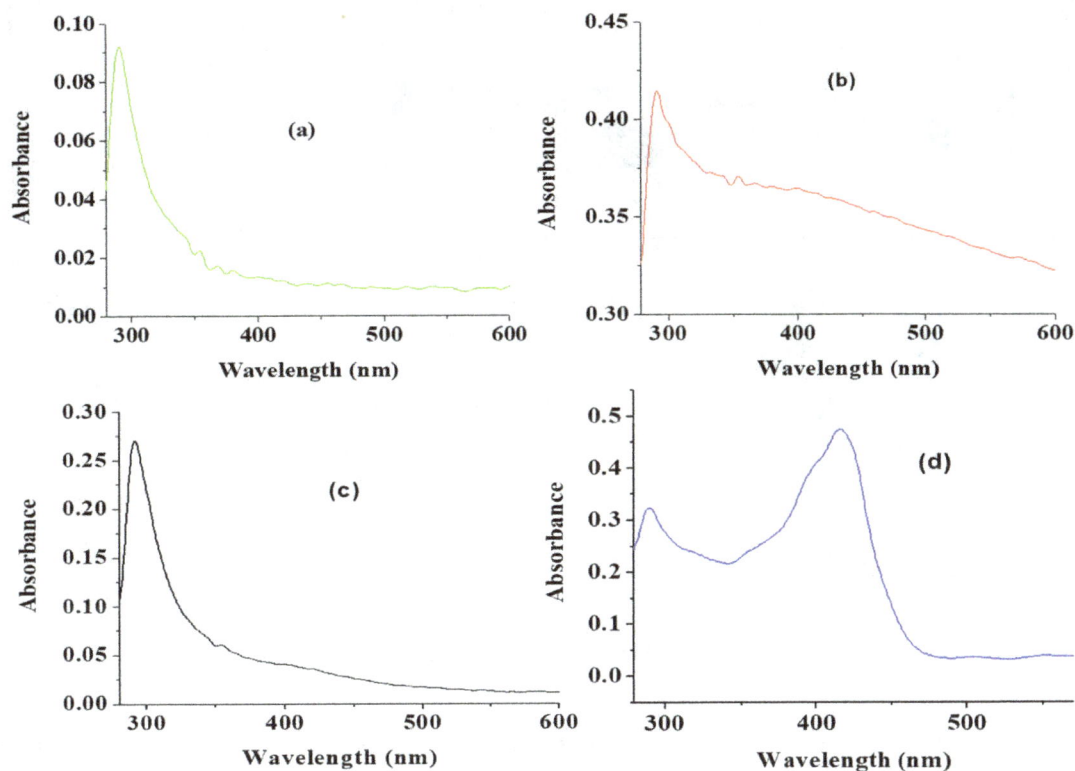

Figure 1. UV-vis spectra of (**a**) CuO obtained using traditional chemical means (cCuO), (**b**) CuO obtained using deionized water (wCuO), (**c**) CuO obtained using acetone (aCuO) and (**d**) CuO obtained using ethanol (eCuO).

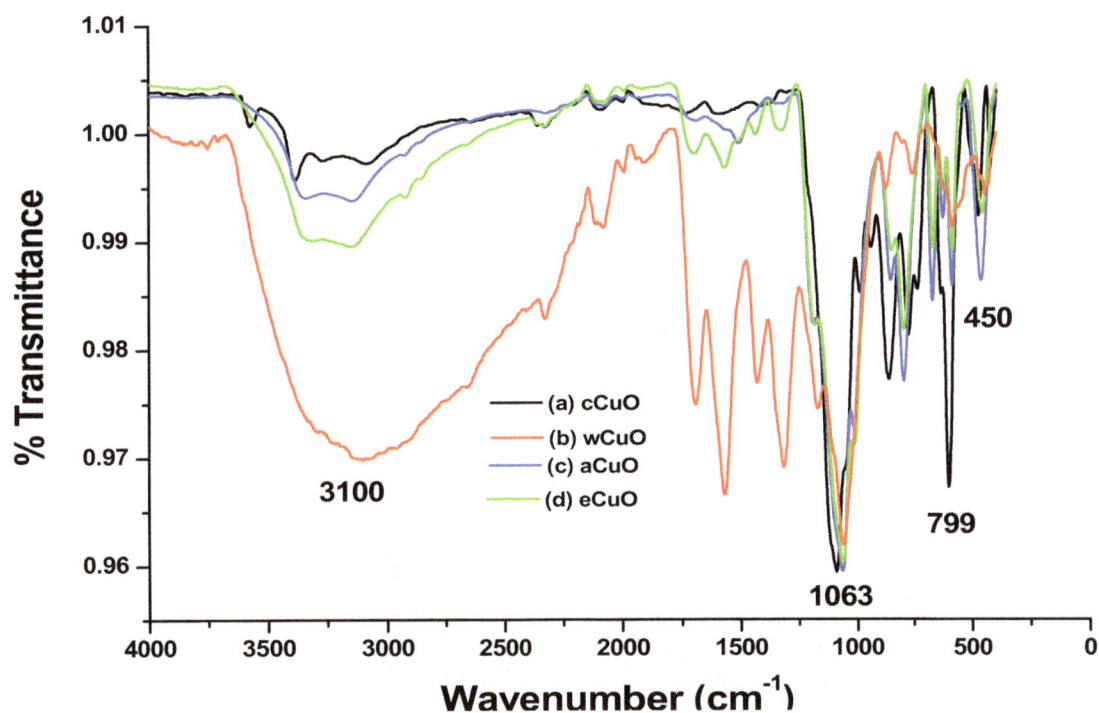

Figure 2. FTIR spectra of CuO from (**a**) cCuO, (**b**) wCuO, (**c**) aCuO and (**d**) eCuO.

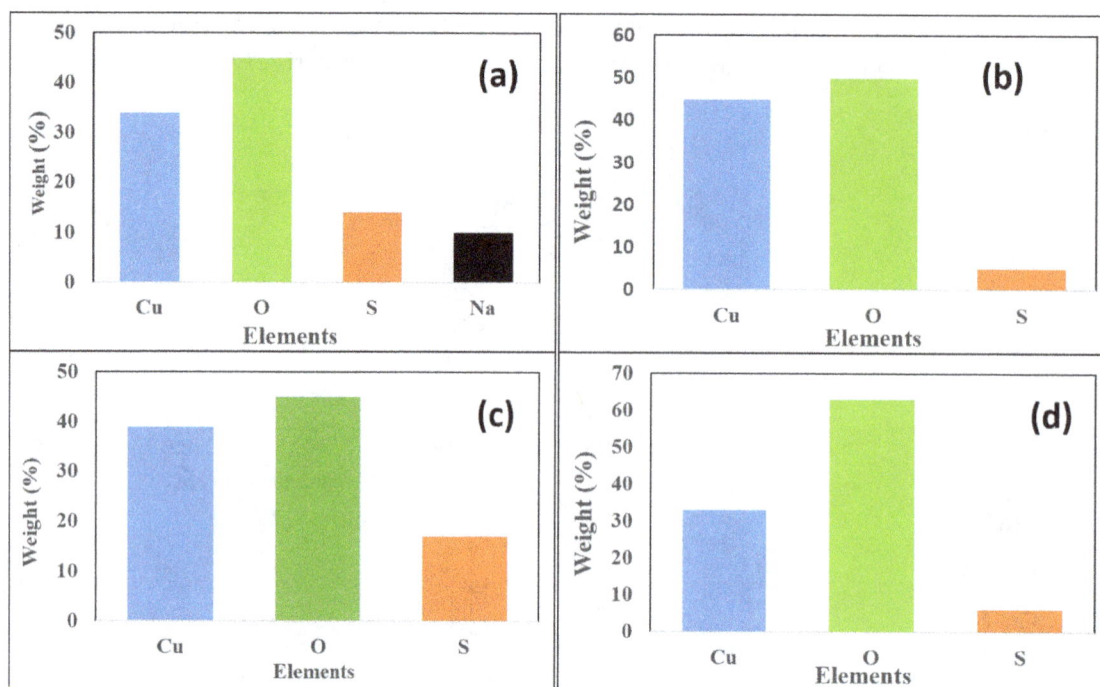

Figure 3. EDX plots of (**a**) cCuO, (**b**) wCuO, (**c**) aCuO and (**d**) eCuO

This confirms the action of ethanol as co-surfactant in the preparation of CuO from plant extract using ethanol as solvent as suggested by the SEM micrograph. Higher percentage of oxygen was evident in all the green mediated copper nanoparticles (Figure 4b–d) relative to the traditional chemical synthesis. The standard deviation for the percentage composition of oxygen in the nanoparticles are 0.87, 4.37, 2.0 and 1.84 for cCuO, wCuO, eCuO and aCuO, respectively. This can be attributed to the high oxygen content of the oxygen containing functionalities in the plant extract.

Figure 4. *Cont.*

Figure 4. SEM images of (**a**) cCuO, (**b**) wCuO, (**c**) aCuO and (**d**) eCuO.

3.1.4. X-ray Diffraction (XRD) Studies

The XRD diffractogram gives an insight to the crystalline structure of a compound. Crystallite size was calculated using X-ray diffraction pattern of CuO nanoparticles (Figure 5). Diffraction peaks for chemically synthesized cCuO (Figure 4a) were noticed at 2θ values of Brags angle for 22.85° (020), 28.00° (021), 30.58° (110), 33.42° (002) 35.67° (111), 41.43° (131), 52.45° (113) and 59.16° (200) and for aqueous extract of plant (Figure 5B) at 22.9° (020), 27.96° (021), 30.60° (110), 33.44° (002) 35.67° (111), 41.34° (131), 52.69° (113) and 60.1° (200) with their corresponding lattice/Miller indices in parentheses. XRD pattern of CuO synthesized from plants extract via acetone and methanol aCuO and eCuO were deficient of peaks (graphs not shown) which possibly could have been from impurities from the organic solvent used in the preparation of the extract. Inter planar d-spacing of cCuO and wCuO were calculated applying Bragg's law equation: [43].

$$2dSin\theta = n\lambda \tag{1}$$

where d is interplanar spacing, θ is Brag's angle of diffraction = 2θ/2, λ is X-ray wavelength (0.154 nm) and n = 1 as represented in Table 1.

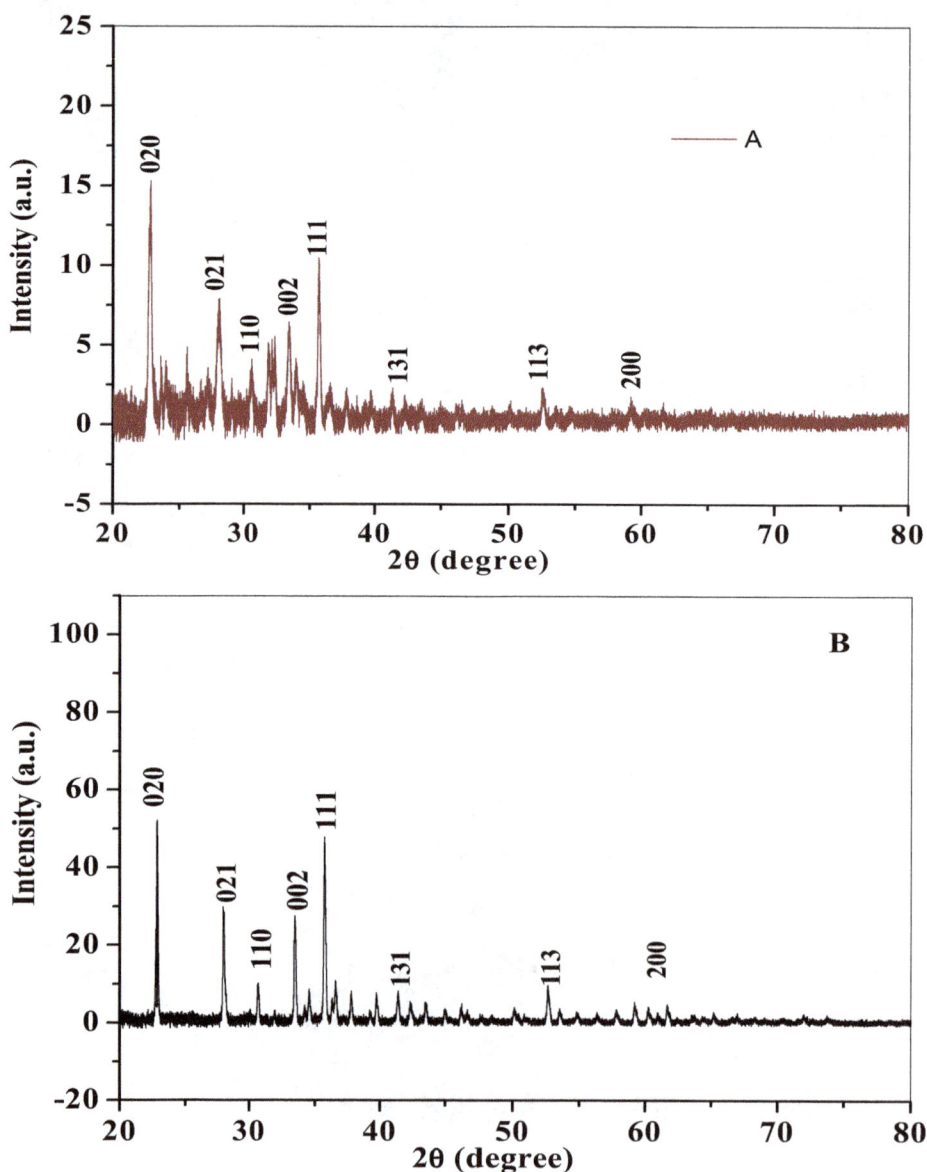

Figure 5. X-ray diffractograms of (**A**) cCuO and (**B**) wCuO.

Table 1. d-inter planar spacing calculations for CuONP.

Peaks (2θ)		θ		Sin θ		d (nm)	
cCuO	**wCuO**	**cCuO**	**wCuO**	**cCuO**	**wCuO**	**cCuO**	**wCuO**
22.85	22.91	11.42	11.45	0.1980	0.1985	0.3887	0.3879
28.00	27.96	14.00	13.98	0.2419	0.2415	0.3183	0.3187
30.58	30.60	15.29	15.30	0.2637	0.2638	0.2919	0.2637
33.42	33.44	16.71	16.72	0.2875	0.2876	0.2678	0.2875
35.66	35.67	16.83	17.83	0.2895	0.3061	0.2659	0.2895
41.43	41.34	20.71	20.67	0.3536	0.3529	0.2177	0.3536
52.45	52.69	26.22	26.35	0.4418	0.4438	0.1742	0.4420
59.16	60.18	29.58	30.05	0.4936	0.5013	0.1559	0.4939

Crystallite size calculated from the diffraction peak corresponding to most intense peak (θ/2) for cCuO (22.85°) and wCuO (22.9°) using Debye–Scherrer's formula were found to be 17.8 nm and 24 nm for cCuO and wCuO, respectively, which are close to reported values in the range of 14–25 nm. [44,45] Differences in the size could, possibly, be attributed to different route of synthesis, imperfect crystallization and strain.

3.2. Zeta Potential Studies

The size of the zeta potential gives information about the particle synthesized stability. For example, particles with high zeta potential exhibit increased stability due to larger electrostatic repulsion between particles [46]. While cCuO show a charge of −5.60 mV, wCuO, aCuO and eCuO had corresponding charges of −22.3, −16.5 and −15.6 mV, respectively (Figure 6). All the CuO synthesized had negative surface charges. As highlighted by Aparna et al., [47] negative zeta potential obtained for nanoparticles might be an indication that CuO are negative particles and moderately stable. The obtained zeta potentials for green synthesis method were about three times greater than that of traditional chemical method which suggests that CuO synthesized by green method are more stable than those from chemical method. Higher zeta potentials were obtained from the CuO prepared from wCuO (Figure 6b) and aCuO (Figure 6c) compared to the cCuO (Figure 6a) and eCuO (Figure 6d). This suggests lesser agglomeration of the particles which manifested in a better porosity as seen in Figure 4b–c. Zeta potential close to that of wCuO (−28.9 mV) has earlier been reported by Sankar et al. [37] for green CuO prepared from the leaf extract of *Carica papaya*.

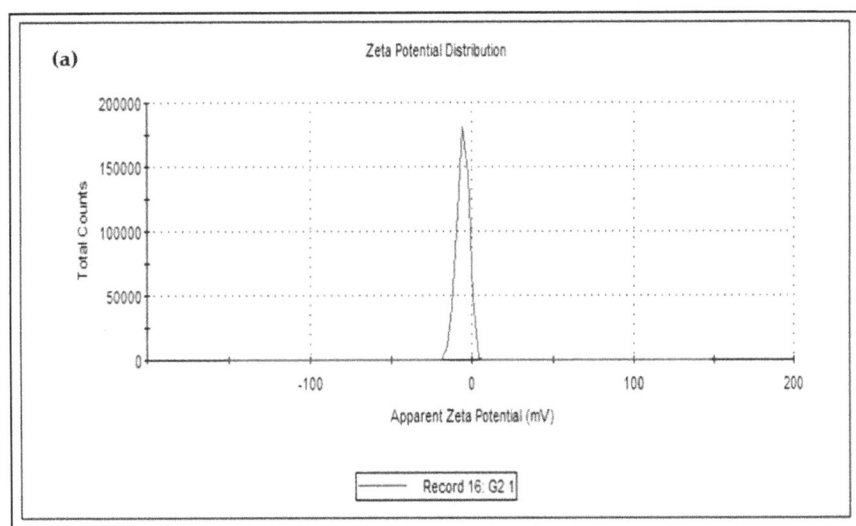

(a) Zeta Potential Distribution

Figure 6. *Cont.*

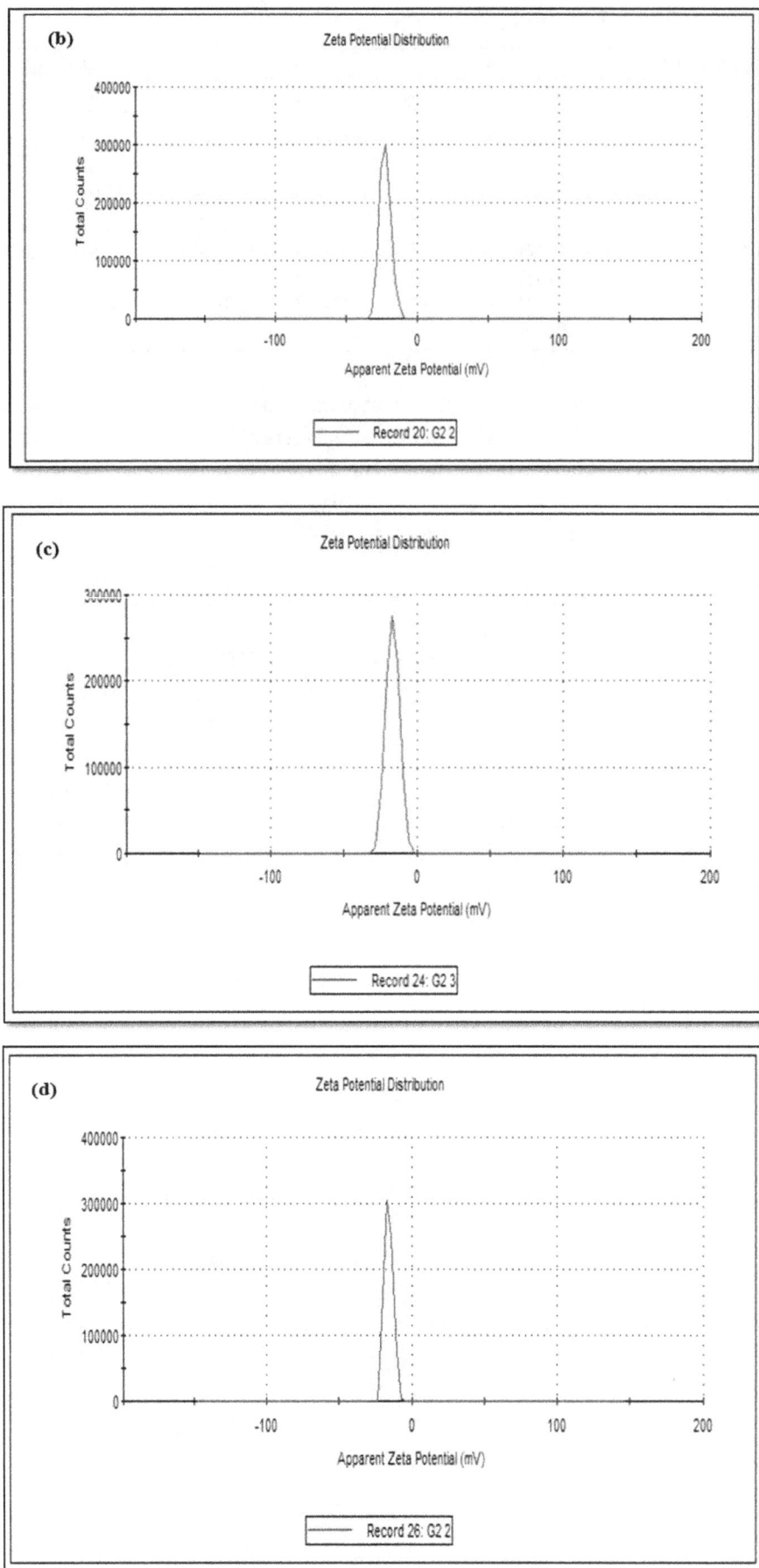

Figure 6. Zeta potential of (**a**) cCuO, (**b**) wCuO, (**c**) aCuO and (**d**) eCuO.

3.3. Electrochemical Study Using Cyclic Voltammetry

Electrochemical comparative studies of CuO was studied using cyclic voltammetry at a scan rate of 50 mV/s in 10 mM $[Fe(CN)_6]^{4-}/[Fe(CN)_6]^{3-}$ solution prepared in 0.1 M PBS. The screen print carbon electrode was modified with the CuO nanoparticle synthesized from chemical and the green method. The screen print electrodes were modified with the CuO nanoparticles and denoted as SPC/cCuO, SPC/wCuO, SPC/aCuO and SPC/eCuO, representing chemically prepared CuO nanoparticles and the green mediated CuO from water, acetone and ethanol solvents, respectively (Figure 7). The voltammogram of the modified SPC electrodes of SPC/cCuO, SPC/wCuO, SPC/aCuO and SPC/eCuO showed redox peaks, an anodic peak at about 0.25 V and cathodic peaks at reduction potential of 0.6 V for the $[Fe(CN)_6]^{4-}/[Fe(CN)_6]^{3-}$ probe. Another observable anodic peak ranging from observed 1.0 V to 1.5 V in Figure 7 is attributed to the CuO on the SPC electrodes. The SPC/cCuO gave better current response compared to the SPC/wCuO, SPC/aCuO and SPC/eCuO modified electrodes (Figure 7) which can be attributed to its lower band-gap between the valence and the conduction band as seen in the UV study also the XRD showed that cCuO has smaller particle size as compared to that of the green mediated CuO nanoparticles. This better current response remained conspicuous with varying scan rates as evident in Figure 8a–d. This could be ascribed to the better catalytic activity of the chemically synthesized CuO towards the redox probe. The anodic peak current of the CuO follows the order; SPC/cCuO > SPC/wCuO > SPC/eCuO > SPC/aCuO. This result confirms that chemical synthesis enhances the electroactivity of the synthesized nanoparticles better than the green mediated nanoparticle. However, the electroactivity of the green mediated CuO can be enhanced by modification with conducting materials such as carbon nanotubes, quantum dots and graphene or graphene oxide for enhance electroactivity.

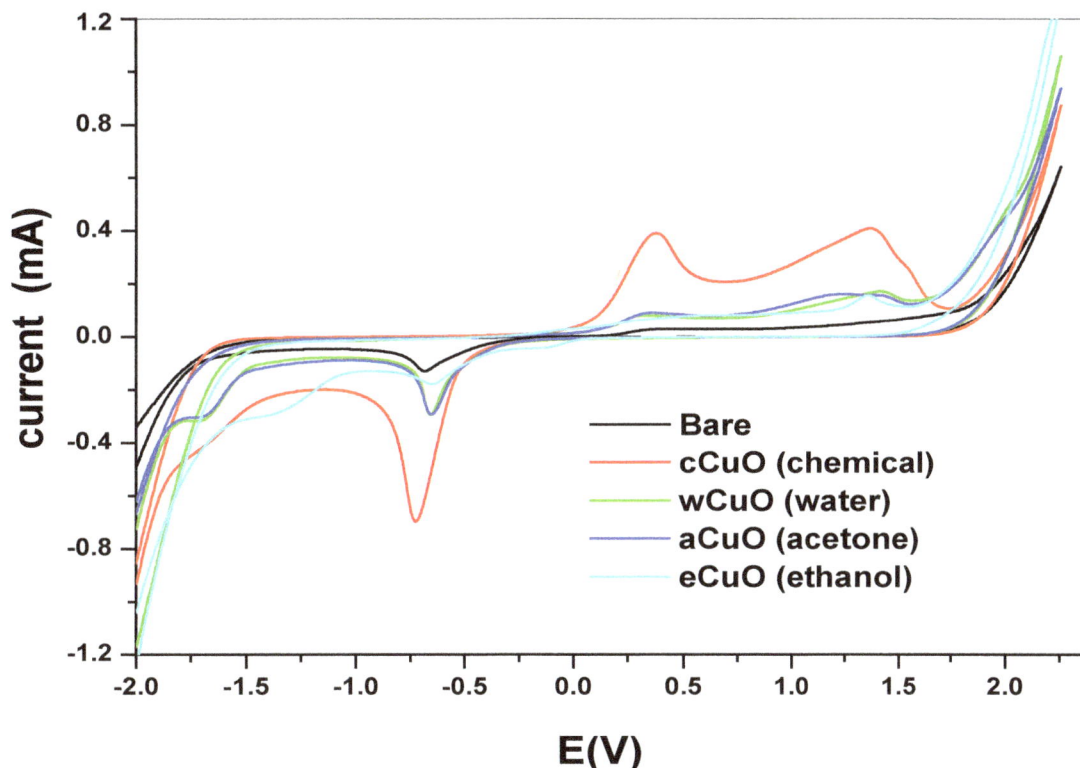

Figure 7. Cyclic voltammograms of the bare screen printed carbon electrode (SPC) and SPC modified with chemical and green synthesis using different solvent in 10 mM $[Fe(CN)_6]^{4-}/[Fe(CN)_6]^{3-}$ solution prepared in 0.1 M phosphate buffer saline (PBS) at a scan rate of 50 mV/s.

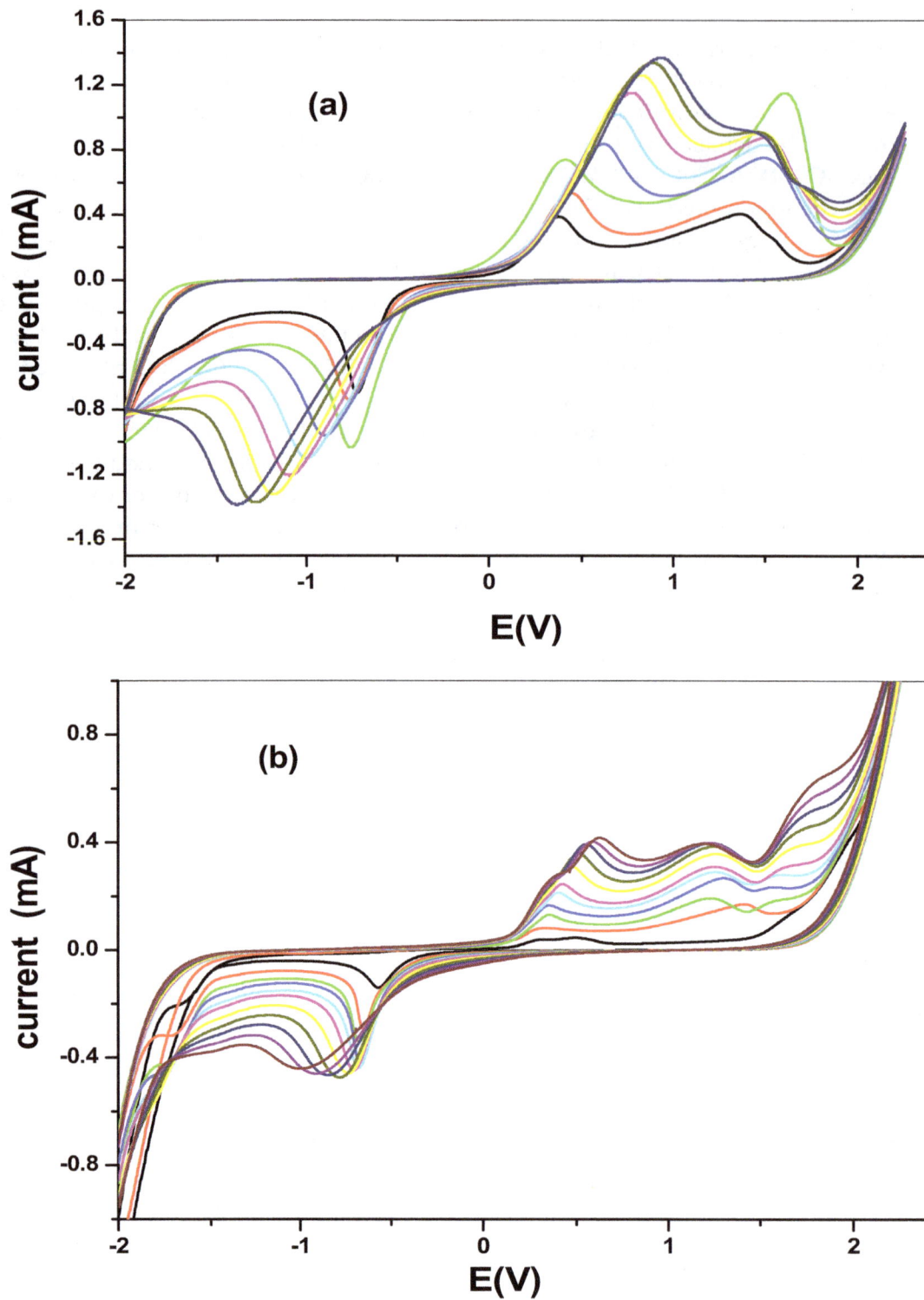

Figure 8. *Cont.*

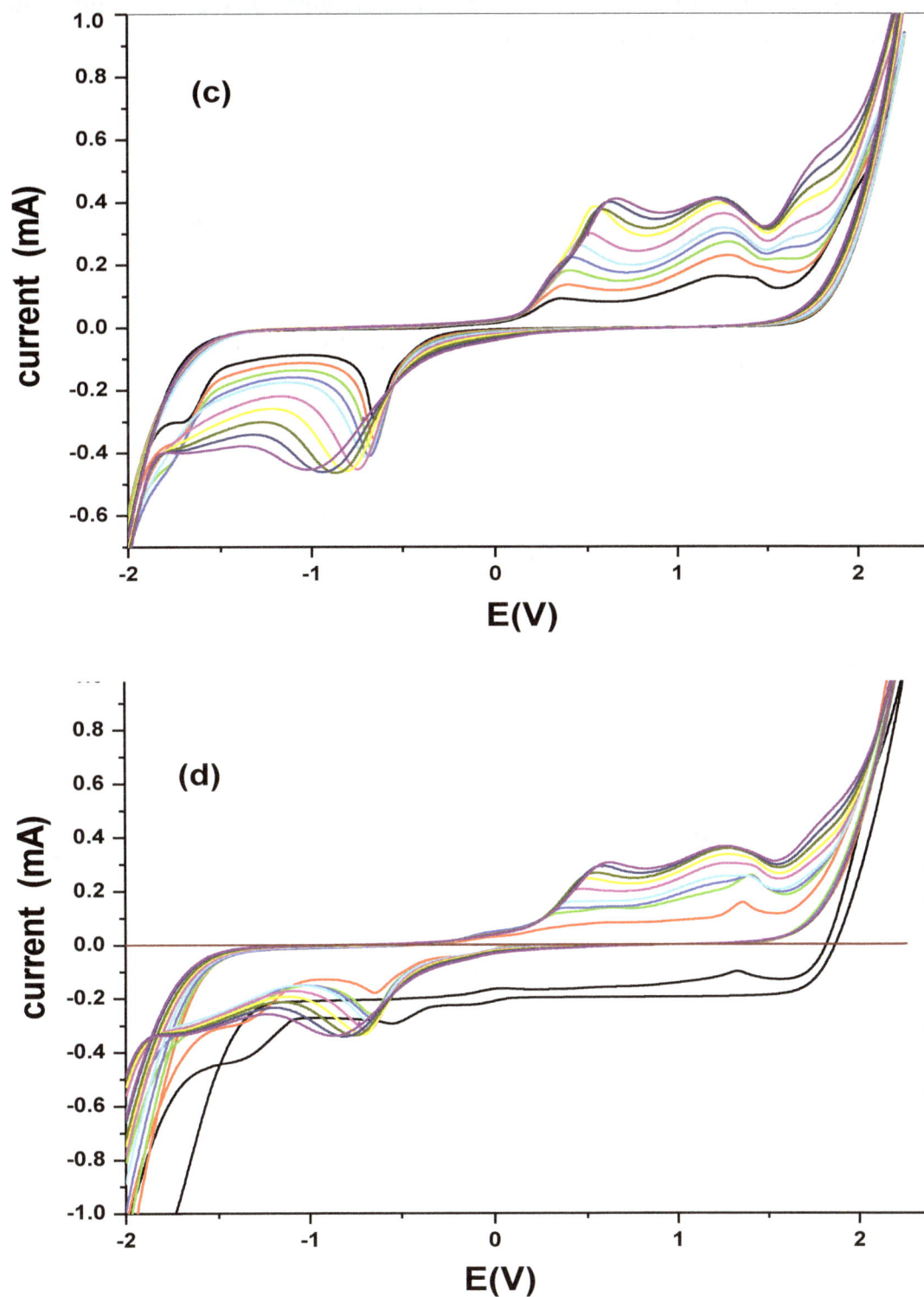

Figure 8. Scan rate voltammogram of (**a**) SPC/cCuO, (**b**) SPC/wCuO, (**c**) SPC/aCuO, and (**d**) SPC/eCuO nanoparticles in 10 mM $[Fe(CN)_6]^{4-}/[Fe(CN)_6]^{3-}$ solution prepared in 0.1 M PBS at varying scan rates (50–400 mV/s).

The effect of the scan rate was also studied for different CuO that are synthesized as seen in Figure 8a–d. The results indicated a direct proportional relationship between scan rate and the current. Increase in scan rate resulted in an increase in current. Linear plots for the comparative are shown in Figure 9 with regression values of approximately 0.9, for SPC/cCuO, SPC/wCuO,

SPC/aCuO and SPC/eCuO modified SPC electrodes, respectively. The anodic and cathodic plots suggest diffusion-controlled processes at the electrodes.

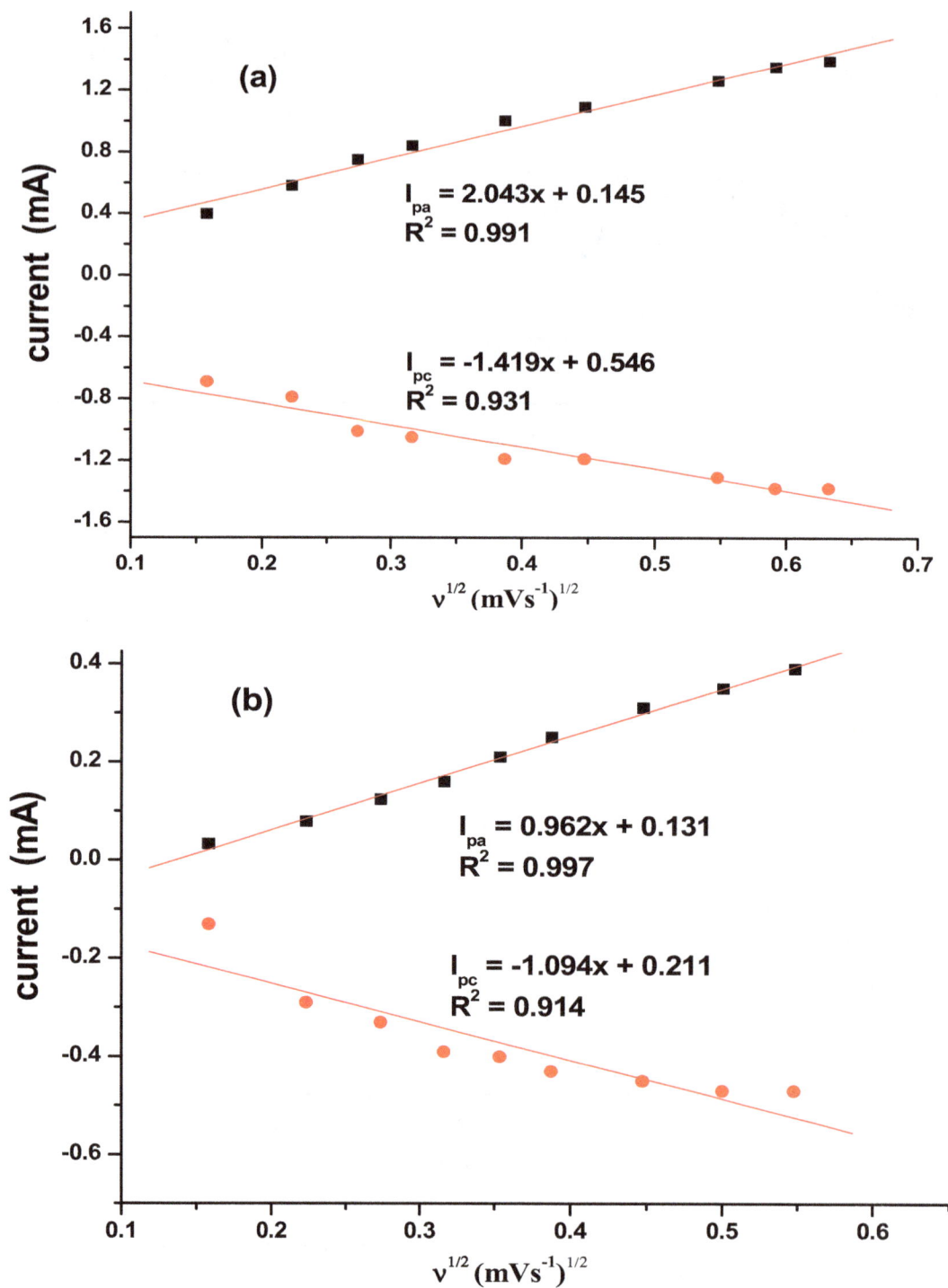

(a)

$$I_{pa} = 2.043x + 0.145$$
$$R^2 = 0.991$$

$$I_{pc} = -1.419x + 0.546$$
$$R^2 = 0.931$$

(b)

$$I_{pa} = 0.962x + 0.131$$
$$R^2 = 0.997$$

$$I_{pc} = -1.094x + 0.211$$
$$R^2 = 0.914$$

Figure 9. *Cont.*

$I_{pa} = 0.601x + 0.023$
$R^2 = 0.996$

$I_{pc} = -0.476x + 0.326$
$R^2 = 0.918$

$I_{pa} = 0.583x + 0.089$
$R^2 = 0.997$

$I_{pc} = -0.474x + 0.073$
$R^2 = 0.932$

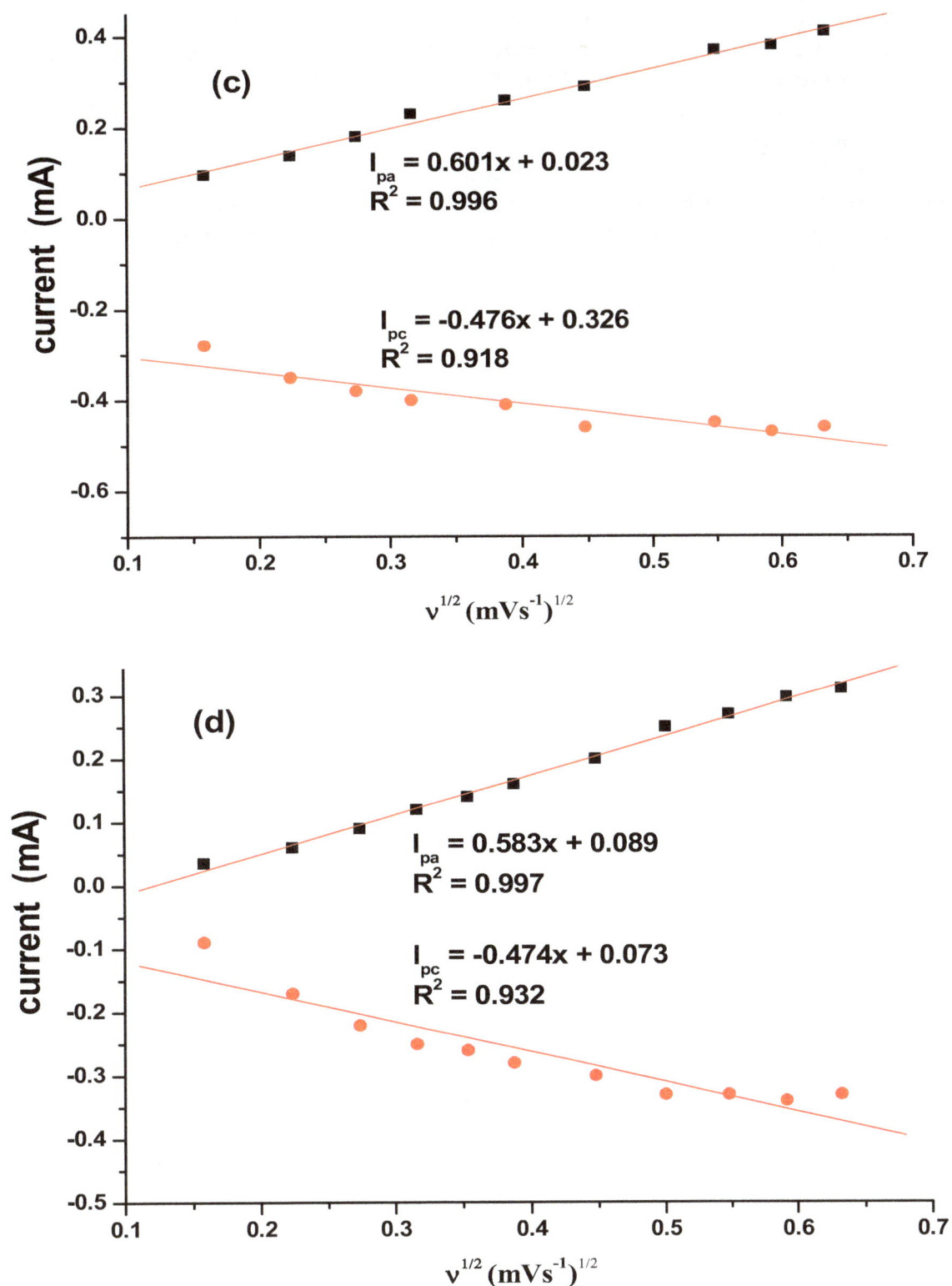

Figure 9. Corresponding linear plots of $V^{1/2}$ versus currents for (**a**) SPC/cCuO, (**b**) SPC/wCuO, (**c**) SPC/aCuO and (**d**) SPC/eCuO varying scan rates (50–400 mV/s).

Table 2 summarizes the electrochemical behavior of the bare electrode and the SPC modified electrodes of SPC/cCuO, SPC/wCuO, SPC/aCuO and SPC/eCuO. From this table, it is apparent that the best current response (anodic peak current, I_{pa}) to the redox probe was obtained from SPC/cCuO. This was followed by the current response obtained at SPC/aCuO. Basically, the current response obtained from the electrodes followed the order SPC/cCuO > SPC/aCuO > SPC/wCuO > SPC/eCuO >

bare SPC. The reversibility of the redox reaction at an electrode can be estimated from the difference of the anodic (E_{pa}) and the cathodic peak potentials (E_{pc}) (ΔE_p). The closer this difference is to 59/n mV, the more reversible the process [48]. The reaction of the redox probe at the surface of the bare electrode, the electrode modified with chemically synthesized nanoparticles (NPs) and the green NPs are irreversible or quasi-reversible at best. This inference was drawn from the ΔE_p values of 1.04, 1.03, 1.02, 1.00 and 1.19 obtained at bare SPC, SPC/wCuO, SPC/aCuO, SPC/eCuO and SPC/cCuO, respectively. Similarly, the extent of departure of the ratio I_{pa}/I_{pc} of all the electrodes from unity further confirms the irreversibility of the redox reaction at the surface of the bare electrode and the modified electrodes.

$$i_p = 2.69 \times 10^5 \, n^{2/3} A_{eff} D^{1/2} C V^{1/2} \tag{2}$$

Table 2. Electrochemical parameters of bare SPC and electrodes modified with CuO nanoparticles.

Electrodes	E_{pa}	E_{pc}	I_{pa}	I_{pc}	A_{eff} (cm^2)	I_{pa}/I_{pc}	ΔE_p
Bare SPC	0.32	−0.72	6.87×10^{-5}	-1.76×10^{-4}	6.90×10^{-5}	3.90×10^{-1}	1.04
SPC/wCuO	0.36	−0.67	1.66×10^{-4}	-4.12×10^{-4}	0.12	4.03×10^{-1}	1.03
SPC/aCuO	0.37	−0.65	1.76×10^{-4}	-3.94×10^{-4}	0.08	4.47×10^{-1}	1.02
SPC/eCuO	0.34	−0.66	1.27×10^{-4}	-2.60×10^{-4}	0.061	4.88×10^{-1}	1.00
SPC/cCuO	0.44	−0.75	7.32×10^{-4}	-1.03×10^{-3}	0.196	7.11×10^{-1}	1.19

Using the Randles–Sevcik equation (Equation (2)) where i_p, n, A_{eff}, D, C and v represent the peak current (A), number of electrons transferred, effective surface area of the electrode (cm^2), diffusion coefficient of the redox probe (cm^2 s^{-1}), concentration of the analyte (mol cm^{-3}) and the scan rate (V s^{-1}), respectively. From the plot of the anodic peak potential (I_{pa}) against the square root of the scan rate ($V^{\frac{1}{2}}$) (Figure 8a–d), the slope which is equal to $2.69 \times 10^5 \, n^{2/3} A_{eff} D^{1/2} C$ was used to calculate A_{eff} given the value of D, n and C for the redox probe as 7.6×10^{-6} cm^2 s^{-1}, 1 and 10 mM, respectively [49]. The A_{eff} of the bare SPC, SPC/wCuO, SPC/aCuO, SPC/eCuO and SPC/cCuO were calculated as 0.69×10^{-4}, 0.275, 0.129, 0.081 and 0.079, respectively. Again, the highest surface area was obtained for the electrode modified with cCuO while the best surface area from the electrode modified with the green CuO was obtained at SPC/aCuO. The extremely small surface area obtained for the bare SPC is an indication that the modification of the bare electrode with the green mediated CuO has affected the surface area of the electrode in no small measures. As a result, the incorporation of the CuO into other materials with large surface area could create a composite with improved surface area and electrocatalytic activity towards an analyte of interest. This possibility has precedence in literature where green mediated nanoparticles have been combined with materials such as polypyrrole [50] and graphene oxide [51–53] with the nanoparticles bringing forth improved electrocatalytic activity to the resultant composite. Interestingly, green-mediated CuO have also been individually used for the modification of bare glassy carbon electrode for the detection of biomolecules [51,54].

4. Conclusions

CuO were synthesized using a chemical and green method for a comparative study. The morphological studies revealed that the chemically synthesized CuO have similar morphology with that of the copper oxide prepared from plant extract using ethanol as solvent. Similar conclusion could be drawn from the SEM micrograph of green CuO prepared from plant extracts obtained with water and acetone as solvents. The electrochemical activity of the chemically prepared CuO towards ferrocyanide redox probe was much better than that of the green CuO. Further studies on the modification of the CuO with other conducting materials could increase their electrocatalytic activity towards the analytes of interest. On this basis, the green CuO have the prospect of application in electrochemical sensing among other industrial applications, as with the chemically synthesized CuO.

Author Contributions: Conceptualization, O.E.F.; methodology, O.E.F., O.P.K., and A.O.A.; software, O.E.F., O.P.K., and S.E.E.; validation, O.P.K., A.O.A., and O.E.F.; formal analysis, O.E.F., O.P.K., and A.O.A.; investigation, O.E.F., O.P.K., and A.O.A.; resources, A.O.A. and O.E.F.; data curation, O.P.K., A.O.A., S.E.E. and O.E.F.; writing—original draft preparation, O.P.K., A.O.A., S.E.E. and O.E.F.; writing—review and editing, O.P.K., A.O.A., S.E.E. and O.E.F.; supervision, O.E.F.; project administration, O.E.F.; funding acquisition, O.E.F. All authors have read and agreed to the published version of the manuscript.

Acknowledgments: O.E.F., O.P.K. and A.O.A. thank the North-West University and MaSIM for their financial support and research facilities. O.E.F. acknowledges the FRC of North-West University and the National Research Foundation of South Africa for Thuthuka funding for Researchers (UID: 117709).

References

1. Chen, H.; Roco, M.C.; Li, X.; Lin, Y.-L. Trends in nanotechnology patents. *Nat. Nanotechnol.* **2008**, *3*, 123–125. [CrossRef] [PubMed]

2. Usman, M.S.; Ibrahim, N.A.; Shameli, K.; Zainuddin, N.; Yunus, W.M.Z.W. Copper Nanoparticles Mediated by Chitosan: Synthesis and Characterization via Chemical Methods. *Molecules* **2012**, *17*, 14928–14936. [CrossRef] [PubMed]

3. Iravani, S.; Korbekandi, H.; Mirmohammadi, S.; Zolfaghari, B. Synthesis of silver nanoparticles: Chemical, physical and biological methods. *Res. Pharm. Sci.* **2015**, *9*, 385–406.

4. Altikatoglu, M.; Attar, A.; Erci, F.; Cristache, C.M.; Isildak, I. Green synthesis of copper oxide nanoparticles using Ocimum basilicum extract and their antibacterial activity. *Fresenius Environ. Bull.* **2017**, *25*, 7832–7837.

5. Sathiyavimal, S.; Vasantharaj, S.; Bharathi, D.; Saravanan, M.; Manikandan, E.; Kumar, S.S.; Pugazhendhi, A. Biogenesis of copper oxide nanoparticles (CuONPs) using Sida acuta and their incorporation over cotton fabrics to prevent the pathogenicity of Gram negative and Gram positive bacteria. *J. Photochem. Photobiol. B Biol.* **2018**, *188*, 126–134. [CrossRef] [PubMed]

6. Khatoon, U.T.; Mantravadi, K.M.; Rao, G.V.S.N. Strategies to synthesise copper oxide nanoparticles and their bio applications—A review. *Mater. Sci. Technol.* **2018**, *34*, 2214–2222. [CrossRef]

7. Mohan, S.; Singh, Y.; Verma, D.K.; Hasan, S.H. Synthesis of CuO nanoparticles through green route using Citrus limon juice and its application as nanosorbent for Cr(VI) remediation: Process optimization with RSM and ANN-GA based model. *Process. Saf. Environ. Prot.* **2015**, *96*, 156–166. [CrossRef]

8. IAARD—International Association of Advances in Research and Development. *Int. J. Nano Sci. Nano Technol.* **2016**, *13*, 19–52. [CrossRef]

9. Guzmán, M.G.; Dille, J.; Godet, S. Synthesis of silver nanoparticles by chemical reduction method and their antibacterial activity. *Int. J. Chem. Biomol. Eng.* **2009**, *2*, 104–111.

10. Sumitha, S.; Vidhya, R.; Lakshmi, M.S.; Prasad, K.S. Leaf extract mediated green synthesis of copper oxide nanoparticles using Ocimum tenuiflorum and its characterization. *Int. J. Chem. Sci.* **2016**, *14*, 435–440.

11. Kavitha, K.; Baker, S.; Rakshith, D.; Kavitha, H.; Yashwantha Rao, H.; Harini, B.; Satish, S. Plants as green source towards synthesis of nanoparticles. *Int. Res. J. Biol. Sci.* **2013**, *2*, 66–76.

12. Kumar, B.V.; Naik, H.B.; Girija, D. ZnO nanoparticle as catalyst for efficient green one-pot synthesis of coumarins through Knoevenagel condensation. *J. Chem. Sci.* **2011**, *123*, 615–621. [CrossRef]

13. Zhang, D.; Ni, X.; Zheng, H.; Li, Y.; Wang, G.; Yang, Z. Synthesis of needle-like nickel nanoparticles in water-in-oil microemulsion. *Mater. Lett.* **2005**, *59*, 2011–2014. [CrossRef]

14. Yuvakkumar, R.; Hong, S. Green Synthesis of Spinel Magnetite Iron Oxide Nanoparticles. *Adv. Mater. Res.* **2014**, *1051*, 39–42. [CrossRef]

15. Nasrollahzadeh, M.; Sajadi, S.M. Green synthesis of copper nanoparticles using Ginkgo biloba L. leaf extract and their catalytic activity for the Huisgen (3 + 2) cycloaddition of azides and alkynes at room temperature. *J. Colloid Interface Sci.* **2015**, *457*, 141–147. [CrossRef]

16. Din, M.I.; Rehan, R. Synthesis, Characterization, and Applications of Copper Nanoparticles. *Anal. Lett.* **2017**, *50*, 50–62. [CrossRef]

17. Song, J.Y.; Jang, H.-K.; Kim, B.S. Biological synthesis of gold nanoparticles using Magnolia kobus and Diopyros kaki leaf extracts. *Process. Biochem.* **2009**, *44*, 1133–1138. [CrossRef]

18. Roy, A.; Bulut, O.; Some, S.; Mandal, A.K.; Yilmaz, M.D. Green synthesis of silver nanoparticles: Biomolecule-nanoparticle organizations targeting antimicrobial activity. *RSC Adv.* **2019**, *9*, 2673–2702. [CrossRef]

19. Seydi, N.; Saneei, S.; Jalalvand, A.R.; Zangeneh, M.M.; Zangeneh, A.; Tahvilian, R.; Pirabbasi, E. Synthesis of titanium nanoparticles using Allium eriophyllum Boiss aqueous extract by green synthesis method and evaluation of their remedial properties. *Appl. Organomet. Chem.* **2019**, *33*. [CrossRef]

20. Hameed, S.; Abbasi, B.A.; Ali, M.; Khalil, A.T.; Abbasi, B.A.; Numan, M.; Shinwari, Z.K. Green synthesis of zinc nanoparticles through plant extracts: Establishing a novel era in cancer theranostics. *Mater. Res. Express* **2019**, *6*, 102005. [CrossRef]

21. Bose, S.; Ganayee, M.A.; Mondal, B.; Baidya, A.; Chennu, S.; Mohanty, J.S.; Pradeep, T. Synthesis of Silicon Nanoparticles from Rice Husk and their Use as Sustainable Fluorophores for White Light Emission. *ACS Sustain. Chem. Eng.* **2018**, *6*, 6203–6210. [CrossRef]

22. Arsiya, F.; Sayadi, M.H.; Sobhani, S. Green synthesis of palladium nanoparticles using Chlorella vulgaris. *Mater. Lett.* **2017**, *186*, 113–115. [CrossRef]

23. Suresh, Y.; Annapurna, S.; Singh, A.; Bhikshamaiah, G. Green synthesis and characterization of tea decoction stabilized copper nanoparticles. *Int. J. Innov. Res. Sci. Eng. Technol.* **2014**, *3*, 11265–11270.

24. Khanna, P.; Gaikwad, S.; Adhyapak, P.; Singh, N.; Marimuthu, R. Synthesis and characterization of copper nanoparticles. *Mater. Lett.* **2007**, *61*, 4711–4714. [CrossRef]

25. Wang, G.; Wang, G.; Liu, X.; Wu, J.; Li, M.; Gu, J.; Liu, H.; Fang, B. Different CuO Nanostructures: Synthesis, Characterization, and Applications for Glucose Sensors. *J. Phys. Chem. C* **2008**, *112*, 16845–16849. [CrossRef]

26. Humplik, T.; Lee, J.; O'Hern, S.C.; Fellman, B.A.; Baig, M.A.; Hassan, S.F.; Atieh, M.A.; Rahman, F.; Laoui, T.; Karnik, R.; et al. Nanostructured materials for water desalination. *Nanotechnology* **2011**, *22*, 292001. [CrossRef] [PubMed]

27. Naika, H.R.; Lingaraju, K.; Manjunath, K.; Kumar, D.; Nagaraju, G.; Suresh, D.; Nagabhushana, H. Green synthesis of CuO nanoparticles using *Gloriosa superba* L. extract and their antibacterial activity. *J. Taibah Univ. Sci.* **2015**, *9*, 7–12. [CrossRef]

28. Narayanan, R.; El-Sayed, M.A. Effect of Catalysis on the Stability of Metallic Nanoparticles: Suzuki Reaction Catalyzed by PVP-Palladium Nanoparticles. *J. Am. Chem. Soc.* **2003**, *125*, 8340–8347. [CrossRef]

29. Uschakov, A.; Karpov, I.; Karpov, I.V.; Petrov, M. Plasma-chemical synthesis of copper oxide nanoparticles in a low-pressure arc discharge. *Vacuum* **2016**, *133*, 25–30. [CrossRef]

30. Swarnkar, R.K.; Singh, S.C.; Gopal, R.; Singh, M.R.; Lipson, R.H. Synthesis of Copper/Copper-Oxide Nanoparticles: Optical and Structural Characterizations. In *Transport and Optical Properties of Nanomaterials: Proceedings of the International Conference—ICTOPON-1*; AIP Publishing LLC: Beijing, China, 2009; pp. 205–210. [CrossRef]

31. Honary, S.; Barabadi, H.; Gharaei-Fathabad, E.; Naghibi, F. Green synthesis of copper oxide nanoparticles using Penicillium aurantiogriseum, Penicillium citrinum and Penicillium waksmanii. *Dig. J. Nanomater. Bios.* **2012**, *7*, 999–1005.

32. Hasan, S.S.; Singh, S.; Parikh, R.Y.; Dharne, M.S.; Patole, M.S.; Prasad, B.L.V.; Shouche, Y.S. Bacterial Synthesis of Copper/Copper Oxide Nanoparticles. *J. Nanosci. Nanotechnol.* **2008**, *8*, 3191–3196. [CrossRef] [PubMed]

33. Kumar, P.P.N.V.; Shameem, U.; Kollu, P.; Kalyani, R.L.; Pammi, S.V.N. Green Synthesis of Copper Oxide Nanoparticles Using Aloe vera Leaf Extract and Its Antibacterial Activity Against Fish Bacterial Pathogens. *BioNanoScience* **2015**, *5*, 135–139. [CrossRef]

34. Singh, J.; Kumar, V.; Kim, K.-H.; Rawat, M. Biogenic synthesis of copper oxide nanoparticles using plant extract and its prodigious potential for photocatalytic degradation of dyes. *Environ. Res.* **2019**, *177*, 108569. [CrossRef] [PubMed]

35. Ijaz, F.; Shahid, S.; Khan, S.A.; Ahmad, W.; Zaman, S. Green synthesis of copper oxide nanoparticles using Abutilon indicum leaf extract: Antimicrobial, antioxidant and photocatalytic dye degradation activitie. *Trop. J. Pharm. Res.* **2017**, *16*, 743. [CrossRef]

36. Awwad, A.; Albiss, B.; Salem, N. Antibacterial activity of synthesized copper oxide nanoparticles using Malva sylvestris leaf extract. *SMU Med. J.* **2015**, *2*, 91–101.

37. Sankar, R.; Manikandan, P.; Malarvizhi, V.; Fathima, T.; Shivashangari, K.S.; Renu, S. Green synthesis of colloidal copper oxide nanoparticles using Carica papaya and its application in photocatalytic dye degradation. *Spectrochim. Acta Part A Mol. Biomol. Spectrosc.* **2014**, *121*, 746–750. [CrossRef]

38. Sutradhar, P.; Saha, M.; Maiti, D. Microwave synthesis of copper oxide nanoparticles using tea leaf and coffee powder extracts and its antibacterial activity. *J. Nanostruct. Chem.* **2014**, *4*, 1–6. [CrossRef]

39. Vishveshvar, K.; Krishnan, M.V.A.; Haribabu, K.; Vishnuprasad, S. Green Synthesis of Copper Oxide Nanoparticles Using Ixiro coccinea Plant Leaves and its Characterization. *BioNanoScience* **2018**, *8*, 554–558. [CrossRef]

40. Saif, S.; Tahir, A.; Asim, T.; Chen, Y. Plant Mediated Green Synthesis of CuO Nanoparticles: Comparison of Toxicity of Engineered and Plant Mediated CuO Nanoparticles towards Daphnia magna. *Nanomaterials* **2016**, *6*, 205. [CrossRef]

41. Sawicki, B.; Tomaszewicz, E.; Piątkowska, M.; Gron, T.; Duda, H.; Górny, K. Correlation between the Band-Gap Energy and the Electrical Conductivity in $MPr_2W_2O_{10}$ Tungstates (Where M = Cd, Co, Mn). *Acta Phys. Pol. A* **2016**, *129*, 94–96. [CrossRef]

42. Prakash, V.; Diwan, R. Characterization of synthesized copper oxide nanopowders and their use in nanofluids for enhancement of thermal conductivity. *Indian J. Pure Appl. Phys.* **2015**, *53*, 753–758.

43. Suresh, S.; Karthikeyan, S.; Jayamoorthy, K. FTIR and multivariate analysis to study the effect of bulk and nano copper oxide on peanut plant leaves. *J. Sci. Adv. Mater. Devices* **2016**, *1*, 343–350. [CrossRef]

44. Syame, S.M.; Mohamed, W.S.; Mahmoud, R.K.; Omara, S.T. Synthesis of Copper-Chitosan Nanocomposites and its Application in Treatment of Local Pathogenic Isolates Bacteria. *Orient. J. Chem.* **2017**, *33*, 2959–2969. [CrossRef]

45. Tamuly, C.; Saikia, I.; Hazarika, M.; Das, M.R. Reduction of aromatic nitro compounds catalyzed by biogenic CuO nanoparticles. *RSC Adv.* **2014**, *4*, 53229–53236. [CrossRef]

46. Anandhavalli, N.; Mol, B.; Manikandan, S.; Anusha, N.; Ponnusami, V.; Rajan, K. Green synthesis of cupric oxide nanoparticles using the water extract of Murrya koenigi and its photocatalytic activity. *Asian J. Chem.* **2015**, *27*, 2523–2526. [CrossRef]

47. Aparna, Y.; Rao, K.V.; Subbarao, P.S. Preparation and characterization of CuO Nanoparticles by novel sol-gel technique. *J. Nano-Electron. Phys.* **2012**, *4*, 03005–03009.

48. Thomas, D.; Rasheed, Z.; Jagan, J.S.; Kumar, K.G. Study of kinetic parameters and development of a voltammetric sensor for the determination of butylated hydroxyanisole (BHA) in oil samples. *J. Food Sci. Technol.* **2015**, *52*, 6719–6726. [CrossRef]

49. Siswana, M.P.; Ozoemena, K.I.; Nyokong, T. Electrocatalysis of asulam on cobalt phthalocyanine modified multi-walled carbon nanotubes immobilized on a basal plane pyrolytic graphite electrode. *Electrochim. Acta* **2006**, *52*, 114–122. [CrossRef]

50. Uwaya, G.E.; Fayemi, O.E. Electrochemical detection of serotonin in banana at green mediated PPy/Fe_3O_4NPs nanocomposites modified electrodes. *Sens. Bio-Sensing Res.* **2020**, *28*, 100338. [CrossRef]

51. Pourbeyram, S.; Abdollahpour, J.; Soltanpour, M. Green synthesis of copper oxide nanoparticles decorated reduced graphene oxide for high sensitive detection of glucose. *Mater. Sci. Eng. C* **2019**, *94*, 850–857. [CrossRef]

52. Nayak, S.P.; Ramamurthy, S.S.; Kumar, J.K.K. Green synthesis of silver nanoparticles decorated reduced graphene oxide nanocomposite as an electrocatalytic platform for the simultaneous detection of dopamine and uric acid. *Mater. Chem. Phys.* **2020**, *252*, 123302. [CrossRef]

53. Naghdi, S.; Sajjadi, M.; Nasrollahzadeh, M.; Rhee, K.Y.; Sajadi, S.M.; Jaleh, B. Cuscuta reflexa leaf extract mediated green synthesis of the Cu nanoparticles on graphene oxide/manganese dioxide nanocomposite and its catalytic activity toward reduction of nitroarenes and organic dyes. *J. Taiwan Inst. Chem. Eng.* **2018**, *86*, 158–173. [CrossRef]

54. Sundar, S.; Venkatachalam, G.; Kwon, S.J. Biosynthesis of Copper Oxide (CuO) Nanowires and Their Use for the Electrochemical Sensing of Dopamine. *Nanomaterials* **2018**, *8*, 823. [CrossRef] [PubMed]

Flower-Based Green Synthesis of Metallic Nanoparticles: Applications beyond Fragrance

Harsh Kumar [1] , Kanchan Bhardwaj [2] , Kamil Kuča [3,*] , Anu Kalia [4], Eugenie Nepovimova [3], Rachna Verma [2] and Dinesh Kumar [1,*]

[1]   School of Bioengineering & Food Technology, Shoolini University of Biotechnology and Management Sciences, Solan-173229, H. P., India; microharshs@gmail.com

[2]   School of Biological and Environmental Sciences, Shoolini University of Biotechnology and Management Sciences, Solan-173229, H. P., India; kanchankannu1992@gmail.com (K.B.); rachnaverma@shooliniuniversity.com (R.V.)

[3]   Department of Chemistry, Faculty of Science, University of Hradec Kralove, Hradec Kralove 50003, Czech Republic; eugenie.nepovimova@uhk.cz

[4]   Electron Microscopy and Nanoscience Laboratory, Punjab Agricultural University, Ludhiana-141004, Punjab, India; kaliaanu@pau.edu

*   Correspondence: kamil.kuca@uhk.cz or kamil.kuca@fnhk.cz (K.K.); dineshkumar@shooliniuniversity.com (D.K.).

Abstract: Green synthesis has gained wide attention as a sustainable, reliable, and eco-friendly approach to the synthesis of a variety of nanomaterials, including hybrid materials, metal/metal oxide nanoparticles, and bioinspired materials. Plant flowers contain diverse secondary compounds, including pigments, volatile substances contributing to fragrance, and other phenolics that have a profound ethnobotanical relevance, particularly in relation to the curing of diseases by 'Pushpa Ayurveda' or floral therapy. These compounds can be utilized as potent reducing agents for the synthesis of a variety of metal/metal oxide nanoparticles (NPs), such as gold, silver, copper, zinc, iron, and cadmium. Phytochemicals from flowers can act both as reducing and stabilizing agents, besides having a role as precursor molecules for the formation of NPs. Furthermore, the synthesis is mostly performed at ambient room temperatures and is eco-friendly, as no toxic derivatives are formed. The NPs obtained exhibit unique and diverse properties, which can be harnessed for a variety of applications in different fields. This review reports the use of a variety of flower extracts for the green synthesis of several types of metallic nanoparticles and their applications. This review shows that flower extract was mainly used to design gold and silver nanoparticles, while other metals and metal oxides were less explored in relation to this synthesis. Flower-derived silver nanoparticles show good antibacterial, antioxidant, and insecticidal activities and can be used in different applications.

Keywords: flower extract; green synthesis; nanoparticles; phytochemicals; antibacterial; antioxidants; catalytic; insecticidal

1. Introduction

The theoretical concept of nanotechnology was first described in 1959 by the physicist, Richard Feynman [1]. Nanotechnology is defined as understanding, controlling, and manipulating matter at the level of individual atoms and molecules [2]. Metal nanoparticles (NPs) with distinct physico-chemical properties have gained considerable attention in the last few decades [3]. Due to their ultra-small size and large surface area to volume ratio, a great interest in the use of NPs—which display variations in both physical and chemical properties, as compared to the bulk of similar chemical compositions—has developed [4–6]. As a result of their unique optoelectronic and physico-chemical properties, NPs have a number of applications, including their use as catalysts, electronic components, and chemical sensors in medical diagnostic imaging, medical treatment protocols, and pharmaceutical products [7].

Nanoparticles can be synthesized using two different fundamental approaches (top down and bottom up methods) to obtain nanomaterials with a desired shape, size, and functionality [8]. The former involves the generation of nanomaterials/nanoparticles using diverse synthesis approaches, like ball milling, lithographic techniques, etching, and sputtering [9]. The bottom-up approach usually used to synthesize nanoparticles normally involves aggressive reducing agents (hydrazine and sodium borohydride), along with a capping agent and volatile solvent, like chloroform and toluene. These methods are effective in synthesizing well-defined and pure metallic nanoparticles, but their production cost remains the main hinderance [10]. Therefore, there is a need for the development of a cost-effective and environmentally friendly alternative, which would allow an eco-friendly reducing agent, environmentally compatible solvents and nonhazardous capping agents to be used for the synthesis of nanoparticles. All these criteria have been proposed as the primary prerequisite for green nanoparticle synthesis [11].This review focuses on the use of flower extracts for the green synthesis of several types of nanoparticles and their applications. It also highlights the key challenges of green flower-mediated nanoparticles.

2. Importance of Flowers in Daily Life

There is a special association between humans and flowers, and the aesthetic appeal of flowers triggered humans to cultivate flowers and propagate them, just like insects do with pollen [12,13]. Flowers have an attractive visual quality, and vision is a multimodal procedure that activates visual regions of the brain, as well as the viscera-motor, sensory-motor and affective cerebral circuits. Various parts of the brain are activated by flowers, creating an interesting perceptual experience [14]. Flowers also induce a multisensory experience, as observed while watching flowers sway in the wind and their use in perfume [15,16]. Additionally, in the Ayurveda and Siddha systems, some flowers have been reported to possess distinct medicinal properties [17]. In rasayana medicines, about 18,000 kinds of flowers have been mentioned [18]. An ayurvedic text,"Kaiyadevanighantu", mainly describes the flowers of many medicinal plants as having therapeutic benefits [19].

Due to the vast and ancient knowledge of health care, the contemporary medical challenges can possibly be tackled through research on the phytochemical constituents present in flowers and their pharmacological properties. The phytochemical analysis of the *Hibiscus rosa-sinensis* flower shows the presence of constituents, such as indole alkaloids, saponins, reducing sugars, tannins, and terpenoids;while their aqueous extracts may contain cardiac glycosides and flavonoids, such as cyanidin, quercetin, and saponins [20,21]. Most of these secondary metabolites are responsible for antibacterial activities or possess haemo-protective properties [22–25]. The flower of *Mimusops elengi* contains 74 different compounds belonging to flavonoids, alkaloids, phenolics, and tannins, which can be isolated using various extraction methods [26–28]. Methanolic extract has been reported to inhibit the growth of a number of bacterial pathogens [29,30]. An anti-malarial compound, cyclohexyl ethanoid (rengyolone), isolated from the ethanolic extract of the *Nyctanthes arbor-tristis* flower, has been reported to be effective against *Plasmodium falciparum* [31]. Another compound, benzofuranone, 3, 3a, 7, 7a-tetrahydro-3a hydroxy-6(2H)-benzofuranone, was also isolated from this flower and exhibits a significant antibacterial activity against both Gram-negative and Gram-positive bacteria [32]. Furthermore, there are also reports indicating that the antidiabetic activity of the *Nyctanthes arbor-tristis* flower extract is more effective than the leaf extract [33]. *Tussilago farfara* flower buds yield two flavonoids, namely, quercetin 3-O-beta-D-glucopyranoside and quercetin 3-O-beta-L-arabinopyranoside, with a higher antioxidative activity than their aglycone and quercetin, as shown by a nitro blue tetrazolium (NBT) superoxide scavenging assay [34]. The diverse compounds present in various flower extracts can act as oxidizing/reducing agents or as biotemplates to aid in the green synthesis of NPs, particularly metal/metal oxide NPs.

3. Green Synthesis of Nanoparticles (NPs)

Green-synthesized NPs can be obtained through an easy, efficient, economical and eco-friendly biological synthesis approach [35]. Metallic nanoparticles can be obtained from cell or cell-free extracts of a variety of biological resources, as shown in Figure 1. The key factor that should be considered during the nanoparticle preparation is that it should be evaluated against green chemistry principles, like the selection of a solvent medium, eco-friendly reducing agent, and non-toxic material for nanoparticle stabilization [36]. Furthermore, compounds like peptides, polyphenolics, sugars, vitamins, and water from coffee and tea extracts were found to be appropriate for the synthesis of nanoparticles [37–42]. As compared to microbial NPs, plant-based NPs are more stable and monodispersed, and plant extract takes less time to reduce metal ions. Microbial synthesis is one of the approaches to the synthesis of nanomaterials.

Prokaryotic bacterial cell/cell extracts have been reported in relation to the synthesis of a variety of NPs, including cadmium sulfide (CdS), gold (Au), silver (Ag), silver oxide (AgO), and titanium dioxide (TiO_2) [43–49]. Some fungi have also been used for the synthesis of CdS, Ag, and TiO_2 NPs [45,47,50–53]. Recently, gold, iron oxide, silver, and zinc oxide NPs have been synthesized using algae [54–59]. Likewise, leaf, seed, and root extracts, latex and bulbs of plants have also been utilized for the synthesis of Ag, palladium (Pd), and Au NPs [60–69]. Other materials of a biological origin, such as honey, can also synthesize carbon, Ag, Au, Pd, and platinum (Pt) nanoparticles [70–74].

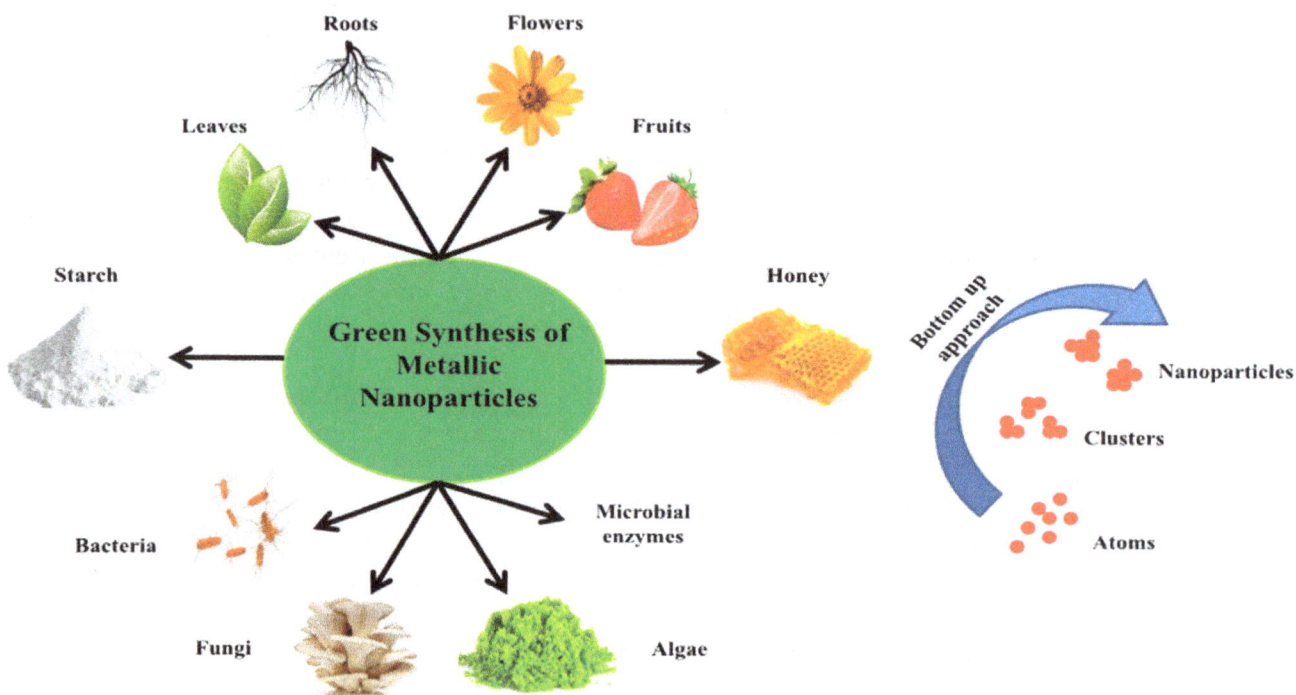

Figure 1. Different types of green synthesis used for the preparation of metal nanoparticles.

4. Green Synthesis of Nanoparticles Mediated by Flowers

Flowers have unique chemical properties that can be useful for nanoparticle synthesis. The synthesis of flower-mediated NPs is advantageous, as compared with other biological NPs synthesis methods, particularly the one mediated through microorganisms, as microorganisms need to be maintained or cultured under aseptic and pure culture conditions. It is a difficult task to separate nanoparticles during the downstream processing of microbial broth cultures. Furthermore, it takes more time to convert soluble metallic salts to elemental or element oxide NPs. A generalized mechanism (Figure 2) for the biosynthesis of different nanoparticles using flower extracts has been summarized in Table 1. The various types of nanoparticles derived from different flower extracts are discussed in the following sections.

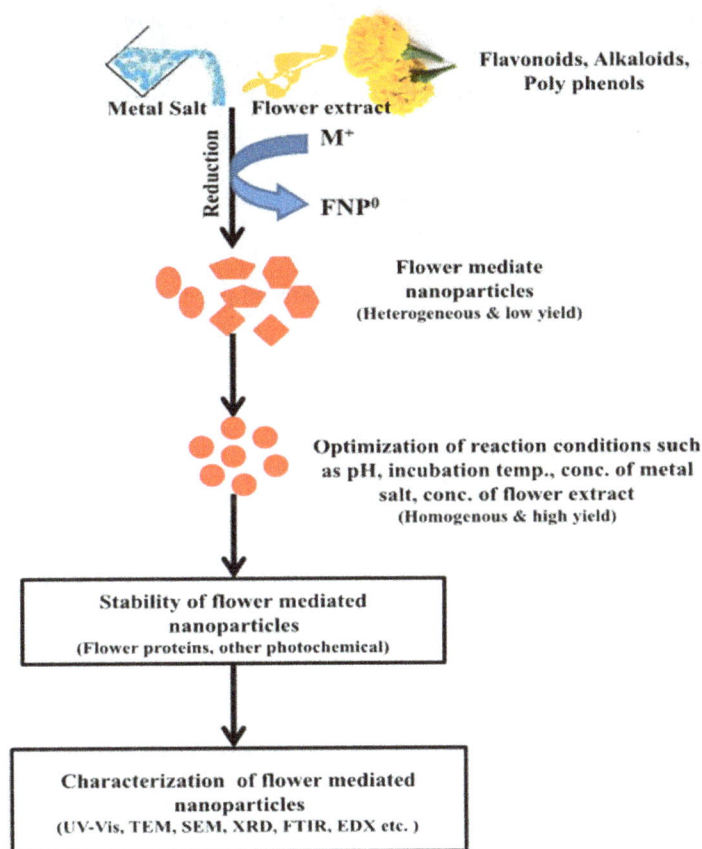

Figure 2. A general mechanism of nanoparticle formation from flower extract. M^+ (metal); FNP^0 (flower nanoparticles).

Table 1. Different types of reducing and stabilizing agents used in the flower-mediated biosynthesis of nanoparticles.

Nanoparticle Types	Reducing Agent	Stabilizing Agent	Specific Temperature	Ref
	chlorine ions	proteins present in the flower	RT	[75,76]
Silver	water-soluble heterocyclic components, polyols, and certain proteins	flower	RT	[77]
	flower	flower	40 °C	[78]
	flower	flower	60 °C	[79]
	sesquiterpenoids	DMEM + FBS	80°C	[80]
	sesquiterpenoids	DMEM + FBS	80 °C	[80]
Gold	flower	flower	40 °C	[81]
	polyphenols and flavonols	flower	25–60 °C	[82]
Zinc	flower	flower	microwave irradiation	[83]
Cadmium	tannins, flavonoids, alkaloids, and carotenoids	flower	RT	[84]
Titanium	flower	flower	60 °C	[85]
Magnesium	flower	flower	70 °C	[86]
Iron	flower	flower	RT	[87]

RT—room temperature; DMEM—Dulbecco's modified eagle medium; FBS—fetal bovine serum.

4.1. Silver Nanoparticles (AgNPs)

Silver nanoparticles (AgNPs) show a considerably large surface area, which leads to a significant biochemical reactivity, catalytic action, and atomic behavior, when compared with large particles with an identical chemical configuration [88]. The synthesis of noble AgNPs is a two-step procedure that first involves the reduction of Ag^+ ions to Ag^0, and after this agglomeration and stabilization is completed, the synthesis involves the development of oligomeric clusters of colloidal AgNPs [89]. The reduction procedure occurs in the presence of biological catalysts. The flower-derived AgNPs have shown numerous applications, which are given in Table 2.

Table 2. Applications of silver NPs synthesized from various flower varieties.

Family	Flower Variety	Applications	Ref
Fabaceae	*Lablab purpureus*	Antibacterial activityagainst *Escherichia coli* and *Staphylococcus aureus*	[90]
Apocynaceae	*Plumeria rubra*	Antibacterial activity against *Escherichia coli* and *Bacillus sp.*	[91]
Apocynaceae	*Catharanthus roseus*	Antibacterial activity against *Escherichia coli,* *Pseudomonas putida, Staphylococcus aureus, Klebsiella pneumoniae,* and *Bacillus subtilus*	[79]
Fabaceae	*Cassia angustifolia*	Antioxidant and cytotoxicity activity	[92]
Apocynaceae	*Allamanda cathartica*	Antioxidant activity and antibacterial activity against *Salmonella typhimurium, Staphylococcus aureus, Escherichia coli,*and *Klebsiella pneumoniae*	[93]
Malvaceae	*Malva sylvestris*	Antibacterial activity against *Escherichia coli, Staphylococcus aureus,* and *Streptococcus pyogenes*	[76]
Fabaceae	*Caesalpinia pulcherrima*	Antibacterial activity against *Staphylococcus aureus*; antifungal activity against *Candida glabrata*; antioxidant activity; cytotoxicity activity	[94]
Asteraceae	*Tussilago farfara*	Antibacterial activity against *Enterococcus faecium*; cyrotoxicity activity	[80]
Asteraceae	*Tagetes erecta*	Antibacterial activity against *Escherichia coli* and *Pseudomonas aeruginosa*; antifungal activity against *Candida albicans*	[95]
Sapotaceae	*Madhuca longifolia*	Antibacterial activity against *Bacillus cereus* and *Staphylococcus saprophyticus*	[78]
Malvaceae	*Hibiscus rosa-sinensis*	antibacterial activity against *Aeromonas hydrophila*	[77]
Convolvulaceae	*Ipomoea digitata* Linn	Antibacterial activity against *Staphylococcus epidermidis*; catalytic activity against methylene blue	[96]
Asteraceae	*Chrysanthemum indicum* L.	Larvicidal and pupicidalactivity against *Anopheles stephenis*	[97]

4.2. Gold Nanoparticles (AuNPs)

Extensive attention has been paid to gold nanoparticles (AuNPs) due to their good shape, size, optical characteristics, and biocompatibility [98]. AuNPs of several sizes and morphologies have gained significant attention in relation to applications in the field of medicine, i.e., as carriers for drugs, such as paclitaxel, tumor-detectors, photothermal agents, or radiotherapy dose enhancers [99–103]. Flower-mediated AuNPs have also shown antimicrobial and catalytic activity, which is shown in Table 3.

Table 3. Applications of gold NPs synthesized from various flower varieties.

Family	Flower Variety	Applications	Ref
Apocynaceae	*Plumeria alba* Linn	Antibacterial activity against *Escherichia coli*	[104]
Thymelaeaceae	*Gnidia glauca*	Chemocatalytic activity against 4-nitrophenol	[81]
Anacardiaceae	*Mangifera indica*	Catalytic activity against 4-nitrophenol	[82]
Asteraceae	*Tussilago farfara*	Antibacterial activity against *Enterococcus faecium*; cyrotoxicity activity	[80]

4.3. Other Nanoparticles

Metal nanoparticles based on titanium (Ti), cadmium (Cd), copper (Cu), iron (Fe), zinc (Zn), and magnesium (Mg), etc., have been emerging as a new class, owing to their exclusive applications in research (Table 4). Rosemary extract (*Rosmarinus officinalis* L.) was used in MgO nano-flower synthesis in a stirring situation at 70 °C for 4 h [86]. Marigold flower (*Tageta sp.*) petal extract was used in the synthesis of cadmium nanoparticles (CdNPs) [84]. In this study, a solution of cadmium chloride (88 ml) and petal extract (12 ml) was mixed, which resulted in a yellow nanoparticle solution with a sphereshape, as observed under a fluorescent microscope. In the combustion method, zinc nitrate [$Zn(NO_3)_2 \cdot 6H_2O$] was used as a substrate to synthesize ZnO NPs using *Syzygium aromaticum* bud and flower extract. The solution was poured into a China dish and stirred for 5–10 min at a constant temperature of 400± 10°C in a muffle furnace for 4 min to complete the entire combustion process. The synthesized SaZnO NPs of an off-white color were obtained as the final product [105]. An aqueous flower extract of *Piliostigma thonningii* was also used in the synthesis of iron nanoparticles by reacting the flower extract with a ferrous chloride solution. Reductants already present in the flower extract functioned both as reducing and stabilizing agents [87]. A *Calotropis gigantean* flower extract was used for the synthesis of titanium dioxide nanoparticles (TiO$_2$ NPs) [85]. *Mimusops elengi* flower powder was used to synthesize CuNPs and showed good antibacterial, anti-coagulant, antifungal, and anti-larval activities [106].

Table 4. Applications of other types of NPs synthesized from various flower varieties.

Family	Flower Variety	Types of Nanoparticles Synthesized	Applications	Ref
Sapotaceae	*Mimusops elengi*	Copper	Antibactrial activity against *Escherichia coli*, *Streptococcus*, *Staphylococcus*, *Pseudomonas*, and *Bacillus subtilis*; antifungal activity *Aspergillus flavus*, *Candida albicans*, *Penicillium* and *Aspergillus fumigates*; antioxidant activity; thrombolytic activity; anti-larval activity; cytotoxicity activity; heavy metals removal	[106]
Fabaceae	*Piliostigma thonningii*	Iron	Antibacterial activity against *Escherichia coli* and *Staphylococcus aureus*	[87]
Oleaceae	*Nyctanthes arbor-tristis*	Zinc	Antifungal activity against *Alternaria alternate*, *Aspergillus niger*, *Botrytis cinerea*, *Fusarium oxysporum*, and *Penicillium expansum*	[107]
Myrtaceae	*Syzygium aromaticum*	Zinc	Antifungal activity against *Fusarium graminearum*	[105]
Bignoniaceae	*Jacaranda mimosifolia*	Zinc	Antibacterial activity against *Enterococcus faecium*	[83]
Asteraceae	*Tagetes sp.*	Cadmium	Larvicidal activity against *Aedes albopictus*	[84]
Apocynaceae	*Calotropis gigantean*	Titanium	Acaricidal activity against *Rhipicephalus microplus* and *Haemaphysalis bispinosa*	[85]
Lamiaceae	*Rosmarinus officinalis* L.	Magnesium	Antibacterial activity against *Xanthomonas oryzae* pv. *oryzae*	[86]

5. Approaches Used in the Characterization of Nanoparticles

Metallic nanoparticles synthesized from extracts of several flowers of a diverse size, shape, and surface areas are categorized using different approaches, as shown in Table 5. The composition, size, structure, and crystal phase of the synthesized nanoparticles are deduced using UV–vis, XRD, FT-IR, DLS, EDS, and Raman spectroscopy. The range of the UV spectra wavelength, from 300 to 800 nm, illustrates the existence of several metallic nanoparticles of a size ranging from 2 nm to 100 nm. Usually, the detection of gold nanoparticles is conducted using UV spectroscopy in the range of 500 and 580 nm [108]. Estimation of the size of the synthesized nanoparticles, along with the quantification of the charges on the surface of the nanoparticles, is conducted using DLS analysis. The composition of the element is determined through EDAX analysis [109]. XRD is performed to recognize the size of the crystallite. FT-IR spectroscopy is used to detect the residues on the surface and the functional groups—such as flavonoids, phenols, and hydroxyls—which bond with the surface of the nanoparticles throughout the process of the synthesis for an effective reduction and stabilization.

Table 5. Synthesis and characterization of metallic NPs from various flower varieties.

Family	Flower Variety	Types of Nanoparticles Synthesized	UV-vis	TEM	SEM	FT-IR	XRD	EDX	DLS	Zeta Potential	HRTEM	AFM	GC-MS	Size	Morphology	Ref
Fabaceae	*Lablab purpureus*	Silver	✓	-	✓	✓	✓	-	-	-	-	-	-	5–50 nm	Spherical	[90]
Sapotaceae	*Mimusopselengi*	Copper	✓	-	✓	✓	✓	-	-	-	-	-	-	42–90 nm	Rod and spherical	[106]
Fabaceae	*Piliostigma thonningii*	Iron	✓	-	✓	✓	✓	-	-	-	-	-	-	20–100 μm	Rod and spherical	[87]
Oleaceae	*Nyctanthes arbor-tristis*	Zinc	✓	✓	-	-	✓	-	✓	-	-	-	-	12–32 nm	Aggregate	[107]
Apocynaceae	*Plumeria rubra*	Silver	✓	✓	-	-	-	-	-	-	-	-	-	20–80 nm	Spherical and irregular	[91]
Apocynaceae	*Catharanthus roseus*	Silver	✓	✓	✓	✓	-	-	-	-	-	-	-	6–25 nm	spherical	[79]
Fabaceae	*Cassia angustifolia*	Silver	✓	-	✓	✓	✓	✓	-	-	-	-	-	10–80 nm	Spherical	[92]
Apocynaceae	*Plumeria alba* Linn	Gold	✓	-	-	-	-	-	-	-	✓	-	-	20–30 and 80–150 nm	Spherical	[104]
Myrtaceae	*Syzygium aromaticum*	Zinc	✓	✓	✓	✓	✓	-	-	-	-	-	-	30–40 nm	Triangular and hexagonal	[105]
Thymelaeaceae	*Gnidia glauca*	Gold	✓	✓	✓	✓	✓	-	✓	-	✓	-	-	50–150 nm	Spherical	[81]
Apocynaceae	*Allamanda cathartica*	Silver	✓	✓	✓	✓	✓	-	-	-	-	-	-	39 nm	Spherical	[93]
Malvaceae	*Malva sylvestris*	Silver	✓	✓	✓	✓	-	✓	-	-	-	✓	-	20–40 nm	Spherical	[76]
Fabaceae	*Caesalpinia pulcherrima*	Silver	✓	✓	✓	✓	✓	-	-	-	-	-	-	12 nm	Spherical	[94]
Asteraceae	*Tussilago farfara*	Silver and Gold	✓	-	-	-	✓	-	-	✓	-	✓	-	13.57 and 18.20 nm	Spherical	[80]
Anacardiaceae	*Mangifera indica*	Gold	✓	-	-	-	✓	-	-	-	✓	-	-	10–60 nm	Spherical	[82]
Asteraceae	*Tagetes erecta*	Silver	✓	-	✓	✓	✓	✓	-	-	-	-	-	10–90 nm	Spherical, hexagonal, and irregular	[95]
Sapotaceae	*Madhuca longifolia*	Silver	✓	✓	✓	✓	✓	✓	-	✓	-	-	-	30–50 nm	Spherical and oval	[78]
Bignoniaceae	*Jacaranda mimosifolia*	Zinc	✓	-	-	✓	✓	-	-	-	✓	-	✓	2–4 nm	Spherical	[83]
Malvaceae	*Hibiscus rosa-sinensis*	Silver	✓	-	✓	✓	✓	✓	-	-	-	-	-	5–40 nm	Spherical	[77]
Convolvulaceae	*Ipomoea digitata* Linn	Silver	✓	-	✓	✓	✓	✓	-	-	-	-	-	111 nm	Spherical	[96]
Asteraceae	*Tagetes sp.*	Cadmium	✓	-	✓	✓	✓	-	-	-	-	-	-	50 μm	Spherical	[84]
Apocynaceae	*Calotropis gigantean*	Titanium	-	-	✓	✓	✓	✓	-	-	-	-	-	160–220 nm	Spherical	[85]
Lamiaceae	*Rosmarinus officinalis* L.	Magnesium	✓	✓	✓	-	✓	-	-	-	-	-	-	20 nm	Spherical	[86]
Asteraceae	*Chrysanthemum indicum* L.	Silver	✓	✓	-	-	✓	✓	-	-	-	-	-	25–59 nm	Spherical	[97]

UV-vis–Ultraviolet-visible spectroscopy; TEM–Transmission electron microscopy; SEM–Scanning electron microscopy; FT-IR–Fourier-transform infrared spectroscopy; XRD–X-ray powder diffraction; EDX–Energy dispersive X-ray spectroscopy; DLS–Dynamic light scattering; HRTEM–High-resolution transmission electron microscopy; AFM–Atomic force microscopy; GC-MS–Gas chromatography-mass spectroscopy.

6. Antibacterial Activity of Flower-Derived NPs

NPs should come in contact with the bacterial cells to show the antibacterial function. The NPs pass through the membrane of the bacteria, add up along the pathway for metabolism, and influence the activity of a cell [110]. Subsequently, NPs associate with the elementary components of the bacterial cell, like DNA, lysosomes, ribosomes, and enzymes, and result in oxidative stress, heterogeneous altera tions, variations in the permeability of the cell membrane, disorders related to the balance of electrolytes, an inhibition of enzymes, a deactivation of proteins, and variations in the expression of the gene.

Cell walls and membranes are significant protective checkpoints for bacterial resistance to the outside environment, and the cell wall of the bacteria plays a vital function in sustaining the bacteria's normal shape. The parts of the cell membrane of both Gram-positive and Gram-negative bacteria use diverse pathways for the adsorption of NPs [111]. Lipopolysaccharides (LPS) are an exclusive structure of the Gram-negative bacteria cell wall, which offers an area that is negatively charged for attracting NPs. On the other hand, the presence of teichoic acid is noted in Gram-positive bacteria cell walls; hence, NPs circulate throughout the phosphate molecular chain and avoid aggregation. NPs are more effective against Gram-positive than against Gram-negative bacteria, as their cell wall is made up of LPS, lipoproteins, and phospholipids, which produce a barrier that only permits the entry of macromolecules. On the other hand, cell membrane damage and cell death occur in Gram-positive bacteria, as its cell wall contains a thin sheet of peptidoglycan, teichoic acid, and ample pores, which permit the entry of foreign molecules [110].

The synthesis of AuNPs from *Plumeria alba* flower extract was conducted by adding 5 mL of flower extract to 45 mL of 0.002 M $AuCl_4$ solution [104]. The process was continued for 3–4 h in the dark, until a pale-yellow solution was obtained. Synthesized AuNPs exhibited a higher antibacterial activity, performing a synergistic interaction with antibiotics—such as imipenem, vancomycin, and norfloxacin—against *Escherichia coli*. However, AuNPs synergistic to vancomycin and norfloxacin showed more antifungal activity against *Aspergillus flavus*. Iron nanoparticles were found to be efficient for the inhibition of bacterial growth, and the maximum zone of inhibition was observed for *E. coli* (21.8±0.2 mm), followed by *Staphylococcus aureus* (20.2 ± 0.3 mm) [87]. The *Catharanthus roseus* flower has been used for the synthesis of AgNPs and showed a potential antibacterial activity against *Bacillus subtilis*, *E. coli*, *Klebsiella pneumoniae*, *Pseudomonas putida*, and *S. aureus* [79]. Padalia et al. (2014) found that AgNPs formed from the flower extract of *Tagetes erecta* showed more antibacterial activity against *S. aureus* than against *Bacillus cereus* [95]. Lee et al. (2019) reported that *Tussilago farfara* flower bud extract, containing sesquiterpenoids, was efficiently utilized as a reducing agent for AgNPs synthesis [80]. The surface plasmon resonance peak of these silver NPs was observed at 416 nm on a UV–vis spectrophotometer, and TEM images revealed the shape of these nanoparticles as spherical, with a mean size of 13.57± 3.26 nm. These AgNPs displayed a better antibacterial activity in both Gram-negative and Gram-positive bacteria than the extract, and the maximum recorded antibacterial activity was against vancomycin-resistant enterococci (Van-A type), i.e., *Enterococcus faecium*. Sharma et al. (2016) synthesized spherical zinc oxide nanoparticles (ZnO NPs), with a size of 2–4 nm, from fallen *Jacaranda mimosifolia* flower aqueous extract (JMFs) [83]. In GC-MS analysis, the oleic acid in the flower extract was found to act as a reducing and capping agent, and the presence of oleic acid stabilized ZnO NPs, showing antibacterial activity against both Gram-positive *E. faecium* and Gram-negative *E. coli* bacteria. Abdallah et al. (2019) found that magnesium oxide (MgO) NPs formed from the flower extract of *Rosmarinus officinalis* L. showed a strong inhibitory effect against biofilms of a rice pathogen, *Xanthomonas oryzae* pv. *oryzae* strain GZ 0005 [86].

7. Antioxidant Potentials of Flower-Derived NPs

The presence of a variety of phytochemicals in flowers allows their extracts to contain antioxidant properties. A flower extract of *Cassia angustifolia* contains carbonyls, phenols, nitro compounds, alkane compounds, aromatics, alkyl halides, and many other aromatic phyto-compounds, which may act as reducing, capping, and stabilizing agents for AgNPs synthesis [92]. The DPPH potential of the synthesized AgNPs showed an IC_{50} value of 47.24µg/mL. On the other hand, theAgNPs H_2O_2-IC_{50} value was found to be 78.10µg/mL, while the FRAP-IC_{50} value was recorded to be 63.21µg/mL. The phyto-synthesized AgNPs induced 50% (IC_{50}) of the anti-cancer activity against MCF 7 cells at a concentration of 73.82µg/mL, and *C. angustifolia* flower aqueous extract exhibited only a moderate activity against the tested cell line [92].

The AgNPs synthesized from flower extract of *Caesalpinia pulcherrima* were evaluated for antioxidant activity by ABTS cation radical scavenging activity, DPPH-free radical, FRAP, super oxide anion radical and reducing power assessment and showed that AgNPs were more effective in scavenging a variety of reactive oxygen species (ROS) [94]. However, at higher concentrations, the AgNPs resulted in decreased the cell viability of the HeLa cell line, and at a concentration of 200 µg/mL, AgNPs exhibited their maximum inhibition (18%), while at a concentration of 50 µg/mL, the cell viability was 23%. Whereas, an *in vivo* genotoxicity study showed that at a lower concentration, AgNPs do not cause any visibly harmful effects.

8. Catalytic Properties of Flower-Derived NPs

Generally, 4-nitrophenol and its derivatives are used in the production of herbicides, insecticides, and synthetic dyestuffs, and they can badly harm the ecosystem as a general organic pollutant of wastewater [8]. As a result of its toxic and inhibitory nature, 4-nitrophenol is considered to have a huge risk to the environment. Therefore, the reduction of these pollutants must be crucial. The 4-nitrophenol reduction product (i.e., 4-aminophenol)has been used as a mediator for paracetamol, sulfur dyes, rubber antioxidants, the making of black/white film developers, corrosion inhibitors, and precursors in antipyretic and analgesic drugs [112,113]. The use of $NaBH_4$ as a reductant and a metal catalyst for Au NPs, AgNPs, CuO NPs, and Pd NPs is the easiest and most effective approach to reduce 4-nitrophenol [114–117]. Methylene blue (MB), which is the member of the thiazine class of dyes, is another heterocyclic aromatic industrial pollutant [118]. The ingestion of MB in human body has been reported to restrict oxidase enzymes in the body, which may lead to grave disorders, i.e., toxicity of the central nervous system, gastrointestinal infections and decolorization of the brain parenchyma [119,120]. During the reduction of MB, $NaBH_4$ acts as a reducing agent, and NPs act as an absorbent [121].

Nayan et al. (2018) synthesized AuNPs using *Mangifera indica* flower extract (MIFE) [82]. In the aqueous phase, these AuNPs showed a high nano-catalysis to reduce 4-nitrophenol to 4-aminophenol through the use of $NaBH_4$ at room temperature. *Ipomoea digitata* (ID) flower extract-synthesized AgNPs were studied to check their catalytic activity against MB dye, with $NaBH_4$ as a model reducing agent [96]. The oxidized state of MB (blue color) becomes colorless when reduced to leuco-methylene blue (LMB). The addition of ID-AgNPs to the reaction mixture resulted in the formation of an intermediate between MB dye and BH_4 ions. This study showed a good catalytic ability of synthesized nanoparticles, as the catalytic reduction of MB dye by $NaBH_4$ was completed within 15 min, indicating a prospective application of the ID-AgNPsfor environmental remediation.

9. Insecticidal Properties of Flower-Derived NPs against Parasites

The insecticidal potential of flower-derived nanoparticles has also been identified. Cadmium nanoparticles (CdNPs) (10 ppm) synthesized from marigold petal extract showed a mortality rate of 68.9% against *Aedes albopictus*, while at the same concentration, CdNPs showed a 100% mortality rate after 72 h of treatment against *A. albopictus* [84]. Similarly, another study reported the killing potential of *Chryasanthemum indicum* L. floral extract-derived AgNPs [97]. In this study, different concentrations of *C. indicum* aqueous extract and synthesized AgNPs were tested against *Anopheles stephensi* mosquito larvae and pupae, and the maximum mortality was observed with the synthesized AgNPs against the vector, *A. stephensi* (LC_{50} = 5.07, 10.35, 14.19, 22.81, and 35.05 ppm; LC_{90} = 29.18, 47.15, 65.53, 87.96, and 115.05 ppm). *Calotropis gigantean* flower extract-derived TiO_2 NPs showed LC_{50} values of 9.15 mg/L and 5.43 mg/L against the larvae of *Haemaphysalis bispinosa* and against *Rhipicephalus microplus*, respectively [85].

10. Challenges in the Use of Flower-Mediated Nanoparticles

Technical barriers are the obstructions that are involved during the synthesis of flower-mediated nanoparticles. While green nanoscience has gained significant attention, efforts are still being made to standardized the protocols for the synthesis of uniform nanoparticles. Further advancements involving the use of tools and techniques for the scaled-up production of NPs through green synthesis need to be identified to design commercially feasible production technology at the industrial scale. Another pivotal issue regarding the large-scale use of green synthesized nanoparticles is nano-toxicity, which has to be addressed stringently. The toxicology and analysis protocols have to be developed and updated constantly to reflect the need of the application. Furthermore, the uncertainty and ambiguity associated with the regulatory bodies and laws has to be clearly understood to allow for the use and commercialization of ecologically safe nano-based products. The end market demands need to be made clear, as there are only limited numbers of commercial grade products that can be compared to conventional materials in terms of performance [122].

A unique idea, which still needs to be developed and established, is the use of flowers in the green synthesis of nanoparticles, as this research is still restricted to the synthesis of Au and Ag NPs. To further strengthen this field, it is important to create monodispersed nanoparticles—such as CdS, ZnO, TiO_2, and Fe_3O_2. More studies are required to recognize the various components that may lead to the reduction of metal ions. In the literature, it has been reported that proteins are responsible for the equilibrium, but it is very difficult to recognize the proteins responsible for the functionalization of these nanoparticles [123].

11. Conclusions

The use of biological materials for the production of nanoparticles has a great potential as a cost-effective and eco-friendly synthesis method for novel and innovative nanomaterials. Non-hazardous biological wastes also play a crucial role in green synthetic protocols for the generation of nanoparticles [36]. The green chemistry approach is completely different from the conventional physical and chemical processes, which frequently utilize environmentally corrosive agents with the ability to cause cytotoxicity, environmental toxicity, and carcinogenicity. On the other hand, the flower-mediated green synthesis of NPs is a vigorous method that does not require any specific isolation and maintenance procedures, which are needed in bacteria-, fungi-, or algae-based nanoparticle synthesis approaches. Flower-induced nanoparticles can exhibit specialized properties, including antimicrobial, antioxidant, catalytic, and cytotoxic activities.The present study intends to highlight the potential of flower-derived metallic nanoparticles. Of all the studied nanoparticles, Au and AgNPs were shown to be the best potential nanoparticles in terms of their effective antibacterial, antioxidant, and insecticidal activities. Bio-accumulation and toxicity are the two challenges associated with green metallic nanoparticles that prevent their use as therapeutic agents in humans and that need to be resolved through scientific intervention. With further improvement, the flower-mediated green synthesis of nanoparticles may offer important, ecofriendly end products, with wide applications, as compared to the harsh and lethal procedures used at present for the synthesis of nanoparticles.

Author Contributions: Conceptualization, A.K., K.K., and D.K.; Manuscript writing, H.K. and K.B.; Manuscript editing, E.N. and R.V.; Critical revising, A.K., K.K., E.N., R.V., and D.K. All authors have read and agreed to the published version of the manuscript.

Acknowledgments: We acknowledge university of Hradec Kralove (Faculty of Science, VT2019-2021) for financial support.

References

1. Bhattacharyya, D.; Singh, S.; Satnalika, N. Nanotechnology, big things from a tiny world: A review. *Int. J. u-e- Serv. Sci. Technol.* **2009**, *2*, 29–38.

2. Goddard, W.A., III; Brenner, D.; Lyshevski, S.E.; Iafrate, G.J. *Handbook of Nanoscience, Engineering, and Technology*, 2nd ed.; CRC Press: Boca Raton, FL, USA, 2007.

3. Khandel, P.; Shahi, S.K. Mycogenic nanoparticles and their bio-prospective applications: Current status and future challenges. *J. Nanostruct. Chem.* **2018**, *8*, 369–391. [CrossRef]

4. Bogunia-Kubik, K.; Sugisaka, M. From molecular biology to nanotechnology and nanomedicine. *BioSystem* **2002**, *65*, 123–138. [CrossRef]

5. Daniel, M.C.; Astruc, D. Gold nanoparticles: Assembly, supramolecular chemistry, quantum-size-related properties, and applications toward biology, catalysis, and nanotechnology. *Chem. Rev.* **2004**, *104*, 293–306. [CrossRef]

6. Zharov, V.P.; Kim, J.W.; Curiel, D.T.; Everts, M. Self-assembling nanoclusters in living systems: Application for integrated photothermal nanodiagnostics and nanotherapy. *Nanomedicine* **2005**, *1*, 326–345. [CrossRef] [PubMed]

7. Shah, M.; Fawcett, D.; Sharma, S.; Tripathy, S.K.; Poinern, G.E.J. Green synthesis of metallic nanoparticles via biological entities. *Materials* **2015**, *8*, 7278–7308. [CrossRef] [PubMed]

8. Singh, J.; Dutta, T.; Kim, K.H.; Rawat, M.; Samddar, P.; Kumar, P. 'Green' synthesis of metals and their oxidenanoparticles: Applications for environmental remediation. *J. Nanobiotechnol.* **2018**, *16*, 84. [CrossRef]

9. Cao, G. *Nanastructures and Nanomaterials-Synthesis, Properties and Applications*, 2nd ed.; World Scientific: Singapore, 2004.

10. Varma, R.S. Greener approach to nanomaterials and their sustainable applications. *Curr. Opin. Chem. Eng.* **2012**, *1*, 123–128. [CrossRef]

11. Nadagouda, M.N.; Varma, R.S. Green and controlled synthesis of gold and platinum nanomaterials using vitamin B2: Density assisted self-assembly of nanospheres, wires and rods. *Green Chem.* **2006**, *8*, 516–518. [CrossRef]

12. Comba, L.; Corbet, S.A.; Barron, A.; Bird, A.; Collinge, S.; Miyazaki, C.; Powell, M. Garden flowers: Insect visits and the floral reward of horticulturally-modified variants. *Ann. Bot.* **1999**, *83*, 73–86. [CrossRef]

13. Huss, E.; Yosef, K.B.; Zaccai, Z. Humans' relationship to flowers as an example of the multiple components of embodied aesthetics. *Behav. Sci.* **2018**, *8*, E32. [CrossRef] [PubMed]

14. Varela, F.J.; Thompson, E.; Rosch, E. *The Embodied Mind: Cognitive Science and Human Experience*; MIT Press: Cambridge, MA, USA, 1991.

15. Baron, R.A. The sweet smell of ... helping: Effects of pleasant ambient fragrance on prosocial behavior in shopping malls. *Pers. Soc. Psychol. Bull.* **1997**, *23*, 498–503. [CrossRef]

16. Sarid, O.; Zaccai, M. Changes in mood states are induced by smelling familiar and exotic fragrances. *Front. Psychol.* **2016**, *7*, 1–7. [CrossRef] [PubMed]

17. Shubhashree, M.N.; Shantha, T.R.; Ramarao, V.; Prathapa Reddy, M.P.; Venkateshawarulu, G. A review on therapeutic uses of flowers as depicted in classical texts of Ayurveda and Siddha. *J. Res. Educ. Indian Med.* **2015**, *21*, 1–14.

18. Varadhan, K.P. Introduction to pushpaayurveda. *Anc. Sci. Life.* **1985**, *4*, 153–157.

19. Nishteswar, K. Pushpayurveda (flowers of medicinal plants) delineated in Kaiyadevanighantu. *Punarna V.* **2015**, *2*, 1–10.

20. Puckhaber, L.S.; Stipanovic, R.D.; Bost, G.A. Analyses for flavonoid aglycones in fresh and preserved *Hibiscus* flowers. In *Trends in New Crops and New Uses*; Jules, J., Anna, W., Eds.; ASHS Press: Alexandria, VA, USA, 2002; pp. 56–563.

21. Khan, Z.S.; Shinde, V.N.; Bhosle, N.O.; Nasreen, S. Chemical composition and antimicrobial activity of angiospermic plants. *Middle-East J. Sci. Res.* **2010**, *6*, 56–61.

22. Arullappan, S.; Zakaria, Z.; Basri, D.F. Preliminary screening of antibacterial activity using crude extracts of *Hibiscus rosa-sinensis*. *Trol. Life Sci. Res.* **2009**, *20*, 109–118.

23. Ruban, P.; Gajalakshmi, K. In vitro antibacterial activity of *Hibiscus rosa-sinensis* flower extract against human pathogens. *Asian Pac. J. Trop. Biomed.* **2012**, *2*, 399–403. [CrossRef]

24. Khan, Z.A.; Naqvi, S.A.; Mukhtar, A.; Hussain, Z.; Shahzad, S.A.; Mansha, A.; Mahmood, N. Antioxidant and antibacterial activities of *Hibiscus Rosa-sinensis* Linn flower extracts. *Pak. J. Pharm. Sci.* **2014**, *27*, 469–474.

25. Meena, A.K.; Patidar, D.; Singh, R.K. Ameliorative effect of *Hibiscus rosa-sinensis* on phenylhydrazine induced haematotoxicity. *Int. J. Innov. Res. Sci. Eng. Technol.* **2014**, *3*, 8678–8683.

26. Wong, K.C.; Teng, Y.E. Volatile components of *Mimusops elengi* L. flowers. *J. Essent. Oil. Res.* **1994**, *6*, 453–458. [CrossRef]

27. Rout, P.K.; Sahoo, D.; Misra, L.N. Comparison of extraction methods of *Mimusops elengi* L. flowers. *Ind. Crops. Prod.* **2010**, *32*, 678–680. [CrossRef]

28. Sundari, U.T.; Rekha, S.; Parvathi, A. Phytochemical analysis of some therapeutic medicinal flowers. *Int. J. Pharm.* **2012**, *2*, 583–585.

29. Koppula, S.B. Antimicrobial activity of floral extracts on selected human pathogens. *Int. J. Bio-Pharm. Res.* **2013**, *2*, 141–143.

30. Reddy, L.J.; Jose, B. Evaluation of antibacterial activity of *Mimusops elengi* L. flowers and *Trichosanthes cucumerina* L. fruits from South India. *Int. J. Pharm. Pharm. Sci.* **2013**, *5*, 362–364.

31. Tuntiwachwuttikul, P.; Rayanil, K.; Taylor, W.C. Chemical constituents from the flowers of *Nyctanthes arbortristis*. *Sci. Asia.* **2003**, *29*, 21–30. [CrossRef]

32. Khatune, N.A.; Mossadik, M.A.; Rahman, M.M.; Khondkar, P.; Haque, M.E.; Gray, A.I. A benzofuranone from the flowers of *Nyctanthes arbortristis* and its antibacterial and cytotoxic activities. *Dhaka Univ. J. Pharm. Sci.* **2005**, *4*, 33–37. [CrossRef]

33. Nanu, R.; Raghuveer, I.; Chitme, H.; Chandra, R. Antidiabetic activity of *Nyctanthes arbortristis*. *Pharmacogn. Mag.* **2008**, *4*, 335–340.

34. Kim, M.R.; Lee, J.Y.; Lee, H.H.; Aryal, D.K.; Kim, Y.G.; Kim, S.K.; Woo, E.R.; Kang, K.W. Antioxidative effects of quercetin-glycosides isolated from the flower buds of *Tussilago farfara* L. *Food Chem. Toxicol.* **2006**, *44*, 1299–1307. [CrossRef]

35. Maurya, S.; Bhardwaj, A.K.; Gupta, K.K.; Agarwal, S.; Kushwaha, A.; Chaturvedi, V.K.; Pathak, R.K.; Gopal, R.; Uttam, K.N.; Soingh, A.K.; et al. Green synthesis of silver nanoparticles using *Pleurotus* and bactericidal activity. *Cell Mol. Biol.* **2016**, *62*, 131.

36. Baruwati, B.; Varma, R.S. High value products from waste: Grape pomace extract- a three-in-one package for the synthesis of metal nanoparticles. *ChemSusChem* **2009**, *2*, 1041–1044. [CrossRef] [PubMed]

37. Nadagouda, M.N.; Varma, R.S. A greener synthesis of core (Fe, Cu)-shell (Au, Pt, Pd, and Ag) nanocrystals using aqueous vitamin C. *Cryst. Growth Des.* **2007**, *7*, 2582–2587. [CrossRef]

38. Nadagouda, M.N.; Varma, R.S. Microwave-assisted shape-controlled bulk synthesis of noble nanocrystals and their catalytic properties. *Cryst. Growth Des.* **2007**, *7*, 686–690.

39. Baruwati, B.; Polshettiwara, V.; Varma, R.S. Glutathione promoted expeditious green synthesis of silver nanoparticles in water using microwaves. *Green Chem.* **2009**, *11*, 926–930. [CrossRef]

40. Polshettiwar, V.; Baruwati, B.; Varma, R.S. Self-assembly of metal oxides into three-dimensional nanostructures: Synthesis and application in catalysis. *ACS Nano* **2009**, *3*, 728–736. [CrossRef]

41. Baruwati, B.; Nadagouda, M.N.; Varma, R.S. Bulk synthesis of monodisperse ferrite nanoparticles at water–organic interfaces under conventional and microwave hydrothermal treatment and their surface functionalization. *J. Phys. Chem. C* **2008**, *112*, 18399–18404. [CrossRef]

42. Nadagouda, M.N.; Varma, R.S. Green synthesis of silver and palladium nanoparticles at room temperature using coffee and tea extract. *Green Chem.* **2008**, *10*, 859–862. [CrossRef]

43. Klaus, T.; Joerger, R.; Olsson, E.; Granqvist, C. Silver-based crystalline nanoparticles, microbially fabricated. *Proc. Natl. Acad. Sci. USA* **1999**, *96*, 13611–13614. [CrossRef]

44. Ahmad, A.; Senapati, S.; Khan, M.I.; Kumar, R.; Ramani, R.; Srinivas, V.; Sastry, M. Intracellular synthesis of gold nanoparticles by a novel alkalotolerant actinomycete, *Rhodococcus* species. *Nanotechnology* **2003**, *14*, 824–828. [CrossRef]

45. Jha, A.K.; Prasad, K.; Kulkarni, A.R. Synthesis of TiO_2 nanoparticles using microorganisms. *Colloids Surf. B Biointerfaces.* **2009**, *71*, 226–229. [CrossRef] [PubMed]

46. Saifuddin, N.; Wong, C.W.; Yasumira, A.A.N. Rapid biosynthesis of silver nanoparticles using culture supernatant of bacteria with microwave irradiation. *E J. Chem.* **2009**, *6*, 61–70. [CrossRef]

47. Prasad, K.; Jha, A.K. Biosynthesis of CdS nanoparticles: An improved green and rapid procedure. *J. Colloid. Interface Sci.* **2010**, *342*, 68–72. [CrossRef] [PubMed]

48. Dhoondia, Z.H.; Chakraborty, H. *Lactobacillus* mediated synthesis of silver oxide nanoparticles. *Nanomater. Nanotechno.* **2012**, *2*, 1–7. [CrossRef]

49. Wadhwani, S.A.; Shedbalkar, U.U.; Singh, R.; Karve, M.S.; Chopade, B.A. Novel polyhedral gold nanoparticles: Green synthesis, optimization and characterization by environmental isolate of *Acinetobacter* sp. SW30. *World J. Microbiol. Biotechnol.* **2014**, *30*, 2723–2731. [CrossRef]

50. Kowshik, M.; Ashtaputre, S.; Kharrazi, S.; Vogel, W.; Urban, J.; Kulkarni, S.K.; Paknikar, K.M. Extracellular synthesis of silver nanoparticles by a silver-tolerant yeast strain MKY3. *Nanotechnology* **2003**, *14*, 95–100. [CrossRef]

51. Li, G.; He, D.; Qian, Y.; Guan, B.; Gao, S.; Cui, Y.; Yokoyama, K.; Wang, L. Fungus-mediated green synthesis of silver nanoparticles using *Aspergillus terreus*. *Int. J. Mol Sci.* **2012**, *13*, 466–476. [CrossRef]

52. Korbekandi, H.; Ashari, Z.; Iravani, S.; Abbasi, S. Optimization of biological synthesis of silver nanoparticles using *Fusarium oxysporum*. *Iran. J. Pharm. Res.* **2013**, *12*, 289–298.

53. Gholami-Shabani, M.; Akbarzadeh, A.; Norouzian, D.; Amini, A.; Gholami-Shabani, Z.; Imani, A.; Chiani, M.; Riazi, G.; Shams-Ghahfarokhi, M.; Razzaqhi-Abyaneh, M. Antimicrobial activity and physical characterization of silver nanoparticles green synthesized using nitrate reductase from *Fusarium oxysporum*. *Appl. Biochem. Biotechnol.* **2014**, *172*, 4084–4098. [CrossRef]

54. Singaravelu, G.; Arockiamary, J.S.; Kumar, V.G.; Govindaraju, K. A novel extracellular synthesis of monodisperse gold nanoparticles using marine alga, *Sargassum wightii* Greville. *Colloids Surf. B Biointerfaces.* **2007**, *57*, 97–101. [CrossRef]

55. Venkatpurwar, V.; Pokharkar, V. Green synthesis of silver nanoparticles using marine polysaccharide: Study of in-vitro antibacterial activity. *Mater. Lett.* **2011**, *65*, 999–1002. [CrossRef]

56. Rajeshkumar, S.; Kannan, C.; Annadurai, G. Green synthesis of silver nanoparticles using marine brown algae *Turbinaria conoides* and its antibacterial activity. *Int. J. Pharm. Bio Sci.* **2012**, *3*, 502–510.

57. El-Rafie, H.M.; El-Rafie, M.H.; Zahran, M.K. Green synthesis of silver nanoparticles using polysaccharides extracted from marine macro algae. *Carbohydr. Polym.* **2013**, *96*, 403–410. [CrossRef] [PubMed]

58. Mahdavi, M.; Namvar, F.; Ahmad, M.B.; Mohamad, R. Green biosynthesis and characterization of magnetic iron oxide (Fe_3O_4) nanoparticles using seaweed *(Sargassum muticum)* aqueous extract. *Molecules* **2013**, *18*, 5954. [CrossRef] [PubMed]

59. Azizi, S.; Ahmad, M.B.; Namvar, F.; Mohamad, R. Green biosynthesis and characterization of zinc oxide nanoparticles using brown marine macroalga *Sargassum muticum* aqueous extract. *Mater. Lett.* **2014**, *116*, 275–277. [CrossRef]

60. Shankar, S.S.; Ahmad, A.; Sastry, M. *Geranium* leaf assisted biosynthesis of silver nanoparticles. *Biotechnol. Program* **2003**, *19*, 1627–1631. [CrossRef] [PubMed]

61. Chandran, S.P.; Chaudhary, M.; Pasricha, R.; Ahmad, A.; Sastry, M. Synthesis of gold nanotriangles and silver nanoparticles using *Aloe vera* plant extract. *Biotechnol. Prog.* **2006**, *22*, 577–583. [CrossRef]

62. Bar, H.; Bhui, D.K.; Sahoo, G.P.; Sarkar, P.; Pyne, S.; Misra, A. Green synthesis of silver nanoparticles using seed extract of *Jatropha curcas*. *Colloids Surf. A Physicochem. Eng. Asp.* **2009**, *348*, 212–216. [CrossRef]

63. Bar, H.; Bhui, D.K.; Sahoo, G.P.; Sarkar, P.; De, S.P.; Misra, A. Green synthesis of silver nanoparticles using latex of *Jatropha curcas*. *Colloids Surf. A Physicochem. Eng. Asp.* **2009**, *339*, 134–139. [CrossRef]

64. Dubey, S.P.; Lahtinen, M.; Sillanpää, M. Green synthesis and characterizations of silver and gold nanoparticles using leaf extract of *Rosa rugosa*. *Colloids Surf. A Physicochem. Eng. Asp.* **2010**, *364*, 34–41. [CrossRef]

65. Krishnaraj, C.; Jagan, E.G.; Rajasekar, S.; Selvakumar, P.; Kalaichelvan, P.T.; Mohan, N. Synthesis of silver nanoparticles using *Acalypha indica* leaf extracts and its antibacterial activity against water borne pathogens. *Colloids Surf. B Biointerfaces.* **2010**, *76*, 50–56. [CrossRef] [PubMed]

66. Singh, A.; Jain, D.; Upadhyay, M.K.; Khandelwal, N.; Verma, H.N. Green synthesis of silver nanoparticles using *Argemone mexicana* leaf extract and evaluation of their antimicrobial activities. *Dig. J. Nanomater. Biost.* **2010**, *5*, 483–489.

67. Yang, X.; Li, Q.; Wang, H.; Huang, J.; Lin, L.; Wang, W.; Sun, D.; Su, Y.; Opiyo, J.B.; Hong, L.; et al. Green synthesis of palladium nanoparticles using broth of *Cinnamomum camphora* leaf. *J. Nanopart Res.* **2010**, *12*, 1589–1598. [CrossRef]

68. Kumar, V.G.; Gokavarapu, S.D.; Rajeswari, A.; Dhas, T.S.; Karthick, V.; Kapadia, Z.; Shrestha, T.; Barathy, I.A.; Roy, A.; Sinha, S. Facile green synthesis of gold nanoparticles using leaf extract of antidiabetic potent *Cassia auriculata*. *Colloids Surf. B Biointerfaces* **2011**, *87*, 159–163. [CrossRef] [PubMed]

69. Zargar, M.; Hamid, A.A.; Bakar, F.A.; Shamsudin, M.N.; Shameli, K.; Jahanshiri, F.; Farahani, F. Green synthesis and antibacterial effect of silver nanoparticles using *Vitex negundo* L. *Molecules* **2011**, *16*, 6667. [CrossRef]

70. Philip, D. Honey mediated green synthesis of gold nanoparticles. *Spectrochim. Acta A Mol. Biomol. Spectrosc.* **2009**, *73*, 650–653. [CrossRef]

71. Venu, R.; Ramulu, T.S.; Anandakumar, S.; Rani, V.S.; Kim, C.G. Bio-directed synthesis of platinum nanoparticles using aqueous honey solutions and their catalytic applications. *Colloids Surf. A Physicochem. Eng. Asp.* **2011**, *384*, 733–738. [CrossRef]

72. Reddy, S.M.; Datta, K.K.R.; Sreelakshmi, C.; Eswaramoorthy, M.; Reddy, B.V.S. Honey mediated green synthesis of Pd nanoparticles for suzuki coupling and hydrogenation of conjugated olefins. *Nanosci. Nanotechnol. Lett.* **2012**, *4*, 420–425. [CrossRef]

73. Haiza, H.; Azizan, A.; Mohidin, A.H.; Halin, D.S.C. Green synthesis of silver nanoparticles using local honey. *Nano Hybrids.* **2013**, *4*, 87–98. [CrossRef]

74. Wu, L.; Cai, X.; Nelson, K.; Xing, W.; Xia, J.; Zhang, R.; Stacy, A.J.; Luderer, M.; Lanza, G.M.; Wang, L.V.; et al. A green synthesis of carbon nanoparticles from honey and their use in real-time photoacoustic imaging. *Nano Res.* **2013**, *6*, 312–325. [CrossRef]

75. Chidambaram, J.; Saritha, K.; Maheshwari, R.; Muzammil, M.S. Efficacy of green synthesis of silver nanoparticles using flowers of *Calendula officinalis*. *Chem. Sci. Trans.* **2014**, *3*, 773–777.

76. Esfanddarani, H.M.; Kajani, A.A.; Bordbar, A.K. Green synthesis of silver nanoparticles using flower extract of *Malva sylvestris* and investigation of their antibacterial activity. *IET Nanobiotechnol.* **2018**, *12*, 412–416. [CrossRef]

77. Surya, S.; Kumar, G.D.; Rajakumar, R. Green synthesis of silver nanoparticles from flower extract of *Hibiscus rosa-sinensis* and its antibacterial activity. *Int. J. Innov Res. Sci. Eng. Technol.* **2016**, *5*, 5242–5247.

78. Patil, M.P.; Singh, R.D.; Koli, P.B.; Patil, K.T.; Jagdale, B.S.; Tipare, A.R.; Kim, G.D. Antibacterial potential of silver nanoparticles synthesized using *Madhuca longifolia* flower extract as a green resource. *Micro Pathog.* **2018**, *121*, 184–189. [CrossRef] [PubMed]

79. Manisha, D.R.; Alwala, J.; Kudle, K.R.; Rudra, M.P.P. Biosynthesis of silver nanoparticles using flower extracts of *Catharanthus roseus* and evaluation of its antibacterial efficacy. *World J. Pharm. Pharm. Sci.* **2014**, *3*, 877–885.

80. Lee, Y.J.; Song, K.; Cha, S.H.; Cho, S.; Kim, Y.S.; Park, Y. Sesquiterpenoids from *Tussilago farfara* flower bud extract for the eco-friendly synthesis of silver and gold nanoparticles possessing antibacterial and anticancer activities. *Nanomaterials* **2019**, *9*, E819. [CrossRef] [PubMed]

81. Ghosh, S.; Patil, S.; Ahire, M.; Kitture, R.; Gurav, D.D.; Jabgunde, A.M.; Kale, S.; Pardesi, K.; Shinde, V.; Bellare, J.; et al. *Gnidia glauca* flower extract mediated synthesis of gold nanoparticles and evaluation of its chemocatalytic potential. *J. Nanobiotechnol.* **2012**, *10*, 17. [CrossRef] [PubMed]

82. Nayan, V.; Onteru, S.K.; Singh, D. *Mangifera indica* flower extract mediated biogenic green gold nanoparticles: Efficient nanocatalyst for reduction of 4-nitrophenol. *Environ. Prog. Sustain. Energy* **2018**, *37*, 283–294. [CrossRef]

83. Sharma, D.; Sabela, M.I.; Kanchi, S.; Mdluli, P.S.; Singh, G.; Stenström, T.A.; Bisetty, K. Biosynthesis of ZnO nanoparticles using *Jacaranda mimosifolia* flowers extract: Synergistic antibacterial activity and molecular simulated facet specific adsorption studies. *J. Photoc. Photobiol. B Biol.* **2016**, *162*, 199–207. [CrossRef]

84. Hajra, A.; Dutta, S.; Mondal, N.K. Mosquito larvicidal activity of cadmium nanoparticles synthesized from petal extracts of marigold (*Tagetes sp.*) and rose (*Rosa sp.*) flower. *J. Parasit Dis.* **2016**, *40*, 1519–1527. [CrossRef]

85. Marimuthu, S.; Rahuman, A.A.; Jayaseelan, C.; Kirthi, A.V.; Santhoshkumar, T.; Velayutham, K.; Bagavan, A.; Kamaraj, C.; Elango, G.; Iyappan, M.; et al. Acaricidal activity of synthesized titanium dioxide nanoparticles using *Calotropisgigantea* against *Rhipicephalus microplus* and *Haemaphysalis bispinosa*. *Asian J. Trop. Med.* **2013**, *6*, 682–688. [CrossRef]

86. Abdallah, Y.; Ogunyemi, S.O.; Abdelazez, A.; Zhang, M.; Hong, X.; Ibrahim, E.; Hossain, A.; Fouad, H.; Li, B.; Chen, J. The green synthesis of MgO nano-flowers using *Rosmarinus officinalis* L. (Rosemary) and the antibacterial activities against *Xanthomonas oryzae* pv. *Oryzae*. *BioMed. Res. Int.* **2019**, *2019*, 5620989. [CrossRef]

87. Igwe, O.U.; Nwamezie, F. Green synthesis of iron nanoparticles using flower extract of *Piliostigma thonningii* and their antibacterial activity evaluation. *Chem. Int.* **2018**, *4*, 60–66.

88. Xu, Z.P.; Zeng, Q.H.; Lu, G.Q.; Yu, A.B. Inorganic nanoparticles as carriers for efficient cellular delivery. *Chem. Eng. Sci.* **2006**, *61*, 1027–1040. [CrossRef]

89. Mashwani, Z.R.; Khan, M.A.; Khan, T.; Nadhman, A. Applications of plant terpenoids in the synthesis of colloidal silver nanoparticles. *Adv. Colloid Interface Sci.* **2016**, *234*, 132–141. [CrossRef]

90. Muruganantham, N.; Govindharaju, R.; Anitha, P.; Anusuya, V. Synthesis and Characterization of silver nanoparticles using *Lablab purpureus* flowers (Purple colour) and its anti-microbial activities. *Int. J. Sci. Res. Biol. Sci.* **2018**, *5*, 1–7. [CrossRef]

91. Mandal, P. Biosynthesis of silver nanoparticles by *Plumeria rubra* flower extract: Characterization and their antimicrobial activities. *Int. J. Eng. Sci. Inv.* **2018**, *7*, 1–6.

92. Bharathi, D.; Bhuvaneshwari, V. Evaluation of the cytotoxic and antioxidant activity of phyto-synthesized silver nanoparticles using *Cassia angustifolia* flowers. *BioNanoScience* **2018**, *9*, 155–163. [CrossRef]

93. Karnuakaran, G.; Jagathambal, M.; Gusev, A.; Kloesnikov, E.; Mandal, A.R.; Kuznestov, D. *Allamanda cathartica* flower's aqueous extract-mediated green synthesis of silver nanoparticles with excellent antioxidant and antibacterial potential for biomedical application. *MRS Commun.* **2016**, *6*, 41–46. [CrossRef]

94. Moteriya, P.; Chanda, S. Synthesis and characterization of silver nanoparticles using *Caesalpinia pulcherrima* flower extract and assessment of their in vitro antimicrobial, antioxidant, cytotoxic, and genotoxic activities. *Artif. Cells Nanomed. Biotechnol.* **2016**, *45*, 1556–1567. [CrossRef]

95. Padalia, H.; Moteriya, P.; Chanda, S. Green synthesis of silver nanoparticles from marigold flower and its synergistic antimicrobial potential. *Arab. J. Chem.* **2014**, *8*, 732–741. [CrossRef]

96. Varadavenkatesan, T.; Selvaraj, R.; Vinayagam, R. Dye degradation and antibacterial activity of green synthesized silver nanoparticles using *Ipomoea digitata* Linn. flower extract. *Int. J. Environ. Sci. Te.* **2019**, *16*, 2395–2404. [CrossRef]

97. Arokiyaraj, S.; Kumar, V.D.; Elakya, V.; Kamala, T.; Park, S.K.; Saravanan, M.; Bououdina, M.; Arasu, M.V.; Kovendan, K.; Vincent, S. Biosynthesized silver nanoparticles using floral extract of *Chrysanthemum indicum* L.-potential for malaria vector control. *Environ. Sci. Pollut. Res.* **2015**, *22*, 9759–9765. [CrossRef] [PubMed]

98. Elia, P.; Zach, R.; Hazan, S.; Kolusheva, S.; Porat, Z.; Zeiri, Y. Green synthesis of gold nanoparticles using plant extracts as reducing agents. *Int. J. Nanomed.* **2014**, *9*, 4007–4021.

99. Hainfeld, J.F.; Slatkin, D.N.; Smilowitz, H.M. The use of gold nanoparticles to enhance radiotherapy in mice. *Phys. Med. Biol.* **2004**, *49*, N309–N315. [CrossRef] [PubMed]

100. Gibson, J.D.; Khanal, B.P.; Zubarev, E.R. Paclitaxel-functionalized gold nanoparticles. *J. Am. Chem. Soc.* **2007**, *129*, 11653–11661. [CrossRef]

101. Qian, X.; Peng, X.H.; Ansari, D.O.; Yin-Goen, Q.; Chen, G.Z.; Shin, D.M.; Yang, L.; Young, A.N.; Wang, M.D.; Nie, S. *In vivo* tumor targeting and spectroscopic detection with surface-enhanced Raman nanoparticle tags. *Nat. Biotechnol.* **2008**, *26*, 83–90. [CrossRef]

102. Hainfeld, J.F.; Dilmanian, F.A.; Zhong, Z.; Slatkin, D.N.; Kalef-Ezra, J.A.; Smilowitz, H.M. Gold nanoparticles enhance the radiation therapy of a murine squamous cell carcinoma. *Phys. Med. Biol.* **2010**, *55*, 3045–3059. [CrossRef]

103. McMahon, S.J.; Hyland, W.B.; Muir, M.F.; Coulter, J.A.; Jain, S.; Butterworth, K.T.; Schettino, G.; Dickson, G.R.; Hounsell, A.R.; O'Sullivan, J.M.; et al. Biological consequences of nanoscale energy deposition near irradiated heavy atom nanoparticles. *Sci. Rep.* **2011**, *1*, 18. [CrossRef]

104. Nagaraj, B.; Malakar, B.; Divya, T.K.; Krishnamurthy, N.B.; Liny, P.; Dinesh, R. Environmental benign synthesis of gold nanoparticles from the flower extracts of *Plumeria alba* Linn. (Frangipani) and evaluation of their biological activities. *Int. J. Drug Dev. Res.* **2012**, *4*, 144–150.

105. Lakshmeesha, T.R.; Kalagatur, N.K.; Mudili, V.; Mohan, C.D.; Rangappa, S.; Prasad, B.D.; Ashwini, B.S.; Hashem, A.; Alqarawi, A.A.; Malik, J.A.; et al. Biofabrication of zinc oxide nanoparticles with *Syzygium aromaticum* flower buds extract and finding its novel application in controlling the growth and mycotoxins of *Fusarium graminearum*. *Front. Microbiol.* **2019**, *10*, 1244. [CrossRef] [PubMed]

106. Sarah, S.L.R.; Iyer, P.R. Green synthesis of copper nanoparticles from the flowers of *Mimusops elengi*. *Int. J. Recent. Sci. Res.* **2019**, *10*, 32956–32963.

107. Jamdagni, P.; Khatri, P.; Rana, J.S. Green synthesis of zinc oxide nanoparticles using flower extract of *Nyctanthes arbor-tristis* and their antifungal activity. *J. King Saud. Univ. Sci.* **2016**, *30*, 168–175. [CrossRef]

108. Mishra, A.; Tripathy, S.K.; Yun, S.I. Fungus mediated synthesis of gold nanoparticles and their conjugation with genomic DNA isolated from *Escherichia coli* and *Staphylococcus aureus*. *Process. Biochem.* **2012**, *47*, 701–711. [CrossRef]

109. Jiang, J.; Oberdörster, G.; Biswas, P. Characterization of size, surface charge, and agglomeration state of nanoparticles dispersions for toxicological studies. *J. Nanopart Res.* **2009**, *11*, 77–89. [CrossRef]

110. Wang, L.; Hu, C.; Shao, L. The antimicrobial activity of nanoparticles: Present situation and prospects for the future. *Int. J. Nanomed.* **2017**, *12*, 1227–1249. [CrossRef] [PubMed]

111. Lesniak, A.; Salvati, A.; Santos-Martinez, M.J.; Radomski, M.W.; Dawson, K.A.; Åberg, C. Nanoparticle adhesion to the cell membrane and its effect on nano particle uptake efficiency. *J. Am. Chem. Soc.* **2013**, *135*, 1438–1444. [CrossRef]

112. Panigrahi, S.; Basu, S.; Praharaj, S.; Pande, S.; Jana, S.; Pal, A.; Ghosh, S.K.; Pal, T. Synthesis and size-selective catalysis by supported gold nanoparticles: Study on heterogeneous and homogeneous catalytic process. *J. Phys. Chem. C* **2007**, *111*, 4596–4605. [CrossRef]

113. Woo, Y.; Lai, D.Y. Aromatic amino and nitro-amino compounds and their halogenated derivatives. In *Patty's Toxicology*; Bingham, E., Cohrssen, B., Powell, C.H., Eds.; Wiley: Hoboken, NJ, USA, 2012.

114. Sharma, J.K.; Akhtar, M.S.; Ameen, S.; Srivastva, P.; Singh, G. Green synthesis of CuO nanoparticles with leaf extract of *Calotropis gigantea* and its dye-sensitized solar cells applications. *J. Alloys Compd.* **2015**, *632*, 321–325. [CrossRef]

115. Lim, S.H.; Ahn, E.Y.; Park, Y. Green synthesis and catalytic activity of gold nanoparticles synthesized by *Artemisia capillaries* water extract. *Nanoscale Res. Lett.* **2016**, *11*, 474. [CrossRef]

116. Rostami-Vartooni, A.; Nasrollahzadeh, M.; Alizadeh, M. Green synthesis of perlite supported silver nanoparticles using *Hamamelis virginiana* leaf extract and investigation of its catalytic activity for the reduction of 4-nitrophenol and Congo red. *J. Alloys Compd.* **2016**, *680*, 309–314. [CrossRef]

117. Gopalakrishnan, R.; Loganathan, B.; Dinesh, S.; Raghu, K. Strategic green synthesis, characterization and catalytic application to 4-nitrophenol reduction of palladium nanoparticles. *J. Clust. Sci.* **2017**, *28*, 2123–2131. [CrossRef]

118. Senobari, S.; Nezamzadeh-Ejhieh, A. A comprehensive study on the enhanced photocatalytic activity of CuO-NiO nanoparticles: Designing the experiments. *J. Mol. Liq.* **2018**, *261*, 208–217. [CrossRef]

119. Chen, A.; Chen, W.; Latham, P. 10 Fatal Cholangiocarcinoma in the setting of treatment-resistant hepatitis C virus infection. *Am. J. ClinPathol.* **2018**, *149*, S4. [CrossRef]

120. Sultan, M.; Waheed, A.; Bibi, I.; Islam, A. Ecofriendly reduction of methylene blue with polyurethane catalyst. *Int. J. Polym Sci.* **2019**, *2019*, 3168618. [CrossRef]

121. Begum, R.; Najeeb, J.; Sattar, A.; Naseem, K.; Irfan, A.; Al-Sehemi, A.G.; Farooqi, Z.H. Chemical reduction of methylene blue in the presence of nanocatalysts: A critical review. *Rev. Chem. Eng.* **2019**. [CrossRef]

122. Matus, K.J.M.; Hutchison, J.E.; Peoples, R.; Rung, S.; Tanguay, R.L. Green Nanotechnology Challenges and Opportunities. Available online: https://greennano.org/sites/greennano2.uoregon.edu/files/GCI_WP_GN10.pdf (accessed on 2 November 2019).

123. Balasooriya, E.R.; Jayasinghe, C.D.; Jayawardena, U.A.; Ruwanthika, R.W.D.; de Silva, R.M.; Udagama, P.V. Honey mediated green synthesis of nanoparticles: New era of safe nanotechnology. *J. Nanomater.* **2017**, *2017*, 5919876. [CrossRef]

Green Synthesis of Gold and Silver Nanoparticles using Leaf Extract of *Clerodendrum inerme*; Characterization, Antimicrobial and Antioxidant Activities

Shakeel Ahmad Khan [1,*] ⓘ, **Sammia Shahid** [2] ⓘ and **Chun-Sing Lee** [1,*]

1   Center of Super-Diamond and Advanced Films (COSDAF) and Department of Chemistry,
     City University of Hong Kong, 83 Tat Chee Avenue, Kowloon 999077, Hong Kong
2   Department of Chemistry, School of Science, University of Management and Technology, Lahore 54770,
     Pakistan; sammia.shahid@umt.edu.pk
*   Correspondence: shakilahmadkhan56@gmail.com (S.A.K.); apcslee@cityu.edu.hk (C.-S.L.)

Abstract: Due to their versatile applications, gold (Au) and silver (Ag) nanoparticles (NPs) have been synthesized by many approaches, including green processes using plant extracts for reducing metal ions. In this work, we propose to use plant extract with active biomedical components for NPs synthesis, aiming to obtain NPs inheriting the biomedical functions of the plants. By using leaves extract of *Clerodendrum inerme* (*C. inerme*) as both a reducing agent and a capping agent, we have synthesized gold (CI-Au) and silver (CI-Ag) NPs covered with biomedically active functional groups from *C. inerme*. The synthesized NPs were evaluated for different biological activities such as antibacterial and antimycotic against different pathogenic microbes (*B. subtilis, S. aureus, Klebsiella,* and *E. coli*) and (*A. niger, T. harzianum,* and *A. flavus*), respectively, using agar well diffusion assays. The antimicrobial propensity of NPs further assessed by reactive oxygen species (ROS) glutathione (GSH) and FTIR analysis. Biofilm inhibition activity was also carried out using colorimetric assays. The antioxidant and cytotoxic potential of CI-Au and CI-Ag NPs was determined using DPPH free radical scavenging and MTT assay, respectively. The CI-Au and CI-Ag NPs were demonstrated to have much better antioxidant in terms of %DPPH scavenging (75.85% ± 0.67% and 78.87% ± 0.19%), respectively. They exhibited excellent antibacterial, antimycotic, biofilm inhibition and cytotoxic performance against pathogenic microbes and MCF-7 cells compared to commercial Au and Ag NPs functionalized with dodecanethiol and PVP, respectively. The biocompatibility test further corroborated that CI-Ag and CI-Au NPs are more biocompatible at the concentration level of 1–50 μM. Hence, this work opens a new environmentally-friendly path for synthesizing nanomaterials inherited with enhanced and/or additional biomedical functionalities inherited from their herbal sources.

Keywords: green synthesis; *C. inerme*; gold; silver; antibacterial; antimycotic; antioxidant

1. Introduction

Development of multi-drug resistance (MDR) in bacterial strains, including *Enterococci, Staphylococci, Klebsiella, Acinetobacter, Pseudomonas, Enterobacter* species, etc., has become a severe challenge [1]. These bacteria are displaying resistance to world-leading antibiotics, including Cephalosporins, Carbapenems, Vancomycin, and Methicillin [2]. Furthermore, some fungi, for example, *Candida* species, are also showing resistance to various antifungal drugs, such as azole [3]. These pathogenic microbes are causing life-threatening diseases, such as aspergillosis, candidiasis, pneumocystis pneumonia, sepsis, osteomyelitis, meningitis, cholecystitis, severe bacteremia, diarrhea, tuberculosis [4,5].

Numerous antimicrobial drugs have been developed; however, because of the emergence of MDR in pathogens, clinical efficiency of existing drugs is vulnerable. Microbial spices are showing resistance to antimicrobial medicines by enzymatic deactivation and altering the drug target sites, decreasing antibiotics cell wall permeability, and displaying efflux mechanisms [6]. According to the World Health Organization (WHO), current deaths due to microbial diseases are ~0.7 million per year, if we could not develop efficient drugs to control or destroy these pathogenic microbes, the death caused by microbial diseases may rise to ~10 million by 2050 [7]. Therefore, this has become obligatory to find out alternate routes to tackle these MDR pathogenic microbes.

Nanotechnology has gained much attention for confronting these challenges. Nanomaterials of metal (Au, Ag, Se, etc.) and their oxides (CuO, ZnO, NiO, MnO, etc.) have been exploited as antibacterial agents, targeted drug delivery vehicles, antimycotic agents, antioxidant agents, anticancer agents, etc. [8–14]. Among them, Au and Ag NPs are of high significance due to their unique properties. They are extensively employed as anti-inflammatory, antibacterial, and antifungal agents in the coating of catheters, disinfecting medical devices, antimicrobial filters, dental hygiene, eye treatments, and wound dressings [8,9]. With nanometer sizes, these NPs can easily penetrate cell walls and cell membranes of pathogenic microbes in comparison to conventional antibacterial and antifungal drugs. This is a critical factor for their superior antimicrobial properties.

Numerous approaches have been used for preparing these metal NPs. These include physical (laser ablation, arc discharging, photolithography, ball milling, etc.), chemical (sol-gel, solvothermal, co-precipitation, pyrolysis, chemical redox reaction, etc.) and biological (plants, fungi, bacteria, virus, yeast, etc.) methods [15]. Physical and chemical methods often involve the use of toxic chemicals and solvents, which could have a harmful impact on the environment. In fact, the presence of residual hazardous chemical species on the surface of the synthesized NPs cannot be removed easily and could prohibit their biological and clinical applications. Moreover, their production often demands more energy and is not easily scalable [16]. Therefore, the use of biological methods for synthesizing NPs has gained much consideration as an alternative because it uses natural resources and is believed to be more biocompatible [17]. Synthesis of NPs using plants has gained tremendous attention over the last five years. It has eradicated complex steps, including maintenance of microbial cell culture, prolonged incubation time, several purification steps, etc., required for NPs synthesis using microorganisms such as fungi, bacteria, and yeast. Moreover, the usage of plants is considered more effective, easily scalable, and economical than other biological methods [16,17].

So far, uses of different plants for the synthesis metal nanoparticles are employed mostly for the lower manufacturing cost, easy scalability, and environmental friendliness. However, it should be noted that many plants have intrinsic biomedical applications that stem from their biologically active components, including polyphenols, alkaloids, saponins, terpenoids, flavonoids, etc. Here, we propose that if plants with intrinsic biomedical applications are used for preparing metal NPs, the obtained NPs might be capped with some of the biologically active components and thus inheriting their biomedical functions. To implement this concept, we have chosen leaves of C. inerme, which has been widely used for treating venereal infections, cough, fever, skin diseases, microbial infections, rheumatism, leprosy, etc., as the raw materials for preparing Au and Ag NPs [18–20]. By reducing Ag^+ and Au^{3+} using C. inerme extract, Ag and Au NPs capped with various functional groups are obtained. It was found that these C. inerme –derived NPs show much better performance compared to commercial as well as other reported plant-derived Au and Ag NPs in terms of antimicrobial, antioxidant, ROS generation performance. To the best of our knowledge, it is the first demonstration that gold and silver NPs prepared with a C. inerme extract can inherit active biomedical components of the plant.

2. Materials and Methods

2.1. Chemicals

Analytical grade chemicals and reagents were purchased from Sigma-Aldrich or Merck. Leaves of the plant *"Clerodendrum inerme"* were collected from the hall area of City University Hong Kong on 10 September 2019. *C. inerme* leaves were identified and authenticated by taxonomist Professor Mansoor Hamid, Department of Botany, University of Agriculture Faisalabad, Pakistan.

2.2. Plant Extract Preparation

C. inerme leaves were first gently washed with deionized (DI) water to remove dust particles. Washed leaves were placed under shady areas for drying at room temperature (25–30 °C). Dried leaves were blended into powder with a commercial blender. After removing larger particles with a 200-mesh sieve (pore diameter of 0.074 mm), 10 g of the sifted leaves powder was added to 100 mL of DI water. The mixture was then heated to boil with continuous stirring for 5 min. The boiled mixture was further filtered with a sintered glass crucible to obtain a yellowish colored extract. The leaves extract of *C. inerme* was then stored in an airtight glass bottle placed in a refrigerator at 4 °C until further use (Figure 1).

Figure 1. Schematic showing extract preparation and green synthesis of CI-Au and CI-Ag nanoparticles (NPs) using aqueous leaves extract of *Clerodendrum inerme*.

2.3. Synthesis of CI-Au and CI-Ag NPs

The leaf extract was then used as a reducing and capping agent to convert gold and silver salts to their metallic forms (Figure 1). A total of 1 mM of, respectively, $HAuCl_4.3H_2O$ and $AgNO_3$ were added into 25 mL of leaves extract of *C. inerme*. The mixtures of gold and silver salts were both heated at 80 °C for 65 min with continuous stirring to obtain a ruby red and a dark brown dispersion of CI-Au and CI-Ag NPs, respectively (Figure 1). After centrifugation at 15,000 rpm for 15 min, the obtained CI-Au and CI-Ag NPs were washed with DI water three times and dried in an oven at 70 °C.

2.4. Characterizations

Standard transmission electron microscopy (TEM), X-ray diffraction (XRD), FTIR and UV-Vis spectroscopies, dynamic light scattering measurements were used for characterizing the compositions and structures of the synthesized CI-Au and CI-Ag NPs. Their performance in terms of antioxidant, antibacterial and antimycotic activities, and capability of reactive oxygen generation, as well as biocompatibility, were evaluated as described below. In addition to the CI-Au and CI-Ag NPs, for comparison, we also carried out the same measurements for the leaf extract, standard Au NPs functionalized with dodecanethiol (catalog no. 660434), and Ag NPs functionalized with PVP (catalog no. AGPB5-1M) purchased from Sigma-Aldrich and nanoComposix, respectively, as well as prototypical standard compounds including butylated hydroxytoluene (BHT, an antioxidant), Cephradine (an antibiotic drug), terbinafine hydrochloride (an antifungal medicine).

2.5. Antioxidant Activity

Antioxidant activity of the CI-Au NPs, the CI-Ag NPs, were determined by using a DPPH free radical scavenging assay [12]. In a typical experiment, 0.1 mM solution of DPPH was prepared in ethanol. Aqueous solution/dispersion of each sample with concentrations from 125 to 1000 µg/mL was prepared separately in order to evaluate concentration-dependent antioxidant potential. Each sample solution/dispersion was respectively mixed with a DPPH solution. The resultant reaction mixtures were stirred for 10 min at room temperature and set aside for 1 h. The antioxidant activity in terms of DPPH scavenging was then determined via the optical absorbance (As) at 517 nm measured with a UV-Visible spectrophotometer. Percentage of DPPH free radical scavenging was calculated by using the following equation:

$$\% \text{ DPPH free radical scavenging} = [(A_c - A_s)/A_c] \times 100 \tag{1}$$

where, A_s is the absorbance of the sample, and A_c is the absorbance of the control (only DPPH solution).

2.6. Antibacterial Propensity

The antibacterial propensity of all samples against two Gram-positive bacterial strains (*B. subtilis* ATCC 6051 and *S. aureus* ATCC 15564) and two Gram-negative bacterial strains (*Klebsiella* ATCC 13883 and *E. coli* ATCC BAA-196) were evaluated with the standard well diffusion method [21]. In a typical experiment, by means of a sterilized cotton-swab, a suspension of bacterial strains at a concentration of 5×10^5 CFU/mL was swabbed onto Mueller-Hinton agar plates. Each sample of 50 µL with a concentration of 250 µg/mL was separately added into the wells of 6 mm in diameter. After hatching for 24 h at 37 °C, the sizes of bacteria inhibition zones on the plates were measured.

2.7. Minimum Inhibitory Concentrations

The minimum inhibitory concentration (MICs) of the green synthesized CI-Au NPs, and CI-Ag NPs was determined in comparison with other samples following the protocol reported by [9] with slight modifications. In a typical procedure, the bacteriological strains at a concentration of 5×10^5 CFU/mL were inoculated into 96-well plates. After that, 100 µL of Mueller Hinton broth containing different concentrations (40, 35, 30, 25, 12.5, 6.25, 3.125, 1.562, and 0.781 µg/mL) of each sample was serially diluted into a well of 96-well plates and incubated for 24 h at 37 °C. The 10 µL of 0.5% freshly prepared MTT (3-(4,5-Dime-thylthiazol-2-yl)-2,5-Diphenyltetrazolium Bromide) was added and incubated for 2 h in the dark. The 100 µL of DMSO (0.5%) was then added to solubilize the crystals of formazan and kept in the dark for 30 min. Finally, the optical density (OD) was measured at 595 nm wavelength to determine the percentage of bacterial cell death.

$$\text{Percentage of bacterial inhibition} = [(OD_{control} - OD_{treatment})/OD_{control}] \times 100 \tag{2}$$

The MIC of each sample was determined based on the lowest concentration of the sample required to prohibit the 80% growth of the bacteriological strains.

2.8. Antimycotic Activity

The green synthesized CI-Au NPs, and CI-Ag NPs were assessed for their antimycotic activity against three pathogenic fungal strains (*Aspergillus niger*, *Aspergillus flavus*, and *Trichoderma harzianum*) using the standard agar well diffusion assay [22]. In a typical procedure, a freshly prepared autoclaved solution of potato dextrose (25 mL) was transferred to autoclaved petri dishes. Then 1 mL inoculum of already cultured fungal strains was transferred to each petri dish. Petri dishes were put aside for a while to allow solidification of the whole medium. Wells of 2 mm were bored from the solidified agar gel at four peripheral positions of each petri dish with a sterilized hollow iron tube. A total of 50 μL of each sample at 250 μg/mL was respectively added to the four wells on one petri dish. The Petri dishes were set aside for 1 h. Finally, the sizes of the fungal inhibition zone were measured after incubation at 25 °C for 24 h.

2.9. Biofilm Inhibition Activity

The green synthesized CI-Au NPs and CI-Ag NPs were assessed for their biofilm inhibition activity against pathogenic bacterial (*B. subtilis*, *S. aureus*, *Klebsiella*, and *E. coli*) and fungal strains (*Aspergillus niger*, *Aspergillus flavus*, and *Trichoderma harzianum*). In brief, biofilms of the microbial strains were developed using suitable media (TSB for bacteria and RPMI for fungi) in the 96-microtiter plate (10^7 cells/well) at 37 °C for 24 h. The planktonic cells were then separated, and each well washed three times with PBS (phosphate buffer saline). After 50 μL of each sample at different concentration levels (0–120 μg/mL) was added separately into each well of 96-well plates. The 96-microtiter well plate was further incubated at 37 °C for 24 h. After drug treatment, the biofilms containing wells were gently rewashed with PBS. The staining agent, 90 μL of XTT, and 10 μL of phenazine methosulfate were added in each well and incubated in the dark at 37 °C for 4 h. The optical density (OD) was measured at 492 nm wavelength. The percentage of biofilm inhibition was calculated using the following equation

$$\text{Percentage of biofilm inhibition} = [(OD_{control} - OD_{treatment})/OD_{control}] \times 100 \qquad (3)$$

The minimum biofilm inhibitory concentration (MBIC) was determined as the lowest concentration of the drug molecule at which no biofilm formation of pathogenic microbes occur.

2.10. FT-IR Analysis of Bacterial and Fungal Strains

FTIR analysis of bacterial and fungal strains was carried out to identify molecular functionalities changes after their treatment with the green synthesized CI-Au NPs and CI-Ag NPs using FTIR spectrophotometer. At first, the greens synthesized NPs at the concentration of 250 μg/mL were employed to treat *E. coli* and *A. flavus* (5×10^5 CFU/mL). Afterward, the microbial cells were obtained upon centrifugation (10,000 rpm, 15 min), and then their pellets formation (control and treated) was achieved upon treatment with KBr (1:100 ratios). Finally, they were analyzed using an FT-IR spectrophotometer.

2.11. Intracellular Reactive Oxygen Species (ROS) Analysis

Capability for producing intracellular ROS was investigated by employing 2′,7′-dichlorodihydrofluorescein diacetate (H_2-DCFDA) as a probe, as described in the literature [23]. In a typical procedure, for microbial cells (E. coli and A. flavus) (10^5 CFU/mL) incubation with a probe at 37 °C, 200 μM concentration of DCFH-DA was employed, followed by adding 50 μL of each sample at 250 μg/mL. After that, the incubation of microbial cells was further continued for 4 h at 37 °C. The results of intracellular ROS generation were recorded by measuring fluorescence emission at 523 nm and excitation at 503 nm using a Varian Eclipse spectrofluorometer. The results of NPs treated microbial cells were compared with 1 mM H_2O_2 treated (positive control) and untreated cells (negative control) to determine the ROS production capability.

2.12. Intracellular Glutathione (GSH) Investigation

The investigation of intracellular GSH production was performed following the procedure, as stated by Park et al., with minor amendments [24]. In a typical process, the microbial cells (10^5 CFU/mL) were treated with 50 μL of each sample at 250 μg/mL, and subsequently, by using 5% TCA (trichloroacetic acid), they were lysed on ice for 15 min. After, 100 μL of cell lysate was treated with 900 μL of Tris–HCl (pH 8.3) and 100 μL of 1 mg/mL o-phthaldialdehyde solution. The resultant reaction mixtures were then subjected for incubation at 30 °C in the dark for 1:30 h. The fluorescence intensity of each sample with emission and excitation wavelengths of 420 and 350 nm, respectively, was recorded by employing the Varian Eclipse spectrofluorometer. The results of NPs treated microbial cells were compared with 1 mM H_2O_2 treated (Positive control) and untreated cells (negative control).

2.13. Cytotoxicity Activity

The cytotoxicity activity of the green synthesized CI-Au, and CI-Ag NPs was determined against the MCF-7 cancerous cell lines compared to plant extract, Au, and Ag NPs following the MTT colorimetric protocol. The MCF-7 cancerous cells were placed in Dulbecco's Modified Eagle's Medium (DMEM) provided with streptomycin (100 μg/mL), penicillin (100 U/mL), and 10% FBS (fetal bovine serum) in a humidified atmosphere consisting of 5% CO_2 and 95% air at 37 °C. The MCF-7 cancerous cells were cultured in 150 μL of DMEM in a 96-microtiter plate for 24 h at 37 °C in 5% CO_2 to get cell-confluency up to 5×10^5 cells/well. After 50 μL of NPs, plant extract, and standard drug at the concentration of 100 μg/mL were added separately in each well containing cultured cancerous cells, and the plate was then incubated for 24 h at 37 °C. Afterward, the cells were subjected to centrifugation for removing the supernatant, and then cells were washed with PBS solution. The 10 μL of MTT at 0.6 mg mL^{-1} concentration was added to each well, and the plate was further incubated again at 37 °C for 4 h. The DMSO at 100 μL volume was then transferred to each well for solubilizing the un-dissolved formazan crystals and placed the well plate on the shaker for 20 min. After, formazan's absorption spectrum at 570 nm with reference at 655 nm was determined in each well employing the Varian Eclipse spectrophotometer and the cell viability (%) was calculated using the given formula:

$$\text{Cell viability (\%)} = (\text{OD}_{\text{value of treated cells}})/(\text{OD}_{\text{value of negative control}}) \times 100 \qquad (4)$$

Treatment of cells with the standard drug (doxorubicin) was named as the positive control while cancerous cells without any treatment served as the negative control.

2.14. Biocompatibility

Hemolytic activity was carried out to determine the biocompatibility of the green synthesized CI-Au and CI-Ag NPs in comparison to the purchased metal NPs via the standard protocol, as reported by Khan et al. [18].

2.15. Statistical Analysis

All the experiments were repeated triplicates, and the results were presented as mean ± standard deviation. To determine the statistical difference, we have performed ANOVA analysis at a fixed significance level (0.05). Moreover, pairs Tukey's test carried out to find out the significant pairs.

3. Results

3.1. Compositions and Structures Analysis

XRD patterns of the green synthesized CI-Au NPs, and CI-Ag NPs using C. inerme leaves extract are shown in Figure 2. The peak positions match well to those of metallic gold and silver, respectively, and show no observable impurity.

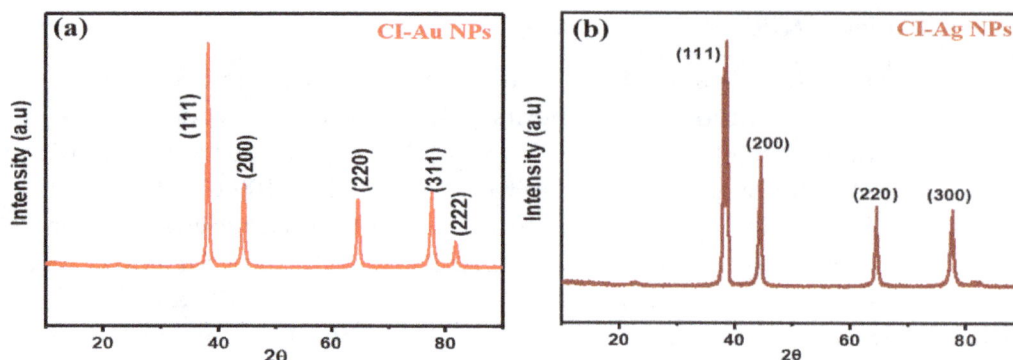

Figure 2. XRD patterns of green synthesized (**a**) CI-Au NPs and (**b**) CI-Ag NPs using leaves extract of *C. inerme*.

TEM images (Figure 3a,b) of the samples show that both samples are spherical in morphology, and they have average sizes 5.82 (CI-Au NPs) and 5.54 nm (CI-Ag NPs), respectively (Figure 3c). Further, the DLS particle size distribution of green synthesized CI-Au NPs, and CI-Ag NPs was also verified with the histogram generated by TEM (Figure S1). The composition of the NPs was measured using an energy dispersive X-ray spectrometer attached to the TEM. EDX spectra (Figure 3d,e) confirm that the two samples consist mainly of gold and silver, respectively. Peaks from C, O, and N are attributed to signal from the surface functional groups (e.g., polyphenols, flavonoids, proteins, etc.) on the nanoparticles as well as the holey carbon film, which holds the nanoparticle samples.

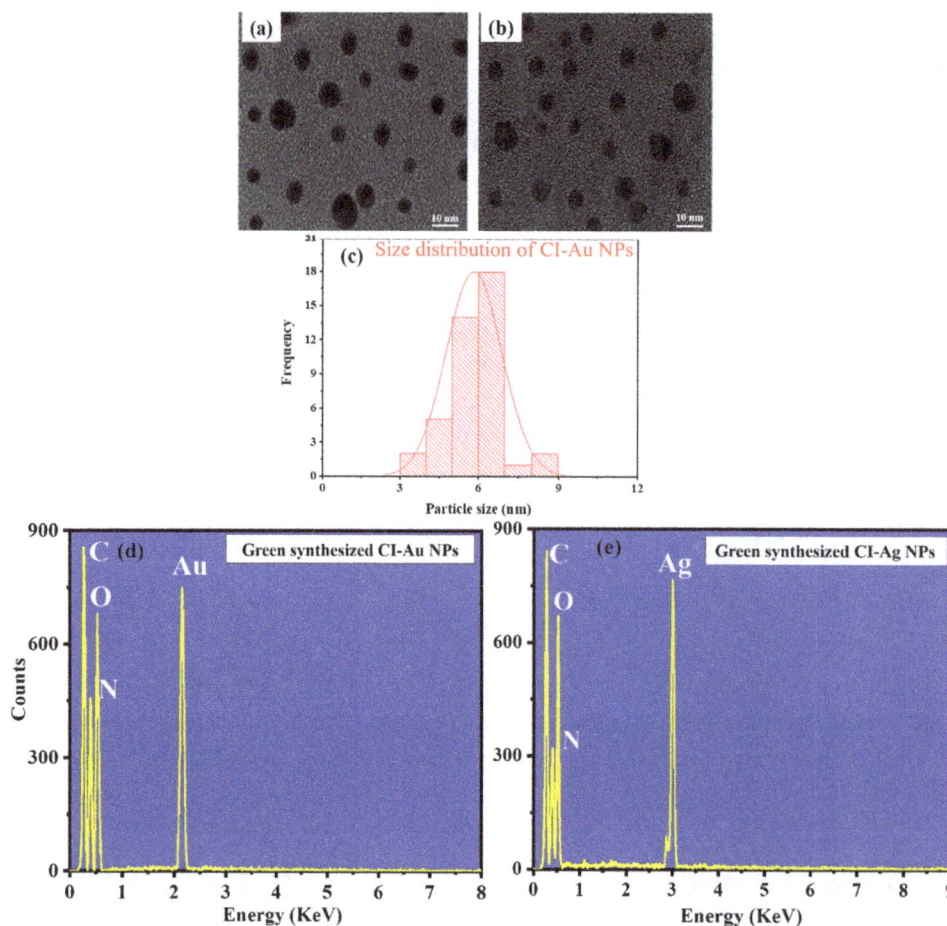

Figure 3. (**a**,**b**) TEM, (**c**) DLS, and (**d**,**e**) EDX spectra of green synthesized CI-Au and CI-Ag NPs, respectively, using leaves extract of *C. inerme*.

Absorption spectra of the leaf extract and the nanoparticle are shown in Figure 4. Due to the surface plasmonic resonance phenomenon, the maximum absorption bands were observed at 534 nm for CI-Au NPs (Figure 4c) while for CI-Ag NPs at 412 nm (Figure 4b). Moreover, the leaf extract of *C. inerme* exhibited an absorption band at 380 nm (Figure 4a), which can be attributed to the absorption of polyphenols and flavonoids [9,25,26]. Alfuraydi et al. and Latha et al. reported the similar absorption spectrum for the green synthesized silver and gold NPs respectively [27,28].

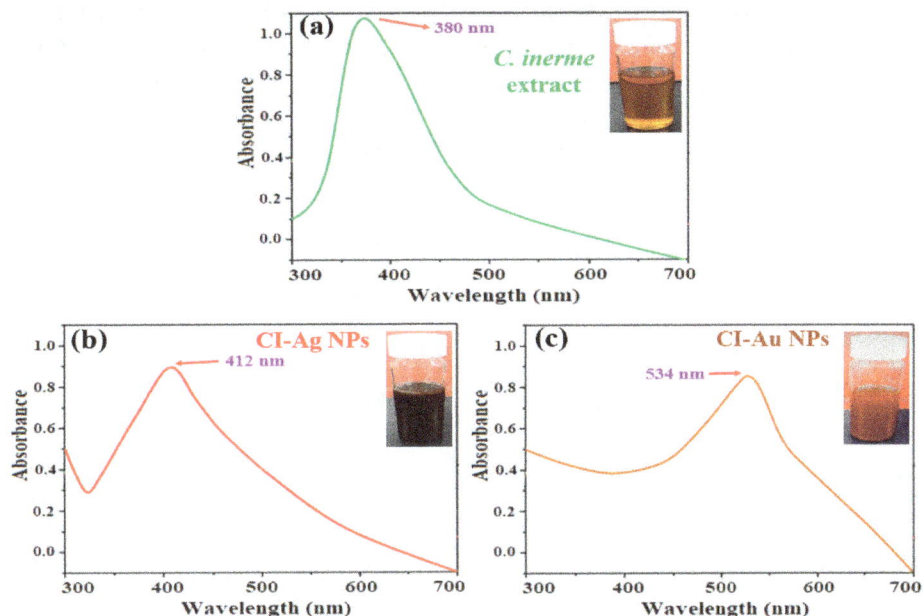

Figure 4. Absorption spectra of (**a**) extract of *C. inerme* leaves, (**b**) CI-Au NPs and (**c**) CI-Ag NPs.

FT-IR spectral study was carried out to characterize surface functional groups on the CI-Au NPs and the CI-Ag NPs. It can be seen from Figure 5 that types of nanoparticles show a FTIR signal corresponding to aromatic C=C (1520–1590 cm^{-1}), C-H (2750–2860 cm^{-1}), O-H (3310–3390 cm^{-1}), N-H (1415–1490 cm^{-1}), C–O–C (1025–1195 cm^{-1}), C-N (2310–2350 cm^{-1}), C=O (1690–1740 cm^{-1}), O-H (1250–1310 cm^{-1}), and aromatic compounds (675–815 cm^{-1}). In fact, these match well to those in the FTIR spectrum of the *C. inerme* leaves extract. This suggests that many of the organic functional units in the leaf's extracts are actually left on the surface of the CI-Au NPs and the CI-Ag NPs.

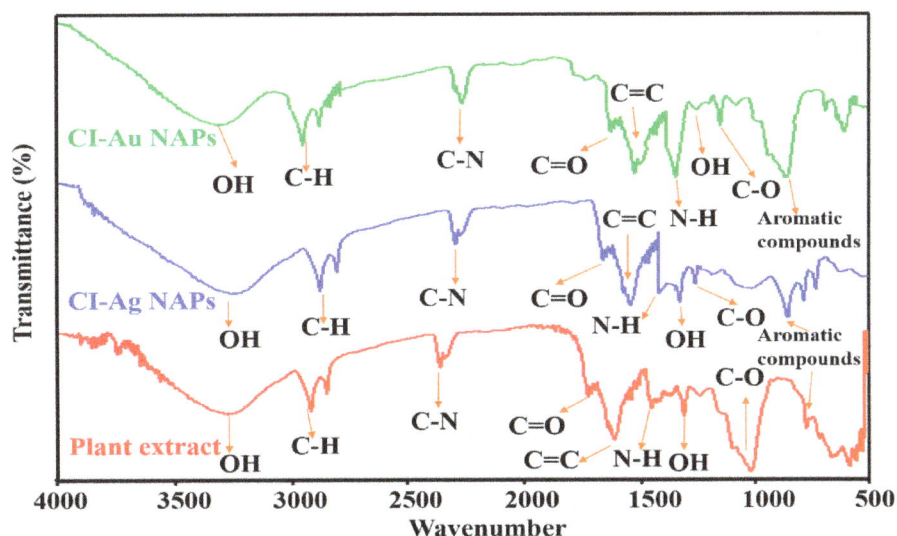

Figure 5. FT-IR spectra of green synthesized CI-Au NPs, CI-Ag NPs, and the *C. inerme* leaves extract.

3.2. Antioxidant, Antibacterial, and Antimycotic Performance

Antioxidant potentials of the CI-Au NPs and the CI-Ag NPs were determined by measuring their abilities to scavenge DPPH free radicals. Among all samples, the standard antioxidant BHT has the most substantial antioxidant capability (Figure 6a). The CI-Ag NPs and the CI-Au NPs show the second and third highest antioxidant capability, which are only slightly lower than that of BHT. It is interesting that the leaves extract shows higher antioxidant strength than the commercial Ag and Au NPs. These results suggest that the good antioxidant power of the CI-Au NPs and the CI-Ag NPs are likely to be associated with the leaves extract. We performed ANOVA test on DPPH scavenging results of six groups against different concentrations of 125, 250, 500, and 1000 µg/mL and the results revealed the statistical difference ($F_{(5,12)}$ = 2043.11, $p < 0.005$), ($F_{(5,12)}$ = 2081.81, $p < 0.001$), ($F_{(5,12)}$ = 1755.24, $p < 0.001$), and ($F_{(5,12)}$ = 1429.78, $p < 0.001$), respectively.

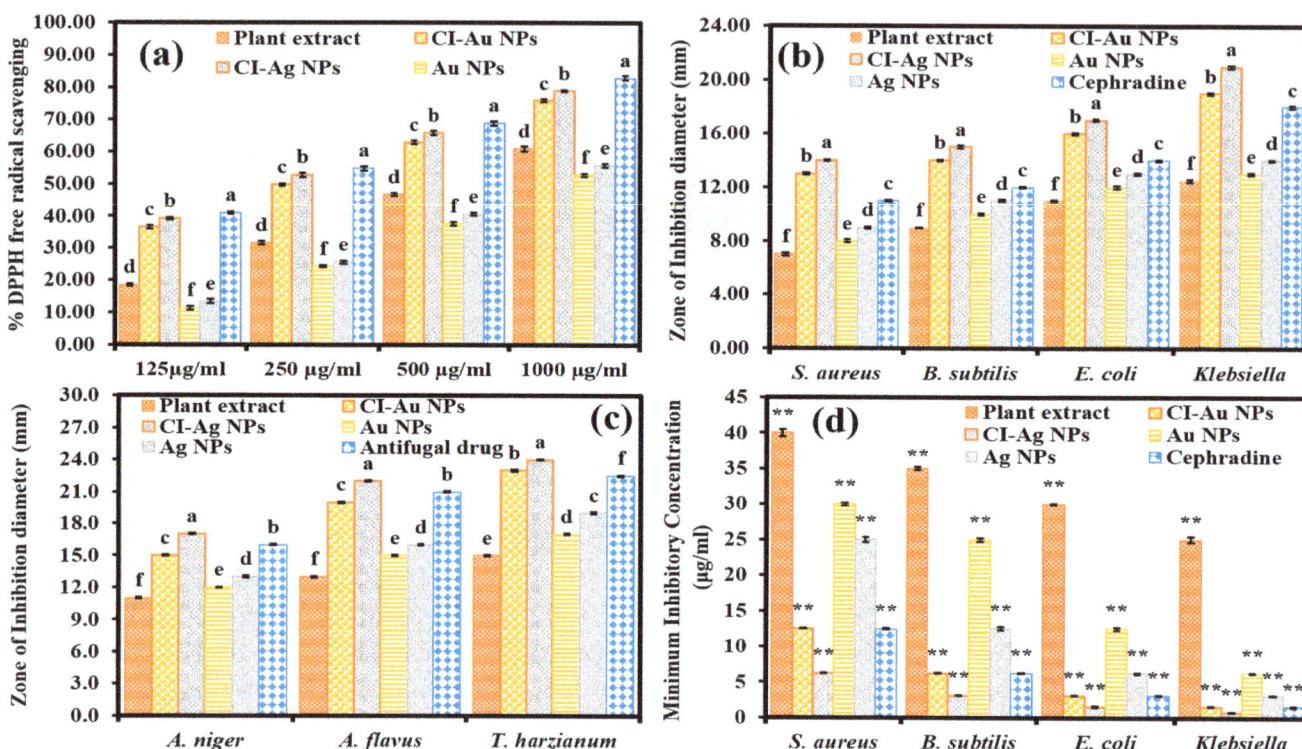

Figure 6. (**a**) Antioxidant, (**b**) antibacterial, (**c**) antimycotic activities, and (**d**) MICs results of green synthesized CI-Au NPs, and CI-Ag NPs in comparison to *C. inerme* leaves extract, Au NPs, Ag NPs, and standards (BHT, antibacterial and antifungal drugs). (Note; Tukey based heterogeneous lower-case letters represent significant statistical pairs). (** $p < 0.01$).

Antibacterial activity of all samples was compared via their zone of bacteria inhibition (ZOIs), as described in the experimental section. Figure 6b shows that for all the bacteria strains employed here, CI-Ag NPs and CI-Au NPs have the highest and the second uppermost antibacterial activities, respectively. It is impressive that their antibacterial performance is even better than the standard antibacterial drug Cephradine and the commercial Ag and Au NPs. Furthermore, we performed an ANOVA test on ZOIs results of six groups against each bacterial strain (*S. aureus*, *B. subtilis*, *E. coli*, and *Klebsiella*) and the results revealed the statistical difference ($F_{(5,12)}$ = 3189.19, $p < 0.001$), ($F_{(5,12)}$ = 3604.48, $p < 0.001$), ($F_{(5,12)}$ = 2783.86, $p < 0.001$), and ($F_{(5,12)}$ = 5454.22, $p < 0.001$), respectively.

The antimycotic propensity of green synthesized CI-Au, and CI-Ag NPs was evaluated by using agar well diffusion assays against different pathogenic mycological strains. Figure 6c shows that the CI-Ag NPs have the best antimycotic performance for all the employed fungal strains. The performance of the CI-Au NPs and terbinafine hydrochloride are slightly below.

Again, the antimycotic performance of the present CI-Ag and CI-Au NPs are much better than those of the commercial Ag and Au NPs. We performed an ANOVA test on ZOIs results of six groups against each fungal strain (*A. niger, A. flavus, and T. harzianum*) and the results displayed statistical significance ($F_{(5,12)} = 4097.56$, $p < 0.002$), ($F_{(5,12)} = 10{,}326.18$, $p < 0.001$), and ($F_{(5,12)} = 8930.23$, $p < 0.001$), respectively.

The MICs generally employed to know about the minimum concentration of the drug molecule that required to prohibit the microbial growths. For this, four bacterial strains were evaluated using green synthesized CI-Au and CI-Ag NPs in comparison to the standard drug, commercial Ag, Au NPs, and plant extract. Figure 6d shows that CI-Ag NPs presented the highest antibacterial efficacy in terms of MICs compared to the standard drug, commercially purchased Ag NPs, Au NPs, and plant extract. On the other hand, the second-highest antibacterial performance in terms of MICs was exhibited by CI-Au NPs, which was comparable to standard drug. We carried out an ANOVA test on MICs results of six groups against each bacterial strain (*S. aureus, B. subtilis, E. coli,* and *Klebsiella*) and the results revealed the statistical difference ($F_{(5,12)} = 3368.21$, $p < 0.001$), ($F_{(5,12)} = 3815.39$, $p < 0.005$), ($F_{(5,12)} = 3051.75$, $p < 0.001$), and ($F_{(5,12)} = 5649.33$, $p < 0.002$), respectively.

3.3. Biofilm Inhibition Activity

The biofilm inhibition ability of green synthesized CI-Au, and CI-Ag NPs was evaluated by using colorimetric assays against different pathogenic bacteriological and mycological strains, and their results in terms of MBIC are shown in Figures 7 and 8, respectively. Results demonstrated that CI-Ag NPs and CI-Au NPs had demonstrated the highest and second uppermost biofilm inhibition activity against all the bacteria strains. It is impressive that the biofilm inhibition efficacy of CI-Ag NPs is even better than the standard antibacterial drug and the commercial Ag and Au NPs. While CI-Au NPs demonstrated comparable biofilm inhibition efficacy to the standard antibacterial drug (Figure 7). We performed an ANOVA test on the MBIC results of six groups against each bacterial strain (*S. aureus, B. subtilis, E. coli,* and *Klebsiella*) and the results revealed the statistical difference ($F_{(5,12)} = 4415.21$, $p < 0.001$), ($F_{(5,12)} = 4705.24$, $p < 0.001$), ($F_{(5,12)} = 3817.71$, $p < 0.001$), and ($F_{(5,12)} = 5297.75.22$, $p < 0.001$), respectively.

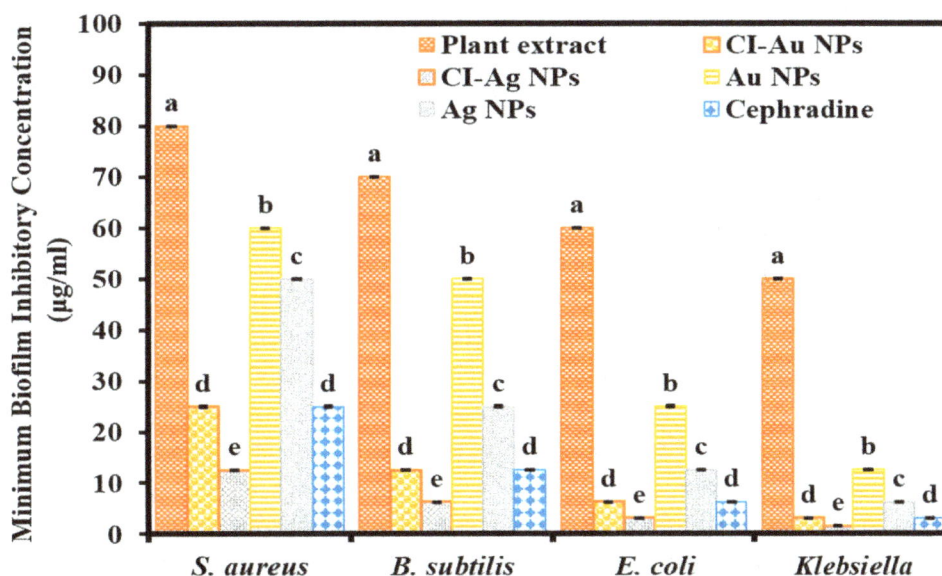

Figure 7. Biofilm inhibition activity of green synthesized CI-Au NPs, and CI-Ag NPs in terms of minimum biofilm inhibitory concentrations (MBICs) against different bacterial strains in comparison to *C. inerme* leaves extract, commercially purchased Au and Ag NPs, as well as Cephradine. (Note: Tukey based heterogenous lower-case letters represent significant statistical pairs).

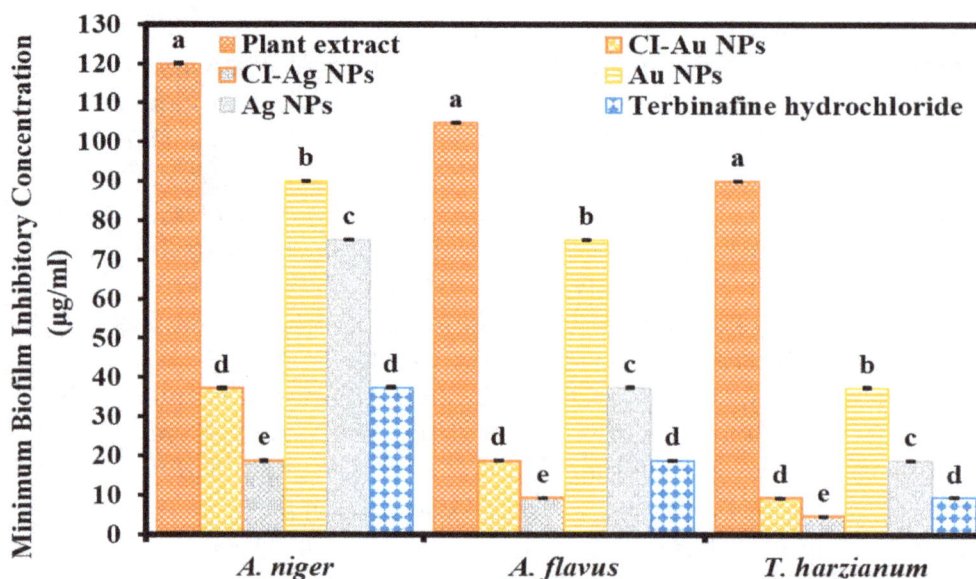

Figure 8. Biofilm inhibition activity of green synthesized CI-Au NPs, and CI-Ag NPs in terms of MBICs against different mycological strains in comparison to *C. inerme* leaves extract, commercially purchased Au and Ag NPs, as well as an antifungal drug. (Note: Tukey based heterogenous lower-case letters represent significant statistical pairs).

Moreover, results were displayed that CI-Ag NPs have demonstrated excellent biofilm inhibition activity in terms of MBIC against all mycological strains than commercial Ag NPs, Au NPs, extract, and standard drug. While CI-Au NPs exhibited comparable biofilm inhibition activity in terms of MBIC to the standard antifungal drug but more significant than other samples (Ag NPs, Au NPs, and extract) (Figure 8). We performed an ANOVA test on MBIC results of six groups against each fungal strain (A. niger, A. flavus, and T. harzianum) and the results displayed statistical significance ($F_{(5,12)} = 3161.91$, $p < 0.001$), ($F_{(5,12)} = 1454.04$, $p < 0.005$), and ($F_{(5,12)} = 2906.33$, $p < 0.001$), respectively. A correlation has been observed in all results of antibacterial, antifungal, and biofilm inhibition activities.

3.4. The Capability of ROS Generation

It has been reported that oxidative stress plays a vital role in the annihilation of microbial strains [29]. Metal NPs interaction with microbial cells often generates ROS, including hydroperoxyl radicals (HO_2^-), hydrogen peroxide (H_2O_2), superoxide ions $O_2^{-\bullet}$, and hydroxyl radicals $OH^\bullet$. They can induce oxidative stress inside the cell, leading to the destruction of various organelles and biomolecules. By using the H_2-DCFDA assay, oxidative stresses in microbial cells (*E. coli* and *A. flavus*) after their treatment with the CI-Au and CI-Ag NPs were evaluated. H_2-DCFDA will be oxidized in the presence of ROS and gives a green fluorescence peak at 503 nm upon photoexcitation. Figure 9 shows the fluorescent spectra of the H_2-DCFDA probe inside *E. coli* (Figure 9a) and *A. flavus* (Figure 9b) cells after incubating with different samples. It can be seen that the ROS generation capabilities of the tested samples are in the order of H_2O_2 > CI-Ag NPs > Ag NPs > CI-Au NPs > plant extract > Au NPs > control (untreated) in both *E. coli* and *A. flavus*. In terms of ROS generation, silver NPs appear to have much better performance than gold NPs. Nevertheless, for both cases, those prepared with the *C. inerme* extract are better than the commercially purchased nanoparticles. This is hinting that the phytochemicals adsorbed on the surfaces of CI-Au and CI-Ag NPs are likely to contribute to the ROS generation.

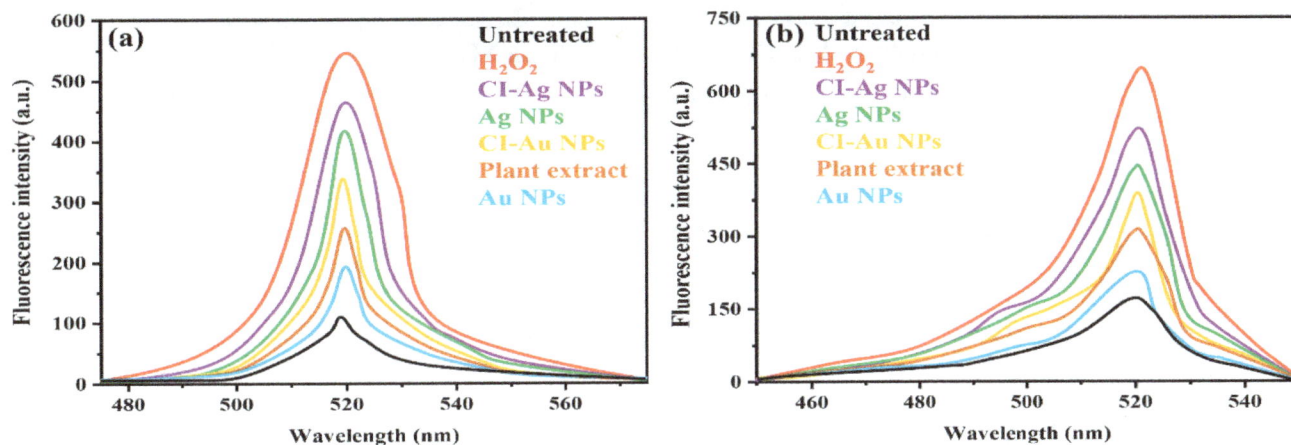

Figure 9. Fluorescent spectra measured with a confocal microscope from the H_2-DCFDA probe inside (**a**) *E. coli* and (**b**) *A. flavus* cells after incubating with different samples.

3.5. Measurement of GSH Concentration

The capability of ROS generation was further corroborated by the quantification of produced intracellular GSH (a thiol-containing tripeptide) in microbial cells in response to oxidative stress results from NPs action. GSH is present in many of the microbial cells in its reduced form up to the concentration of 0.1–10 mM. GSH has the duty to provide protection to microbial cells against oxidative stress results from ROS and to maintain a cellular redox environment [30]. Therefore, it is very important for the survival of microbial cells to preserve naturally occurring antioxidant defense systems based on GSH. Nevertheless, upon GSH subjection to molecular oxygen, it oxidizes extemporaneously to disulfide (GSSG) ($O_2 + 2R–SH \rightarrow RSSR + H_2O_2$; $\Delta G_0 = -96$ kJ/mol) [31], thus upon enough GSH oxidation to GSSG leads to cell demise. Therefore, to understand the cellular oxidative stress in microbial cells upon treatment with the CI-Au NPs and the CI-Ag NPs, the intracellular GSH concentration was examined.

GSH concentrations measured from cells immediately taken out of the CO_2 incubator are considered as 100%. GSH concentrations in cells incubated with different samples for 4 h at 37 °C were measured (Figure 10). The "untreated" samples referred to cells incubated without adding any of the NPs, H_2O_2 nor plant extract, and show GSH concentration of 95% and 94%, respectively, in untreated *E. coli* and *A. flavus*. This shows that under standard laboratory lighting of 4 h, a few percent of GSH were oxidized. The GSH concentration was reduced considerably to 45%, 35%, 20%, 55%, and 25% in E. coli and 50%, 40%, 25%, 60%, and 30% in *A. flavus* upon incubating, respectively, with *C. inerme* leaves extract, CI-Au NPs, CI-Ag NPs, Au NPs, and Ag NPs. As anticipated, the GSH concentration was attenuated remarkably to 15% and 14% in *E. coli* and *A. flavus*, respectively, in the presence of H_2O_2 (1 mM). We performed an ANOVA test on GSH concentration results of six groups against *E. coli* and *A. flavus*, and the results displayed statistical significance ($F_{(5,12)} = 5944.36$, $p < 0.001$), and ($F_{(5,12)} = 4735.37$, $p < 0.001$), respectively. These observations are consistent with the ROS generation results (Figure 9) showing that interaction of CI-Au NPs and CI-Ag NPs with microbial cells has developed the ROS mediated oxidative stress by depleting or destroying the antioxidant defense that leads to cell demise. Banerjee et al. has reported similar results upon E. coli treatment with iodinated chitosan−silver nanoparticle composite [32] and described that those NPs interact with microbial cells produced intracellular ROS mediated oxidative stress in the cell leading to cell membrane impairment and cell demise.

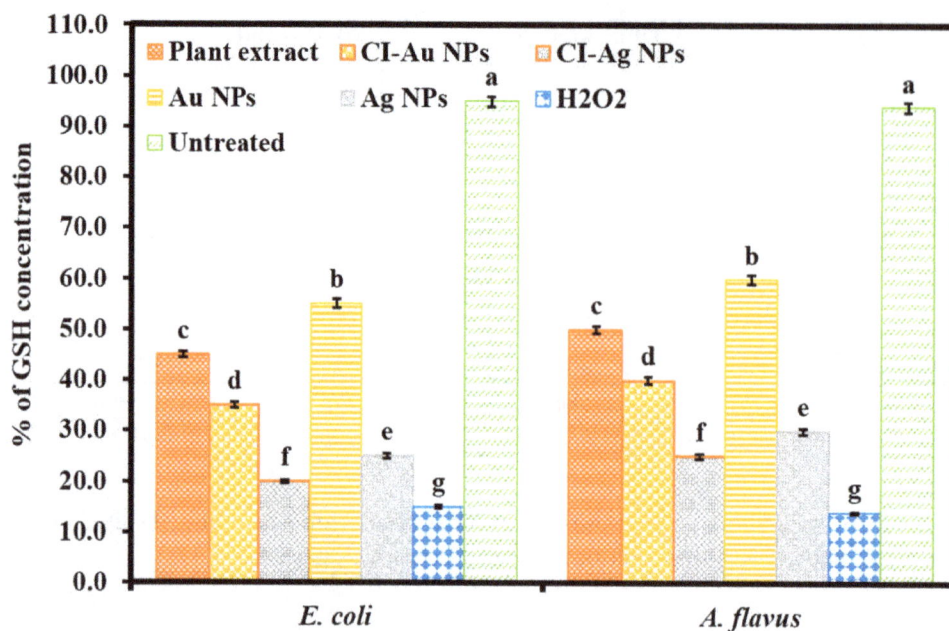

Figure 10. The level of GSH concentration in *E. coli* and *A. flavus* treated with *C. inerme* leaves extract, CI-Au NPs, CI-Ag NPs, Au NPs, and Ag NPs. (Note: Tukey based heterogeneous lower-case letters represent significant statistical pairs).

3.6. FTIR Analysis of Bacteria and Fungi

FTIR analysis of both untreated and treated microbial cells (*E. coli* and *A. flavus*) was performed to investigate further the binding of CI-Au NPs and CI-Ag NPs on their cell surfaces and subsequent changes in molecular functionalities of their cell membrane. Results demonstrated that the untreated bacterial cell (*E. coli*) showed characteristic FT-IR peaks (3335 cm^{-1} for O-H), (3095 cm^{-1} for -COOH), 2930 and 2845 cm^{-1} for C-H), (1580 cm^{-1} for amide I and amide II), (1490, 1445, and 1325 for -CH$_2$), (1275 cm^{-1} for PO$_2^-$) and (1180, 1130, 1040, 890, and 830 cm^{-1} for C-O-C) from different proteins, fatty acids, and polysaccharide molecules present on the cell surface (Figure 11) [33]. The untreated mycological cell (*A. flavus*) showed FT-IR peaks from several functional groups, such as -NH$_2$ (3465 cm^{-1}), -OH (3390 cm^{-1}), -COOH (3065 cm^{-1}), aliphatic -CH (2925 and 2855 cm^{-1}), amide I and amide II (1599 and 1548 cm^{-1}), aromatic C=C (1423 cm^{-1}), PO$_2^-$ (1280 cm^{-1}), C-O-C (1190, 1110 cm^{-1}), glucose ring band (1010 cm^{-1}), and C-Cl (855 cm^{-1}) (Figure 12) [34].

In contrast to untreated *E. coli*, the treated cells demonstrated the noticeable changes in whole FTIR spectral regions (Figure 11). The FTIR peaks shifting and reduction in band intensity were observed in treated microbial cells. The O-H peak at 3335 cm^{-1} in untreated *E. coli* disappeared in CI-Au NPs, and CI-Ag NPs treated bacterial cells. Moreover, the intensity of methylene and -COOH stretching vibrations were reduced in CI-Ag NPs treated and shifted to lower wavenumber, while their peaks were totally disappeared in CI-Au NP-treated *E. coli* cells. These observations demonstrate that fatty acids may undergo a number of significant reductions, which leads the *E. coli* cell membrane to transform from a well-ordered to disordered state [35,36]. The amide I and amide II peaks were shifted to downfield and appear at 1575 and 1570 cm^{-1} in *E. coli* cells treated with CI-Au NPs and CI-Ag NPs, which indicate changes in protein structure. This might be because of the cell membrane lysis of *E. coli*. The mild downfield shifting of -CH$_2$ and C-O-C peaks were also observed in the treated *E. coli* cells. PO$_2^-$ peaks bands appeared strongly shifted in CI-Ag NPs treated cells but completely disappeared in CI-Au NPs treated *E. coli* cells. Both of these results evidently indicated that phospholipid molecules of the *E. coli* cell membrane were denatured upon processing with the CI-Au and CI-Ag NPs.

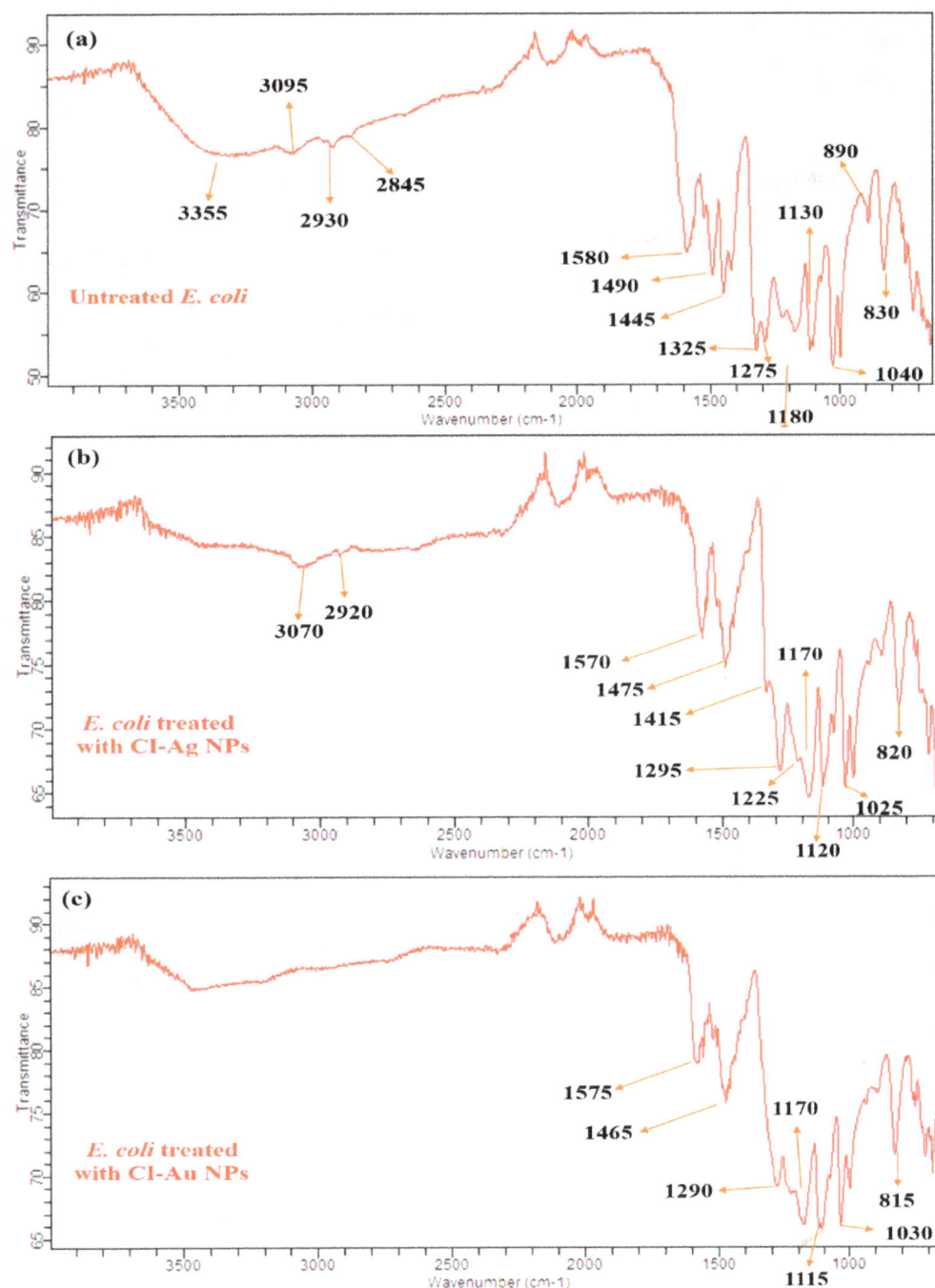

Figure 11. FTIR of (**a**) untreated and treated *E. coli* with (**b**) CI-Ag NPs and (**c**) CI-Au NPs synthesized by using *C. inerme* leaves extract.

Similarly, CI-Au NPs and CI-Ag NPs treated *A. flavus* also showed noticeable changes in its FTIR spectrum (Figure 12). The $-NH_2$, $-OH$, $-COOH$, aliphatic $-CH$, and PO_2^- functional groups peaks appear in untreated *A. flavus* cells were utterly removed upon treatments with CI-Au NPs or CI-Ag NPs. In contrast to untreated *A. flavus* cells, FTIR peaks of amide I, amide II, Aromatic C=C, C-O-C, glucose ring band, and C-Cl are shifted higher and downfield following reduction in their intensity in CI-Au NPs, and CI-Ag NPs treated cells. In addition, some new peaks at 1365 and 1380 cm^{-1} are also observed in the FTIR spectrum of CI-Ag NPs, and CI-Au NPs treated *A. flavus* cells, respectively.

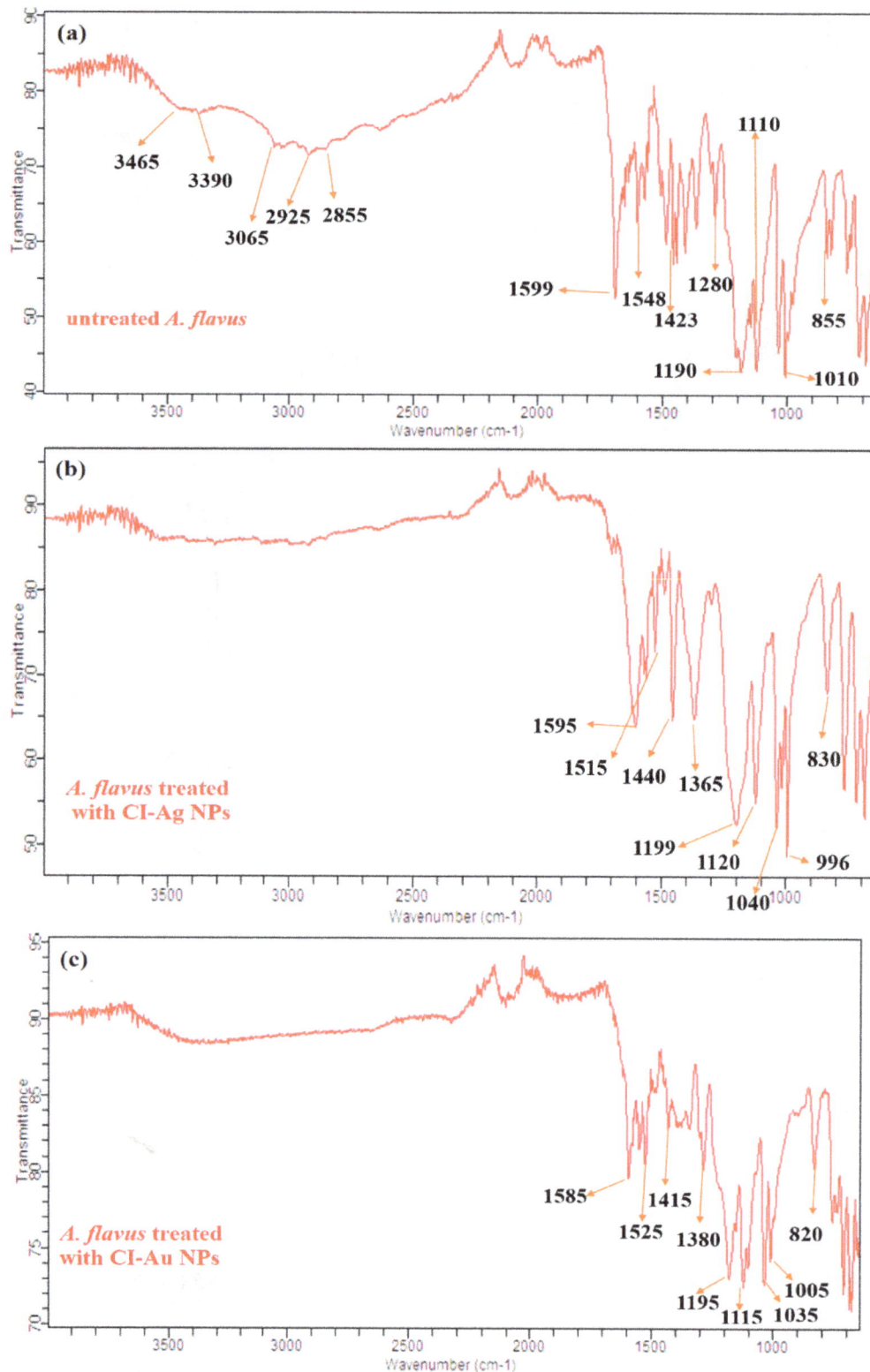

Figure 12. FTIR of (**a**) untreated and treated *A. flavus* with (**b**) CI-Ag NPs and (**c**) CI-Au NPs synthesized by using *C. inerme* leaves extract.

These FTIR results suggest that upon treatment with the CI-Au NPs and CI-Ag NPs, cell membranes microbial cells are destructed with the following changes: (1) well-ordered to disordered state transformation of cell membrane's fatty acids as revealed by peaks intensity reduction and disappearance of OH, methylene, and COOH; (2) changes in the structure of membrane proteins shown by the downfield shifting of amide I and amide II peaks; (3) downfield shifting and disappearance of

PO_2^- peaks displaying possibly denaturation of phospholipid molecules; (4) obliteration of glycoside linkages (C-O-C) of membrane's polysaccharide molecules [36–41].

3.7. Cytotoxicity Study

The green synthesized CI-Au and Cl-Ag NPs were evaluated for their cytotoxic propensity against the MCF-7 cancerous cell line in vitro compared to the *C. inerme* leaf extract, Au, Ag NPs, and standard drug. Figure 13 shows the cytotoxicity results. The results were demonstrated that green synthesized CI-Ag NPs displayed the superior cytotoxic effect on MCF-7 cancerous cells by lowering their cell viability percentage compared to other samples (Au, Ag NPs, and standard drug). While, the green synthesized CI-Au NPs have exhibited slightly lower cytotoxic propensity against cancerous cells compared to the standard drug but was higher than plant extract, Au, and Ag NPs. Interestingly, leaves extract of *C. inerme* also displayed cytotoxic potential against MCF-7 breast cancerous cells. Moreover, the statistical significance in the results of cytotoxicity activity of NPs, plant extract, and the standard drug was also validated with the ANOVA (F value = 3255.924, $p < 0.001$) and Tukey test (Figure 13). These good cytotoxicity activity results of green synthesized CI-Au, and Cl-Ag NPs might be attributed to their physical properties (small size, morphology, and surface area), and their functionalization with the biologically active phytomolecules of the leaves extract of *C. inerme*. Many reports disclosed that leaves extract of *C. inerme* have different phytomolecules that are biological active [18–20].

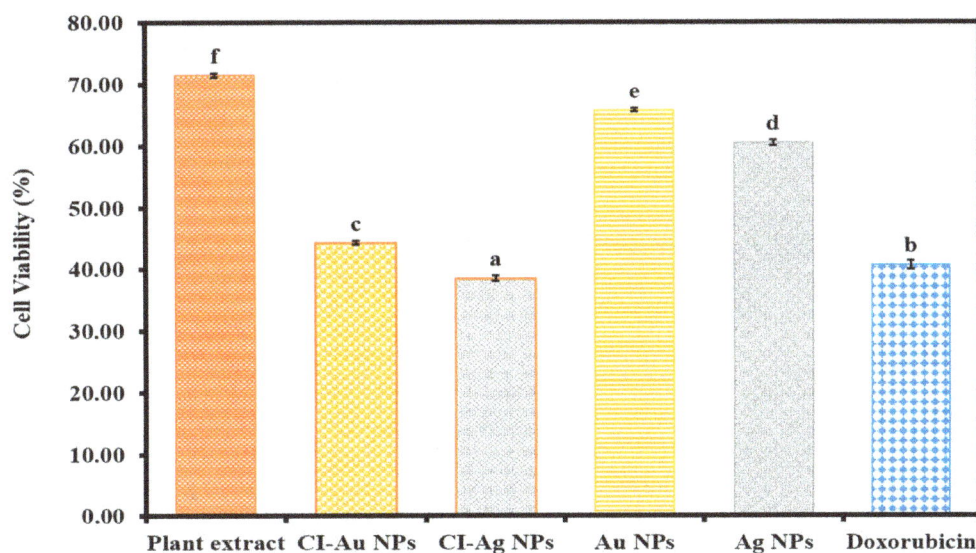

Figure 13. Cytotoxicity activity of green synthesized CI-Au NPs and CI-Ag NPs against the MCF-7 cancerous cells compared to *C. inerme* leaves extract, Au NPs, Ag NPs, and standard drug. (Note: Tukey based heterogeneous lower-case letters represent significant statistical pairs).

3.8. Biocompatibility Analysis

Biocompatibility of CI-Au and Cl-Ag NPs are assessed via their effects on red blood cells (RBCs). The hemolysis induced by the sample is presented in Figure S2. The ASTM international method was followed to determine whether the samples were hemolytic (0–2% non-hemolytic, 2–5% partially hemolytic, and ≥5% hemolytic).

The results show that CI-Au NPs and CI-Ag NPs were non-hemolytic at their lower concentrations (1 µM). However, they are partially hemolytic at higher concentrations (50 µM). On the other hand, the commercial Au and Ag NPs are both partially hemolytic even at a low concentration of 1 µM and fully hemolytic at a higher concentration (50 µM). The *C. inerme* leaves extract was demonstrated to be non-hemolytic at all tested concentrations. These results suggest that adsorbed phytochemicals on the surface of CI-Au NPs and CI-Ag NPs might lower the toxicity of the core metal nanoparticles.

The statistical significance of the hemolytic activity results was further corroborated by the ANOVA ($p < 0.001$, F-value = 117,502.57) and Tukey test (heterogeneous lower-case letters) (Figure S2). Similar results were reported by Parthiban et al. [9].

4. Discussion

It has been reported that *C. inerme* leaves extract possesses numerous biological active phytochemical compounds such as flavonoids, phenolics, alkaloids, terpenoids, anthraquinones, carbohydrates, saponins, and tannins, as shown in Figure 14 [18–20]. In the present synthesis processes, metal ions are reduced to metal nanoparticles with only the leaves extract as the other reactant. This suggests that some components of the leaves extract act as reducing agents in the synthetic reaction. At the same time, the CI-Au and CI-Ag NPs show much better antioxidant, antibacterial and antimycotic activities compared with commercial Au and Ag NPs. Together with the FTIR results, we can conclude that some bioactive components from the leaves extract do remain on the surfaces of the CI-Au and CI-Ag NPs. During synthesis, gold and silver salts in leaves extract first dissociated into their ions, such as Au^{3+} and Ag^{+}. Phytochemicals, such as flavonoids, phenolics, carbohydrates, cardiac glycosides, and anthraquinones, in the leaves extract can first reduce the metal ions into their zero-valent species. While capping agents, such as terpenoids, tannins, saponins, alkaloids, and proteins, can encapsulate the Au and Ag zero-valent species to stabilize them (Figure 15).

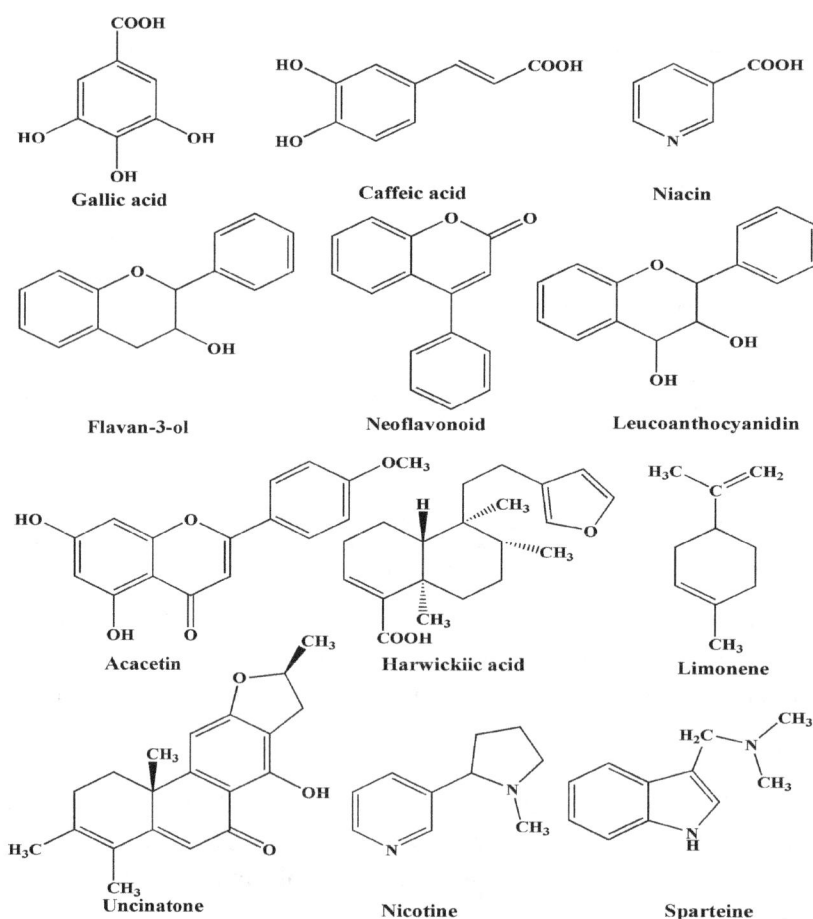

Figure 14. Phytochemicals present in *C. inerme* leaves extract [18–20].

It has been shown that antimicrobial and antioxidant properties can be enhanced by anchoring biocompatible and biologically active molecules to synthesized metal NPs. In the current work, we demonstrate that the benefits of these bioactive components can be simultaneously obtained in a green synthesis process using a plant with bioactive components. In the present case, the adsorbed biologically active phytochemicals are bacteriostatic and fungicidal in nature as they have a substitution of different molecular functionalities (-OH, -NO$_2$, -COOH, -SO$_3$H, -NH$_2$, -CONH$_2$, etc.), which play a vital role in various biological activities (Figure 14) [19].

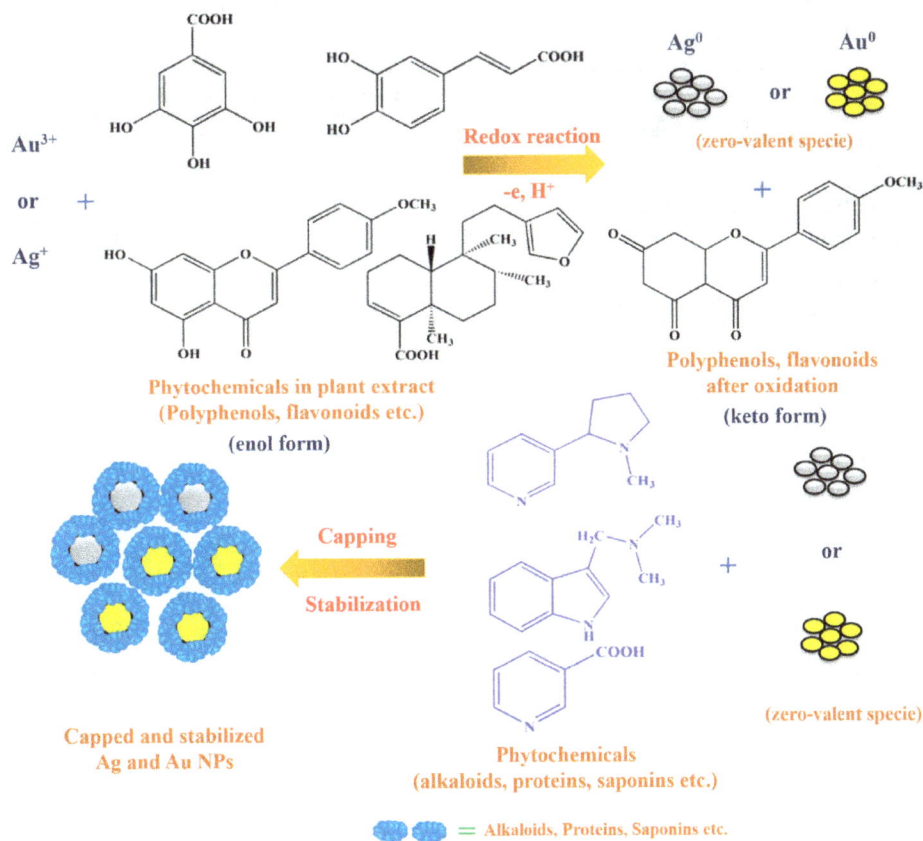

Figure 15. The schematic illustration displays the anticipated mechanism for green synthesis of CI-Au NPs and CI-Ag NPs using leaves extract of *C. inerme*.

Finally, we have compared the antimicrobial potential of our synthesized CI-Ag, and CI-Au NPs at the concentration of 250 µg/mL (equivalent to 2317.643 µM and 1269.250 µM) with Ag and Au NPs, respectively, synthesized with other plants in Tables 1 and 2. It can be seen that the green synthesized CI-Au NPs and CI-Ag NPs are highly effective for prohibiting the growth of both Gram-negative bacteria (*E. coli* and *Klebsiella*) than Gram-positive (*S. aureus* and *B. subtilis*) as well as fungi (*A. flavus* and *A. niger*).

It has been anticipated from the antibacterial results that green fabricated CI-Au NPs and CI-Ag NPs were found to manifest more excellent growth inhibitory action against *E. coli*, and *Klebsiella* (Gram-negative bacterial strains) than *S. aureus* and *B. subtilis* (Gram-positive bacterial strains) (Figure 6b). This attributes to the fact of differences in chemical composition and structure of their cell wall. The cell wall of Gram-negative bacterial strains has a thin layer of peptidoglycan with an extra outer covering layer of lipopolysaccharide called periplasm. On the other hand, the cell wall of Gram-positive bacterial strains has a thick peptidoglycan layer, as shown in Figure 16 [37]. The literature demonstrates that Gram-negative bacterial strain, i.e., *E. coli*, has ~8 nm thick peptidoglycans layer and 1–3 µm thick lipopolysaccharides layer as well in their cell wall. While Gram-positive bacterial strain, i.e., *S. aureus*, has much thick layer (~80 nm) of peptidoglycans with covalently-attached teichuronic and teichoic acid. Due to the thinner layer of peptidoglycans in the cell walls, Gram-negative bacterial strains are highly vulnerable to the penetration of NPs and their antibacterial action than Gram-positive bacteria. Another factor for Gram-negative bacterial strains to exhibit high sensitivity towards NPs is the existence of lipopolysaccharides coatings outside their cell as these coatings are negatively charged. These lipopolysaccharides coatings have a greater affinity towards NPs with positive surface charge [37]. Hence, due to these above factors, green fabricated CI-Au NPs and CI-Ag NPs are proved to demonstrate significant effectiveness against Gram-negative bacterial strains.

Table 1. Comparison of antimicrobial activities of green synthesized CI-Ag NPs with reported Ag NPs synthesized with other plants.

Materials (NPs)	Size (nm)	Plant Used	Antimicrobial Properties			References
			Species	Conc. of NPs	ZOIs	
CI-Ag	2–10	C. inerme	E. coli	250 µg/mL	17	This work
Ag	8–20	D. bulbifera	E. coli	500 µg/mL	15	[42]
Ag	8–50	Allium ampeloprasum	E. coli	300 µg/mL	13	[43]
Ag	20	Umbrella	E. coli	250 µg/mL	16	[44]
Ag	36–74	Trianthema decandra	E. coli	10 mg/mL	15.5	[45]
CI-Ag	2–10	C. inerme	S. aureus	250 µg/mL	14	This work
Ag	10–20	Green and black tea	S. aureus	1 mg/mL	19–21	[46]
Ag	10–20	Zingiber officinale	S. aureus	0.1 mg/mL	6.5	[47]
Ag	8–50	Allium ampeloprasum	S. aureus	300 µg/mL	8	[43]
Ag	20	Umbrella	S. aureus	250 µg/mL	12.7	[44]
Ag	36–74	Trianthema decandra	S. aureus	10 mg/mL	13.5	[45]
CI-Ag	2–10	C. inerme	K. pneumoniae	250 µg/mL	21	This work
Ag	8–20	D. bulbifera	K. pneumoniae	500 µg/mL	15	[42]
Ag	20	Umbrella	K. pneumoniae	250 µg/mL	13.1	[44]
Ag	50	Aesculus hippocastanum	K. pneumoniae	100 µg/mL	12.5	[48]
CI-Ag	2–10	C. inerme	B. subtilis	250 µg/mL	15	This work
Ag	37	E. scaber	B. subtilis	1 mg/mL	16	[49]
Ag	20–25	P. guajava	B. subtilis	300 µg/mL	19	[50]
Ag	10–20	Zingiber officinale	B. subtilis	0.1 mg/mL	0	[47]
Ag	36–74	Trianthema decandra	B. subtilis	10 mg/mL	12	[45]
CI-Ag	2–10	C. inerme	A. flavus	250 µg/mL	22	This work
Ag	37	E. scaber	A. flavus	1 mg/mL	12	[49]
CI-Ag	2–10	C. inerme	A. niger	250 µg/mL	17	This work
Ag	20–25	P. guajava	A. niger	300 µg/mL	18.79	[50]

Antimicrobial activities of Au NPs and Ag NPs have been attributed either by their physical or oxidative vandalization or by both to the microbial cells [36,38]. It has been reported that gold and silver NPs possess a higher affinity to proteins and tend to bind to the surface proteins of cells [38]. As per the hard-soft acid–base theory, Au NPs and Ag NPs possess a higher affinity for phosphorus and sulfur moieties of proteins. In addition, Ag and Au also have a tendency to form bonds with nitrogen (i.e., Ag–N and Au–N bonds) and with oxygen (i.e., Ag–O) moieties of proteins [39]. The drastic changes take place in membrane permeability upon the binding of Au NPs and Ag NPs with the cell surface proteins, which leads to depletion in the level of intracellular ATP and the dissipation of proton motive force that results in microbial cells to demise [36,40]. As well, Ag and Au possess higher redox potential [EH^0 (Ag^+/Ag^0) = 0.8 V] and [EH^0 (Au^+/Au^0) = 1.83 V] respectively. Their higher redox potential does oxidative disintegration of lipopolysaccharides and cell surface proteins, which further leads to cell membrane destruction and pore formation on the cell membrane due to which seepage of intracellular contents occurs [36–39]. Moreover, this pore formation on cell membranes also causes Au and Ag NPs internalization, which vandalizes the intracellular proteins and nucleic acids due to NPs interaction with them [36,41]. Recent reports on antimicrobial mechanism of metal NPs also proposed that Au and Ag NPs demonstrate microbicidal activity due to the production of oxidative stress in microbial cells by them leading to the generation of ROS species, which subsequently cause drastic destruction to cells such as cell membrane impairment, seepage of cellular material, loss of respiratory activity, as well as DNA damage leading to cell demise [36,38].

Table 2. Comparison of antimicrobial activities of green synthesized CI-Au NPs with reported Au NPs synthesized with other plants.

| Materials (NPs) | Size (nm) | Plant Used | Antimicrobial Properties | | | References |
			Species	Conc. of NPs	ZOIs	
CI-Au	3–9	C. inerme	E. coli	250 µg/mL	16	This work
Au	15.6	Plumeria alba	E. coli	400 µg/mL	16	[51]
Au	2.7–38.7	Achillea wilhelmsii	E. coli	300 µg/mL	0	[52]
Au	20–140	Citrullus lanatus	E. coli	1000 µg/mL	9.23	[53]
Au	33–65	Trianthema decandra	E. coli	10 mg/mL	9.5	[45]
Au	40–45	Gundelia tournefortii	E. coli	2 mg/mL	9.8	[54]
Au	40–45	Falcaria vulgaris	E. coli	4 mg/mL	8.6	[55]
Au	40–45	Allium saralicum	E. coli	4 mg/mL	10.8	[56]
CI-Au	3–9	C. inerme	S. aureus	250 µg/mL	13	This work
Au	20–140	Citrullus lanatus	S. aureus	1000 µg/mL	0	[53]
Au	33–65	Trianthema decandra	S. aureus	10 mg/mL	14.5	[45]
Au	40–45	Gundelia tournefortii	S. aureus	2 mg/mL	11.2	[54]
Au	40–45	Falcaria vulgaris	S. aureus	4 mg/mL	13	[55]
Au	40–45	Allium saralicum	S. aureus	4 mg/mL	11.6	[56]
CI-Au	3–9	C. inerme	B. subtilis	250 µg/mL	14	This work
Au	2.7–38.7	Achillea wilhelmsii	B. subtilis	300 µg/mL	11	[52]
Au	33–65	Trianthema decandra	B. subtilis	10 mg/mL	9.5	[45]
Au	40–45	Gundelia tournefortii	B. subtilis	2 mg/mL	14.2	[54]
Au	40–45	Falcaria vulgaris	B. subtilis	4 mg/mL	14	[55]
Au	40–45	Allium saralicum	B. subtilis	4 mg/mL	14.2	[56]
CI-Au	3–9	C. inerme	A. niger	250 µg/mL	15	This work
Au	12–22	Brassica oleracea	A. niger	50 µg/mL	9	[57]
CI-Au	3–9	C. inerme	A. flavus	250 µg/mL	20	This work
Au	12–22	Brassica oleracea	A. flavus	50 µg/mL	9	[57]

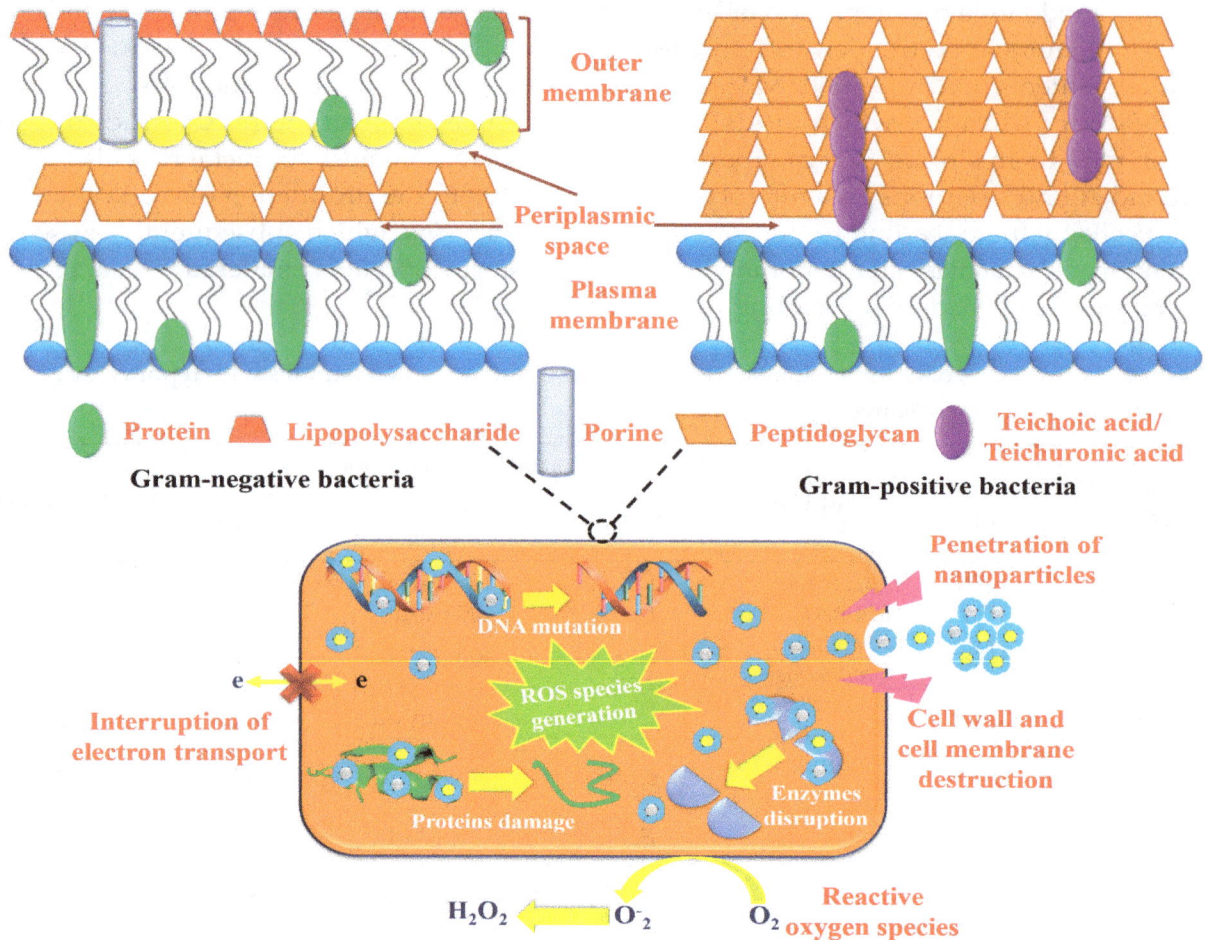

Figure 16. Cell wall comparison of Gram-positive and Gram-negative bacteriological strains.

The outcomes of this research work recommend that green synthesized CI-Au NPs and CI-Ag NPs exhibit excellent antimicrobial activity because of the extraordinary colloidal stability of phytochemicals capped NPs. The intracellular ROS investigations have been affirmed that annihilation of microbial's cell membrane and following cell demise by CI-Au NPs and CI-Ag NPs was caused by the generation of ROS species and membrane permeabilization. Further, FT-IR spectroscopic study disclosed alterations in chemical compositions of cell's biological molecules (fatty acids, carbohydrates, and proteins) during the assassination process of microbial cells. In view of the uniformity and consistency in results we propose that antimicrobial propensity presented by CI-Au NPs and CI-Ag NPs is the combining effect of physical and oxidative destructions with the following cellular changes: (1) binding of CI-Au NPs and CI-Ag NPs with lipopolysaccharides and cell surface proteins leads to the cell membrane destruction; (2) deterioration of microbial cell membrane increased the membrane permeability, consequently instigated the seepage of intracellular biomolecular functionality; (3) over the cellular antioxidant defense system, generation of intracellular ROS species impaired the microbial cells, which leads to cell demise.

5. Conclusions

Current research work demonstrated the successful fabrication of CI-Au and CI-Ag NPs with inherited biomedical functions of *C. inerme* extract via an environment-friendly green approach. Green synthesized NPs were successfully characterized using different characterization techniques such as XRD, UV-Visible, FTIR, TEM, EDX, and DLS. Results demonstrate that green synthesized CI-Au NPs and CI-Ag NPs have much better antioxidant, antibacterial, antimycotic performance comparing to commercial Au and Ag NPs functionalized with dodecanethiol and PVP, respectively.

Further, they appeared more biocompatible than commercial NPs. The synthesized NPs exhibited enhanced biological activities due to the synergetic addition of biologically active adsorbed phytochemicals. Hence, this research work further has proven that environment-friendly and modest green synthesis of CI-Au NPs and CI-Ag NPs with enhanced and/or additional biomedical functionalities employing leaf extract of *C. inerme* would be an economical and viable substitute to conventional chemical procedures.

Author Contributions: Conceptualization, S.A.K.; methodology, S.A.K.; software, S.A.K.; validation, S.A.K.; formal analysis, S.A.K. and S.S.; investigation, S.A.K. and S.S.; resources, S.A.K. and S.S.; data curation, S.A.K.; writing—original draft preparation, S.A.K.; writing—review and editing, S.A.K. and C.-S.L.; visualization, S.A.K. and C.-S.L.; supervision, C.-S.L.; project administration, C.-S.L.; funding acquisition, C.-S.L. All authors have read and agreed to the published version of the manuscript.

References

1. Balabanian, G.; Rose, M.; Manning, N.; Landman, D.; Quale, J. Effect of Porins and bla KPC Expression on Activity of Imipenem with Relebactam in Klebsiella pneumoniae: Can Antibiotic Combinations Overcome Resistance? *Microb. Drug. Resist.* **2018**, *24*, 877–881. [CrossRef]

2. Boucher, H.W.; Talbot, G.H.; Bradley, J.S.; Edwards, J.E.; Gilbert, D.; Rice, L.B.; Scheld, M.; Spellberg, B.; Bartlett, J. Bad bugs, no drugs: No ESKAPE! An update from the Infectious Diseases Society of America. *Clin. Infect. Dis.* **2009**, *48*, 1–2. [CrossRef] [PubMed]

3. Howden, B.P.; Slavin, M.A.; Schwarer, A.P.; Mijch, A.M. Successful control of disseminated Scedosporium prolificans infection with a combination of voriconazole and terbinafine. *Eur. J. Clin. Microbiol. Infect. Dis.* **2003**, *22*, 111–113. [PubMed]

4. Aquino-Andrade, A.; Merida-Vieyra, J.; de la Garza, E.A.; Arzate-Barbosa, P.; Ranero, A.D. Carbapenemase-producing Enterobacteriaceae in Mexico: Report of seven non-clonal cases in a pediatric hospital. *BMC Microbiol.* **2018**, *18*, 38. [CrossRef] [PubMed]

5. Di Pilato, V.; Arena, F.; Tascini, C.; Cannatelli, A.; De Angelis, L.H.; Fortunato, S.; Giani, T.; Menichetti, F.; Rossolini, G.M. mcr-1.2, a new mcr variant carried on a transferable plasmid from a colistin-resistant KPC carbapenemase-producing Klebsiella pneumoniae strain of sequence type 512. *Antimicrob. Agents Chemother.* **2016**, *60*, 5612–5615. [CrossRef]

6. Li, X.Z.; Nikaido, H. Efflux-mediated drug resistance in bacteria. *Drugs* **2009**, *69*, 1555–1623. [CrossRef]

7. Antimicrobial Resistance. Available online: https://www.who.int/news-room/detail/29-04-2019-new-report-calls-for-urgent-action-to-avert-antimicrobial-resistance-crisis (accessed on 9 March 2020).

8. Boomi, P.; Ganesan, R.M.; Poorani, G.; Prabu, H.G.; Ravikumar, S.; Jeyakanthan, J. Biological synergy of greener gold nanoparticles by using Coleus aromaticus leaf extract. *Mater. Sci. Eng. C* **2019**, *99*, 202–210. [CrossRef]

9. Parthiban, E.; Manivannan, N.; Ramanibai, R.; Mathivanan, N. Green synthesis of silver-nanoparticles from Annona reticulata leaves aqueous extract and its mosquito larvicidal and anti-microbial activity on human pathogens. *Biotechnol. Rep.* **2019**, *21*, e00297. [CrossRef]

10. Lian, S.; Diko, C.S.; Yan, Y.; Li, Z.; Zhang, H.; Ma, Q.; Qu, Y. Characterization of biogenic selenium nanoparticles derived from cell-free extracts of a novel yeast Magnusiomyces ingens. *3 Biotech* **2019**, *9*, 221. [CrossRef]

11. Ijaz, F.; Shahid, S.; Khan, S.A.; Ahmad, W.; Zaman, S. Green synthesis of copper oxide nanoparticles using Abutilon indicum leaf extract: Antimicrobial, antioxidant and photocatalytic dye degradation activitie. *Trop. J. Pharm. Res.* **2017**, *16*, 743–753.

12. Khan, S.A.; Noreen, F.; Kanwal, S.; Iqbal, A.; Hussain, G. Green synthesis of ZnO and Cu-doped ZnO nanoparticles from leaf extracts of Abutilon indicum, Clerodendrum infortunatum, Clerodendrum inerme and investigation of their biological and photocatalytic activities. *Mater. Sci. Eng. C* **2018**, *82*, 46–59. [CrossRef] [PubMed]

13. Abbasi, B.A.; Iqbal, J.; Mahmood, T.; Ahmad, R.; Kanwal, S.; Afridi, S. Plant-mediated synthesis of nickel oxide nanoparticles (NiO) via Geranium wallichianum: Characterization and different biological applications. *Mater. Res. Express* **2019**, *6*, 0850a7. [CrossRef]

14. Khan, S.A.; Shahid, S.; Shahid, B.; Fatima, U.; Abbasi, S.A. Green Synthesis of MnO Nanoparticles Using Abutilon indicum Leaf Extract for Biological, Photocatalytic, and Adsorption Activities. *Biomolecules* **2020**, *10*, 785. [CrossRef] [PubMed]

15. Ahmad, S.; Munir, S.; Zeb, N.; Ullah, A.; Khan, B.; Ali, J.; Bilal, M.; Omer, M.; Alamzeb, M.; Salman, S.M.; et al. Green nanotechnology: A review on green synthesis of silver nanoparticles—An ecofriendly approach. *Int. J. Nanomed.* **2019**, *14*, 5087. [CrossRef] [PubMed]

16. Nayak, S.; Bhat, M.P.; Udayashankar, A.C.; Lakshmeesha, T.R.; Geetha, N.; Jogaiah, S. Biosynthesis and characterization of Dillenia indica-mediated silver nanoparticles and their biological activity. *Appl. Organomet. Chem.* **2020**, e5567. [CrossRef]

17. Gharehyakheh, S.; Ahmeda, A.; Haddadi, A.; Jamshidi, M.; Nowrozi, M.; Zangeneh, M.M.; Zangeneh, A. Effect of gold nanoparticles synthesized using the aqueous extract of Satureja hortensis leaf on enhancing the shelf life and removing Escherichia coli O157: H7 and Listeria monocytogenes in minced camel's meat: The role of nanotechnology in the food industry. *Appl. Organomet. Chem.* **2020**, *34*, e5492.

18. Khan, S.A.; Rasool, N.; Riaz, M.; Nadeem, R.; Rashid, U.; Rizwan, K.; Zubair, M.; Bukhari, I.H.; Gulzar, T. Evaluation of antioxidant and cytotoxicity studies of *Clerodendrum inerme. Asian J. Chem.* **2013**, *13*, 7457–7462. [CrossRef]

19. Ba Vinh, L.; Thi Minh Nguyet, N.; Young Yang, S.; Hoon Kim, J.; Thi Vien, L.; Thi Thanh Huong, P.; Van Thanh, N.; Xuan Cuong, N.; Hoai Nam, N.; Van Minh, C.; et al. A new rearranged abietane diterpene from Clerodendrum inerme with antioxidant and cytotoxic activities. *Nat. Prod. Res.* **2018**, *32*, 2001–2007. [CrossRef]

20. Anandhi, K.; Ushadevi, T. Analysis of Phytochemical Constituents and Antibacterial Activities of *Clerodendron inerme* L. Against Some Selected Pathogens. *Int. J. Biotechnol. Allied Fields* **2013**, *1*, 387–393.

21. Shahid, S.; Fatima, U.; Sajjad, R.; Khan, S.A. Bioinspired nanotheranostic agent: Zinc oxide; green synthesis and biomedical potential. *Dig. J. Nanomater Bios.* **2019**, *14*, 1023–1031.

22. Iqbal, A.; Khan, Z.A.; Shahzad, S.A.; Usman, M.; Khan, S.A.; Fauq, A.H.; Bari, A.; Sajid, M.A. Synthesis of E-stilbene azomethines as potent antimicrobial and antioxidant agents. *Turk. J. Chem.* **2018**, *42*, 1518–1533. [CrossRef]

23. Arakha, M.; Saleem, M.; Mallick, B.C.; Jha, S. The effects of interfacial potential on antimicrobial propensity of ZnO nanoparticle. *Sci. Rep.* **2015**, *15*, 9578. [CrossRef] [PubMed]

24. Park, E.J.; Yi, J.; Kim, Y.; Choi, K.; Park, K. Silver nanoparticles induce cytotoxicity by a Trojan-horse type mechanism. *Toxicol. In Vitro* **2010**, *24*, 872–878. [PubMed]

25. Ajitha, B.; Reddy, Y.A.; Reddy, P.S. Green synthesis and characterization of silver nanoparticles using Lantana camara leaf extract. *Mater. Sci. Eng. C* **2015**, *49*, 373–381. [CrossRef] [PubMed]

26. Kasprzak, M.M.; Erxleben, A.; Ochocki, J. Properties and applications of flavonoid metal complexes. *RSC Adv.* **2015**, *5*, 45853–45877.

27. Alfuraydi, A.A.; Devanesan, S.; Al-Ansari, M.; AlSalhi, M.S.; Ranjitsingh, A.J. Eco-friendly green synthesis of silver nanoparticles from the sesame oil cake and its potential anticancer and antimicrobial activities. *J. Photochem. Photobiol. B* **2019**, *192*, 83–89. [CrossRef]

28. Latha, D.; Sampurnam, S.; Arulvasu, C.; Prabu, P.; Govindaraju, K.; Narayanan, V. Biosynthesis and characterization of gold nanoparticle from Justicia adhatoda and its catalytic activity. *Mater. Today Proc.* **2018**, *5*, 8968–8972. [CrossRef]

29. Kim, J.S.; Kuk, E.; Yu, K.N.; Kim, J.H.; Park, S.J.; Lee, H.J.; Kim, S.H.; Park, Y.K.; Park, Y.H.; Hwang, C.Y.; et al. Antimicrobial effects of silver nanoparticles. *Nanomed. Nanotechnol. Biol. Med.* **2007**, *3*, 95–101. [CrossRef]

30. Masip, L.; Veeravalli, K.; Georgiou, G. The many faces of glutathione in bacteria. *Antioxid. Redox Sign.* **2006**, *8*, 753–762. [CrossRef]

31. Vecitis, C.D.; Zodrow, K.R.; Kang, S.; Elimelech, M. Electronic-structure-dependent bacterial cytotoxicity of single-walled carbon nanotubes. *ACS Nano* **2010**, *4*, 5471–5479. [CrossRef]

32. Banerjee, M.; Mallick, S.; Paul, A.; Chattopadhyay, A.; Ghosh, S.S. Heightened reactive oxygen species generation in the antimicrobial activity of a three component iodinated chitosan—Silver nanoparticle composite. *Langmuir* **2010**, *26*, 5901–5908. [PubMed]

33. Das, S.K.; Khan, M.M.; Guha, A.K.; Das, A.R.; Mandal, A.B. Silver-nano biohybride material: Synthesis, characterization and application in water purification. *Bioresour. Technol.* **2012**, *124*, 495–499. [CrossRef] [PubMed]

34. Qayyum, S.; Khan, I.; Bhatti, Z.A.; Tang, F.; Peng, C. Fungal strain Aspergillus flavus F3 as a potential candidate for the removal of lead (II) and chromium (VI) from contaminated soil. *Main Group Met. Chem.* **2016**, *39*, 93–104. [CrossRef]

35. Gué, M.; Dupont, V.; Dufour, A.; Sire, O. Bacterial swarming: A biochemical time-resolved FTIR—ATR study of Proteus mirabilis swarm-cell differentiation. *Biochemistry* **2001**, *40*, 11938–11945. [CrossRef]

36. Ramalingam, B.; Parandhaman, T.; Das, S.K. Antibacterial effects of biosynthesized silver nanoparticles on surface ultrastructure and nanomechanical properties of gram-negative bacteria viz. Escherichia coli and Pseudomonas aeruginosa. *ACS Appl. Mater. Interfaces* **2016**, *8*, 4963–4976. [CrossRef]

37. Slavin, Y.N.; Asnis, J.; Häfeli, U.O.; Bach, H. Metal nanoparticles: Understanding the mechanisms behind antibacterial activity. *J. Nanobiotechnol.* **2017**, *15*, 65.

38. Rizzello, L.; Pompa, P.P. Nanosilver-based antibacterial drugs and devices: Mechanisms, methodological drawbacks, and guidelines. *Chem. Soc. Rev.* **2014**, *43*, 1501–1518. [CrossRef]

39. Bondarenko, O.; Ivask, A.; Käkinen, A.; Kurvet, I.; Kahru, A. Particle-cell contact enhances antibacterial activity of silver nanoparticles. *PLoS ONE* **2013**, *8*, e64060.

40. Lok, C.N.; Ho, C.M.; Chen, R.; He, Q.Y.; Yu, W.Y.; Sun, H.; Tam, P.K.; Chiu, J.F.; Che, C.M. Proteomic analysis of the mode of antibacterial action of silver nanoparticles. *J. Proteome Res.* **2006**, *5*, 916–924. [CrossRef]

41. Parandhaman, T.; Das, A.; Ramalingam, B.; Samanta, D.; Sastry, T.P.; Mandal, A.B.; Das, S.K. Antimicrobial behavior of biosynthesized silica–silver nanocomposite for water disinfection: A mechanistic perspective. *J. Hazard. Mater.* **2015**, *290*, 117–126.

42. Ghosh, S.; Patil, S.; Ahire, M.; Kitture, R.; Kale, S.; Pardesi, K.; Cameotra, S.S.; Bellare, J.; Dhavale, D.D.; Jabgunde, A.; et al. Synthesis of silver nanoparticles using Dioscorea bulbifera tuber extract and evaluation of its synergistic potential in combination with antimicrobial agents. *Int. J. Nanomed.* **2012**, *7*, 483. [PubMed]

43. Jalilian, F.; Chahardoli, A.; Sadrjavadi, K.; Fattahi, A.; Shokoohinia, Y. Green synthesized silver nanoparticle from Allium ampeloprasum aqueous extract: Characterization, antioxidant activities, antibacterial and cytotoxicity effects. *Adv. Powder Technol.* **2020**. [CrossRef]

44. Elemike, E.E.; Onwudiwe, D.C.; Ekennia, A.C. Eco-friendly synthesis of silver nanoparticles using Umbrella plant, and evaluation of their photocatalytic and antibacterial activities. *Nano-Met. Chem.* **2020**, *50*, 389–399. [CrossRef]

45. Geethalakshmi, R.; Sarada, D.V. Gold and silver nanoparticles from Trianthema decandra: Synthesis, characterization, and antimicrobial properties. *Int. J. Nanomed.* **2012**, *7*, 5375.

46. Asghar, M.A.; Zahir, E.; Shahid, S.M.; Khan, M.N.; Asghar, M.A.; Iqbal, J.; Walker, G. Iron, copper and silver nanoparticles: Green synthesis using green and black tea leaves extracts and evaluation of antibacterial, antifungal and aflatoxin B1 adsorption activity. *LWT* **2018**, *90*, 98–107. [CrossRef]

47. Velmurugan, P.; Anbalagan, K.; Manosathyadevan, M.; Lee, K.J.; Cho, M.; Lee, S.M.; Park, J.H.; Oh, S.G.; Bang, K.S.; Oh, B.T. Green synthesis of silver and gold nanoparticles using Zingiber officinale root extract and antibacterial activity of silver nanoparticles against food pathogens. *Bioproc. Biosystems Eng.* **2014**, *37*, 1935–1943.

48. Küp, F.Ö.; Çoşkunçay, S.; Duman, F. Biosynthesis of silver nanoparticles using leaf extract of Aesculus hippocastanum (horse chestnut): Evaluation of their antibacterial, antioxidant and drug release system activities. *Mater. Sci. Eng. C* **2020**, *107*, 110207.

49. Francis, S.; Joseph, S.; Koshy, E.P.; Mathew, B. Microwave assisted green synthesis of silver nanoparticles using leaf extract of elephantopus scaber and its environmental and biological applications. *Artif. Cell Nanomed. B.* **2018**, *46*, 795–804.

50. Wang, L.; Wu, Y.; Xie, J.; Wu, S.; Wu, Z. Characterization, antioxidant and antimicrobial activities of green synthesized silver nanoparticles from Psidium guajava L. leaf aqueous extracts. *Mater. Sci. Eng. C* **2018**, *86*, 1–8. [CrossRef]

51. Mata, R.; Bhaskaran, A.; Sadras, S.R. Green-synthesized gold nanoparticles from Plumeria alba flower extract to augment catalytic degradation of organic dyes and inhibit bacterial growth. *Particuology* **2016**, *24*, 78–86. [CrossRef]

52. Asariha, M.; Chahardoli, A.; Karimi, N.; Gholamhosseinpour, M.; Khoshroo, A.; Nemati, H.; Shokoohinia, Y.; Fattahi, A. Green synthesis and structural characterization of gold nanoparticles from Achillea wilhelmsii leaf infusion and in vitro evaluation. *Bull. Mater. Sci.* **2020**, *43*, 57. [CrossRef]

53. Patra, J.K.; Baek, K.H. Novel green synthesis of gold nanoparticles using Citrullus lanatus rind and investigation of proteasome inhibitory activity, antibacterial, and antioxidant potential. *Int. J. Nanomed.* **2015**, *10*, 7253. [PubMed]

54. Zhaleh, M.; Zangeneh, A.; Goorani, S.; Seydi, N.; Zangeneh, M.M.; Tahvilian, R.; Pirabbasi, E. In vitro and in vivo evaluation of cytotoxicity, antioxidant, antibacterial, antifungal, and cutaneous wound healing properties of gold nanoparticles produced via a green chemistry synthesis using Gundelia tournefortii L. as a capping and reducing agent. *Appl. Organomet. Chem.* **2019**, *33*, e5015. [CrossRef]

55. Zangeneh, M.M.; Saneei, S.; Zangeneh, A.; Toushmalani, R.; Haddadi, A.; Almasi, M.; Amiri-Paryan, A. Preparation, characterization, and evaluation of cytotoxicity, antioxidant, cutaneous wound healing, antibacterial, and antifungal effects of gold nanoparticles using the aqueous extract of Falcaria vulgaris leaves. *Appl. Organomet. Chem.* **2019**, *33*, e5216. [CrossRef]

56. Hu, X.; Ahmeda, A.; Zangeneh, M.M. Chemical characterization and evaluation of antimicrobial and cutaneous wound healing potentials of gold nanoparticles using Allium saralicum RM Fritsch. *Appl. Organomet. Chem.* **2020**, *34*, e5484. [CrossRef]

57. Piruthiviraj, P.; Margret, A.; Krishnamurthy, P.P. Gold nanoparticles synthesized by Brassica oleracea (Broccoli) acting as antimicrobial agents against human pathogenic bacteria and fungi. *Appl. Nanosci.* **2016**, *6*, 467–473.

4

Facile and Robust Solvothermal Synthesis of Nanocrystalline CuInS$_2$ Thin Films

Anna Frank [1], Jan Grunwald [2], Benjamin Breitbach [1] and Christina Scheu [1,3,*]

[1]   Max-Planck-Institut für Eisenforschung GmbH, Max-Planck-Straße 1, 40237 Düsseldorf, Germany;
     frank@mpie.de (A.F.); breitbach@mpie.de (B.B.)

[2]   Ludwig-Maximilians-Universität, Butenandtstraße 5-11, 81377 Munich, Germany; jan_grunwald92@gmx.de

[3]   Materials Analytics, RWTH Aachen University, Kopernikusstraße 10, 52074 Aachen, Germany

*   Correspondence: scheu@mpie.de

Abstract: This work demonstrates that the solvothermal synthesis of nanocrystalline CuInS$_2$ thin films using the amino acid L-cysteine as sulfur source is facile and robust against variation of reaction time and temperature. Synthesis was carried out in a reaction time range of 3–48 h (at 150 °C) and a reaction temperature range of 100–190 °C (for 18 h). It was found that at least a time of 6 h and a temperature of 140 °C is needed to produce pure nanocrystalline CuInS$_2$ thin films as proven by X-ray and electron diffraction, high-resolution transmission electron microscopy, and energy-dispersive X-ray spectroscopy. Using UV-vis spectroscopy, a good absorption behavior as well as direct band gaps between 1.46 and 1.55 eV have been determined for all grown films. Only for a reaction time of 3 h and temperatures below 140 °C CuInS$_2$ is not formed. This is attributed to the formation of metal ion complexes with L-cysteine and the overall slow assembly of CuInS$_2$. This study reveals that the reaction parameters can be chosen relatively free; the reaction is completely nontoxic and precursors and solvents are rather cheap, which makes this synthesis route interesting for industrial up scaling.

Keywords: solvothermal synthesis; CuInS$_2$; TEM

1. Introduction

Due to the growing energy needs of our society, the scarcity of fossil fuels, and threatening greenhouse effect, research on materials that offer appropriate functionalities to overcome these problems is desperately needed; this is therefore a very active research field. Possible applications involve the generation of electricity via solar energy, the production of alternative fuels like hydrogen, and the decomposition of contaminants; but involves also research on how to store the produced energy [1–7]. For example, the photosynthesis of plants is mimicked to split water by light [8,9] or to convert CO$_2$ into less-harmful compounds [10]. To keep the costs economic, the synthesis of green energy materials should also be a green synthesis. Such a synthesis should be feasible without the need for expensive precursors, high pressures, or temperatures (i.e., high energy input), using a route that tolerates deviations in temperature and time, ideally accomplished in only one synthesis step, and should also avoid toxic chemicals during preparation, [11,12]. The use of biomolecules as precursors in chemical reactions and the formation of nanomaterials for diverse applications has been actively investigated in recent times [4,6,12–17].

Copper indium disulfide, CuInS$_2$, is a material suitable for diverse solar-driven applications [18–20]. It offers a band gap of 1.5 eV for the bulk, a high-absorption coefficient ($\alpha = 10^5$ cm^{-1}) [21] and can be used to convert sunlight into electricity or as a photocatalyst. CuInS$_2$ can be fabricated with various techniques which, in most cases, require high temperatures, high pressure, and clean precursor metals, but also with wet-chemical approaches [22–25]. We chose the solvothermal route to prepare CuInS$_2$, as it can be considered a green synthesis route—it only uses simple solvents and metal salts, while achieving a wide variety of nanostructures at low-reaction temperatures [26,27]. Furthermore, the solvothermal growth allows for the direct, one-pot $CuInS_2$ deposition on a suitable substrate—growing (thin) films in-situ without the need to deposit synthesized material afterwards [28].

Peng et al. [28] developed a solvothermal synthesis strategy for growing $CuInS_2$ films directly on fluorine-doped tin oxide (FTO) using simple salts as precursors. This synthesis strategy has been used and slightly modified in our group to prepare $CuInS_2$ thin films as well as microspheres [29,30]. However, this synthesis route involves the carcinogenic substance thioacetamide as sulfur source [31]. Therefore, we recently changed the sulfur source to the natural amino acid L-cysteine to achieve a complete non-toxic, green synthesis pathway towards $CuInS_2$ films on FTO substrates [32]. There, we varied the concentration and ratio of the used precursor salts to investigate the influence on sample morphology and properties, while keeping the reaction conditions constant (150 °C, 18 h). In short, at high sulfur ratios, an additional nanoflake layer of In_2S_3 on top of a compact $CuInS_2$ film was observed [32]. On reviewing literature, it becomes clear that it is possible to synthesize $CuInS_2$ thin films in a wide variety of reaction conditions and with many possible precursor salts and solvents; e.g., Peng et al. [28] used $CuSO_4$, $InCl_3$ and thioacetamide in ethanol at 160 °C for 12 h to produce pure $CuInS_2$. Wochnik et al. [29] were also able to synthesize pure tetragonal $CuInS_2$ films but at 150 °C for 24 h using the same precursors and solvent. Furthermore, it is possible to produce pure and stoichiometric $CuInS_2$ thin films with $CuCl_2$, $In(NO_3)_3$, thiourea, CTAB and oxalic acid in ethanol at 200 °C for 24 h as demonstrated by Xia et al. [33] The synthesis from Cu_2O, $In(OH)_3$, thioacetic acid and ammonia in ethanol at 150 °C for 6 h is also possible, as shown by Liu et al. [34] Solvothermal synthesis of nanostructured $CuInS_2$ using L-cysteine as a sulfur source has already been published in literature [35,36]. Liu et al. [35] reported about the formation of $CuInS_2$ using $CuCl_2$, $InCl_3$ and L-cystine in 1:1 ethylene diamine: water. They kept the autoclave at 200 °C for 12 h, resulting in microspheres and nanoparticles in the tetragonal Chalcopyrite structure without any visible impurities or side products. The composition of their $CuInS_2$ samples was also in a stoichiometric range. Wen et al. [36] synthesized $CuInS_2$ microspheres out of $CuCl_2$, $InCl_2$ and L-cysteine in N,N-dimethylformamide (DMF) as a solvent, also at 200 °C for 12 h. Their product displayed the tetragonal Chalcopyrite modification as well.

The results in literature indicate that pure, crystalline $CuInS_2$ nanostructures can be fabricated within a relative large reaction window; however, systematic studies, where the reaction temperature and time are varied in a broad range while keeping all other parameters constant, are rare. Additionally, many of the already existing synthesis routes involve toxic substances as raw materials.

In the present work, we fill that gap and focus on the influence of the reaction temperature and time on the L-cysteine-assisted solvothermal growth of nanocrystalline $CuInS_2$ films. We show that the solvothermal synthesis of $CuInS_2$ using L-cysteine as sulfur source is not only non-toxic but also extremely robust over a large temperature range from 140 °C to 190 °C as well as less critical on large time variations from 6 to 48 h. Thus, this synthesis pathway is very interesting for possible industrial utilization.

2. Materials and Methods

2.1. Synthesis of CuInS₂ Films

Chemicals were used as-purchased from Sigma-Aldrich (Sigma-Aldrich Chemie Gmbh, Munich, Germany) without further purification. The FTO glass substrates (Sigma-Aldrich) were cut into pieces of 15 mm × 20 mm × 2 mm, cleaned ultrasonically in dilute nitric acid, double-distilled water, acetone and ethanol for 5 min each prior to synthesis. The films were grown with our recently reported synthesis strategy using L-cysteine as sulfur source [32], which is based on the method published by Peng et al. [28] and our group [29] where thioacetamide was used.

The procedure is as follows: $CuSO_4 \cdot 5H_2O$ (0.2 mol, 0.050 g) and $InCl_3$ (0.2 mol, 0.044 g) were weighed out directly into a Teflon liner (20 mL capacity) and dissolved in 10 mL ethanol. The mixture was stirred for 10 min after which L-cysteine (0.5 mol, 0.061 g) was added. After stirring for another

5 min, a piece of FTO was placed inside the Teflon liner, conducting side facing down, the stainless-steel autoclave was sealed and put into an electric oven. There it was kept for 3, 6, 9, 12, 15, 18, 21, 24 and 48 h at a temperature of 150 °C and for 18 h at temperatures of 100, 120, 140, 150, 160, 180 and 190 °C. That means that for a variation of the reaction time, the reaction temperature was kept constant at 150 °C and for a variation of the reaction temperature the reaction time was always 18 h. The ratio between the precursors was kept at Cu:In:S 1:1:2.5. The film grown at this concentration/ratio and at 150 °C for 18 h has been published before [32] and will be referred to as film_S (standard reaction conditions). The other films are named according to their variation in time or temperature as film_time or film_temperature.

2.2. Characterization

To investigate the crystal structure of the synthesized $CuInS_2$ films on a global scale, X-ray diffraction (XRD) was used. To minimize the contribution from the FTO substrate, the measurements were performed under grazing incidence geometry with an incident angle of $\alpha = 2°$ in a Seifert THETA/2THETA X-ray diffractometer. The diffractometer was equipped with a Co source ($\lambda K\alpha = 1.79$ Å), polycapillary beam optics and an energy dispersive point detector. The 2θ values ranging from $10°$ to $140°$ were measured with a step size of $0.05°$/s and a count time of 30 s/step. The X-ray generator was operated at 40 kV and 30 mA. Literature data were used to identify the obtained phases. To calculate the average crystallite size, the Scherrer equation [37] was applied, fitting the most intense $CuInS_2$ peaks (112) and (204) with a Gaussian function.

The morphology of the $CuInS_2$ films was evaluated using scanning electron microscopy (SEM). For this purpose, a ZEISS Merlin, operated at 5.0 kV and a probe current of 2.0 nA, was used. Imaging was performed using the attached InLens® ZEISS standard detector. To analyze the chemical composition, energy-dispersive X-ray (EDX) spectroscopy using the XFlash detector 6 l 30 was done with an acceleration voltage of 20.0 kV and a probe current of 4.0 nA. Quantification was done using the Cliff-Lorimer equation. The intensities of the element-specific X-ray lines were determined by using Gaussian functions. The k-factors were calculated using the Bruker software. The results were normalized to Cu. In the case of thin films on substrates, the spectrum can also contain signals from the substrate as a result of the large interaction volume when using high acceleration voltages in SEM. For example, the In L line from $CuInS_2$ and the Sn L line from the substrate FTO (SnO_2:F) overlap and complicate the quantification of In (compare also our recent publication) [38]. Nevertheless, EDX measurements have been performed in the SEM at 20 kV acceleration voltage but In is not considered for the analysis and only the ratio between Cu and S is given (stoichiometric ratio Cu:S for $CuInS_2$ should be 1:2).

The film thicknesses were measured by focused ion beam (FIB) sectioning on a FEI Helios Nanolab 600. Cuts were performed at sample areas coated with conductive silver paint to avoid destruction of the film surface.

To conduct UV-vis measurements of the $CuInS_2$ films a Perkin Elmer Lambda 800 in transmission mode has been used. Spectral range was from 260 nm to 900 nm with a step size of 1 nm. From the UV-vis data band gaps were calculated using the Tauc method for direct band gap semiconductors [39]. The energy was plotted vs. (energy·absorption)^2 and the first linear slope was fitted and the intersection with the x-axis calculated.

For in-depth characterization of the films (scanning), transmission electron microscopy ((S)TEM) was used. Measurements were performed on a FEI Titan Themis 300 (S)TEM at 300 kV acceleration voltage. The (S)TEM is equipped with a C_S probe corrector, a Gatan Quantum ERS energy filter, and a Super X-EDX detector from Bruker. Electron diffraction data, calibrated with the help of a Si standard, were evaluated by comparing the results to literature data. (S)TEM scratch samples have been prepared to avoid an influence of the sample preparation on the crystallinity and composition of the investigated films (compare a recent publication of our group) [38]. As mentioned for EDX in SEM, quantification was done using the Cliff-Lorimer equation with the help of the Bruker software and normalizing the results relative to Cu. In STEM mode, several EDX maps have been recorded

($\approx$6 maps per sample) and quantification of the Cu:In:S ratio was done on $\approx$10 areas of each map with each area ≈ 100 nm^2 and calculating the average value.

3. Results

3.1. Reaction Time

Top-view secondary electron SEM images of CuInS$_2$ films, solvothermally synthesized for different reaction times at 150 °C, are shown in Figure 1 and in the Supporting Information (Figure S1). At first sight, all the films show a very similar surface topology. Only film_3 h, synthesized with the shortest reaction time of only 3 h at 150 °C, seems to be composed of individual agglomerates that grow on the FTO surface. Between these small agglomerates, the substrate is still visible, indicating an incomplete coverage after 3 h of solvothermal reaction (see Figure 1). With increasing reaction time, the agglomerates seem to grow laterally until they cover the underlying FTO substrate completely and form a compact CuInS$_2$ layer (for cross-sectional views see later images and the supporting information, Figure S2). After the compact layer has formed, more CuInS$_2$ agglomerates deposit on top of it. For some of the films, small nanoflakes growing out of the agglomerates can be observed (exemplarily marked in Figure 1). Nevertheless, large changes in the surface morphology of the films cannot be seen. From 6 h reaction time, the FTO substrate is not visible anymore in top-view and no cracks or delamination of the film is observed in cross-sectional SEM micrographs (see Figure S2 in the supporting information and also Figure 4a,b), implying a good homogeneity and adhesion of the films to the substrate.

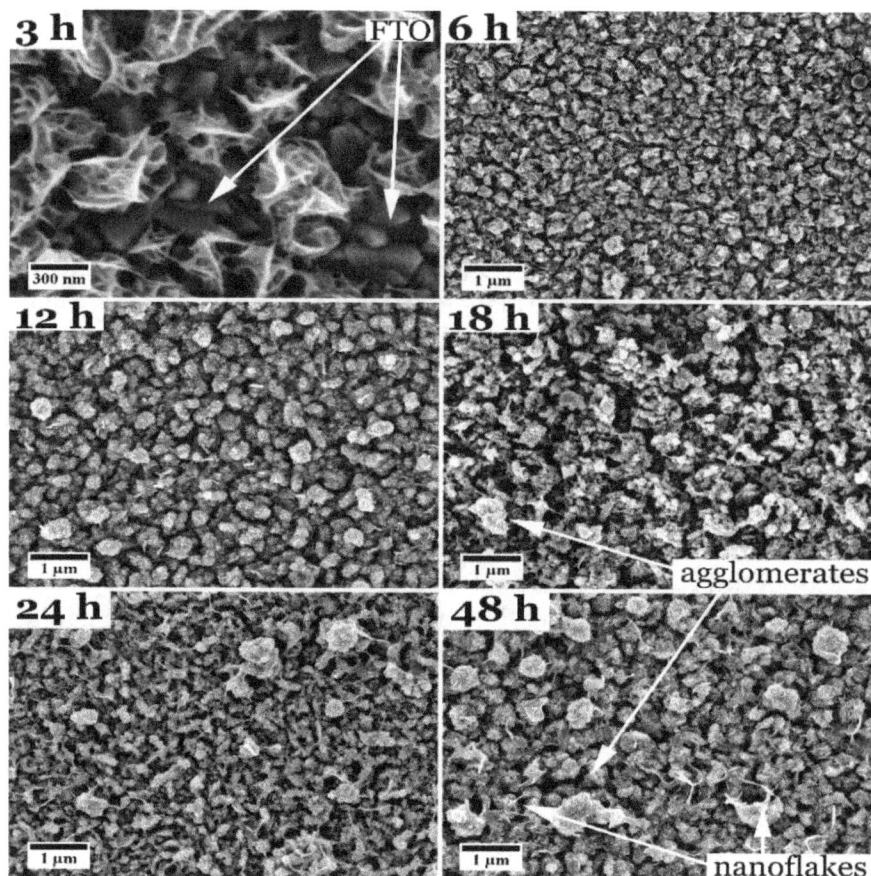

Figure 1. Top-view SEM images of CuInS$_2$ films synthesized with L-cysteine for different reaction times at 150 °C. The time varied between 3 h and 48 h. Please note a different scale bar for a reaction time of 3 h (film_3 h).

Measurement of the film thicknesses of the films via FIB is difficult, as the films consist of a compact $CuInS_2$ layer and outgrowing agglomerates. This problem is demonstrated exemplarily with the help of a FIB cross sectional cut, shown in Figure S2 (can also be seen in Figure 4a,b for a (S)TEM cross sectional sample). However, for all films, the thickness of the compact layer was determined to be in a relatively small range around 400 nm and is therefore in the same size regime.

Figure 2 shows exemplary XRD pattern of the films grown for 3, 12 and 24 h at 150 °C. The XRD pattern of the films synthesized for the other reaction times and of the pure FTO substrate are shown in the supporting information (Figures S3 and S6). Besides the strong reflections of the substrate FTO, (marked with * in Figure 2) all films show some more distinct signals. These reflections can be indexed according to tetragonal $CuInS_2$ in its Chalcopyrite modification (Figure 2, marked with # and compared to literature data [40]) for the films synthesized with a reaction time of at least 6 h. However, Cu^I and Cu^{II} sulfides possess very similar d-values as $CuInS_2$ and can therefore not be excluded by XRD data alone [41–43]. Additionally, amorphous phases could be present. Only film_3 h displays different reflections compared to the other films, which possess pure $CuInS_2$. These reflections (marked with °) can be assigned to cubic CuCl [44] and orthorhombic InS [45]. Due to the EDX measurements, described below, the presence of InS can be excluded as there is nearly no In detectable in the film as described later.

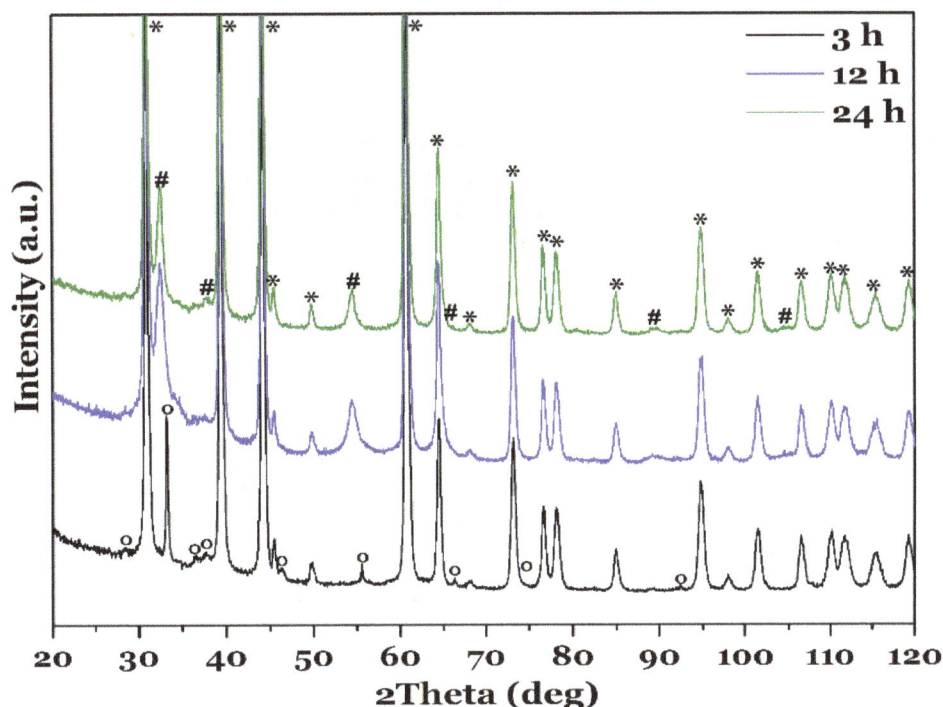

Figure 2. XRD pattern of $CuInS_2$ films synthesized with L-cysteine for different reaction times, 3, 12 and 24 h, at 150 °C. Signals stemming from FTO are marked with *, the ones originating from $CuInS_2$ with #, reflections from CuCl with °. The ° reflections could also stem from InS.

When comparing the sharp FTO reflections with the ones from tetragonal $CuInS_2$ it becomes clear that the latter is rather broad, indicating a small crystal and/or domain size in the films. Applying the Scherrer equation [37] to the most intense reflections (112) and (204) of $CuInS_2$ allows to estimate crystal sizes that are on average 9.0 ± 1.0 nm (Table 1). For film_3 h, in comparison, the distinct signals are relatively sharp, and using the (111) and (220) reflections of CuCl gives a crystal size of ≈ 39 nm for this phase. This value is three times higher than the largest calculated crystallite size of the $CuInS_2$ films.

Table 1. Summary of crystal size, determined with XRD, normalized elemental composition (Cu:S for SEM, Cu:In:S for STEM measurements) and band gap of the $CuInS_2$ films synthesized with L-cysteine for different reaction times at 150 °C. Values of film_3 h are not included in the average value in the last line.

	Crystal Size XRD (nm)	EDX SEM Cu:S (Normalized)	EDX TEM Cu:In:S (Normalized)	Band Gap UV-vis (eV)
film_3 h	39 ± 4	—	Cu:Cl 1.0:0.8 Cu:In:S 1.0:0.1:0.2	1.55
film_6 h	6.6 ± 1.4	1.0:1.3	—	1.54
film_9 h	11.0 ± 0.6	1.0:1.6	—	1.46
film_12 h	7.9 ± 1.0	1.0:2.1	—	1.54
film_15 h	8.5 ± 0.6	1.0:2.4	—	1.50
film_S [32]	9.4 ± 1.0	1.0:2.5	1.0:1.0:2.1	1.47
film_21 h	8.9 ± 1.0	1.0:2.3	—	1.51
film_24 h	11.0 ± 0.5	1.0:1.7	—	1.53
film_48 h	8.7 ± 1.0	1.0:1.9	1.0:1.0:2.2	1.44
Ø	9.0 ± 1.0	$1.0:2.0 \pm 0.4$	—	1.50 ± 0.04

EDX measurements in SEM, as mentioned before, show not only signals from Cu, In and S, but also from the substrate (Sn, Si, O from FTO and glass) due to the large interaction volume. This is discussed in more detail in our previous publication [38]. The overlap between the In L and Sn L line make the quantification of In difficult. This is the reason why for SEM EDX measurements only the Cu:S ratio is given and In is not included. Furthermore, all the SEM EDX spectra show also signals from the conductive coating (Au, Pd) and Cl from the $InCl_3$ precursor.

For film_3 h, a very high amount of Cu and Cl is measured with a lower amount of S and nearly no signal from In. This proves the existence of CuCl as observed from the XRD pattern (see Figure 2). However, an InS phase, which could also explain the reflections in the XRD pattern, is not present due to the very low In amount. Possible other phases might be, e.g., a strongly distorted CuS [41]. This will be described in more detail later. As can be seen (Table 1), the sulfur ratio detected for the films is varying, but always close to the stoichiometric Cu:S ratio of $CuInS_2$ of 1:2, except for reaction times of 3 h and 6 h. Higher amounts of S can be attributed to e.g., incomplete cleaning of the synthesized films with water and therefore remaining precursors/amorphous side products on the film surface. To quantify also the In amount via EDX (S)TEM measurements were performed, which are described later.

All films, synthesized with reaction times from 3 to 48 h, show a similar, strong absorption behavior over the whole visible spectrum, as exemplarily shown for film_3 h, film_12 h and film_24 h in Figure 3 (the UV-vis spectra from the other samples can be found in the supporting information, Figure S4). The absorption is influenced by the film thickness and also by light scattering on film structures or on the interface to the substrate. Since the film thicknesses of our films varies slightly but stays in the same size regime ($\approx$400 nm), this should not influence the absorption drastically. However, our films possess a rough surface structure (compare Figure 1) with larger agglomerates on top, which can scatter the light and 'increase' the absorption. Additionally, in areas with less agglomerates, the absorption is lower due to a smaller effective film thickness. Furthermore, crystal structure and composition can influence the total absorption of the measured films.

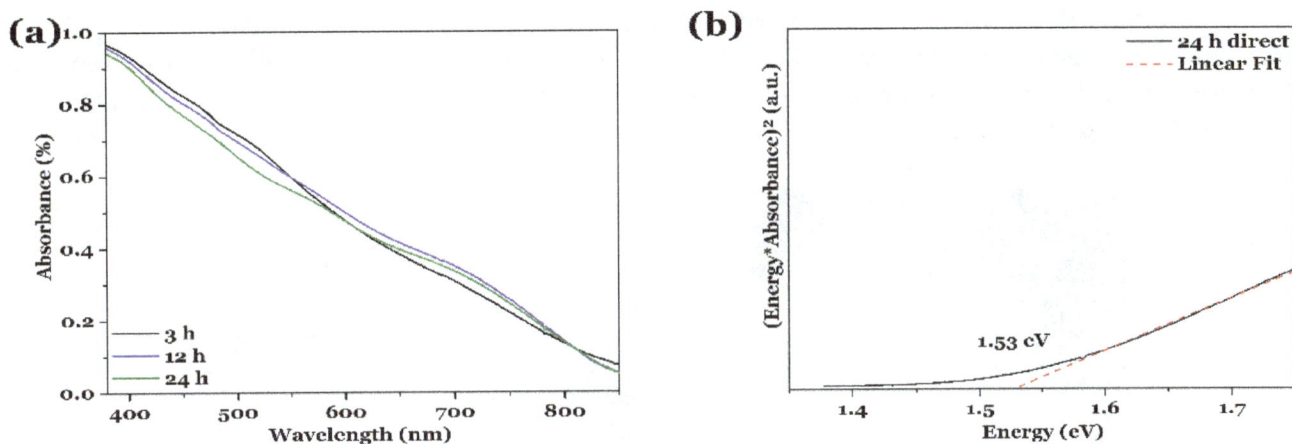

Figure 3. (a) UV-vis spectra of CuInS$_2$ thin films on FTO substrate, synthesized solvothermally with L-cysteine for 3, 12 and 24 h at 150 °C. (b) Exemplary Tauc plot for direct semiconductors for film_24 h, indicating a band gap of 1.53 eV.

The optical band gaps were calculated from the UV-vis data and the values are summarized in Table 1. In Figure 3b, an exemplary Tauc plot for film_24 h is shown. The smallest band gap of 1.44 eV is observed for film_48 h and the largest one with a value of 1.55 eV for film_3 h. All these values are close to the reported band gaps for CuInS$_2$ bulk material and nanostructures [21,46]. Variations in the band gap values can also be affected by the chemical composition, the film thickness and structure, as well as defects in the crystal structure, as mentioned above for the absorption behavior. Due to the solvothermal synthesis of our films and the small crystal size, a large number of defects, i.e., grain boundaries, are present, which could also lead to sub-band gap excitations [47–49]. Although the band gaps seem to vary (compare Table 1) no correlation can be drawn between the band gap value and, e.g., the crystallite size or chemical composition. The average band gap for the films grown for 6 to 48 h is 1.50 ± 0.04 eV. The standard deviation is relatively small, so that the band gaps can be considered as similar for all reaction times. This will be discussed in more detail later.

To investigate and compare the films grown at different reaction times in more detail, (S)TEM investigations have been performed. For the STEM EDX data, the In quantification is not problematic because the influence of the substrate FTO (and therefore the signal from Sn) can be neglected. Out of high-resolution HR TEM images, the crystal size and structure was extracted and compared to the one obtained by XRD. The results of the (S)TEM measurements on film_S [32,38] and film_48 h are shown in Figure 4, a TEM image and according diffraction pattern taken from film_3 h is shown in the supporting information, Figure S5, but also discussed in the following.

Figure 4a,b show the cross-sectional view of a focused-ion beam prepared lamella in STEM, demonstrating the vertical structure of the CuInS$_2$ film grown for 18 h at 150 °C. The film consists of a compact layer close to the FTO substrate with larger agglomerates growing on top of the film [32,38]. As all the other films, grown for different reaction times, display very similar topologies in SEM, it can be concluded that they also display very similar vertical structures. The HR TEM image and electron diffraction pattern in Figure 4c,d, also taken from film_S, prove the good crystallinity of the nanoparticle film and the tetragonal CuInS$_2$ modification [32,38]. The crystal size of the nanoparticles determined from the HR TEM images is 5.3 ± 2.2 nm and therefore smaller than the one determined with XRD (9.4 nm, compare Table 1). EDX quantification for film_S resulted in Cu 24 ± 2 at %, In 25 ± 2 at % and S 51 ± 3 at %, giving a Cu:In:S ratio of 1.0:1.0:2.1 [32]. This ratio is close to the stoichiometric value.

As for film_S, the HR TEM image and electron diffraction pattern for film_48 h reveal a good crystallinity of the solvothermally synthesized CuInS$_2$ film in the tetragonal Chalcopyrite modification (Figure 4e,f). Again, the film is composed of many agglomerated nanoparticles with grain sizes of

5.4 ± 2.3 nm, smaller than the value calculated from the XRD data (Table 1). The reason why the determination of the crystallite size leads to different values in XRD and TEM will be discussed later. EDX measurements on various areas gave Cu 24 ± 2 at %, In 25 ± 1 at % and S 52 ± 1 at %, resulting in a Cu:In:S ratio of 1.0:1.0:2.2. This is also very close to a stoichiometric composition of $CuInS_2$.

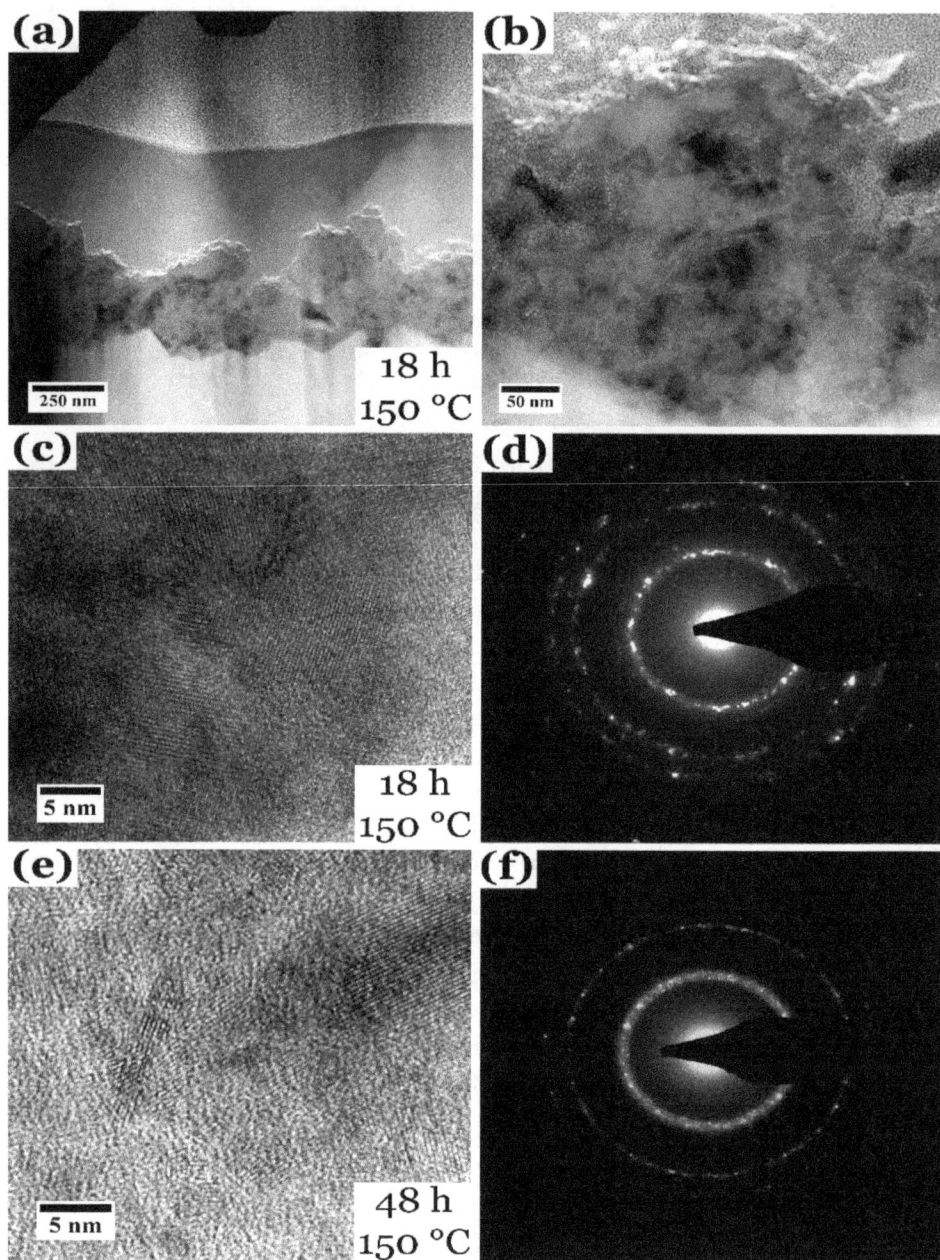

Figure 4. (**a,b**) cross sectional HAADF STEM images of a lamella prepared from film_S [32,38], displaying the vertical structure of the film. (**c,d**) HR TEM image and according electron diffraction pattern of film_S, and (**e,f**) from film_48 h.

Only the film grown for 3 h showed single agglomerates of nanoparticles on the FTO substrate (see Figure 1) and strong reflexes of CuCl in the XRD data. HR TEM and electron diffraction pattern (Figure S5a,b) confirm that these nanoparticles are crystalline. The crystallite sizes determined from HR TEM images resulted in a minimum value of ≈4 nm and a maximum crystal size of 36 nm; the latter was also obtained out of the XRD spectrum. The electron diffraction pattern can be indexed according to cubic CuCl [44] and orthorhombic InS [45]. However, due to very similar d-values of other possible products, e.g., In$_2$S$_3$ or CuS, and the possibility of lattice distortion of these phases caused by intercalation of impurity atoms, it is very difficult to determine the unambiguous phases. EDX measurements and quantification have been performed for Cu, In, S and also Cl. The quantification led to 48 ± 5 at % Cu, 2 ± 2 at % In, 10 ± 8 at % S and 39 ± 6 at % Cl, which gives a ratio of Cu:Cl of 1:0.8. This is close to the stoichiometric Cu:Cl ratio of 1:1 for CuCl. The existence of an InS phase in film_3 h can be excluded because of the very low Indium amount.

3.2. Reaction Temperature

Figure 5 shows an overview over the CuInS$_2$ samples solvothermally grown with L-cysteine at different reaction temperatures for 18 h. Film_100 °C is shown in the supporting information (Figure S6). A reaction temperature of 100 °C seems to be not sufficient to grow a film. Only the pure FTO substrate is found in SEM and XRD (Figure S6). Except for film_100 °C and film_120 °C, all the films have the same appearance when studied in top-view as the films synthesized at 150 °C with different reaction times (see Figure 1). The films consist of a compact nanograined film with agglomerations of nanoparticles on top, which vary in size and density. Film_120 °C, on the other hand, looks different. The film seems to consist of nanoparticles, too, but with a smoother shape and has a white appearance on the FTO substrate when inspected by eye, while all other films are brownish. For film_180 °C, although similar to the other CuInS$_2$ films, a higher number of nanoflakes, which grow out of the film, and agglomerates can be observed. This can also be seen in a more distinct manner for other films (compare Figure 1). The film with the highest reaction temperature (190 °C) displays very large, round-shaped agglomerates. All films again show no cracks or delamination from the substrate.

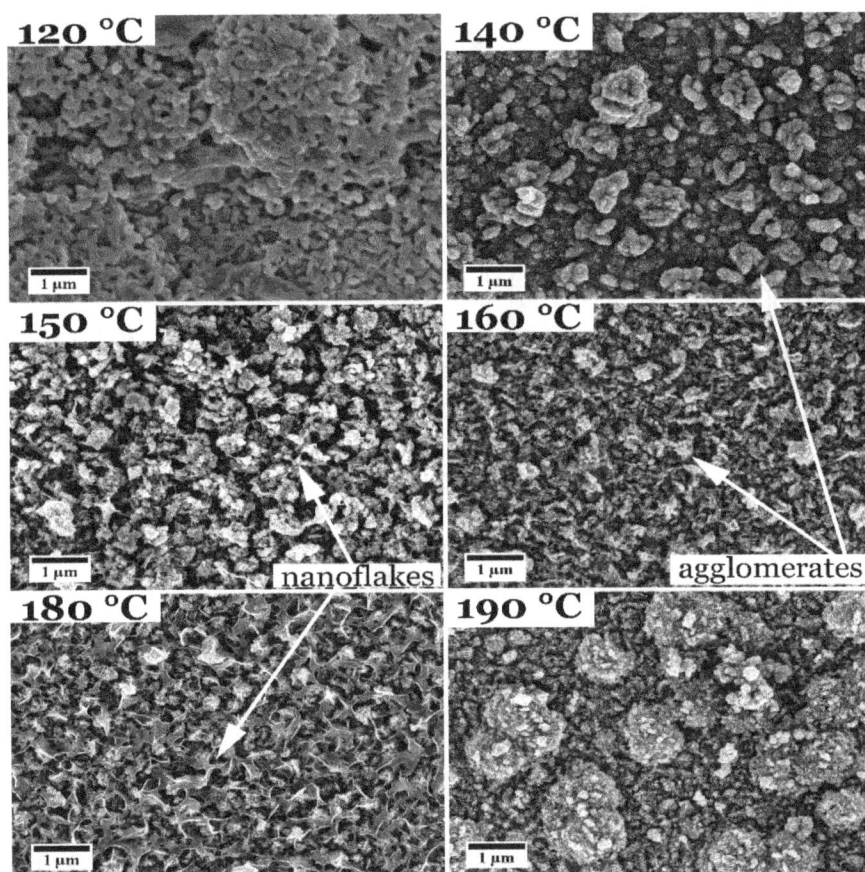

Figure 5. Top-view SEM images of CuInS$_2$ films synthesized with L-cysteine at different reaction temperatures for 18 h. The temperature for the films shown was varied between 120 and 190 °C.

The film thickness of these films has also been investigated by FIB cross sectional cuts. Again, the film thickness of all films is around 400 nm. As mentioned before, the measurement of the film thickness is difficult because of the structure of the film (Figure S2 and HAADF STEM image in Figure 4a).

XRD patterns obtained from the films synthesized at different reaction temperatures for 18 h are shown in Figure 6, exemplarily for 120, 160 and 190 °C, and in the supporting information, Figures S6 and S7, for the other temperatures. For film_100 °C, as already observed in the SEM, only the pure FTO substrate leads to signals in the XRD pattern. All peaks are in accordance to literature data [50]. This implies that a reaction temperature of 100 °C is not sufficient for the growth of $CuInS_2$. In addition, the XRD pattern of the sample grown at 120 °C (Figure 6, black) does not correspond to $CuInS_2$. The pattern shows a lot of reflections, which cannot be indexed unambiguously by one crystalline phase. This means that several different crystalline species are formed for this reaction conditions. It might also be that the film consists of not-reacted precursor salts or preliminary formed complexes. However, with further increase of the reaction temperature, $CuInS_2$ in tetragonal Chalcopyrite modification is formed. As mentioned before, the presence of side products cannot be excluded fully because of very similar d-values of possible compounds, e.g., copper sulfides. Amorphous phases cannot be excluded as well. However, within the detection limit, a pure $CuInS_2$ phase is formed for all films.

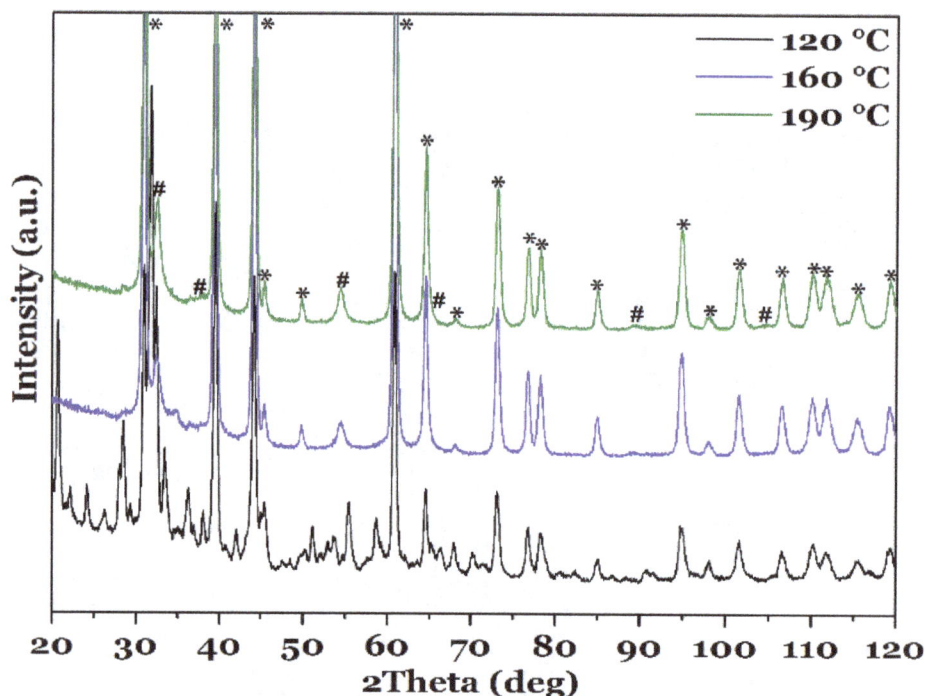

Figure 6. XRD pattern of $CuInS_2$ films synthesized with L-cysteine at different reaction temperatures, 120, 160 and 190 °C, for 18 h. Signals stemming from FTO are marked with *, the ones originating from $CuInS_2$ with #.

The $CuInS_2$ peaks are again rather broad and crystal sizes between 7.9 nm for film_140 °C and 11.4 nm for film_180 °C were estimated from the XRD data. On average, the crystal/domain sizes lay also in the range of 10 nm as observed for the $CuInS_2$ films grown for different reaction times at 150 °C (compare Table 2).

Table 2. Summary of crystal size, determined with XRD, normalized elemental composition (Cu:S for SEM, Cu:In:S for STEM measurements) and band gap of the $CuInS_2$ films synthesized with L-cysteine at different reaction temperatures for 18 h. Values of film_120 °C are not included in the average value in the last row.

	Crystal Size XRD (nm)	EDX SEM Cu:S (Normalized)	EDX TEM Cu:In:S (Normalized)	Band Gap UV-Vis (eV)
film_120 °C	—	1.0:3.9	—	1.51
film_140 °C	7.9 ± 0.5	1.0:1.9	1.0:1.1:2.2	1.52
film_S [32]	9.4 ± 0.6	1.0:2.5	1.0:1.0:2.1	1.47
film_160 °C	9.2 ± 1.0	1.0:1.7	—	1.54
film_180 °C	11.4 ± 1.0	1.0:1.6	—	1.55
film_190 °C	9.8 ± 0.6	1.0:1.4	1.0:1.0:1.9	1.54
Ø	9.5 ± 1.1	1.0:1.8 ± 0.4	—	1.52 ± 0.03

Analogous to the study of different reaction times, also for the films grown at different reaction temperature, EDX measurements have been performed in the SEM but only the Cu:S ratio is given.

As expected, film_100 °C does not give any signal of Cu, In or S in the EDX spectrum, as it is only the plain FTO substrate and no film has grown on top. Film_120 °C shows only a very little amount of copper (Cu:S 1.0:3.9) but large amounts of from Sn and O (FTO substrate). From film_140 °C on the films show a nearly stoichiometric ratio between Cu and S. An increase of the reaction temperature to 180 °C gives a ratio of Cu to S of 1.0:1.6 (Table 2). The decreased amount of sulfur is even more pronounced for a reaction temperature of 190 °C (Cu:S of 1.0:1.4). However, as the XRD pattern in Figure 6 only shows signals from $CuInS_2$ in the tetragonal Chalcopyrite modification, the reduced S amount might be caused by e.g., amorphous side products, which have not been rinsed away.

All the films show a good absorption behavior over the whole visible spectrum (see Figure 7a for film_140 °C, film_160 °C and film_190 °C, the other UV-vis spectra are shown in the supporting information, Figure S8). As described before, fluctuations in the total absorption can be caused by variations of the sample surface, relative thickness, and scattering of light. Because all the films show a similar crystal size (determined with XRD) and film thickness, the absorption is in the same order of magnitude.

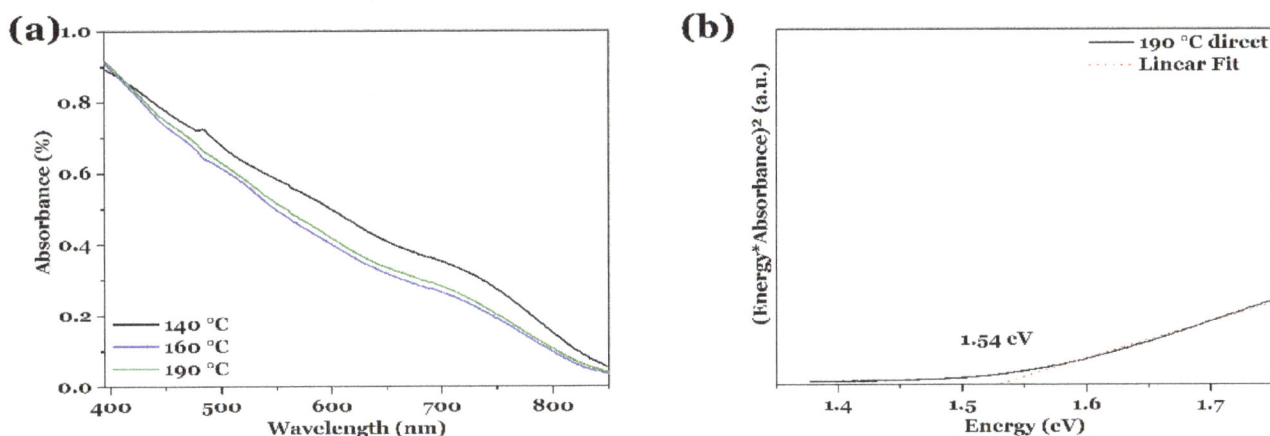

Figure 7. (a) UV-vis spectra of $CuInS_2$ thin films on FTO substrate, synthesized solvothermally with L-cysteine at different reaction temperatures for 18 h. (b) Exemplary Tauc plot for direct semiconductors for film_190 °C, indicating a band gap of 1.54 eV.

Figure 7b shows one exemplary Tauc plot to determine the band gap for film_190 °C; the other calculated band gaps are summarized in Table 2. The smallest band gap is found for film_S (1.47 eV) [32], while the largest one is given for a reaction temperature of 180 °C, with a value of 1.55 eV.

All the other band gap values lay in between these values and are very close to band gaps reported in literature for $CuInS_2$ [21,46]. Again, variations in the band gap can be induced by the chemical composition, the film thickness, and structure, as mentioned already above. The average band gap gives 1.52 ± 0.03 eV, which also displays a small standard deviation and therefore the band gaps are in the same size regime. The impact of these fluctuations in the band gap values will be discussed later.

TEM measurements on film_140 °C show a good crystallinity of the film (Figure 8a,b) and small crystallites with an average size of 5.7 ± 2.0 nm (XRD 7.9 nm, compare Table 2). The size of the crystals is in the same magnitude as for film_S [32] and film_48 h (compare also Figure 4 and Table 1); the deviation between TEM and XRD grain size determination will be discussed later. The electron diffraction pattern in (b) also proves the tetragonal Chalcopyrite modification of $CuInS_2$. STEM EDX quantification gives Cu 23 ± 1 at %, In 25 ± 1 at % and S 52 ± 1 at %, resulting in a Cu:In:S ratio of 1.0:1.1:2.2, which is very close to a stoichiometric $CuInS_2$ ratio.

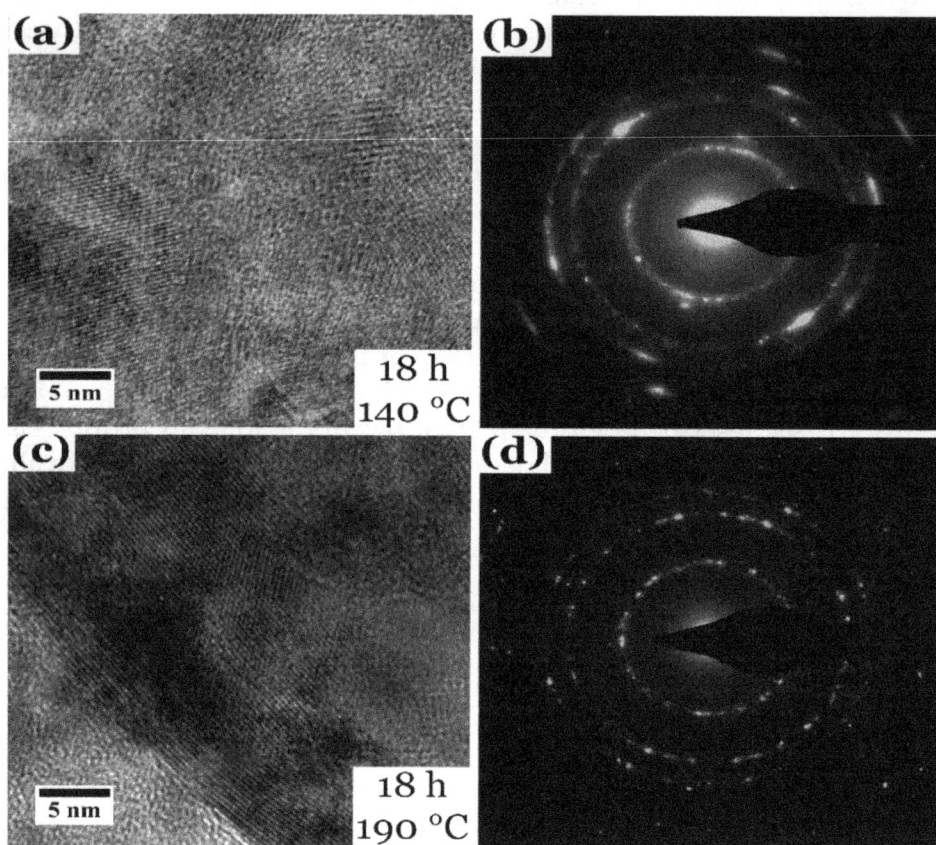

Figure 8. (a) HR TEM image and (b) corresponding electron diffraction pattern of film_140 °C. (c) HR TEM image and (d) corresponding electron diffraction pattern of film_190 °C.

The (S)TEM investigations on film_190 °C (Figure 8c,d) give similar results. The film is also well crystallized and displays the tetragonal $CuInS_2$ modification. Crystallite size was determined to be 10 ± 5.0 nm, which is larger than for the lower temperatures but close to the crystal size determined with XRD (≈ 9.8 nm). Additionally, the reflections in the diffraction pattern of film_190 °C appear much sharper than for film_140 °C (Figure 8b vs. Figure 8d), hinting also at a larger crystallite size. The quantitative EDX measurements in STEM mode show Cu 26 ± 1 at %, In 25 ± 1 at % and S 49 ± 1 at %, resulting in a Cu:In:S ratio of 1.0:1.0:1.9. This is also very close to stoichiometry of $CuInS_2$.

4. Discussion

When comparing the top-view SEM images of the samples synthesized with strongly varying reaction time and temperature, they all appear very similar. Most films are composed of a dense film with outgrowing agglomerates, also built up of nanoparticles, on top. This can also be seen in cross-sectional SEM and STEM images. The film thickness of the different films is roughly in the same size regime of $\approx$400 nm. This can be explained by the use of the same precursor concentrations for all the films, which was found to be the dominating parameter in controlling the film thickness [32]. When all available precursors are consumed, neither time nor temperature have an impact on the thickness of the grown $CuInS_2$ film. Only changing the number of precursor molecules would change the film thickness. However, another possible explanation for this fact is the development of a thermodynamic equilibrium between grown film and free $CuInS_2$ nuclei. Only adding more precursor would shift the equilibrium to an increase in film thickness.

For all synthesis conditions, a pure tetragonal Chalcopyrite phase of $CuInS_2$ can be observed in the XRD data, except for film_3 h, film_100 °C and film_120 °C. Applying the Scherrer equation [37] to the most intense peaks of the pattern gave an average crystal/domain size of $\approx$10 nm. Crystal sizes determined with TEM for film_S [32], film_48 h and film_140 °C give, by contrast, smaller crystal sizes of $\approx$5 nm with the exception of film_190 °C, where both methods led to similar crystal sizes ($\approx$10 nm). This might be related to the fact that the films are buildup of two areas—a dense, nanograined film and overlying agglomerates of nanoparticles. Due to the global nature of XRD, the calculation of the crystallite size takes into account both parts of the film structure, resulting in an overall higher crystal size. In contrast, TEM allows the determination of the crystal size at a very local scale and the here used scratch TEM samples are most likely representing mainly the agglomerates, as they can be removed from the substrate easier. As a consequence, we conclude that our $CuInS_2$ thin films display larger crystal sizes in the more-dense film close to the substrate (measured with XRD), while the agglomerates on top of this film are formed of smaller crystallites (measured with TEM). Only for film_190 °C, the crystal size determined with TEM is close to the XRD measurements, which can mean that higher temperatures lead to larger crystal sizes. Nevertheless, all $CuInS_2$ thin film samples display a good crystallinity and the tetragonal Chalcopyrite modification of $CuInS_2$. Along with EDX measurements, it can be concluded that starting from a reaction time of 6 h and a reaction temperature of 140 °C, pure $CuInS_2$ thin films without any impurities or side products can be grown solvothermally with the help of L-cysteine. The reason for the indeterminable amount of side products like CuS, Cu_2S or In_2S_3 is the use of a high enough L-cysteine to Cu precursor ratio of 2.5:1 [32,51] as well as the complex formation between Cu, In and L-cysteine (and cystine) [35,36,52,53] as will be described below.

All the films show a very good absorption behavior in the visible regime and their band gaps, calculated with the help of the Tauc method, lie in the range between 1.46 and 1.55 eV, which is very close to the band gap for the bulk $CuInS_2$ [21] and has also been reported for $CuInS_2$ nanostructures [46]. As obvious from Tables 1 and 2 and mentioned before, the band gap values of the different films are varying and show no direct relation to the crystal size and/or chemical composition. An average value calculated from the band gaps of film_6 h to film_48 h as well as film_140 °C to film_190 °C results in 1.52 ± 0.04 eV. The small standard deviations show that the band gap variations are small and can be caused by e.g., slight variations in the chemical composition or the local crystallinity, which can be caused e.g., by amorphous side products, etc. Summarized, the absorption properties and band gap values, as well as the good crystallinity, make the $CuInS_2$ thin films a very interesting material for solar driven applications. Changes of the band gaps in the here observed regime should not have any influence on use in applications. A calculation of the theoretical efficiencies based on the band gap values show that, considering single-junction solar cells, a high efficiency of $\approx$30% can be reached in a band gap range from $\approx$1.0–1.6 eV [54,55]. This means that for a solar-driven application, the observed band gap values are excellent. Equally important is the ability to absorb most of the sun light, which is true for $CuInS_2$.

Solvothermal reaction—and chemical reactions in general—are governed by many factors, but most important are the thermodynamic ones, pressure, and temperature, followed by reaction time. In this work, we varied the reaction time at constant temperature, and the reaction temperature for constant time intervals. In the case of a reaction time of 18 h but increasing reaction temperature, the pressure inside the reaction vessel should increase accordingly. However, the consideration in the case of a fixed reaction temperature but changing time is more complex. In general, for an increase of reaction time at fixed temperature, no significant increase in pressure is to be expected—except for the evolution of gaseous products during the reaction, which is not assumed for the $CuInS_2$ reaction as presented here. Nevertheless, for very short reaction times, the temperature inside the autoclave is lower than targeted and a therefore lower pressure is expected.

For film_3 h, mainly a cubic CuCl phase could be observed in the XRD and electron diffraction pattern. The film itself is very thin and the FTO substrate is visible between the agglomerates. EDX measurements in STEM mode reveal mainly Cu and Cl with only little indium and sulfur contents in the film. This shows that a reaction time of 3 h at 150 °C is not sufficient to form $CuInS_2$. The formation of CuCl instead can be explained as follows: Although a reaction temperature of 150 °C is chosen, the autoclave might not reach this temperature during 3 h of reaction time. For a short reaction time (i.e., lower temperature, lower pressure inside the autoclave) the decomposition of L-cysteine and cystine proceeds with a kinetically low rate, leading to (1) the metal ions In^{3+} and Cu^+ still stabilized in complexes and (2) no S^{2-} ions are available. Instead, a lot of free Cl^- ions are accessible due to the solvation of the $InCl_3$ precursor salt. A recrystallization of $InCl_3$ is very unlikely because of an enthalpy of formation of 537 kJ/mol [56], which exceeds the one of In_2S_3 of -346 kJ/mol [56]. However, as mentioned before, due to the low temperature/short reaction time, no sulfur ions are available. Therefore, only CuCl with an enthalpy of formation of -137 kJ/mol [56] is thermodynamically likely to grow.

For film_100 °C, only the pure FTO substrate was obtained. It is obvious that a reaction temperature of 100 °C, although kept for 18 h, and an according low pressure inside the autoclave, is not sufficient to grow $CuInS_2$ or any other crystalline compound on the substrate. The formed Cu^+ and In^{3+} complexes with L-cysteine and cystine are still stable at this temperature and a thermal decomposition is not taking place. Therefore, no S^{2-} and metal ions are released and ready to react. Additionally, the overall solubility of the precursor salts is not promoted at low temperature and pressure, and only a little amount is available in the solution.

When increasing the reaction temperature to 120 °C, a film forms, consisting of nanoparticles, probably embedded in an amorphous matrix. The XRD data show a lot of signals, arising from crystalline compounds, which means that the reaction conditions are sufficient to form crystalline compounds. However, an unambiguously assignment of all emerging signals is not achievable. Possible compounds are L-cystine, L-cysteine, their Cu and In complexes, Cu sulfides like Cu_2S, $Cu_{1.8}S$ or CuS, indium sulfides, e.g., In_2S_3 or modifications of these phases. It is also possible that unchanged precursor materials like $InCl_3$ or $CuSO_4 \cdot 5H_2O$ are still present for this reaction conditions. As observed for film_3 h, the formation of CuCl is also possible.

The reaction mechanism of $CuInS_2$ out of $CuSO_4 \cdot 5H_2O$, $InCl_3$ and L-cysteine can be described as follows: as known, L-cysteine is oxidized to cystine while Cu^{2+} is reduced to Cu^+ [32,51,53]. This Cu^+ and In^{3+} can be coordinated and stabilized by the chelating agents L-cysteine and cysteine [57]. The release of the metal ions is therefore very slow. As can be learned from a reaction at 100 °C for 18 h, at the chosen precursor concentrations (Cu 0.2 mol, In 0.2 mol, L-cysteine 0.5 mol, precursor ratio Cu:In:S 1:1:2.5) no free metal or sulfur ions exist in the solution. However, reaching a certain temperature/pressure (>120 °C), the organic molecules L-cysteine and L-cystine begin to decompose and release S^{2-}. When the decomposition of the sulfur source starts and S^{2-} is released, $(InS_2)^-$ can be formed [35,36,52], but also copper sulfides are possible side products. From previous investigations [32] with higher L-cysteine contents (Cu:In:S precursor ratio 1:1:4), it is known that during a reaction at 150 °C for 18 h, a compact $CuInS_2$ bottom layer with an outgrowing In_2S_3 top layer is formed.

Accordingly, for a short reaction time (3 h) at 150 °C, In_2S_3 should be formed, which is not the case as described above. Here, the reason is a thermodynamically favored formation and stabilization of Cu^+ in a chloride as no sulfur ions are present. Additionally, this gives rise to the conclusion that the combination between $(InS_2)^-$ and Cu^+ to form $CuInS_2$ is the main bottleneck in the reaction and rather slow, which hints at a higher importance of the reaction time compared to the reaction temperature. However, as obvious from film_100 °C and film_120 °C, the reaction temperature is important in the formation of $CuInS_2$. If the temperature of the reaction is not high enough, no reaction will occur (film_100 °C) or a mixture of compounds will grow (film_120 °C). However, temperatures above 120 °C allow the formation of pure $CuInS_2$ without obvious side products, as described above.

Our results are different from Kharkwal et al. [52] They synthesized $CuInS_2$ nanoparticles solvothermally using CuCl, $InCl_3$ and thiourea at 150 °C for different reaction times (2 to 48 h). They observed small nanoparticles of ≈5 nm for 2 h reaction time and larger nanoparticles (up to ≈27 nm) for 48 h. All nanoparticles showed a pure tetragonal $CuInS_2$ and with an increase in the particle size, the band gap decreased. Furthermore, with increasing reaction time, their nanoparticles evolved from agglomerates to flower-shaped structures. Kharkwal et al. ascribed the phase purity of their nanoparticles to the formation of Cu and In complexes with thiourea as we ascribe them to the formation of L-cysteine complexes. However, in our case, the crystallite size does not change with reaction time and a reaction time of 3 h is not sufficient to produce $CuInS_2$. This can be related to the different sulfur and copper precursors. In addition, in our case, the Cu^{2+} of the precursor has to be reduced to Cu^+ before it is usable in the synthesis of $CuInS_2$, while Kharkwal et al. [52] directly used a Cu^+ salt.

Zhuang et al. [58] prepared $CuInS_2$ thin films on FTO substrates as well. They used $CuCl_2$, $In(NO_3)_2$, thiourea, oxalic acid and hexadecyl trimethyl ammonium bromide (CTAB) in ethanol at 200 °C and reaction time was varied (1, 4, 8 and 20 h). For a low thiourea concentration, they observed an increase in film thickness with increasing reaction time (saturation after 8 h of reaction) and an evolution in the surface topology. For a higher concentration, the films are always composed of a close packed microsphere layer, while only the diameter and packing density changed with time. All the samples displayed the tetragonal Chalcopyrite modification of $CuInS_2$. In our synthesis, the influence of the reaction time on film thickness or morphology is neglectable.

To summarize the comparison with literature, it seems that our synthesis procedure using L-cysteine as sulfur source is a very robust route, as it yields $CuInS_2$ films on FTO substrates with excellent properties over a wide reaction temperature and time range.

5. Conclusions

To conclude, we were able to demonstrate that the green synthesis of pure, nanocrystalline $CuInS_2$ thin films on FTO substrates via a solvothermal, non-toxic L-cysteine assisted synthesis approach is possible and stable over a wide reaction time and temperature range. Our systematic study showed that the synthesis of $CuInS_2$ films with a thickness of ≈400 nm, a good crystallinity and band gaps in the range of 1.46 to 1.55 eV is independent of the exact reaction temperature and time, as long as the reaction temperature is above 140 °C and the reaction time longer than 6 h. All the results can be explained by the presented, refined reaction mechanism and the stability of initially formed precursor complexes of Cu and In complexes with L-cysteine and cystine, which provides the basis for the advantage of using L-cysteine compared to other sulfur sources. The facile and robust solvothermal synthesis of $CuInS_2$ using L-cysteine is therefore suggested as a possible route for up-scaling. Due to their excellent properties, the films are also viewed as possible candidates for solar driven applications like solar cells or water splitting.

Supplementary Materials:
Figure S1, Top-view SEM images of $CuInS_2$ films synthesized with L-cysteine for different reaction times at 150 °C. Shown are film_9 h, film_15 h and film_21 h. Figure S2, SE image of an exemplary FIB cross section for the film thickness determination of film_140 °C is shown, displaying large fluctuations in the film thickness due to the

agglomerates. Figure S3, XRD pattern of $CuInS_2$ films synthesized with L-cysteine for different reaction times, 6, 9, 15, 18, 21 and 48 h, at 150 °C. Signals stemming from FTO are marked with *, the ones originating from $CuInS_2$ with #. Figure S4, UV-vis spectra of $CuInS_2$ thin films on FTO substrate, synthesized solvothermally with L-cysteine grown for different reaction times at 150 °C. Figure S5, (a) HR TEM image and (b) electron diffraction pattern of a $CuInS_2$ film synthesized with L-cysteine for 3 h at 150 °C (film_3 h). Figure S6, (a) SEM image and (b) XRD pattern of film_100 °C. Only pure FTO can be observed in SEM and XRD. Figure S7, XRD pattern of $CuInS_2$ films synthesized with L-cysteine at different reaction temperatures, 100, 140, 150, and 180 °C, for 18 h. Signals stemming from FTO are marked with *, the ones originating from $CuInS_2$ with #. Figure S8, UV-vis spectra of $CuInS_2$ thin films on FTO substrate, synthesized solvothermally with L-cysteine grown for 18 h at different reaction temperatures.

Author Contributions: The syntheses were conceived and performed by A.F. and J.G. SEM, TEM and UV-vis experiments and their analyses was done by A.F. XRD measurements and analyses was carried out by B.B. C.S. supervised the experiments and discussed the results. The article was written by A.F. and revised by all co-authors.

Acknowledgments: This research did not receive any specific grant from funding agencies in the public, commercial, or not-for-profit sectors.

References

1. Yamada, Y.; Yamada, T.; Kanemitsu, Y. Free Carrier Radiative Recombination and Photon Recycling in Lead Halide Perovskite Solar Cell Materials. *Bull. Chem. Soc. Jpn.* **2017**, *90*, 1129–1140. [CrossRef]

2. Anasori, B.; Lukatskaya, M.R.; Gogotsi, Y. 2D metal carbides and nitrides (MXenes) for energy storage. *Nat. Rev. Mater.* **2017**, *2*, 16098. [CrossRef]

3. Abe, H.; Liu, J.; Ariga, K. Catalytic nanoarchitectonics for environmentally compatible energy generation. *Mater. Today* **2016**, *19*, 12–18. [CrossRef]

4. Teimouri, M.; Khosravi-Nejad, F.; Attar, F.; Saboury, A.A.; Kostova, I.; Benelli, G.; Falahati, M. Gold nanoparticles fabrication by plant extracts: Synthesis, characterization, degradation of 4-nitrophenol from industrial wastewater, and insecticidal activity—A review. *J. Clean. Prod.* **2018**, *184*, 740–753. [CrossRef]

5. Yu, L.; Zhang, Y.; He, J.; Zhu, H.; Zhou, X.; Li, M.; Yang, Q.; Xu, F. Enhanced photoelectrochemical properties of α-Fe_2O_3 nanoarrays for water splitting. *J. Alloys Compd.* **2018**, *753*, 601–606. [CrossRef]

6. Shahriary, M.; Veisi, H.; Hekmati, M.; Hemmati, S. In situ green synthesis of Ag nanoparticles on herbal tea extract (*Stachys lavandulifolia*)-modified magnetic iron oxide nanoparticles as antibacterial agent and their 4-nitrophenol catalytic reduction activity. *Mater. Sci. Eng. C* **2018**, *90*, 57–66. [CrossRef] [PubMed]

7. Feurer, T.; Bissig, B.; Weiss, T.P.; Carron, R.; Avancini, E.; Löckinger, J.; Buecheler, S.; Tiwari, A.N. Single-graded CIGS with narrow bandgap for tandem solar cells. *Sci. Technol. Adv. Mater.* **2018**, *19*, 263–270. [CrossRef] [PubMed]

8. Bard, A.J.; Fox, M.A. Artificial Photosynthesis: Solar Splitting of Water to Hydrogen and Oxygen. *Acc. Chem. Res.* **1995**, *28*, 141–145. [CrossRef]

9. Gunawan; Haris, A.; Widiyandari, H.; Septina, W.; Ikeda, S. Surface modifications of chalcopyrite $CuInS_2$ thin films for photochatodes in photoelectrochemical water splitting under sunlight irradiation. *IOP Conf. Ser. Mater. Sci. Eng.* **2017**, *172*, 012021.

10. Yang, S.-W.; Pan, G.-T.; Yang, T.C.K.; Chen, C.-C.; Chiang, H.-C. The Photosynthesis of Methanol on 1D Ordered Zn:$CuInS_2$ Nanoarrays. *J. Taiwan Inst. Chem. Eng.* **2014**, *45*, 1509–1515. [CrossRef]

11. Benelli, G. Green synthesized nanoparticles in the fight against mosquito-borne diseases and cancer—A brief review. *Enzym. Microb. Technol.* **2016**, *95*, 58–68. [CrossRef] [PubMed]

12. Benelli, G. Plant-borne compounds and nanoparticles: Challenges for medicine, parasitology and entomology. *Environ. Sci. Pollut. Res.* **2018**, *25*, 10149–10150. [CrossRef] [PubMed]

13. Yue, W.; Wei, F.; He, C.; Wu, D.; Tang, N.; Qiao, Q. L-cysteine assisted Synthesis of 3D In_2S_3 for 3D $CuInS_2$ and its Application in Hybrid Solar Cells. *RSC Adv.* **2017**, *7*, 37578–37587. [CrossRef]

14. Yue, W.; Wei, F.; Li, Y.; Zhang, L.; Zhang, Q.; Qiao, Q.; Qiao, H. Hierarchical $CuInS_2$ synthesized with the induction of histidine for polymer/$CuInS_2$ solar cells. *Mater. Sci. Semicond. Process.* **2018**, *76*, 14–24. [CrossRef]

15. Liu, Y.; Jin, X.; Chen, Z. The formation of iron nanoparticles by Eucalyptus leaf extract and used to remove Cr(VI). *Sci. Total Environ.* **2018**, *627*, 470–479. [CrossRef] [PubMed]

16. Molnár, Z.; Bódai, V.; Szakacs, G.; Erdélyi, B.; Fogarassy, Z.; Sáfrán, G.; Varga, T.; Kónya, Z.; Tóth-Szeles, E.; Szűcs, R.; et al. Green synthesis of gold nanoparticles by thermophilic filamentous fungi. *Sci. Rep.* **2018**, *8*, 3943. [CrossRef] [PubMed]

17. Fathalipour, S.; Pourbeyram, S.; Sharafian, A.; Tanomand, A.; Azam, P. Biomolecule-assisted synthesis of Ag/reduced graphene oxide nanocomposite with excellent electrocatalytic and antibacterial performance. *Mater. Sci. Eng. C* **2017**, *75*, 742–751. [CrossRef] [PubMed]

18. Kazmerski, L.L.; Sanborn, G.A. CuInS$_2$ Thin-Film Homojunction Solar Cells. *J. Appl. Phys.* **1977**, *48*, 3178–3180. [CrossRef]

19. Zheng, L.; Xu, Y.; Song, Y.; Wu, C.; Zhang, M.; Xie, Y. Nearly Monodisperse CuInS$_2$ Hierarchical Microarchitectures for Photocatalytic H$_2$ Evolution under Visible Light. *Inorg. Chem.* **2009**, *48*, 4003–4009. [CrossRef] [PubMed]

20. Yuan, J.; Hao, C. Solar-Driven Photoelectrochemical Reduction of Carbon Dioxide to Methanol at CuInS$_2$ Thin Film Photocathode. *Sol. Energy Mater. Sol. Cells* **2013**, *108*, 170–174. [CrossRef]

21. Tell, B.; Shay, J.L.; Kasper, H.M. Electrical Properties, Optical Properties, and Band Structure of CuGaS$_2$ and CuInS$_2$. *Phys. Rev. B* **1971**, *4*, 2463–2471. [CrossRef]

22. Hollingsworth, J.A.; Banger, K.K.; Jin, M.H.C.; Harris, J.D.; Cowen, J.E.; Bohannan, E.W.; Switzer, J.A.; Buhro, W.E.; Hepp, A.F. Single Source Precursors for Fabrication of I–III–VI$_2$ Thin-Film Solar Cells Via Spray CVD. *Thin Solid Films* **2003**, *431–432*, 63–67. [CrossRef]

23. Kuranouchi, S.I.; Nakazawa, T. Study of One-Step Electrodeposition Condition For Preparation of CuIn(Se,S)$_2$ Thin Films. *Sol. Energy Mater. Sol. Cells* **1998**, *50*, 31–36. [CrossRef]

24. Liu, H.; Gu, C.; Xiong, W.; Zhang, M. A Sensitive Hydrogen Peroxide Biosensor Using Ultra-Small CuInS$_2$ Nanocrystals as Peroxidase Mimics. *Sens. Actuators B Chem.* **2015**, *209*, 670–676. [CrossRef]

25. Amerioun, M.H.; Ghazi, M.E.; Izadifard, M.; Bahramian, B. Preparation and Characterization of CuInS$_2$ Absorber Layers by Sol-Gel Method for Solar Cell Applications. *Eur. Phys. J. Plus* **2016**, *131*, 113. [CrossRef]

26. Gorai, S.; Bhattacharya, S.; Liarokapis, E.; Lampakis, D.; Chaudhuri, S. Morphology Controlled Solvothermal Synthesis of Copper Indium Sulphide Powder and Its Characterization. *Mater. Lett.* **2005**, *59*, 3535–3538. [CrossRef]

27. Yu, C.; Yu, J.C.; Wen, H.; Zhang, C. A Mild Solvothermal Route for Preparation of Cubic-Like CuInS$_2$ Crystals. *Mater. Lett.* **2009**, *63*, 1984–1986. [CrossRef]

28. Peng, S.; Cheng, F.; Liang, J.; Tao, Z.; Chen, J. Facile Solution-Controlled Growth of CuInS$_2$ Thin Films on FTO and TiO$_2$/FTO Glass Substrates for Photovoltaic Application. *J. Alloys Compd.* **2009**, *481*, 786–791. [CrossRef]

29. Wochnik, A.; Heinzl, C.; Auras, F.; Bein, T.; Scheu, C. Synthesis and Characterization of CuInS$_2$ Thin Film Structures. *J. Mater. Sci.* **2012**, *47*, 1669–1676. [CrossRef]

30. Wochnik, A.S.; Frank, A.; Heinzl, C.; Häusler, J.; Schneider, J.; Hoffmann, R.; Matich, S.; Scheu, C. Insight into the Core–Shell Structures of Cu-In-S Microspheres. *Solid State Sci.* **2013**, *26*, 23–30. [CrossRef]

31. Kuroda, K.; Terao, K.; Akao, M. Inhibitory Effect of Fumaric Acid on Hepatocarcinogenesis By Thioacetamide in Rats. *J. Natl. Cancer Inst.* **1987**, *79*, 1047–1051. [PubMed]

32. Frank, A.; Wochnik, A.S.; Bein, T.; Scheu, C. A Biomolecule-Assisted, Cost-Efficient Route for Growing Tunable CuInS$_2$ films for Green Energy Application. *RSC Adv.* **2017**, *7*, 20219–20230. [CrossRef]

33. Xia, J.; Liu, Y.; Qiu, X.; Mao, Y.; He, J.; Chen, L. Solvothermal Synthesis of Nanostructured CuInS$_2$ Thin Films on FTO Substrates and Their Photoelectrochemical Properties. *Mater. Chem. Phys.* **2012**, *136*, 823–830. [CrossRef]

34. Liu, Y.; Xie, Y.; Cui, H.; Zhao, W.; Yang, C.; Wang, Y.; Huang, F.; Dai, N. Preparation of Monodispersed CuInS$_2$ Nanopompons and Nanoflake Films and Application in Dye-Sensitized Solar Sells. *Phys. Chem. Chem. Phys.* **2013**, *15*, 4496–4499. [CrossRef] [PubMed]

35. Liu, H.-T.; Zhong, J.-S.; Liu, B.-F.; Liang, X.-J.; Yang, X.-Y.; Jin, H.-D.; Yang, F.; Xiang, W.-D. L-cystine-Assisted Growth and Mechanism of CuInS$_2$ Nanocrystallines via Solvothermal Process. *Chin. Phys. Lett.* **2011**, *28*, 057702. [CrossRef]

36. Wen, C.; Weidong, X.; Juanjuan, W.; Xiaoming, W.; Jiasong, Z.; Lijun, L. Biomolecule-Assisted Synthesis of Copper Indium Sulfide Microspheres with Nanosheets. *Mater. Lett.* **2009**, *63*, 2495–2498. [CrossRef]

37. Scherrer, P. Bestimmung der Grösse und der Inneren Struktur von Kolloidteilchen Mittels Röntgenstrahlen. *Nachrichten von der Gesellschaft der Wissenschaften zu Göttingen Mathematisch-Physikalische Klasse* **1918**, *26*, 98–100.

38. Frank, A.; Changizi, R.; Scheu, C. Challenges in TEM sample preparation of solvothermally grown $CuInS_2$ films. *Micron* **2018**, *109*, 1–10. [CrossRef] [PubMed]

39. Tauc, J.; Menth, A. States in The Gap. *J. Non-Cryst. Solids* **1972**, *8–10*, 569–585. [CrossRef]

40. Ho, C.H.; Pan, C.C.; Cai, J.R.; Huang, G.T.; Dumcenco, D.O.; Huang, Y.S.; Tiong, K.K.; Wu, C.C. Structural and Band-Edge Properties of $Cu(Al_xIn_{1-x})S_2$ ($0 \leq x \leq 1$) Series Chalcopyrite Semiconductors. *Solid State Phenom.* **2013**, *194*, 133–138. [CrossRef]

41. Fjellvag, H.; Gronvold, F.; Stolen, S.; Andresen, A.F.; Mueller Kaefer, R.; Simon, A. Low-Temperature Structural Distortion in CuS. *Z. Kristallogr.* **1988**, *184*, 111–121. [CrossRef]

42. Lukashev, P.; Lambrecht, W.R.L.; Kotani, T.; van Schilfgaarde, M. Electronic and Crystal Structure of Cu_2S: Full-Potential Electronic Structure Calculations. *Phys. Rev. B* **2007**, *76*, 195202. [CrossRef]

43. Janosi, A. La Structure du Sulfure Cuivreux Quadratique. *Acta Crystallogr.* **1964**, *17*, 311–312. [CrossRef]

44. Hull, S.; Keen, D.A. High-Pressure Polymorphism of The Copper(I) halides: A Neutron-Diffraction Study to ~10 GPa. *Phys. Rev. B* **1994**, *50*, 5868–5885. [CrossRef]

45. Schwarz, U.; Hillebrecht, H.; Syassen, K. Effect of Hydrostatic Pressure on The Crystal Structure of InS. *Z. Kristallogr.* **1995**, *210*, 494–497. [CrossRef]

46. Benchikhi, M.; El Ouatib, R.; Er-Rakho, L.; Durand, B. Synthesis and Characterization of $CuInS_2$ Nanocrystals Prepared by Solvothermal/Molten Salt Method. *Ceram. Int.* **2016**, *42*, 11303–11308. [CrossRef]

47. Das, K.; Panda, S.K.; Gorai, S.; Mishra, P.; Chaudhuri, S. Effect of Cu/In Molar Ratio on The Microstructural and Optical Properties of Microcrystalline $CuInS_2$ Prepared by Solvothermal Route. *Mater. Res. Bull.* **2008**, *43*, 2742–2750. [CrossRef]

48. Werner, J.H.; Mattheis, J.; Rau, U. Efficiency Limitations of Polycrystalline Thin Film Solar Cells: Case of $Cu(In,Ga)Se_2$. *Thin Solid Films* **2005**, *480–481*, 399–409. [CrossRef]

49. Siebentritt, S. What Limits the Efficiency of Chalcopyrite Solar Cells? *Sol. Energy Mater. Solar Cells* **2011**, *95*, 1471–1476. [CrossRef]

50. McCarthy, G.J.; Welton, J.M. X-Ray Diffraction Data for SnO_2. An Illustration of The New Powder Data Evaluation Methods. *Powder Diffr.* **1989**, *4*, 156–159. [CrossRef]

51. Li, B.; Xie, Y.; Xue, Y. Controllable Synthesis of CuS Nanostructures from Self-Assembled Precursors With Biomolecule Assistance. *J. Phys. Chem. C* **2007**, *111*, 12181–12187. [CrossRef]

52. Kharkwal, A.; Sharma, S.N.; Jain, K.; Singh, A.K. A Solvothermal Approach for the Size-, Shape- and Phase-Controlled Synthesis and Properties of $CuInS_2$. *Mater. Chem. Phys.* **2014**, *144*, 252–262. [CrossRef]

53. Cavallini, D.; De Marco, C.; Duprè, S.; Rotilio, G. The Copper Catalyzed Oxidation of Cysteine to Cystine. *Arch. Biochem. Biophys.* **1969**, *130*, 354–361. [CrossRef]

54. Nelson, J. *The Physics of Solar Cells*; Imperial College Press: London, UK, 2007.

55. Saunders, B.R. Hybrid Polymer/Nanoparticle Solar Cells: Preparation, Principles and Challenges. *J. Colloid Interface Sci.* **2012**, *369*, 1–15. [CrossRef] [PubMed]

56. Hollemann; Wiberg, E.; Wiberg, N. *Lehrbuch der Anorganischen Chemie*; Walter de Gruyter: Berlin, Germany, 2007; Volume 102.

57. Dokken, K.M.; Parsons, J.G.; McClure, J.; Gardea-Torresdey, J.L. Synthesis and Structural Analysis of Copper(II) Cysteine Complexes. *Inorg. Chim. Acta* **2009**, *362*, 395–401. [CrossRef]

58. Zhuang, M.X.; Wei, A.X.; Zhao, Y.; Liu, J.; Yan, Z.Q.; Liu, Z. Morphology-Controlled Growth of Special Nanostructure $CuInS_2$ Thin Films on an FTO Substrate and Their Application in Thin Film Solar Cells. *Int. J. Hydrog. Energy* **2015**, *40*, 806–814. [CrossRef]

5

Green Synthesis of High Temperature Stable Anatase Titanium Dioxide Nanoparticles using Gum Kondagogu: Characterization and Solar Driven Photocatalytic Degradation of Organic Dye

Kothaplamoottil Sivan Saranya [1], Vinod Vellora Thekkae Padil [2,*], Chandra Senan [3], Rajendra Pilankatta [4], Kunjumon Saranya [1], Bini George [1,*], Stanisław Wacławek [2] and Miroslav Černík [2,*]

[1] Department of Chemistry, School of Physical Sciences, Central University of Kerala, Kerala 671316, India; sharanyacks@gmail.com (S.K.S.-K.S.S.); kunjumonsaranya916@gmail.com (S.K.-K.S.)

[2] Institute for Nanomaterials, Advanced Technologies and Innovation (CXI), Technical University of Liberec (TUL), Studentská 1402/2, 46117 Liberec 1, Czech Republic; stanislaw.waclawek@tul.cz

[3] Centre for Water Soluble Polymers, Applied Science, Faculty of Arts, Science and Technology, Wrexham Glyndwr University, Wrexham LL11 2AW, Wales, UK; c.senan@glyndwr.ac.uk

[4] Department of Biochemistry and Molecular Biology, School of Biological Sciences, Central, University of Kerala, Kerala 671316, India; praj74@gmail.com

[*] Correspondence: vinod.padil@tul.cz (V.V.T.P.); binigeorgek@gmail.com (B.G.); miroslav.cernik@tul.cz (M.Č.)

Abstract: The present study reports a green and sustainable method for the synthesis of titanium dioxide (TiO_2) nanoparticles (NPs) from titanium oxysulfate solution using Kondagogu gum (*Cochlospermum gossypium*), a carbohydrate polymer, as the NPs formation agent. The synthesized TiO_2 NPs were categorized by techniques such as X-Ray Diffraction (XRD), Fourier transform infrared (FTIR) spectroscopy analysis, Raman spectroscopy, scanning electron microscope- Energy-dispersive X-ray spectroscopy (SEM-EDX), Transmission electron microscopy (TEM), High-resolution transmission electron microscopy (HR-TEM), UV-visible spectroscopy, Brunauer-Emmett-Teller (BET) surface area and particle size analysis. Additionally, the photocatalytic actions of TiO_2 NPs were assessed with regard to their ability to degrade an organic dye (methylene blue) from aqueous solution in the presence of solar light. Various parameters affecting the photocatalytic activity of the TiO_2 NPs were examined, including catalyst loading, reaction time, pH value and calcination temperature of the aforementioned particles. This green synthesis method involving TiO_2 NPs explores the advantages of inexpensive and non-toxic precursors, the TiO_2 NPs themselves exhibiting excellent photocatalytic activity against dye molecules.

Keywords: titanium dioxide nanoparticles; green synthesis; gum kondagogu; methylene blue; photocatalysis

1. Introduction

Nanoparticles (metal and metal oxides) of various types have been widely employed via physical and chemical methods. Although these systems have resulted in the formation of numerous extremely diverse nanostructures, many environmental toxicity issues have emerged [1]. Metal oxides have recently been widely explored because of the huge variety of structural, material and functional properties exhibited by their nanoparticles. Transition metal oxide nanoparticles have generated great scientific interest owing to their unusual properties compared with their corresponding bulk metals. Moreover, they have numerous industrial applications [2]. Metal oxide nanoparticles with disparate morphologies and sizes have been synthesized using different synthetic routes. These include hydrothermal, solvothermal, microwave, vapour deposition, spray pyrolysis and wet-chemical methods [3–5]. However, the usage of solvents in the chemical synthesis route adopted pose limitations for the application of NPs in medicine, pharmacy and other areas, primarily due to the toxicity caused by the solvents. Thus, there is an urgent need for the development of alternative, novel nanoparticle synthesis processes that could be exploited (at both the industrial and commercial level) in order to introduce cleaner, safer and smarter products suitable for application in communication technologies, medicine, pharmacy, agriculture and other industries. The introduction of environmentally benign approaches for designing NPs provides solutions to mounting tasks related to ecological concerns. Use of greener, safe and environmentally non-hazardous chemicals and green protocols for the synthesis of nanoparticles decreases the expense of synthesis while minimizing the utilization of harmful substances and their subsequent disposal [6].

Titanium dioxide (TiO_2) is regarded as an extremely promising metal oxide that can perform a key role in solving the global energy crunch, as well as serving to assuage environmental concerns [7,8]. Nanoparticles of TiO_2 have featured in numerous photovoltaic applications and in various sectors including cosmetic, pharmaceutical and skin care products. These versatile NPs can also confer whiteness to paints, plastics, papers, inks, food colorants and toothpaste while also finding usage in cancer treatment [9,10]. The synthesis of TiO_2 NPs, which possess controlled crystal phases and morphologies that make them eminently suitable in diverse applications requiring high performance, has proven to be fundamentally challenging for the scientific community [11].

Carbohydrate polymers isolated from trees [12], are an innovative type of potentially cost-effective and biologically favorable biomaterials that display highly particular and selective characteristics towards the design of nanostructures for unique applications. They are natural biopolymers based on plant exudates and possess interesting properties e.g., reducing, stabilizing, suspending, gelling etc. [13–15]. Tree exudates (gums) serve as 'green' media and harbor numerous hydroxyl, carbonyl and carboxylic functional groups. Such functional groups can act as good chemical reductants and their presence in these gum hydrocolloid materials facilitates the formation of metal or metal oxide NPs. This is achieved in two ways: either through the reduction of metal ions or by the gum molecules behaving as a stabilizing mediator (to prevent nanoparticle agglomeration). Hence, the essential criterion for designing nanomaterials with desirable properties is fulfilled [16,17].

The Kondagogu (*Cochlospermum gossypium*, KG) is a native tree growing in the forests of India and its exudates have been categorized as being substituted rhamnogalacturonans [16]. KG is a complex and acidic polysaccharide with high solution viscosity and gelation characteristics. KG contains sugars such as arabinose, rhamnose, glucose, galactose, mannose, glucuronic acid and galacturonic acid. The structural features assigned to KG are (1→2) β-D-Gal *p*, (1→6) β-D-Gal *p*, (1→4) β-D-Glc*p* A, 4-*O*-Me-α-D-Glc*p* A and (1→2) α-L-Rha [18,19]. It has a high content of uronic acid and diverse functional groups (hydroxyl, acetyl, carbonyl and carboxylic) [18,20]. The morphological, physicochemical and structural analysis of this biopolymer is currently being extensively researched [18,19].

Organic dyes represent major clusters of pollutants flushed out into wastewaters from textiles and other industrialized procedures. Due to their potential injuriousness and presence in external waters, their elimination and removal has been a matter of considerable importance, since even small amounts of released dyes can discolor surface waters and impact negatively on the otherwise aesthetic and pristine surroundings [21]. The dyes altered the absorption and reflection of sunlight entering the water, which in turn affected bacterial growth and subsequently the level of biological impurities in the water [22]. Wastewater purification is discernible as one of the most serious environmental challenges of the present day. To this end, solid catalysts have been widely used in various water treatment technologies, both in processing and industry. Nanoparticles, courtesy of their small size and high surface-to-volume ratios, display high absorbing, interacting and reactive capabilities, underscoring their value in wastewater remediation [23].

Photocatalysis is an environmentally benign process involving the conversion of light energy into chemical energy under ambient conditions. The outstanding performance of solar driven photocatalytic processes in solving environmental problems has gained much attention in recent years, effluents from the textile and paint industries being the main sources of environmental organic contaminants. Thus, it is essential that the latter is degraded and converted into harmless mineral compounds. The photocatalyst can be harnessed for environmental remediation, which includes abstraction of pesticides, fungicides and fertilizers from wastewater and the eradication of organic pollutants from the air [24]. Activation of a semiconductor photocatalyst is achieved by the absorption of a photon, which results in the transfer of an electron from the valence band to the conduction band by creation of a hole (h^+) in the valence band [25]. These photo-generated charge carriers cause redox reactions on the surface of the photocatalyst, i.e., any contaminant that is adsorbed onto the photocatalyst surface will undergo reduction or oxidation by the electron-hole pair respectively. Generally, the metal oxide photocatalyst surface acts as an active center in photocatalysis, either by the generation of OH· radicals (by oxidation of OH^- anions) or by the generation of O_2^- radicals (by the reduction of O_2.) Subsequently, these photo-generated radicals and anions react with the adsorbed organic contaminants, degrading or mineralizing them into less harmful by-products. The photocatalytic reaction can be employed to bring about the transformation of highly toxic chemicals into less noxious or non-toxic products such as CO_2 and H_2O [26,27]. Consequently, metal oxides can be employed as prime candidates for the effective photocatalytic degradation of environmental contaminants.

In this study, we have focused on the green synthesis of TiO_2 NPs and their effective application in the photocatalytic degradation of a commercially used organic dye, methylene blue (MB). The influence of various solution pH values on photocatalytic efficiency was also studied.

2. Materials and Methods

2.1. Materials

Titanium oxysulfate and methylene blue were purchased from Sigma Aldrich and HiMedia Chemicals, Mumbai, India, respectively. KG samples were obtained from the Girijan Co-operative Corporation Ltd. (GCC), Hyderabad, India. All the other chemicals and solvents used were of analytical grade.

2.2. Fabrication of TiO_2 Nanoparticles

Titanium dioxide NPs were synthesized using a typical procedure as described here. The procedure involved adding KG (50 mg) to 10 mL of titanium oxysulfate (0.1 M) and stirring vigorously (750 rpm) on a magnetic stirrer at 90–95 °C. Later, the product was centrifuged, washed and dried. The dried sample was calcined at 500 °C for 4 h, then pulverized and stored until further use.

2.3. Characterization of TiO_2 Nanoparticles

2.3.1. X-Ray Diffraction (XRD) Analysis

X-Ray Diffraction (XRD) configurations of the calcined samples were obtained with the diffraction angle range (2θ) set between 20° and 90° using a diffractometer (Rigaku Miniflex 600, Tokyo, Japan) with nickel filtered Cu Kα (λ = 1.54 Å) radiation and a liquid nitrogen cooled, germanium solid state detector. The spectral plots were compared with details obtained from Joint Committee on Powder Diffraction Standards (JCPDS), data files for analytical purposes.

2.3.2. FTIR Analysis

For the Fourier transform infrared spectroscopy (FTIR) analysis, a spectrometer (Perkin-Elmer FTIR Spectrum Two, Singapore) in attenuated total reflection (ATR) mode and with the spectral range set between 4000 and 400 cm^{-1} and a resolution of 4 cm^{-1} was used.

2.3.3. Raman Spectra

A Raman microscope (NICOLET DXR, Thermo Scientific, Waltham, MA, USA), equipped with an optical microscope, was used. An argon-ion (532 nm) or helium-neon (632.8 nm) laser was used for the excitation of the Raman signal with appropriate holographic notch filters to eliminate the laser line after excitation. Spectral analysis and curve fitting were performed using GRAMS/AI 8.00 Spectroscopy software (Alfasoft GmbH, Frankfurt, Germany)

2.3.4. Scanning electron microscope- Energy-dispersive X-ray spectroscopy (SEM-EDX) Analysis

The elemental composition and morphology of TiO_2 nanoparticles were determined using a scanning electron microscope (ZEISS, Ultra/Plus, Potsdam, Germany). Energy-dispersive X-ray spectroscopy (EDX) measurements were carried out using a scanning electron microscope equipped with an EDX attachment (JSM-6390 172, JEOL, Tokyo, Japan).

2.3.5. TEM and High-resolution transmission electron microscopy (HR-TEM) Analysis

The nanoparticles and particle distributions determined were recorded by a transmission electron microscopy (TEM) (JEOL, JEM-2100,Tokyo, Japan). The samples were prepared on standard copper TEM grids covered with thin carbon foil. Drops of TiO_2 nanoparticles were dispersed in 1 mL of isopropanol using ultrasound for 10 min. and a drop of the resulting solution was gently spread onto the upper surface of the carbon covered copper TEM grid.

2.3.6. Particle Size Analysis

Nanoparticle size distributions were measured by centrifugal particle sedimentation (CPS) using the Disc Centrifuge technique (DC24000UHR, CPS Instruments Inc., Prairieville, LA, USA).

2.3.7. BET Surface Area

The surface area of the TiO_2 NPs was analyzed using the Brunauer–Emmett–Teller (BET) technique (Autosorb iQ, Quantachrome Instruments, Boynton Beach, FL, USA).

2.3.8. Thermal Stability

Thermal properties of KG and green-synthesized TiO_2 nanoparticles were studied using thermogravimetric analysis (TGA) by means of a Perkin Elmer STA 6000 Thermal Analyzer (Singapore) instrument.

2.3.9. Optical Properties

Optical properties were determined using an ultraviolet (UV)-visible spectrophotometer (Perkin Elmer Lambda 35, Singapore) over the spectral region 200–800 nm.

2.4. Photocatalytic Degradation of Methylene Blue

The photocatalytic activity of green-synthesized TiO_2 nanoparticles was scrutinized by the degradation of MB under sunlight. Photocatalytic activity in the presence of sunlight was determined under direct normal sunlight at an intensity of 100,000 Lux and the solar intensity measurement was carried out throughout the experiment at different time intervals. The degradation of dye was examined by collecting 5 mL aliquots from the reaction mixture at different intervals of time.

In addition, these aliquots were centrifuged at 7000 rpm for 15 min. Photocatalytic degradation of the dye was monitored by measuring the absorbance spectra of the supernatants using a UV−visible spectrophotometer over the wavelength range 200–800 nm.

2.4.1. Photocatalytic Studies Based on Catalyst Concentration

In order to study the effect of catalyst loading on solar light driven photocatalysis of methylene blue, different amounts (1–15 mg) of green-synthesized titanium dioxide nanoparticles calcined at 500 °C were added to 50 mL of dye solution. The reaction suspension was mixed thoroughly using a magnetic stirrer for 90 min in the presence of sunlight. The photocatalytic degradation of the dye was monitored by measuring the absorbance of the solution at regular intervals using a UV-visible spectrophotometer.

2.4.2. Photocatalytic Studies Based on Time

In a typical experiment, 50 mL of dye solution (1.0×10^{-5} M) was mixed with 10 mg of TiO_2 nanoparticles. The mixture was stirred continuously at 600 rpm in the presence of sunlight. The rate of degradation was monitored by measuring the absorbance of the solution (over the 200 to 800 nm wavelength range) with a UV-visible spectrophotometer by removing and monitoring 5 mL aliquots at defined time intervals. The process was continued for 90 min.

2.4.3. Photocatalytic Studies Based on pH

The influence of pH on solar driven photocatalytic activity of the catalyst was studied by conducting the experiment at different pH values (4–9) of dye solution. Green-synthesized titanium dioxide nanoparticles (10 mg) calcined at 500 °C were added to 50 mL of the dye solution having different pH values. The resulting suspension was thoroughly mixed using a magnetic stirrer for 90 min in the presence of sunlight and the absorbance of the solution measured with a UV-visible spectrophotometer.

2.4.4. Photocatalytic Studies Based on Temperature

The green-synthesized TiO_2 nanoparticles were calcined at different temperatures (500–900 °C) and each sample (10 mg) was added to 50 mL of dye solution. The resulting suspension was thoroughly mixed using a magnetic stirrer for 90 min, in the presence of sunlight. The absorbance of the solution was measured with a UV-visible spectrophotometer.

3. Results and Discussion

3.1. Mechanism of TiO_2 NPs Formation Via Green Synthesis

Natural tree based carbohydrate polymers—an environmentally benign medium—contain extensive numbers of hydroxyl, carbonyl and carboxylic groups which can act as good chemical reductants. The presence of these functional groups in the gum hydrocolloid material facilitates the formation of metal nanoparticles. When a metal oxide precursor is introduced into a well dissolved KG homogeneous solution or hydrogel, the polyhydroxylated macromolecules inherent in the gum matrix are able to absorb metal cations. For example, the Ti^{3+} ion could be chelated with KG by means of –OH and –COOH groups. The sequestration of cations $[M^{n+}]$ or hydroxylated cations $[M(OH)]^{m+}$ that can undergo nucleation or growth processes is accelerated by the highly reactive hydroxyl groups present in KG. The last step in the TiO_2 nanoparticle synthesis is the calcination process at 500 °C which removes the gum template via combustion of their organic scaffold, giving rise to the formation of dispersed oxides [11,28]. While XRD and Raman spectroscopy confirmed the formation of anatase as the only distinguished mineral phase; FTIR, TGA and EDX collectively proved that no organic compounds (polysaccharides, hydroxyl, carboxylate and other oxidic groups) were present in the final nanoparticles.

3.2. XRD Analysis

The XRD pattern of TiO$_2$ NPs calcined at 500 °C showed a dominant peak at the 2θ value of 25.25° (Figure 1). This matches the (101) crystallographic plane of the TiO$_2$ anatase structure, indicating that the crystal composition is predominantly anatase. The other characteristic diffraction peaks were at the following 2θ values: 37.75, 48.02, 53.86, 54.95, 62.63, 68.88, 70.28, 75.13 and 82.60°. These values correspond to (004), (200), (105), (211), (204), (116), (220), (215) and (224) crystallographic planes of anatase, respectively (JCPDS No. 01-071-1166), thus confirming the formation of TiO$_2$ nanoparticles in the anatase phase [29]. The average crystallite size of the NPs [i.e., the mean size of the ordered (crystalline) domains, which may be smaller or equal to the grain size] was calculated using the Scherrer formula [$d = 0.89\lambda/\beta\cos\theta$] to be 12.58 nm. The crystallinity and higher purity of prepared TiO$_2$ NPs in the anatase form was validated by the presence of sharp peaks while the absence of peaks represented other crystallite forms of TiO$_2$ [30–32].

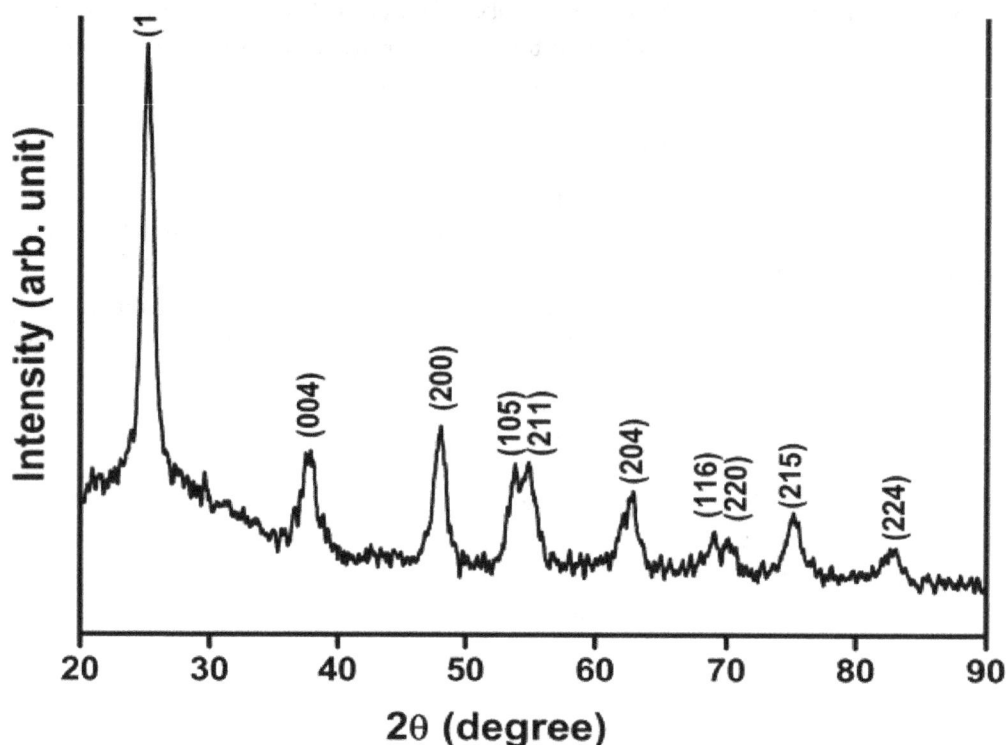

Figure 1. X-ray diffraction analysis (XRD) pattern of titanium dioxide (TiO$_2$) nanoparticles calcined at 500 °C.

XRD patterns of TiO$_2$ nanocatalysts calcined at various temperatures are shown in Figure 2. When the calcination temperatures did not exceed 700 °C, anatase was the only phase, with Joint Committee on Powder Diffraction Standards (JCPDS) No. 01-071-1166, while phase transformation of TiO$_2$ was observed at 800 °C. The rutile phase appeared in the 800 °C sample and became the dominant phase in the 900 °C sample with peaks at 2θ = 27.4°(110), 36.04°(101), 41.2°(111), 54.3°(211) (JCPDS no. 01-073-2224). Formation of the rutile phase of TiO$_2$ is normally observed above 600 °C with a complete transformation to the rutile form occurring at 800 °C [33]. The present XRD patterns illustrate that the anatase to rutile phase transformation of the synthesized TiO$_2$ first took place at 800 °C and was almost completed at 900 °C, thus revealing the formation of a high temperature, stable anatase phase of TiO$_2$ nanocatalysts through the green synthesis method adopted.

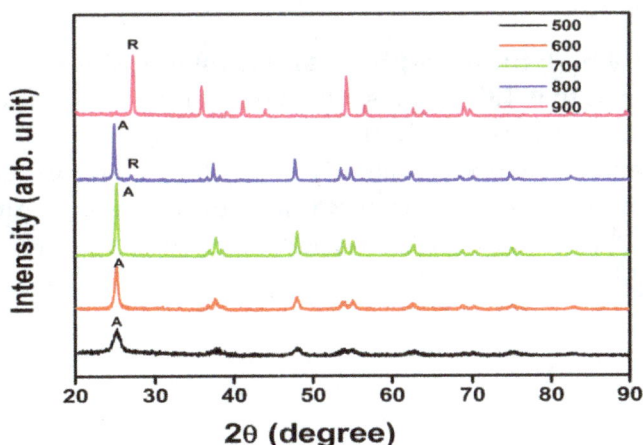

Figure 2. XRD pattern of TiO_2 nanoparticles calcined at different temperatures (R denotes rutile and A denotes anatase phases of TiO_2 nanoparticles).

While anatase and rutile are both mineral forms of titanium dioxide (possessing tetragonal crystal systems), in terms of optical activity, anatase is optically negative whereas rutile is positive. Furthermore, the luster exhibited by anatase is more strongly adamantine or metallic-adamantine than that shown by rutile.

3.3. Fourier Transform Infrared Spectroscopy Analysis

FTIR spectroscopy was employed to identify different functional groups present in KG and on the surface of the formed NPs (Figure 3). The major stretching frequencies in the spectrum for KG were observed at 3368, 1719, 1609, 1417, 1247, 1145 and 1035 cm^{-1}. The band observed at 3368 cm^{-1} suggests the presence of hydroxyl groups while those noted at 1719 cm^{-1} and 1609 cm^{-1} were ascribed to carbonyl stretching vibrations and the asymmetric stretching of carboxylate, respectively. The band seen at 1417 cm^{-1} was due to the symmetrical stretching of the carboxylate group present in the gum's uronic acid. The presence of an acetyl group was inferred by the band appearing at 1247 cm^{-1} while the bands registered at 1145 and 1035 cm^{-1} were indicative of C–O stretching vibrations of ether and alcohol groups, respectively [34]. The absence of these characteristic peaks in the FTIR spectra of TiO_2 nanoparticles promoted by KG may be a consequence of the higher purity of prepared nanoparticles in anatase crystal formations on calcination at 500 °C. The characteristic signal for TiO_2 nanoparticles observed below 1000 cm^{-1} in the FTIR spectra was due to Ti-O-Ti vibrations [35].

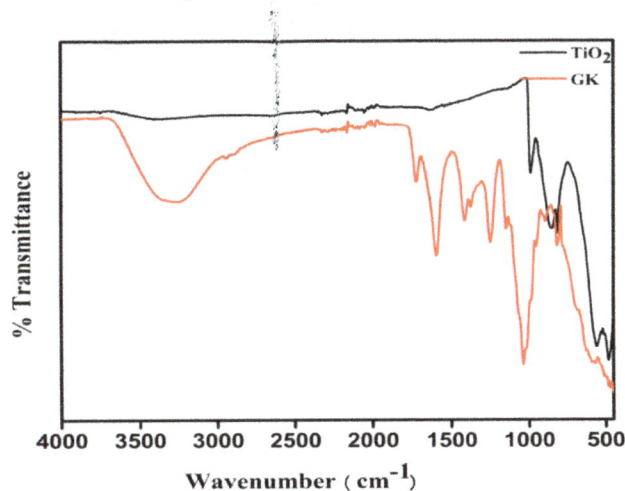

Figure 3. Fourier transform infrared spectroscopy (FTIR) spectra of Kondagogu (KG) and TiO_2 nanoparticles.

3.4. Raman Spectroscopy

To identify and quantify both the amorphous and crystalline TiO_2 phases, Raman spectroscopy was employed. Figure 4 reveals that TiO_2 anatase displays Raman bands at 639, 516, 399 and 197 cm^{-1}, as well as a precise sharp and intense peak at 144 cm^{-1}. According to the literature, the anatase phase of TiO_2 has six Raman active modes, namely $A_{1g} + 2B_{1g} + 3E_g$, determined by group analysis D_{4h}(I4l/amd) [36], and known in the first-order Raman spectrum of single crystal TiO_2 at 144 cm^{-1} (E_g), 197 cm^{-1} (E_g), 399 cm^{-1} (B_{1g}), 516 cm^{-1} ($A_{1g} + B_{1g}$) and 639 cm^{-1} (E_g) [37].

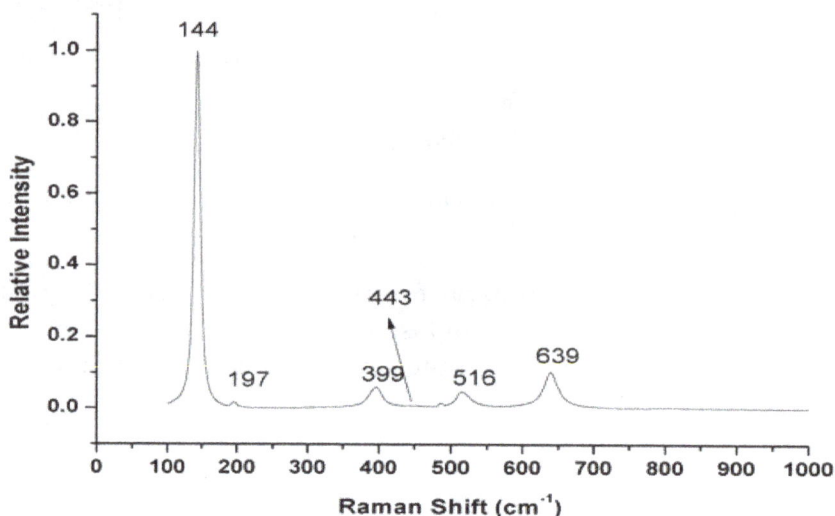

Figure 4. Raman spectrum of TiO_2 nanoparticles.

3.5. Scanning Electron Microscopy (SEM)

The SEM micrograph of TiO_2 nanoparticles is shown in Figure 5a, and the resulting nanoparticles were observed to have a spherical morphology. The chemical composition, analyzed using EDX spectra, confirmed the presence of Ti and O (Figure 5b). Since the sample was coated on a copper grid, peaks corresponding to Cu were also visible in Figure 5b.

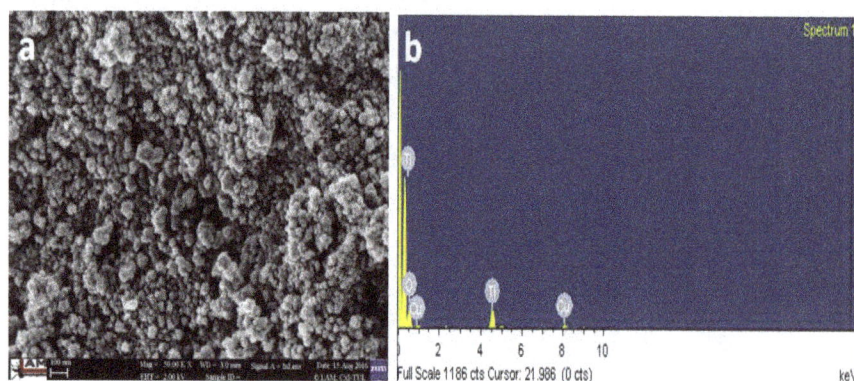

Figure 5. Typical (**a**) scanning electron microscopy (SEM) and (**b**) EDX (Energy Dispersive X-Ray) analysis of TiO_2 nanoparticles.

3.6. TEM and HR-TEM Analysis

Both particle size and morphology of TiO_2 were confirmed by TEM analysis (Figure 6a) which revealed that the particles are monodisperse and spherical in shape. The sizes of particles were in the 8–13 nm range and the selected area electron diffraction (SAED) pattern indicated that the TiO_2 nanoparticles possessed good crystallinity (Figure 6b). HR-TEM observations (Figure 6c) suggest that the TiO_2 nanoparticles have a perfect lattice structure. The particle sizes determined by the TEM image were similar to the reported values obtained by applying the Scherrer equation to the XRD patterns (12.58 nm). The aforementioned equation correlates the size of sub-micrometre particles or crystallites (in a solid) to the broadening of a peak in a diffraction pattern. It is employed for the determination of the size of particles of crystals in the form of powders.

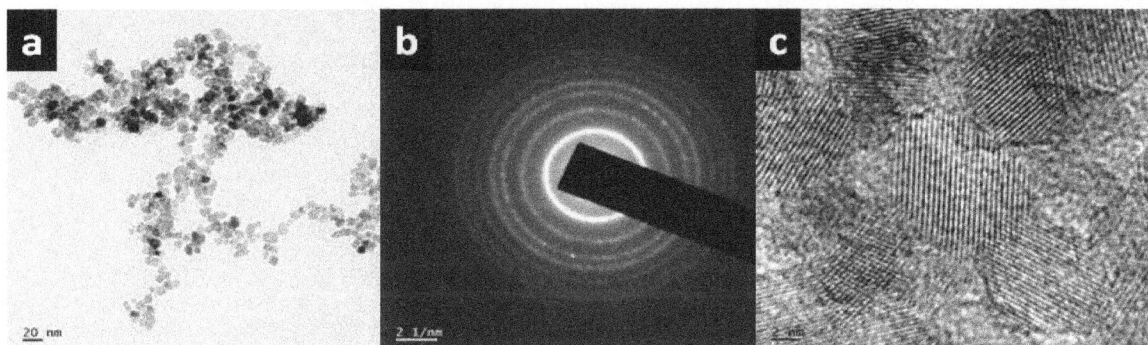

Figure 6. (**a**) TEM (Transmission Electron Microscopy) image, (**b**) SAED pattern and (**c**) HR-TEM (High Resolution TEM) micrograph of TiO_2 nanoparticles.

3.7. Particle Size & BET Analysis

The TiO_2 particle size distribution analyzed by CPS determined the mean size to be 11.2 ± 0.2 nm (Figure 7). Evidently, the TiO_2 nanoparticles appeared to be more stable in the current study involving green synthesis and the particle size measurements corresponded very well with the TEM analytical data (Figure 6). The specific surface area of TiO_2 was determined from the isotherms to be 42.6 m^2/g based on the BET analysis.

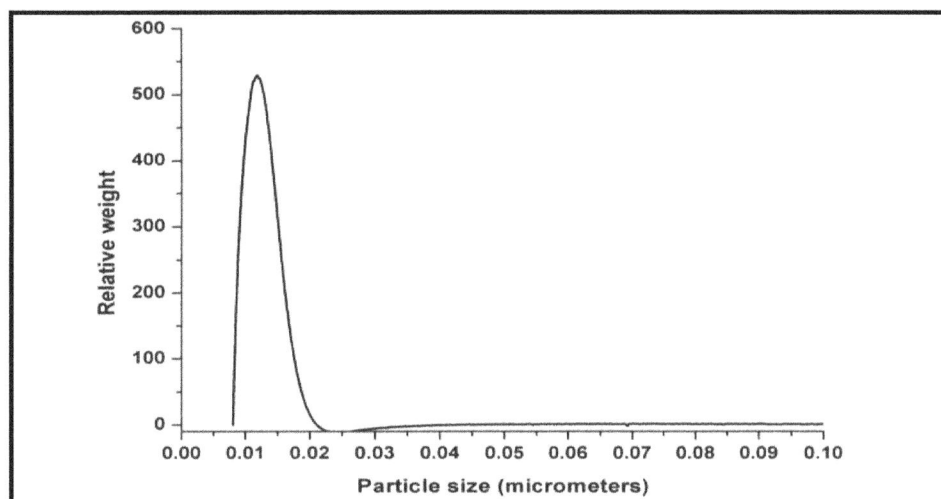

Figure 7. Particle size distributions of TiO_2 nanoparticles as determined by the centrifugal particle sedimentation (CPS) method.

3.8. Thermal Studies

Thermal properties of KG and green-synthesized TiO_2 nanoparticles were studied using thermogravimetric analysis. Depicted in Figure 8 are the TGA curves of KG and TiO_2 nanoparticles heated from 35 °C to 950 °C. In the case of KG, there were two major weight loss events. The first occurrence (observed between 35 °C and 111 °C) of approximately 17% probably represented the loss of adsorbed water - as hydrogen bonded water - from the polysaccharide structure. The second weight loss event of roughly 33% was very significant and was discerned between 232 and 309 °C.

It was ascribed to decomposition of the polysaccharide. A third, far smaller weight loss instance was registered at 590 °C, possibly due to the conversion of the remaining polymer to carbon residue [20].

The TGA curve of TiO_2 (Figure 7b, red curve) did not indicate any weight loss up to 766 °C, a reflection of its high thermal stability, probably due to four hours of calcination at 500 °C. There were small linear weight reductions detected in the range of 766–911 °C, possibly caused by the loss of residual carbon from the gum matrix.

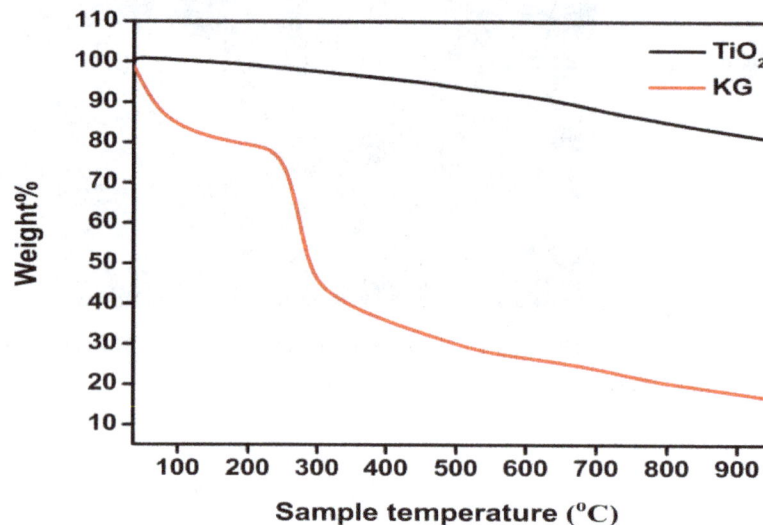

Figure 8. Thermogravimetric analysis (TGA) curves of KG and green-synthesized TiO_2 nanoparticles.

3.9. Optical Properties

The UV-visible absorption spectrum of biosynthesized TiO_2 nanoparticles is shown in Figure 8. The observed absorption spectrum matches those obtained with TiO_2 nanoparticles produced by chemical methods, displaying a broad absorption band in the UV region, up to 380 nm [38].

The UV-visible spectrum was utilized to deduce the optical absorption properties of green-synthesized TiO_2 nanoparticles (Figure 9). The band gap energy of green-synthesized TiO_2 nanomaterials was found to be 3.13 eV from the Tauc plot used to determine the optical bandgap in semiconductors as shown in Figure 10.

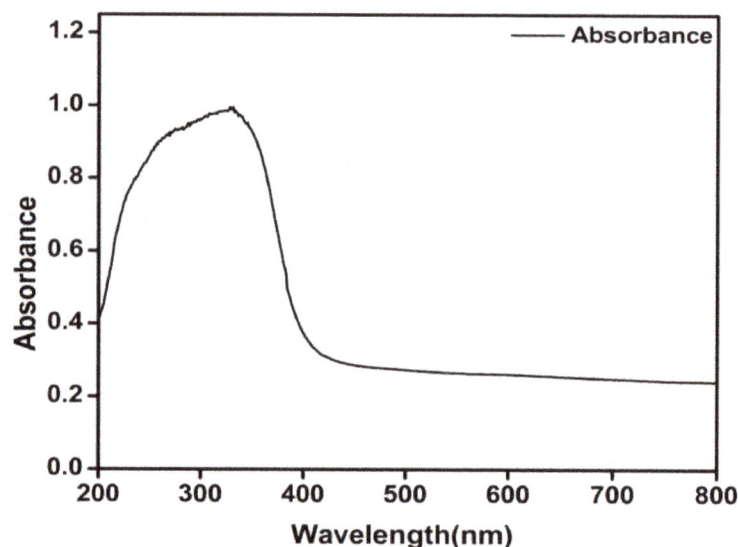

Figure 9. Ultraviolet (UV)-visible absorption spectrum of green-synthesized TiO_2 nanoparticles.

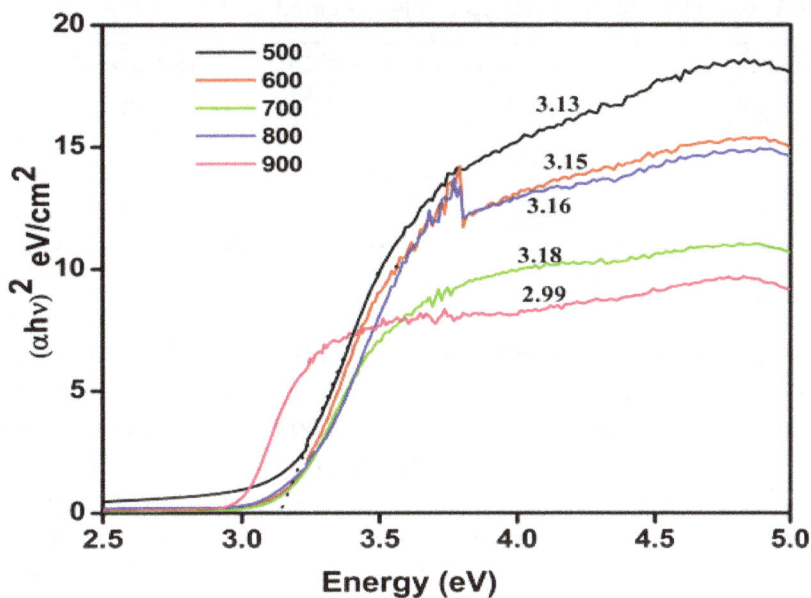

Figure 10. UV-Tauc plot of TiO$_2$ nanoparticles calcined at different temperatures.

After the bulk, TiO$_2$ had a band gap energy of 3.2 eV; the marginal reduction of 0.07 eV in this parameter being attributed to particle size dependence [39]. The absorbance spectra and the corresponding Tauc plot of TiO$_2$ nanoparticles calcined at different temperatures were as given in Figures 10 and 11, respectively.

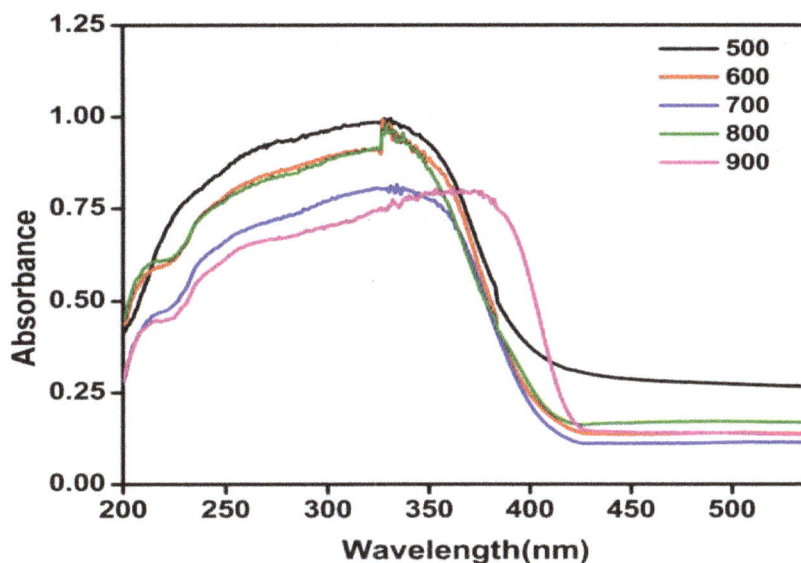

Figure 11. UV-Visible spectra of TiO$_2$ nanoparticles calcined at different temperatures.

The calculated band gap of TiO$_2$ nanoparticles, calcined at different temperatures was as shown in Table 1. It is also evident from the UV data (Table 1) that the band gap increased with the elevation of calcination temperatures from 500 to 800 °C, before decreasing gradually at 900 °C. This variation can be expected and is plausible, given that with an increase in the TiO$_2$ calcination temperature, the anatase phase became endowed with good crystal characteristics. Furthermore, the decrease in band gap at 900 °C was mainly attributed to the complete formation of the rutile phase.

Table 1. Band gap of TiO2 nanoparticles calcined at different temperatures.

Calcination Temperatures of TiO_2 (°C)	Band Gap (eV)
500	3.13
600	3.15
700	3.16
800	3.18
900	3

3.10. Photocatalytic Activity

The photocatalytic activity of green-synthesized TiO_2 NPs was demonstrated by using an organic dye (methylene blue) under solar light, the dye degradation being initially identified by color change (Scheme 1). Additionally, we monitored the intensity of solar light throughout our experiment and confirmed it was in the range 100,000 Lux (see supplementary information, Figure S1). Furthermore, solar UV-radiation, as a function of time, and the UV intensity were measured using a Lux meter and UV filtering goggles. This information has been given in the supplementary document (Figures S2 and S3, respectively). The UV intensity from solar light was ascertained by calculating the difference between the intensity of solar radiation and the radiation through UV filtering goggles.

The dye displayed a distinct absorbance peak in visible light at a wavelength of 663 nm, where absorbance was at a maximum. This peak was used to monitor dye concentration in the solution over time. From the absorbance spectra corresponding to dye degradation, it was indisputable that the presence of titanium dioxide nanocatalysts resulted in a linear increase of the percentage of dye degradation with time, reflected by the decrease in absorbance. This finding indicated that when the time of irradiation was prolonged, the percent degradation increased and reached a maximum after 90 min of solar irradiation. When TiO_2 nanocatalysts, dispersed in the dye solution, were irradiated with solar radiation, photo-generated charge carriers induced redox reactions on the surface of the photocatalyst. Essentially, any contaminant adsorbed onto the photocatalyst surface, by virtue of electron-hole pair generation, will undergo either reduction or oxidation, respectively. Generally, the TiO_2 photocatalyst surface behaves as an active center in photocatalysis, either through the generation of OH radicals (by oxidation of OH^- anions) or by the generation of O_2^- radicals (via the reduction of O_2 molecules) [40]. Subsequently, these photo-generated radicals and anions react with the adsorbed organic contaminants to degrade or mineralize them into less harmful by-products, such as CO_2 and H_2O (Scheme 1).

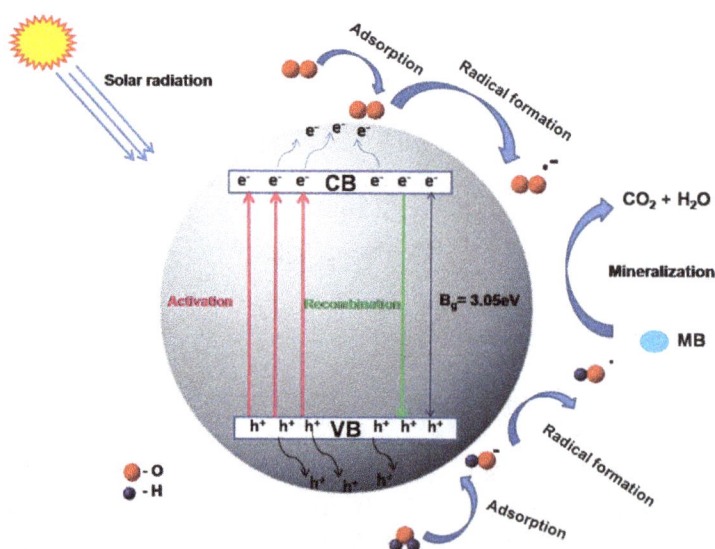

Scheme 1. Photocatalysis mechanism of TiO_2.

As the irradiation time lengthened, absorption of increasingly more light energy impinging on the catalyst surface occurred. This led to the increased formation of photo-excited species, and consequently, enhanced photocatalytic activity. From this study, it was observed that the photocatalytic dye degradation process was enhanced by lengthening exposure time (Figure 12). The Beer-Lambert-Bouguer law—relating the attenuation of light to the properties of the material through which the light is travelling—was used to determine molar concentrations of the degraded dyes.

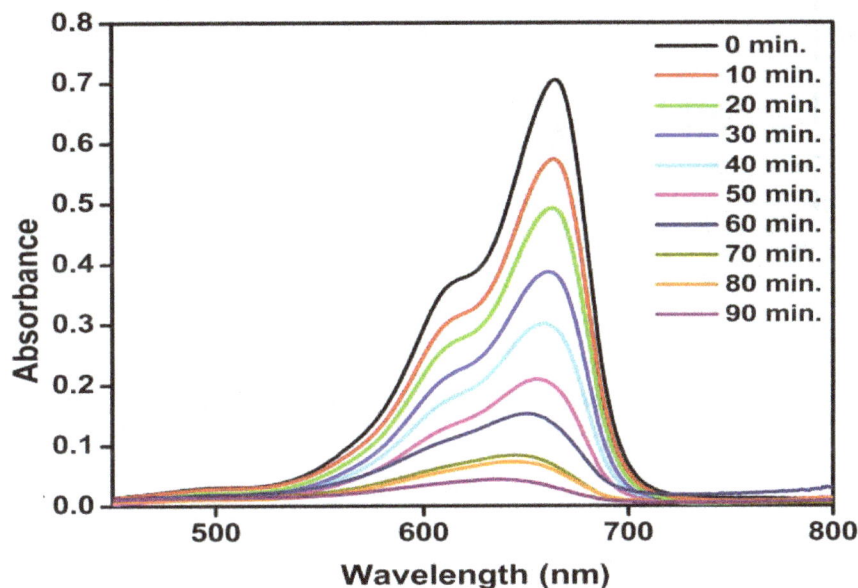

Figure 12. UV-visible spectra of methylene blue (MB) over various time intervals, in the presence of TiO_2 nanocatalysts.

No degradation of dye was discerned in the absence of solar light (supplementary information; Figure S2). The small decrease in absorbance observed was due to the insignificant adsorption of dye molecules on the catalyst surface. We have conducted a control experiment with MB on its own, as shown in supplementary information Figure S4. From this data, it is clear that TiO_2 was activated by solar light and was responsible for dye degradation. Moreover, photocatalytic degradation of dye was confirmed by the supplementary information (Figure S5, i.e., in the absence of sunlight, there is no degradation observed, confirming the major role of sunlight in the activation of the TiO_2 nanocatalyst). From Figures S4 and S5, it was confirmed that dye discoloration was entirely due to photocatalytic degradation, not adsorption.

3.10.1. Effect of Catalyst Concentration on Photocatalytic Activity of TiO_2 Nanoparticles

The effect of catalyst concentration on photocatalytic activity was tested by loading 1 to 15 mg/50 mL of TiO_2 nanocatalyst in methylene blue solution. The photocatalytic degradation of methylene blue was highly influenced by the level of catalyst loading, as evident in Figure 13 which shows that the percent degradation of the dye increased with the amount (from 1 to 15 mg/50 mL) of TiO_2 nanocatalyst loading and remained virtually constant above a certain level. This is because as the amount of catalyst increased, a greater number of active sites on the photocatalyst surface became available. Consequently, more OH radicals were produced, which facilitated their participation in the dye degradation process. However, beyond a certain limiting value of catalyst loading, the solution appeared turbid. As a result, the passage of solar radiation into the reaction mixture (required for the reaction to proceed) was obstructed, and thus, the percent degradation of the dye decreased or remained constant [41].

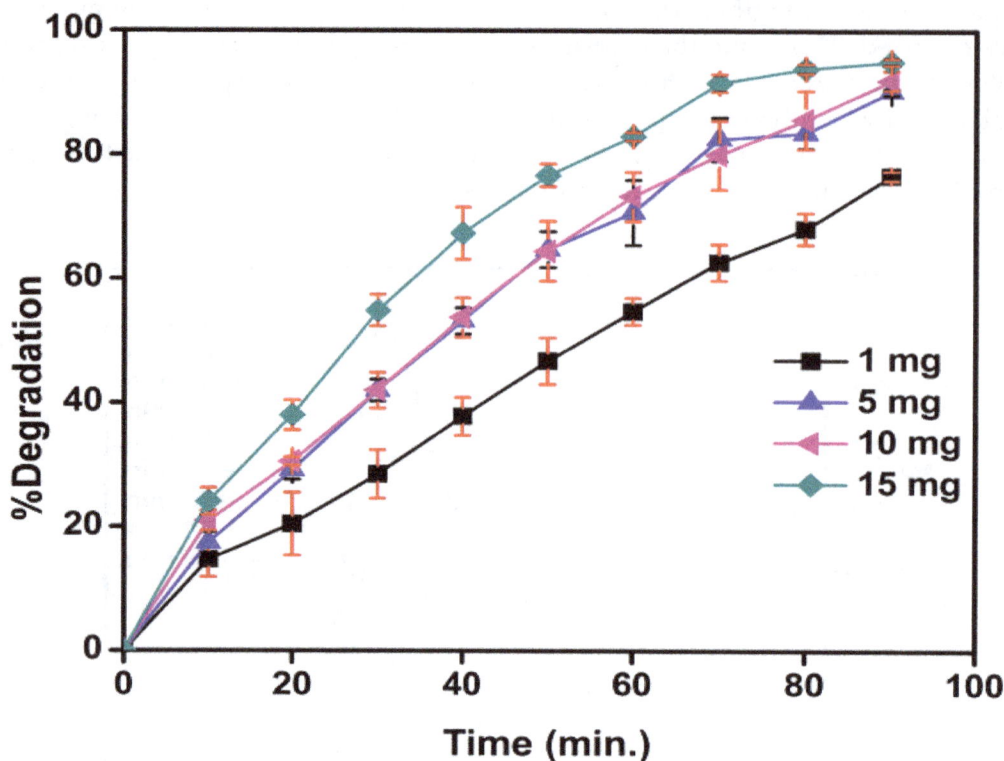

Figure 13. UV-visible spectra of MB dye (10^{-5} M) with different amounts of catalyst loading. Data obtained by degradation versus time are average values of three independent experiments (reported values = mean $\pm$ S.D.)

3.10.2. Effect of pH on Photocatalytic Activity of TiO_2 Nanoparticles

The role of pH on the rate of photocatalytic degradation was studied over the 4–9 pH range and the results are illustrated in Figure 14. It was observed that the percent degradation increased with a rising pH, exhibiting a maximum between the pH 7–9 ranges. This pH variation may bring about changes to the surface charge on the TiO_2 nanoparticles and vary the potential associated with catalytic reactions. With the variation of potential, the extent of dye adsorption on the catalyst surface also fluctuates, culminating in the alteration of reaction velocity. Furthermore, under alkaline conditions, the surface of TiO_2 could acquire a negative charge. Since methylene blue is a cationic dye and the surface of TiO_2 nanoparticles in alkaline media attains a negative charge, the latter can be easily adsorbed onto the catalyst surface. This may lead to enhanced photocatalytic dye degradation under basic conditions [42]. The surface of TiO_2 nanoparticles, in acidic or alkaline circumstances, can be protonated or deprotonated, respectively, according to the following reactions:

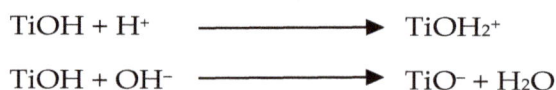

$$TiOH + H^+ \longrightarrow TiOH_2^+$$

$$TiOH + OH^- \longrightarrow TiO^- + H_2O$$

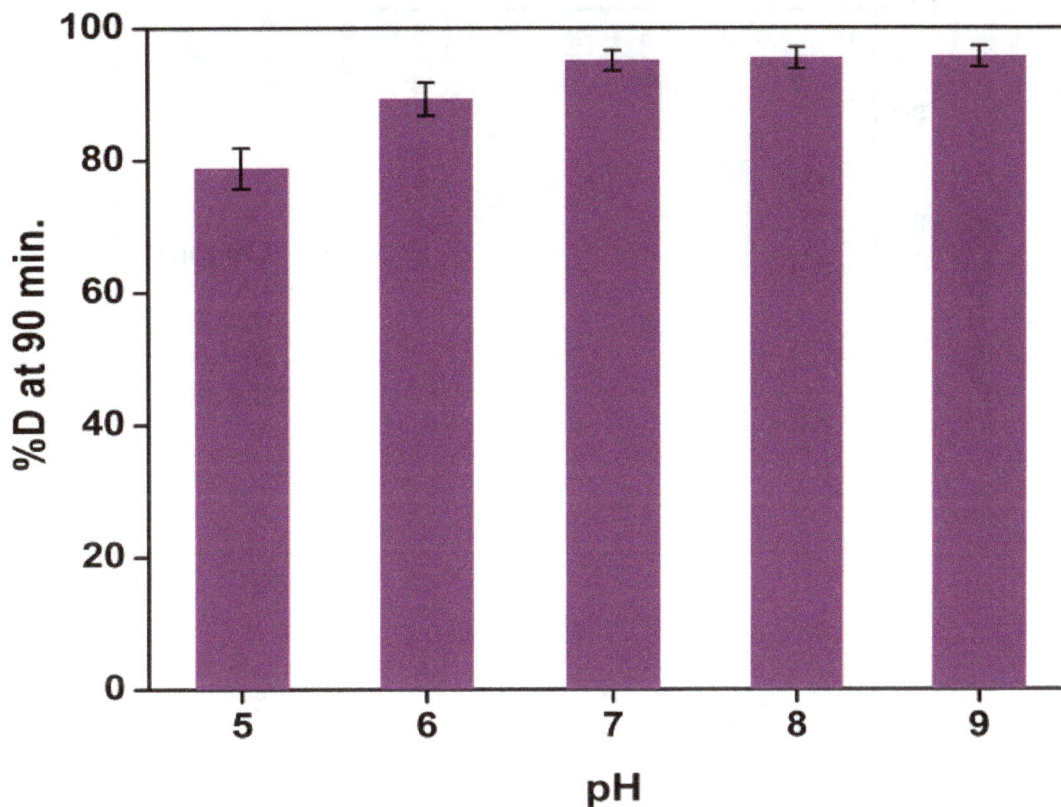

Figure 14. The photocatalytic degradation of MB dye (10^{-5} M) after treatment with photocatalyst TiO_2 (0.2 g/L) in solar light under different pH conditions (reported values are for three independent experiments; average value = mean $\pm$ S.D; n = 3).

3.10.3. Effect of Calcination Temperature on Photocatalytic Activity of TiO_2 Nanoparticles

The effect of calcination temperatures of the TiO_2 nanocatalyst on photocatalytic degradation is depicted in Figure 15. With increasing calcination temperatures of the TiO_2 nanocatalyst, the percentage degradation was found to have risen and attained a maximum of 96.85% at 700 °C. It then decreased gradually at temperatures between 800 °C and 900 °C. This variation can be expected, given that an increase in calcination temperature of the TiO_2 generated an anatase phase possessing good crystal characteristics—an essential criterion for photocatalytic degradation. It is also apparent from the UV data (Table 1) that the band gap increased with rising calcination temperatures, from 500 °C to 800 °C. Hence, the rate of electron-hole recombination decreased, and consequently, photocatalytic degradation increased. The decline in photocatalytic degradation at 800 °C was mainly ascribed to the formation of the rutile phase, which existed as a mixture of both the anatase and rutile forms. Complete formation of the rutile phase transpired at 900 °C, and thus site activity fell and degradation decreased once again.

It is clear that the photocatalytic ability of titanium dioxide NPs is largely dependent on its crystalline form. Thermodynamically, the efficiency corresponding to photo-oxidation by the anatase and rutile phases should be similar. The surface recombination of photo-excited electrons and positive holes was higher in rutile than in anatase, which has a greater free (electron)-carrier mobility. Hence, a range of photoactivities was observed for each crystal form. This showed that variables such as crystal type and particulate sizes, as well as various synthesis routes (temperature, heating time etc.) largely determine its photocatalytic activity [40].

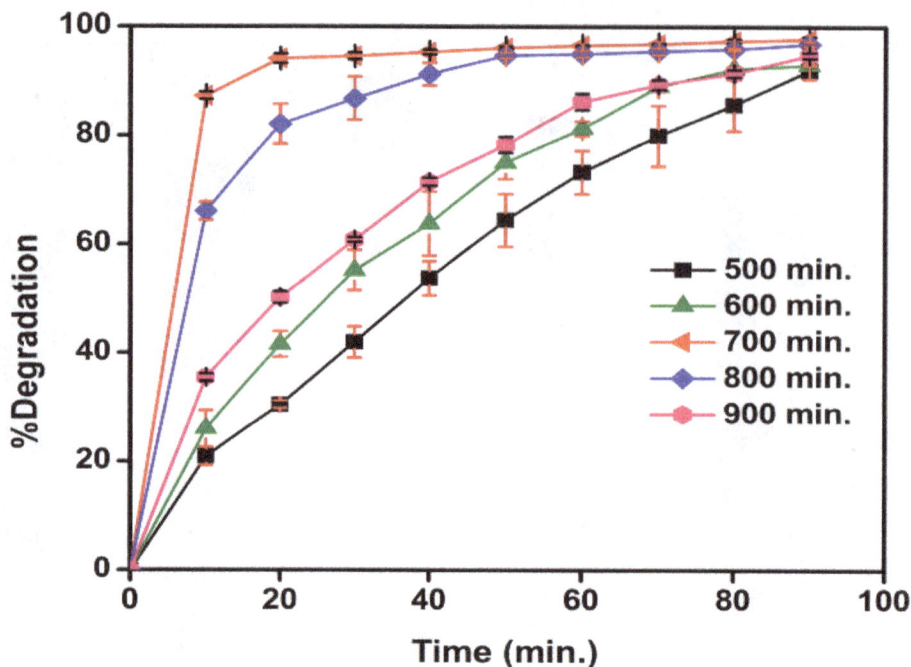

Figure 15. The photocatalytic degradation of MB dye (10^{-5} M) after treatment with photocatalyst TiO_2 (0.2 g/L) calcined at different temperatures under solar light (Values were reported as mean $\pm$ S.D; $n = 3$).

4. Conclusions

Green nanotechnology has, in recent years, been accorded increasing importance for two reasons, namely, its contribution towards the elimination of harmful reagents and its ability to facilitate the synthesis of valuable products in a cost-effective manner. The green synthesis of TiO_2 NPs involves more compatible, eco-friendly, low cost and less time-consuming processes compared to other synthetic methods such as the sol-gel technique, which has been widely used to achieve the same ends. In the present study, titanium dioxide nanoparticles were produced using a natural hydrocolloid, gum Kondagogu (*Cochlospermum gossypium*). The crystallinity and high purity of the synthesized TiO_2 nanoparticles in the anatase form were unambiguously confirmed by the presence of sharp peaks (and the absence of unidentified peaks) in the X-Ray Diffraction patterns, Raman spectroscopic results and TEM images obtained. The absence of the original gum residue on the nanoparticles after calcination for 4 h at 500 °C was confirmed by FTIR, EDX, BET and TGA. The nanoparticles created had a mean particle size of approximately 11 nm, a value which was determined by three independent techniques: Scherrer's formula from XRD (12.58 nm); TEM (8–13 nm) and CPS (11.2 $\pm$ 0.2 nm), the figures in parentheses being the respective particle sizes. The photocatalytic activity of green-synthesized titanium dioxide nanoparticles was evaluated by adopting methylene blue dye as a model system. It is apparent that the photocatalytic effectiveness of titanium dioxide nanoparticles is largely dependent on various factors including its crystalline form, exposure time, extent of catalyst loading and solution pH conditions. The present study has demonstrated that titanium dioxide nanoparticles (synthesized through a green route by the use of natural, renewable and eco-friendly materials) exhibit excellent photocatalytic activity towards organic dye degradation. This system can be employed in water purification and dye effluent treatment.

Supplementary Materials:
Figure S1: The intensity of solar light measured throughout the experiment (range 100,000 Lux); Figure S2: The % of UV intensity in solar radiation at various intervals of time; Figure S3: The solar UV intensity measurement using Lux meter and UV filtering goggles; Figure S4: UV-visible spectra of MB over various time intervals after treatment with TiO2 nanocatalyst under dark condition; Figure S5: UV-visible spectra of MB over various time intervals in the absence of TiO2 nanocatalyst.

Author Contributions: Formal analysis, S.W.; Funding acquisition, M.Č.; Investigation, K.S.; Methodology, C.S.; Project administration, M.Č.; Resources, R.P.; Supervision, B.G. and M.Č.; Writing–original draft, K.S.S.; Writing–review & editing, V.V.T.P.

Acknowledgments: The research reported in this paper was financially supported by the Ministry of Education, Youth and Sports in the framework OPR&DI project 'Extension of CxI facilities' (CZ.1.05/2.1.00/19.0386). The authors' also acknowledge the assistance provided by the Research Infrastructure NanoEnviCz, under Project No. LM2015073 and Ministry of Education, Youth and Sports of the Czech Republic and the European Union—European Structural and Investment Funds in the frames of Operational Program Research, Development and Education—project Hybrid Materials for Hierarchical Structures (HyHi, Reg. No. CZ.02.1.01/0.0/0.0/16_019/0000843). The authors also gratefully acknowledge financial support from the Council for Scientific and Industrial Research, India.

References

1. Makarov, V.V.; Love, A.J.; Sinitsyna, O.V.; Makarova, S.S.; Yaminsky, I.V.; Taliansky, M.E.; Kalinina, N.O. "Green" Nanotechnologies: synthesis of metal nanoparticles using plants. *Acta Naturae* **2014**, *6*, 35–44. [PubMed]

2. Jana, A.; Scheer, E.; Polarz, S. Synthesis of graphene–transition metal oxide hybrid nanoparticles and their application in various fields. *Beilstein J. Nanotechnol.* **2017**, *8*, 688–714. [CrossRef] [PubMed]

3. Kim, C.S.; Moon, B.K.; Park, J.H.; Choi, B.C.; Seo, H.J. Solvothermal synthesis of nanocrystalline TiO_2 in toluene with surfactant. *J. Cryst. Growth* **2003**, *257*, 309–315. [CrossRef]

4. Gotić, M.; Musić, S. Synthesis of nanocrystalline Iron oxide particles in the Iron(iii) acetate/alcohol/acetic acid system. *Eur. J. Inorg. Chem.* **2008**, *6*, 966–973. [CrossRef]

5. Hayashi, H.; Hakuta, Y. Hydrothermal synthesis of metal oxide nanoparticles in supercritical water. *Materials* **2010**, *3*, 3794–3817. [CrossRef] [PubMed]

6. Sharma, D.; Kanchi, S.; Bisetty, K. Biogenic synthesis of nanoparticles: A review. *Arab. J. Chem.* 2015. [CrossRef]

7. Lin, H.; Li, L.; Zhao, M.; Huang, X.; Chen, X.; Li, G.; Yu, R. Synthesis of high-quality brookite TiO_2 single-crystalline nanosheets with specific facets exposed: tuning catalysts from inert to highly reactive. *J. Am. Chem. Soc.* **2012**, *134*, 8328–8331. [CrossRef] [PubMed]

8. Diebold, U. The surface science of titanium dioxide. *Surf. Sci. Rep.* **2003**, *48*, 53–229. [CrossRef]

9. Fernandez-Garcia, M.; Belver, C.; Hanson, J.C.; Wang, X.; Rodriguez, J.A. Anatase-TiO_2 nanomaterials: analysis of key parameters controlling crystallization. *J. Am. Chem. Soc.* **2007**, *129*, 13604–13612. [CrossRef]

10. Fernandez-Garcia, M.; Rodriguez, J.A. *Metal Oxide Nanoparticles*; Brookhaven National Laboratory: Upton, NY, USA, 2007.

11. Bao, S.J.; Lei, C.; Xu, M.W.; Cai, C.J.; Cheng, C.J.; Li, C.M. Environmentally-friendly biomimicking synthesis of TiO_2 nanomaterials using saccharides to tailor morphology, crystal phase and photocatalytic activity. *CrystEngComm* **2013**, *15*, 4694–4699. [CrossRef]

12. Vinod, V.T.P.; Sashidhar, R.B.; Černík, M. Morphology and metal binding characteristics of a natural polymer-kondagogu (*Cochlospermum gossypium*) gum. *Molecules* **2013**, *18*, 8264–8274. [CrossRef] [PubMed]

13. Vinod, V.T.P.; Wacławek, S.; Černík, M.; Varma, R.S. Tree gum-based renewable materials: Sustainable applications in nanotechnology, biomedical and environmental fields. *Biotechnol. Adv.* **2018**, *36*, 1984–2016.

14. Vinod, V.T.P.; Sashidhar, R.B. Surface morphology, chemical and structural assignment of gum Kondagogu (*Cochlospermum gossypium* DC.): An exudate tree gum of India. *Indian J. Nat. Prod. Resour.* **2010**, *1*, 181–192.

15. Vinod, V.T.P.; Sashidhar, R.B.; Sukumar, A.A. Competitive adsorption of toxic heavy metal contaminants by gum kondagogu (*Cochlospermum gossypium*): a natural hydrocolloid. *Colloids Surf. B.* **2010**, *75*, 490–495. [CrossRef] [PubMed]

16. Vinod, V.T.P.; Saravanan, P.; Sreedhar, B.; Devi, D.K.; Sashidhar, R.B. A facile synthesis and characterization of Ag, Au and Pt nanoparticles using a natural hydrocolloid gum kondagogu (*Cochlospermum gossypium*). *Colloids Surf. B.* **2011**, *83*, 291–298. [CrossRef] [PubMed]

17. Padil, V.V.T.; Cernik, M. Green synthesis of copper oxide nanoparticles using gum karaya as a biotemplate and their antibacterial application. *Int. J. Nanomed.* **2013**, *8*, 889–898. [CrossRef]

18. Vinod, V.T.P.; Sashidhar, R.B.; Suresh, K.I.; Rao, B.R.; Vijaya Saradhi, U.V.R.; Rao, T.P. Morphological, physico-chemical and structural characterization of gum kondagogu (*Cochlospermum gossypium*): A tree gum from India. *Food Hydrocoll.* **2008**, *22*, 899–915. [CrossRef]

19. Vinod, V.T.P.; Sashidhar, R.B.; Sarma, V.U.M.; Vijaya Saradhi, U.V.R. Compositional analysis and rheological properties of gum kondagogu (*Cochlospermum gossypium*): a tree gum from India. *J. Agric. Food. Chem.* **2008**, *56*, 2199–2207. [CrossRef]

20. Naidu, V.G.M.; Madhusudhana, K.; Sashidhar, R.B.; Ramakrishna, S.; Khar, R.K.; Ahmed, F.J.; Diwan, P.V. Polyelectrolyte complexes of gum kondagogu and chitosan, as diclofenac carriers. *Carbohydr. Polym.* **2009**, *76*, 464–471. [CrossRef]

21. Khade, G.V.; Suwarnkar, M.B.; Gavade, N.L.; Garadkar, K.M. Green synthesis of TiO_2 and its photocatalytic activity. *J. Mater. Sci. Mater. Electron.* **2015**, *26*, 3309–3315. [CrossRef]

22. Kant, R. Textile dyeing industry an environmental hazard. *Nat. Sci.* **2012**, *4*, 22–26. [CrossRef]

23. Neyaz, N.; Siddiqui, W.A.; Nair, K.K. Application of surface functionalized Iron oxide nanomaterials as a nanosorbents in extraction of toxic heavy metals from ground water: a review. *Int. J. Environ. Sci.* **2014**, *4*, 472–483. [CrossRef]

24. Byrne, C.; Fagan, R.; Hinder, S.; McCormack, D.E.; Pillai, S.C. New approach of modifying the anatase to rutile transition temperature in TiO_2 photocatalysts. *RSC Adv.* **2016**, *6*, 95232–95238. [CrossRef]

25. Carp, O.; Huisman, C.L.; Reller, A. Photoinduced reactivity of titanium dioxide. *Prog. Solid State Chem.* **2004**, *32*, 33–177. [CrossRef]

26. Mahlambi, M.M.; Ngila, C.J.; Mamba, B.B. Recent developments in environmental photocatalytic degradation of organic pollutants: the case of titanium dioxide nanoparticles—A review. *J Nanomater.* **2015**, *2015*, 1–29. [CrossRef]

27. Ahmed, M.A.; El-Katori, E.E.; Gharni, Z.H. Photocatalytic degradation of methylene blue dye using Fe_2O_3/TiO_2 nanoparticles prepared by sol–gel method. *J. Alloys Compd.* **2013**, *553*, 19–29. [CrossRef]

28. Boury, B.; Plumejeau, S. Metal oxides and polysaccharides: An efficient hybrid association for materials chemistry. *Green Chem.* **2015**, *17*, 72–88. [CrossRef]

29. Dong, Y.; Wang, Y.; Cai, T.; Kou, L.; Yang, G.; Yan, Z. Preparation and nitrogen-doping of three-dimensionally ordered macroporous TiO_2 with enhanced photocatalytic activity. *Ceram. Int.* **2014**, *40*, 11213–11219. [CrossRef]

30. Filippo, E.; Carlucci, C.; Capodilupo, A.L.; Perulli, P.; Conciauro, F.; Corrente, G.A.; Gigli, G.; Ciccarella, G. Enhanced photocatalytic activity of pure anatase TiO_2 and Pt-TiO_2 nanoparticles synthesized by green microwave assisted route. *Mater. Res.* **2015**, *18*, 473–481. [CrossRef]

31. Haque, F.Z.; Nandanwar, R.; Singh, P. Evaluating photodegradation properties of anatase and rutile TiO_2 nanoparticles for organic compounds. *Optik* **2017**, *128*, 191–200. [CrossRef]

32. Ba-Abbad, M.M.; Kadhum, A.A.H.; Mohamad, A.B.; Takriff, M.S.; Sopian, K. Synthesis and catalytic activity of TiO_2 nanoparticles for photochemical oxidation of concentrated chlorophenols under direct solar radiation. *Int. J. Electrochem. Sci.* **2012**, *7*, 4871–4888.

33. Periyat, P.; Pillai, S.C.; McCormack, D.E.; Colreavy, J.; Hinder, S.J. Improved high-temperature stability and sun-light-driven photocatalytic activity of sulfur-doped anatase TiO_2. *J. Phys. Chem. C* **2008**, *112*, 7644–7652. [CrossRef]

34. Reddy, G.B.; Madhusudhan, A.; Ramakrishna, D.; Ayodhya, D.; Venkatesham, M.; Veerabhadram, G. Green chemistry approach for the synthesis of gold nanoparticles with gum kondagogu: Characterization, catalytic and antibacterial activity. *J. nanostructure chem.* **2015**, *5*, 185–193. [CrossRef]

35. Chellappa, M.; Anjaneyulu, U.; Manivasagam, G.; Vijayalakshmi, U. Preparation and evaluation of the cytotoxic nature of TiO_2 nanoparticles by direct contact method. *Int. J. Nanomed.* **2015**, *10*, 31–41. [CrossRef]

36. Ma, W.; Lu, Z.; Zhang, M. Investigation of structural transformations in nanophase titanium dioxide by Raman spectroscopy. *Appl. Phys. A* **1998**, *66*, 621–627. [CrossRef]

37. Ohsaka, T. Temperature dependence of the raman spectrum in anatase TiO_2. *J. Phys. Soc. Jpn.* **1980**, *48*, 1661–1668. [CrossRef]

38. Li, Y.; Qin, Z.; Guo, H.; Yang, H.; Zhang, G.; Ji, S.; Zeng, T. Low-temperature synthesis of anatase TiO_2 nanoparticles with tunable surface charges for enhancing photocatalytic activity. *PloS one* **2014**, *9*, 1–19. [CrossRef]

39. Auvinen, S.; Alatalo, M.; Haario, H.; Jalava, J.P.; Lamminmäki, R.J. Size and shape dependence of the electronic and spectral properties in TiO_2 nanoparticles. *J. Phys. Chem. C* **2011**, *115*, 8484–8493. [CrossRef]

40. Rauf, M.A.; Ashraf, S.S. Fundamental principles and application of heterogeneous photocatalytic degradation of dyes in solution. *Chem. Eng. J.* **2009**, *151*, 10–18. [CrossRef]

41. Wang, C.C.; Lee, C.K.; Lyu, M.D.; Juang, L.C. Photocatalytic degradation of C.I. Basic Violet 10 using TiO_2 catalysts supported by Y zeolite: An investigation of the effects of operational parameters. *Dyes. Pigm.* **2008**, *76*, 817–824. [CrossRef]

42. Lakshmi, S.; Renganathan, R.; Fujita, S. Study on TiO_2-mediated photocatalytic degradation of methylene blue. *J. Photochem. Photobiol. A* **1995**, *88*, 163–167. [CrossRef]

Development of Effective Lipase-Hybrid Nanoflowers Enriched with Carbon and Magnetic Nanomaterials for Biocatalytic Transformations

Renia Fotiadou [1], Michaela Patila [1], Mohamed Amen Hammami [2], Apostolos Enotiadis [2], Dimitrios Moschovas [3], Kyriaki Tsirka [3], Konstantinos Spyrou [3], Emmanuel P. Giannelis [2], Apostolos Avgeropoulos [3], Alkiviadis Paipetis [3], Dimitrios Gournis [3] and Haralambos Stamatis [1,*]

[1] Biotechnology Laboratory, Department of Biological Applications and Technologies, University of Ioannina, 45110 Ioannina, Greece; renia.fotiadou@gmail.com (R.F.); mpatila@cc.uoi.gr (M.P.)

[2] Department of Materials Science and Engineering, Cornell University, Ithaca, NY 14853, USA; mah424@cornell.edu (M.A.H.); ae276@cornell.edu (A.E.); epg2@cornell.edu (E.P.G.)

[3] Department of Materials Science and Engineering, University of Ioannina, 45110 Ioannina, Greece; dmoschov@cc.uoi.gr (D.M.); tsirka.kyriaki@gmail.com (K.T.); konstantinos.spyrou1@gmail.com (K.S.); aavger@uoi.gr (A.A.); paipetis@uoi.gr (A.P.); dgourni@uoi.gr (D.G.)

* Correspondence: hstamati@uoi.gr

Abstract: In the present study, hybrid nanoflowers (HNFs) based on copper (II) or manganese (II) ions were prepared by a simple method and used as nanosupports for the development of effective nanobiocatalysts through the immobilization of lipase B from *Pseudozyma antarctica*. The hybrid nanobiocatalysts were characterized by various techniques including scanning electron microscopy (SEM), energy dispersion spectroscopy (EDS), X-ray diffraction (XRD), Raman spectroscopy, and Fourier transform infrared spectroscopy (FTIR). The effect of the addition of carbon-based nanomaterials, namely graphene oxide and carbon nanotubes, as well as magnetic nanoparticles such as maghemite, on the structure, catalytic activity, and operational stability of the hybrid nanobiocatalysts was also investigated. In all cases, the addition of nanomaterials during the preparation of HNFs increased the catalytic activity and the operational stability of the immobilized biocatalyst. Lipase-based magnetic nanoflowers were effectively applied for the synthesis of tyrosol esters in non-aqueous media, such as organic solvents, ionic liquids, and environmental friendly deep eutectic solvents. In such media, the immobilized lipase preserved almost 100% of its initial activity after eight successive catalytic cycles, indicating that these hybrid magnetic nanoflowers can be applied for the development of efficient nanobiocatalytic systems.

Keywords: hybrid nanoflowers; lipase; magnetic nanomaterials; biocatalysis; enzyme immobilization

1. Introduction

Over the last decades, the immobilization of enzymes onto nanostructured supports has been extensively used and has facilitated their applications, owing to their easy handling and operational stability, as well as facile recovery and reusability of the biocatalysts, leading to more efficient bioprocesses [1,2]. Various nanostructured composite materials with extensive active surface areas and desirable pore sizes, such as nanoporous supports, nanofibers, nanoparticles, and carbon-based nanomaterials (e.g., nanotubes and graphene) have been proven to be effective in manipulating the nanoscale environment of biomolecules [3–5] and, as a consequence, their biological function and stability.

Organic-inorganic hybrid nanomaterials (nanoflowers) are a recently developed group of nanoparticles that schematically resemble plant flowers in a nanoscale range [6]. Hybrid nanoflowers

(HNFs) have attracted a lot of interest over the last years as host platforms for immobilizing enzymes, owing to the higher surface-to-volume ratio compared to spherical nanoparticles, as well as to their simple, eco-friendly, and cost-effective synthesis [7,8]. Nanoflowers containing various enzymes have been usually prepared as enzyme-$Cu_3(PO_4)_2 \cdot 3H_2O$ hybrids by combining copper sulfate ($CuSO_4$) with enzymes in phosphate-buffer saline (PBS). Different HNFs mainly based on copper (II) and calcium (II) ions have been used to form complexes with enzymes and other proteins [9–11]. Moreover, the development of protein-embedded HNFs based on other metal ions, such as zinc (II), cobalt (II), and iron (II), was recently reported [12–14]. The formation of HNFs comprises the following steps: the nucleation and formation of primary crystals, the growth of these aggregates, and the complete formulation of nanoflowers [8]. During the first step, protein molecules form complexes with metal ions, primarily through the coordination between nitrogen atoms of the amide groups present in the protein backbone and the metal ion. These complexes provide sites for nucleation. The intramolecular interactions between the metal ion and the protein promote the anisotropic growth of nano-petals (step 2) and, consequently, the formation of a flower-like structure in which proteins serve as the glue that binds the petals together (step 3). The formation of these enzyme-embedded HNFs do not require harsh conditions and toxic reactants for their self-assembly; thus, the immobilization procedure is facilitated with biomolecules in a one-step process. Moreover, the incorporated enzyme is subjected to minor manipulation in comparison with other conventional immobilization procedures, thus retaining its biocatalytic activity [8].

The selection of the enzyme and the metal ions—as well as the pH, the temperature, and the incubation time—plays an essential role for the configuration and the catalytic efficiency of the enzyme-containing nanoflowers [8,15]. A variety of enzymes with biotechnological interest have been encapsulated in HNFs and successfully applied in dye decolorization [11,16], the production of esters [17], the detection of phenols or glucose [9,18], the degradation of pollutants [19], and the development of biosensors [20,21] in which two or more enzymes were successfully encapsulated in the same nanoflower.

The enhanced activity of enzymes that is observed in various HNFs is mainly attributed to their high surface area, which decreases mass-transfer limitations, along with the specific interactions of the enzyme molecules and metal ions [8,22,23]. However, the biocatalytic activity and stability of some HNFs are reduced by the interactions between metal ions and proteins [24]. Recently, it was proposed that the incorporation of surfactants [25], biopolymers such as chitosan [26], and carbon-based nanomaterials [27,28] could enhance the catalytic properties as well as the mechanical strength of enzyme-containing nanoflowers, leading to stable nanohybrids.

Herein, we describe the preparation of novel hybrid nanoflowers consisting of copper (II) or manganese (II) ions, combined with magnetic nanoparticles and carbon-based nanomaterials, and we investigate their use as versatile host platforms for the development of sufficient systems for the immobilization of enzymes. The addition of carbon-based nanomaterials, namely graphene oxide and multi-walled carbon nanotubes, in the preparation of nanoflowers is expected to provide high surface area and extraordinary mechanical properties, whereas the incorporation of magnetic nanoparticles, such as maghemite, allows the easy and quick separation of the nanoflowers by the application of an external magnetic force. The use of these novel HNFs as host platforms for the immobilization of lipase B from *Pseudozyma antarctica,* an enzyme with numerous biotechnological applications, was investigated. The novel nanobiocatalysts were characterized by scanning electron microscopy (SEM), energy dispersion spectroscopy (EDS), X-ray diffraction (XRD), Raman spectroscopy, and Fourier transform infrared spectroscopy (FTIR), while the effect of the composition of nanoflowers on the catalytic activity, thermal activity, and operational stability of the immobilized enzyme was investigated. Moreover, the ability of the lipase-based nanoflowers to catalyze the synthesis of lipophilic derivatives of phenolic antioxidants, such as tyrosol, in non-aqueous media, as well as in environmental-friendly ionic solvents, was also investigated.

2. Materials and Methods

2.1. Materials

Lipase B from *Pseudozyma antarctica* (formerly Candida antarctica, CaLB) was purchased from Novozymes A/S (Bagsværd, Denmark) and was utilized without further purification. 4-nitrophenyl butyrate (*p*-NPB), 4-nitrophenol (*p*-NP), copper (II) sulfate pentahydrate, manganese (II) sulfate, tyrosol, and dimethyl sulfoxide were obtained from Sigma–Aldrich (St. Louis, MO, USA). Vinyl butyrate was obtained from Fluka. The ionic liquid (IL) 1-Butyl-3-methylimidazolium hexafluorophosphate ([BMIM]PF$_6$) with a purity of 97.0% was purchased from Sigma–Aldrich (St. Louis, MO, USA). Choline chloride (ChCl) and urea (U) were obtained from Sigma-Aldrich (St. Louis, MO, USA) and used for the preparation of deep eutectic solvents (DES), according to a previous work [17]. All organic solvents used were of analytical grade.

2.2. Preparation of CaLB Nanoflowers

The CaLB hybrid nanoflowers were prepared according to Ge et al. [8]. Typically, 0.42 mL of CuSO$_4$ or MnSO$_4$ aqueous solutions (120 mM) were added to 50 mL of phosphate buffer saline (PBS 1X, pH 7.4), which contained 0.4 mg mL^{-1} CaLB. The mixtures were placed for incubation at 25 °C for 3 days. The nanoflower precipitates were separated by centrifugation at 4000 rpm for 10 min, washed three times with distilled water, and dried under vacuum over silica gel at room temperature. Nanoflowers were stored at 4 °C until used. The prepared copper- and manganese-based samples are labeled Cu$_3$(PO$_4$)$_2$ and Mn$_3$(PO$_4$)$_2$, respectively.

For the preparation of nanomaterials-modified CaLB nanoflowers, a similar approach was followed. Graphene oxide (GO), oxidized multi-walled carbon nanotubes (CNTs), and maghemite nanoparticles (γ-Fe$_2$O$_3$) were synthesized as reported elsewhere [29–31]. Briefly, 5 mg of GO and 3 mg of oxidized CNTs or γ-Fe$_2$O$_3$ nanoparticles were added in 49 mL of PBS and sonicated for 20 min. After the dispersion of the nanomaterials, 1 mL of CaLB solution and 0.42 mL of CuSO$_4$ or MnSO$_4$ aqueous solutions (120 mM) were added into the mixture. The next steps were the same as those described previously. Nanoflowers containing only GO or CNTs were also prepared. The prepared modified copper-based samples are labeled GO-Cu$_3$(PO$_4$)$_2$, CNTs-Cu$_3$(PO$_4$)$_2$, GO/CNTs-Cu$_3$(PO$_4$)$_2$, and GO/Fe$_2$O$_3$-Cu$_3$(PO$_4$)$_2$, and the prepared modified manganese-based samples are labeled GO-Mn$_3$(PO$_4$)$_2$, CNTs-Mn$_3$(PO$_4$)$_2$, GO/CNTs-Mn$_3$(PO$_4$)$_2$, and GO/Fe$_2$O$_3$-Mn$_3$(PO$_4$)$_2$.

2.3. Characterization of CaLB Nanoflowers

SEM images were acquired from a JEOL JSM-5600 microscope (JEOL Ltd., Tokyo, Japan) with 10 and 25 kV accelerating voltage. Moreover, the surface morphologies of the samples were determined by field emission scanning electron microscopy (FE-SEM) using a SEM Zeiss Gemini 500 (Oberkochen, Germany). Prior to SEM analysis, the nanoflowers were placed in double-sided carbon tape and sputter-coated with gold-platinum. Phase elemental distribution was studied with SEM/EDS (JEOL JSM-6510 LV equipped with an X-Act EDS-detector by Oxford Instruments, Abingdon, Oxfordshire, UK).

The XRD patterns of all CaLB-HNFs were collected on a D8 Advance Bruker diffractometer with Cu Kα radiation (40 kV, 40 mA) and a secondary-beam.

Raman spectrocopy was used to confirm the presence of the carbon nanomaterials in the nanomaterials-modified CaLB-HNFs. The Raman spectra were recorded with the Labram HR system by HORIBA Scientific (HORIBA, Paris, France). The 514.5 green line of an air cooled Ar-Ion Laser was employed for the Raman excitation using a confocal aperture of 100. The laser power at the focal plane of the x100 objective was circa 0.8 mW. Spectral treatment included only a linear baseline subtraction.

FTIR was utilized to confirm the successful immobilization of CaLB in the nanoflower structure. The spectra were recorded in the range of 400–4000 cm^{-1} using a FTIR-8400 infrared spectrometer (Shimadzu, Tokyo, Japan) equipped with a deuterated triglycine sulfate (DTGS) detector. For each sample, a total of 64 scans were averaged, using a 2 cm^{-1} resolution. The samples were prepared using KBr pellets containing a circa 2 wt% sample. The similarity of FTIR spectra in the Amide I region (1600–1700 cm^{-1}) was quantified by calculation of the correlation coefficient, r, using the following equation:

$$r = \frac{\Sigma x_i y_i}{\sqrt{\Sigma x_i^2 \Sigma y_i^2}},$$ (1)

where x and y represent the spectral absorbance values of the reference and sample spectra, respectively, at the ith frequency position [32]. For identical spectra, the r value is equal to 1.0, while spectra that have differences will show lower values.

2.4. Determination of Encapsulation Yield

The amount of the immobilized CaLB was determined by calculating the protein concentration present in the supernatant after the immobilization procedure using the Bradford assay [33]. Enzyme encapsulation was estimated as the difference between the initial amount of the enzyme and the amount of the enzyme in the supernatant after immobilization.

2.5. Activity of CaLB Nanoflowers

The activity of CaLB-HNFs was determined by the hydrolysis of p-NPB. Specifically, 0.5 mg of CaLB nanoflowers was added into 2 mL of phosphate buffer (50 mM, pH 7.5). The reaction was initiated with the addition of 20 µL of a 50 mM p-NPB solution (dissolved in DMSO), and the mixture was incubated for up to 10 min at 40 °C, 650 rpm. The 4-Nitrophenol (p-NP) release was monitored at 405 nm. The activity was estimated by measuring the concentration of p-NP using a standard curve. In this work, one unit of lipase activity was defined as the specific quantity of CaLB nanoflowers required to hydrolyze 1 µmol of p-NPB per reaction minute. Blank measurements without any enzyme were also incubated with the substrate for ten minutes, and their absorbance was measured where no catalytic activity was observed.

2.6. Stability of CaLB Nanoflowers

The thermal stability of free CaLB and CaLB-HNFs was tested at 60 °C for up to 24 h in phosphate buffer (50 mM, pH 7.5). In order to determine the remaining activity of CaLB nanoflowers, aliquots were taken at predetermined interval times for measuring the remaining lipase activity. The remaining hydrolyzing activity was estimated as described before, monitoring the increase in the absorbance of p-NP.

2.7. Transesterification of Tyrosol Catalayzed by CaLB Nanoflowers

The performance of CaLB-HNFs was tested on their ability to synthesize tyrosol esters. Typically, tyrosol (20 mM), vinyl butyrate (100 mM), and 4 mg mL^{-1} of CaLB-HNFs were added in various organic solvents and ionic liquids. The reaction mixtures were incubated for 72 h under stirring at 50 °C. Synthesis reactions were repeated twice, while experiments without nanoflowers were also conducted, and any decrease in the amount of tyrosol was observed for the selected solvents. The concentration of tyrosol in the reaction mixtures was quantified by high performance liquid chromatography (HPLC), equipped with a µBondapack C18 reverse phase column (particle size 10 µm, length 300 mm, diameter 3.9 mm) and a diode array UV detector. The elution was carried out with 40% water (containing 0.1% acetic acid) in methanol at a flow rate of 1 mL min^{-1} for 30 min. Tyrosol and its ester derivative were detected at 280 nm, while the column temperature was set at 35 °C. The conversion yield of the enzymatic transesterification was based on the decrease in the concentration of tyrosol, which was calculated using a tyrosol standard curve.

2.8. Reusability of CaLB Nanoflowers

The reusability of CaLB-HNFs was tested with respect to p-NPB hydrolysis for nine consecutive cycles. After each catalytic cycle, the samples were recovered by centrifugation at 1000 rpm for 2 min and excessively rinsed out three times with phosphate buffer (50 mM, pH 7.5). In the case of GO/Fe$_2$O$_3$-based hybrid nanoflowers, an external magnetic field was applied after each cycle and between washing procedures. The relative activity (%) was defined as the ratio of the remaining activity to the activity of the first cycle.

Magnetic hybrid nanoflowers (GO/Fe$_2$O$_3$-basedHNFs) were tested for their reusability on the transesterification of tyrosol in $tert$-butyl-methylether. Tyrosol (20 mM), vinyl butyrate (100 mM), and 4 mg mL^{-1} of GO/Fe$_2$O$_3$ CaLB-HNFs were added in 1 mL $tert$-butyl-methylether, and the reaction mixture was incubated for 72 h under stirring at 50 °C. The nanobiocatalytic system was separated from the reaction solution by an external magnetic field and washed twice with 1 mL of $tert$-butyl-methylether. The modified nanoflowers were applied to a new reaction solution and tested as described before for eight successive cycles.

3. Results and Discussion

3.1. Morphological and Structural Characterization of CaLB Nanoflowers

In the present work, HNFs based on copper (II) or manganese (II) ions were prepared by a simple method and used as nanosupports for the encapsulation of lipase B from *Pseudozyma antarctica* (CaLB). The effect of the enrichment of HNFs with graphene oxide sheets, oxidized multi-walled carbon nanotubes, and γ-Fe$_2$O$_3$ nanoparticles on the morphological, structural, and catalytic properties of HNFs was investigated.

SEM images of unmodified Cu$_3$(PO$_4$)$_2$ CaLB-HNFs revealed a high quality nanoflower formation with diameters in the range of 15–30 μm (Figure 1a). Moreover, SEM images of unmodified Mn$_3$(PO$_4$)$_2$ CaLB-HNFs (Figure 1b) displayed a flower-like structure, though this structure was not as clear as in the case of Cu$_3$(PO$_4$)$_2$. The nanomaterials-modified CaLB-HNFs, either with GO, CNTs, or both carbon structures together, exhibited different structures and formations, as indicated in Figure 1c–j, while a more detailed analysis can be found in the Supplementary material (Figure S1). It is noteworthy to add that the presence of carbon nanostructures (either GO or CNTs) in manganese-based nanoflowers facilitated the formation of nanoflowers in the final structures compared to the unmodified one (Figure 1g–j). Moreover, the combination of GO and CNTs resulted in the growth of clear crystals forming particular porous flower-like structures.

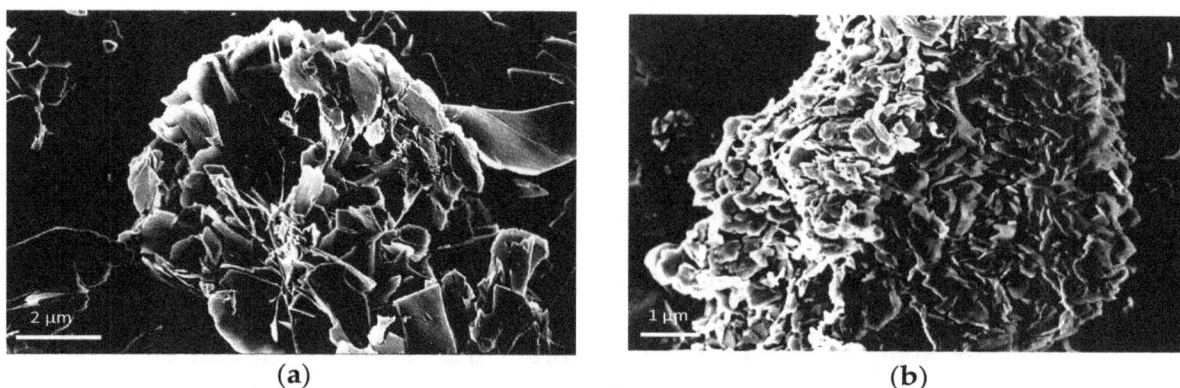

(a)   (b)

Figure 1. *Cont.*

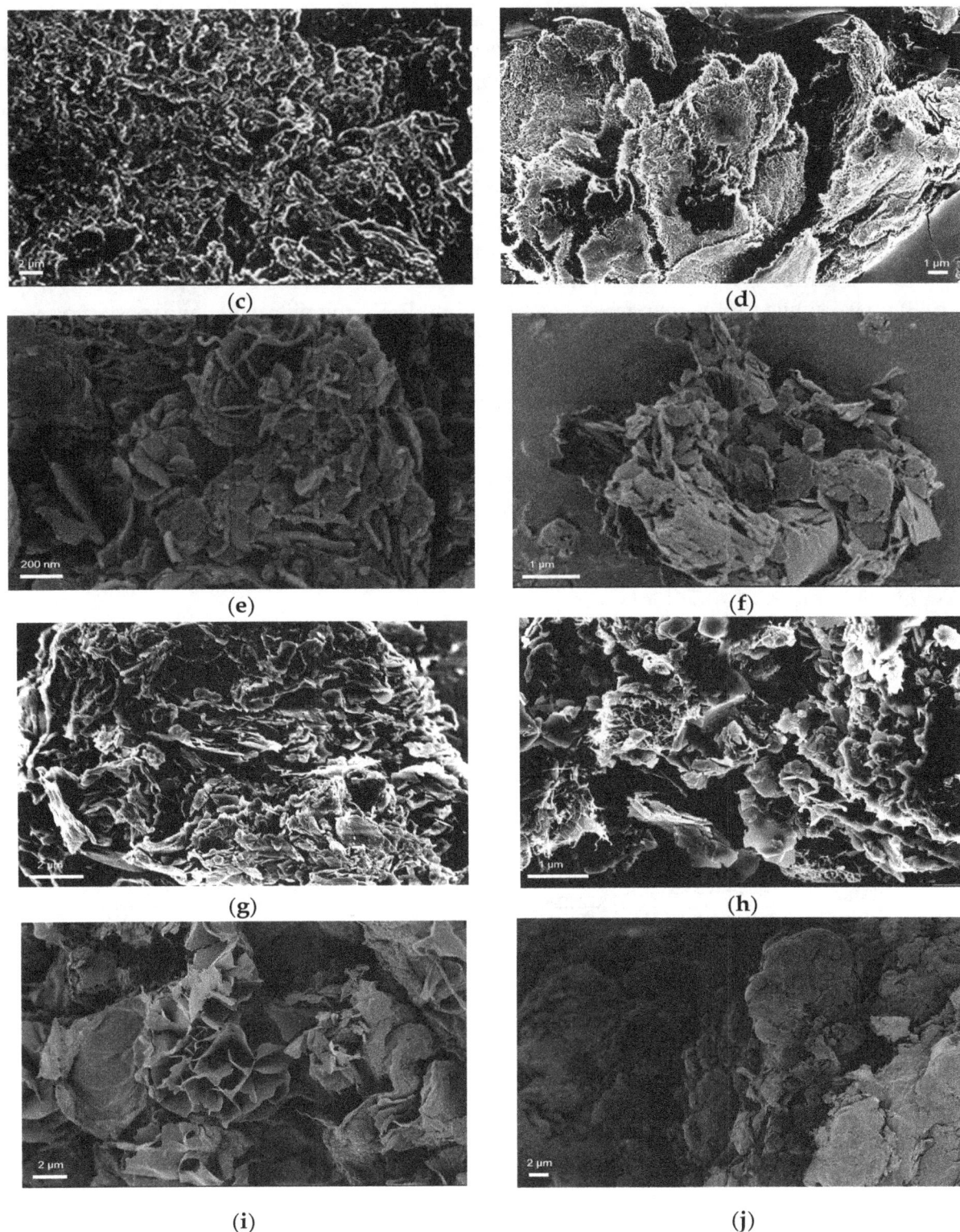

Figure 1. SEM images of: (**a**) unmodified $Cu_3(PO_4)_2$ CaLB-HNFs; (**b**) unmodified $Mn_3(PO_4)_2$ CaLB-HNFs; (**c**) GO-$Cu_3(PO_4)_2$ CaLB-HNFs; (**d**) CNTs-$Cu_3(PO_4)_2$ CaLB-HNFs; (**e**) GO/CNTs-$Cu_3(PO_4)_2$ CaLB-HNFs; (**f**) GO/Fe_2O_3-$Cu_3(PO_4)_2$ CaLB-HNFs; (**g**) GO-$Mn_3(PO_4)_2$ CaLB-HNFs; (**h**) CNTs-$Mn_3(PO_4)_2$ CaLB-HNFs; (**i**) GO/CNTs-$Mn_3(PO_4)_2$ CaLB-HNFs; and (**j**) GO/Fe_2O_3-$Mn_3(PO_4)_2$ CaLB-HNFs.

Modified GO/Fe_2O_3-based HNFs were further elementally analyzed using energy dispersion spectroscopy (EDS) (Figure S2). The peaks of carbon (C) and oxygen (O) were attributed to CaLB and the incorporated nanomaterials, while the presence of nitrogen (N) and sulfur (S) confirmed the successful encapsulation of the enzyme in the nanoflower structure. The appearance of copper (Cu) (Figure S2a), manganese (Mn) (Figure S2b) and phosphate (P) peaks indicated the successful formation of the nanoflowers. Sodium (Na) and chloride (Cl) peaks appeared due to the utilized preparation buffer. Moreover, the confirmation of the presence of iron nanoparticles (Fe) in the hybrid nanoflower structures was also observed—the atomic percentages of iron were 4.34 and 5.42% in the cases of $GO/Fe_2O_3-Cu_3(PO_4)_2$ and $GO/Fe_2O_3-Mn_3(PO_4)_2$ HNFs, respectively (Table S1).

X-ray diffraction (XRD) was used to characterize the unmodified and the nanomaterials-modified CaLB-HNFs, and the XRD patterns are provided in the Supplementary material (Figure S3). For the copper-based CaLB-HNFs, the XRD patterns represented peaks for the $Cu_3(PO_4)_2 \cdot 3H_2O$ (JCPDS 00-022-0548) phase, while for the manganese-based CaLB-HNFs, the phase of manganese changed from $Mn_3(PO_4)_2$ for the unmodified nanoflowers to $Mn_2P_2O_7$ for the nanomaterials-modified HNFs [34].

The presence of the carbon-based nanostructures in the CaLB-HNFs was confirmed with Raman spectroscopy. The Raman spectra of the unmodified $Cu_3(PO_4)_2$ CaLB-HNFs and $Mn_3(PO_4)_2$ CaLB-HNFs, as well as the modified HNFs with GO, CNTs, and γ-Fe_2O_3, are presented in Figure 2. The spectrum of the unmodified $Cu_3(PO_4)_2$ CaLB-HNFs presented several vibrational modes; the most pronounced were located at 645 cm^{-1}, 927 cm^{-1}, and 1147 cm^{-1} and can be attributed to the antisymmetric bending of the PO_4^{3-} ion, the symmetric stretching vibrations of PO_4^{3-} ion, and the antisymmetric stretching vibrations of the PO_4^{3-} ion, respectively [35]. The unmodified $Mn_3(PO_4)_2$ CaLB-HNFs presented a strong vibrational mode at 958 cm^{-1} that can be ascribed to the symmetric stretching mode of the PO_4^{3-} ion [36].

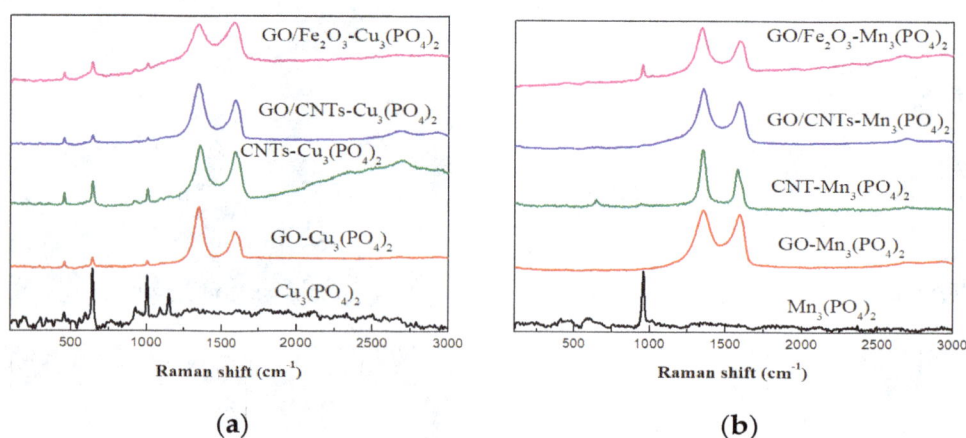

Figure 2. Raman spectra of: (**a**) $Cu_3(PO_4)_2$-based CaLB-HNFs and (**b**) $Mn_3(PO_4)_2$-based CaLB-HNFs.

The preparation of $Cu_3(PO_4)_2$- and $Mn_3(PO_4)_2$-based CaLB-HNFs with GO in both led to the appearance of the characteristic of carbon-based materials vibrational modes, D and G, located at circa 1346 cm^{-1} and 1590 cm^{-1}, respectively [37], while a weak asymmetric 2D vibrational mode was also present in both cases at around 2685 cm^{-1} [38]. Similarly, the preparation of both ion-based CaLB-HNFs with CNTs led to the appearance of the characteristic D and G vibrational modes located at circa 1350 cm^{-1} and 1587 cm^{-1}, respectively, while the 2D vibrational mode of the CNTs was located at circa 2700 cm^{-1}. The more intense D and G vibrational modes were also present when both GO and CNTs were added in the CaLB-HNFs, establishing the successful incorporation of the carbonaceous nanomaterials into the HNFs.

To confirm the successful immobilization of CaLB in the 3D nanostructures, all nanobiocatalytic systems were characterized by FTIR spectroscopy by recording the spectra in the range 400 cm^{-1} to 4000 cm^{-1}. As seen in Figure 3, peaks at the region 950 to 1060 cm^{-1} were associated with the asymmetric stretching vibrations of PO_4^{3-}, while peaks at the region 550 cm^{-1} to 650 cm^{-1} arose from the bending vibrations of bridging phosphate groups, such as O-P-O [39,40]. The presence of CaLB in the nanoflower structure was confirmed by the peak at 1648 cm^{-1}, which arises from the stretching vibrations of C = O of the peptide chain of the enzyme and corresponds to the Amide I band [41,42].

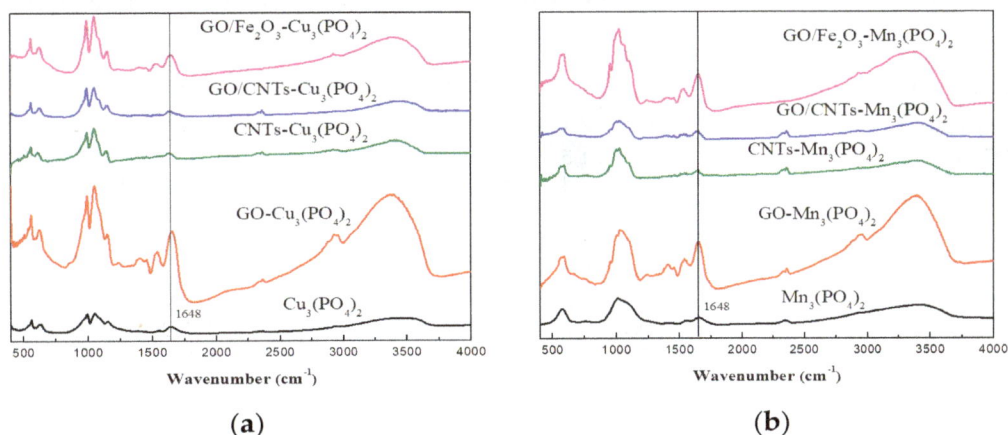

Figure 3. FTIR spectra of: **(a)** Cu$_3$(PO$_4$)$_2$-based CaLB-HNFs and **(b)** Mn$_3$(PO$_4$)$_2$-based CaLB-HNFs.

To better assess the dissimilarities among the spectra of the CaLB-HNFs, we compared the correlation coefficients (r) in the Amide I region (1600–1700 cm^{-1}), according to previously published work [43–45]. As seen from Table 1, for most of the CaLB-HNFs, r was close to 1.0, indicating that CaLB was able to maintain its native secondary structure upon encapsulation in the nanoflower structure. GO has formed a cage-like structure in which lipase was encapsulated, preserving the secondary structure of the enzyme [27]. In contrast, when CaLB-HNFs were prepared with CNTs, especially when combined with GO, the changes in the r value were more pronounced. This result could be attributed to conformational changes occurred during the encapsulation of CaLB in the nanoflower structure. The disorder of the natural conformation of CaLB, may arise from the over-crowded enzyme molecules within the strongly packed GO/CNTs HNF structure.

Table 1. Correlation coefficient (r) between the FTIR spectra of CaLB-HNFs.

Nanoflower	r	Nanoflower	r
Cu$_3$(PO$_4$)$_2$	0.976	Mn$_3$(PO$_4$)$_2$	0.982
GO-Cu$_3$(PO$_4$)$_2$	0.991	GO-Mn$_3$(PO$_4$)$_2$	0.998
CNTs-Cu$_3$(PO$_4$)$_2$	0.899	CNTs-Mn$_3$(PO$_4$)$_2$	0.837
GO/CNTs-Cu$_3$(PO$_4$)$_2$	0.879	GO/CNTs-Mn$_3$(PO$_4$)$_2$	0.801
GO/Fe$_2$O$_3$-Cu$_3$(PO$_4$)$_2$	0.991	GO/Fe$_2$O$_3$-Mn$_3$(PO$_4$)$_2$	0.998

3.2. Biocatalytic Characterization of CaLB Nanoflowers

The encapsulation yield and specific hydrolytic activity of all CaLB-HNFs are presented in Table 2. The protein loading for unmodified Cu$_3$(PO$_4$)$_2$ and Mn$_3$(PO$_4$)$_2$ CaLB-HNFs were 57.6% and 49.0%, respectively, while their specific activity was calculated at 8.3 and 96.7 U g^{-1}, respectively, pointing out that the kind of the metal ion significantly affects the hydrolyzing ability of the immobilized lipase. It has been previously proposed that enzymes provide different binding sites for metal ions, and, as such, nucleation sites are formed in different enzyme regions, affecting the 3D structure and activity of the immobilized biocatalysts [15]. Moreover, in the case of Cu$_3$(PO$_4$)$_2$ CaLB-HNFs, lipase could have been embedded deep inside the flower-like structure, preventing the active sites of the CaLB from interacting with the substrate and thus leading to low catalytic activity, due to steric hindrance phenomena [28,46].

Table 2. Encapsulation yield and specific hydrolytic activity of various CaLB-HNFs.

Nanoflower	Encapsulation Yield (%)	Specific Activity (U g^{-1} Immobilized CaLB)
$Cu_3(PO_4)_2$	57.6 ± 3.1	13.1 ± 0.5
GO-$Cu_3(PO_4)_2$	70.5 ± 1.7	174.4 ± 0.7
CNTs-$Cu_3(PO_4)_2$	57.5 ± 2.1	189.0 ± 3.9
GO/CNTs-$Cu_3(PO_4)_2$	61.6 ± 1.5	167.0 ± 1.7
GO/Fe_2O_3-$Cu_3(PO_4)_2$	59.0 ± 2.4	197.1 ± 2.5
$Mn_3(PO_4)_2$	49.0 ± 1.7	161.2 ± 2.6
GO-$Mn_3(PO_4)_2$	67.1 ± 3.6	284.7 ± 5.2
CNTs-$Mn_3(PO_4)_2$	57.6 ± 1.2	175.6 ± 4.0
GO/CNTs-$Mn_3(PO_4)_2$	65.9 ± 2.5	168.7 ± 1.0
GO/Fe_2O_3-$Mn_3(PO_4)_2$	60.9 ± 2.7	175.9 ± 1.9

In order to provide more binding sites for the formation of CaLB-HNFs, different carbon-based and magnetic nanomaterials were added to the hybrid nanostructures during the preparation procedure. All HNFs enriched with carbon or magnetic nanomaterials exhibited higher encapsulation yields than those without nanomaterials, regardless of the metal ion type. The highest encapsulation yields were observed when GO was used as an additive. For instance, the encapsulation efficiency reached up to 70.5 and 67.1% in the case of GO-$Cu_3(PO_4)_2$ and GO-$Mn_3(PO_4)_2$ CaLB-NHFs, respectively. Similar results have also been reported by Li and co-workers when GO was added in the formation of laccase-based nanoflowers [27]. CNTs also seem to affect the immobilization efficiency of CaLB, as is consistent with previous work [28]. The large surface area of GO and CNTs seems to increase the available binding sites and thus promote enzyme adsorption, in addition to stabilizing the 3D structure of the nanoflower. Moreover, the presence of oxygen-containing groups in the surface of these nanomaterials may result in the formation of electrostatic interactions between those functional groups and the copper cations, stabilizing the nucleation step.

The modification of CaLB-HNFs with carbon or magnetic nanomaterials enhanced the specific hydrolytic activity of the immobilized enzyme. In the case of manganese-based nanoflowers, the specific activity of the enzyme was increased up to around two-times in the presence of nanomaterials. The beneficial effect of the use of nanomaterials was more pronounced in the case of copper-based nanoflowers. More specifically, all the nanomaterials significantly outperformed in terms of activity the unmodified $Cu_3(PO_4)_2$ nanoflowers. GO sheets, CNTs, and γ-Fe_2O_3 nanoparticles, due to randomly distributed oxygen-containing groups on their surface, interact with positively charged metals and amino groups on the enzyme's surface, leading to more stable and active flower-like structures [28]. Such interactions could lead to a more active conformation [47,48]. Compared to each individual nanomaterial, the GO/CNTs hybrid system was not as beneficial as expected, maybe due to the uniform dispersion of lipase within the nanoflower structure or stereochemical hindrance. Moreover, CaLB immobilized on GO/CNTs nanoflowers presented the highest conformational changes (as previously discussed, Table 1), which could result in lower catalytic activity. It is important to mention that the preparation of HNFs containing both carbon nanomaterials and magnetic nanoparticles has not been previously reported. GO/Fe_2O_3-based HNFs reached high encapsulation yields, while GO/Fe_2O_3-$Cu_3(PO_4)_2$ CaLB-HNFs exhibited one of the highest catalytic activities among all nanoflowers.

The thermal stability of the CaLB-HNFs was also investigated. The remaining hydrolytic activity was estimated after incubation of nanoflowers for up to 24 h in phosphate buffer at 60 °C, and is presented in Figure 4.

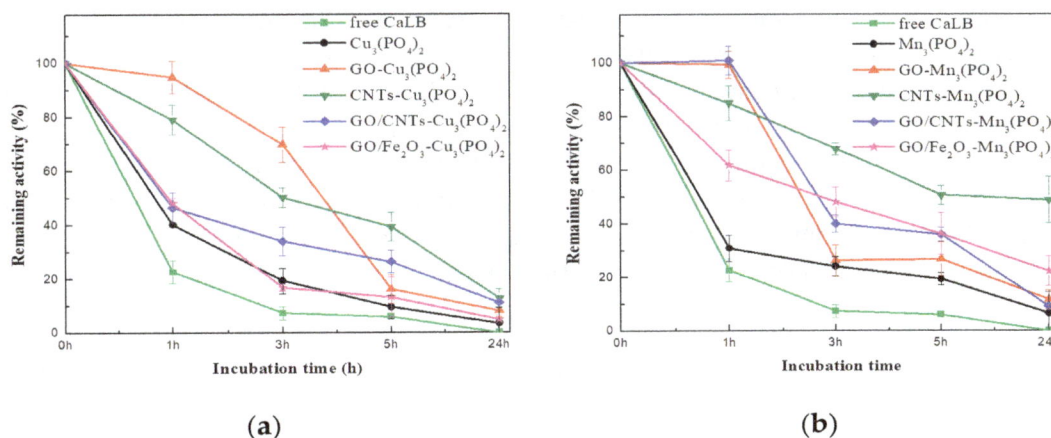

(a)   (b)

Figure 4. Thermal stability of: (**a**) $Cu_3(PO_4)_2$-based CaLB-HNFs and (**b**) $Mn_3(PO_4)_2$-based CaLB-HNFs at 60 °C. The 100% percentage corresponds to the activity at t = 0 min.

As seen in Figure 4, the use of hybrid nanoflowers as supports for the immobilization of CaLB increased the thermal stability of the immobilized enzyme. Specifically, the catalytic activity of free CaLB decreased to 20% after the first hour of incubation, while unmodified CaLB-$Cu_3(PO4)_2$ and $Mn_3(PO_4)_2$ CaLB-HNFs retained up to 40% and 31% of their initial activity, respectively. Moreover, after 5 h of incubation, free CaLB was totally inactivated, while the immobilized lipase on unmodified HNFs retained up to 19.2% of their activity, indicating that the nanoflower 3D structure can protect the active conformation of the enzyme, thus enhancing its stability [49]. Similar observations have also been reported for lipase from the porcine pancreas [50]. The thermal stability of the immobilized CaLB was further improved when HNFs containing carbon and γ-Fe_2O_3 nanomaterials were used as immobilization supports. This observation could be attributed to the protective effect these nanomaterials offer on the stability of protein molecules [51,52]. Amongst the nanomaterials, CNTs stabilized the immobilized enzyme the most for both copper- and manganese-based nanoflowers (12.7% and 49% of enzyme activity, respectively, was retained after 24 h of incubation). Their high surface area, as well as the fact that CNTs are distributed within the petals of the flower-like structure, enables lipase to maintain its stability [27,28]. The conformational changes previously described (Table 1) may lead to a more rigid folding of lipase and thus enhance its stability [51]. Furthermore, in comparing the two inorganic components, it is clear that manganese HNFs exhibited higher remaining activity than copper HNFs, underlining the correlation of the different interactions developed between nanomaterials and each metal ion.

One of the major drawbacks of using soluble enzymes in large-scale reactions is reusability, due to their incapability of maintaining their stability under harsh conditions, and the difficulty of removal from the reaction system, due to their high solubility. Therefore, the immobilization of enzymes enhances their stability and enables their separation and use in successive cycles, making them an asset for industrial applications. In the present study, the operational stability of the CaLB-HNFs was investigated for the hydrolysis of p-NPB, and the results are presented in Figure 5. As seen, unmodified $Cu_3(PO_4)_2$ and $Mn_3(PO_4)_2$ CaLB-HNFs were almost deactivated after the fifth biocatalytic cycle. It is possible that the disruption of the non-covalent bonds between the organic and inorganic parts of the nanoflowers in the aqueous environment accelerated the enzyme leaching or gradual degradation of the flower-like morphology, leading to low enzymatic activity, which is in agreement with that recently reported [40].

In the case of nanomaterials-modified HNFs, the operational stability of CaLB was notably increased. Immobilized CaLB on nanomaterials-based HNFs could be efficiently used for nine consecutive cycles for the hydrolysis of p-NPB. The residual activity of nanomaterials-modified $Cu_3(PO4)_2$ and $Mn_3(PO4)_2$ CaLB-HNFs retained up to 83% even after nine catalytic cycles. These results infer that the presence of nanomaterials in the nanoflowers protects the enzyme configuration, thus enhancing its stability for successive hydrolysis cycles. Similar to the thermal stability studies presented above, CNTs-modified CaLB-HNFs offered the most beneficial impact on the operational stability of CaLB, indicating that the incorporation of CNTs inside the nanoflower structure enables the adoption of a more rigid conformation of CaLB, stabilizing it against repeatable usage [53]. It is interesting to note that, although manganese-based HNFs presented higher thermal stability than copper-based HNFs (as previously discussed), their operational stability was lower compared to copper-based HNFs. It is possible that the enzyme leaching from the manganese-based HNFs is higher compared to copper-based HNFs, resulting in a higher loss of the residual activity of the enzyme.

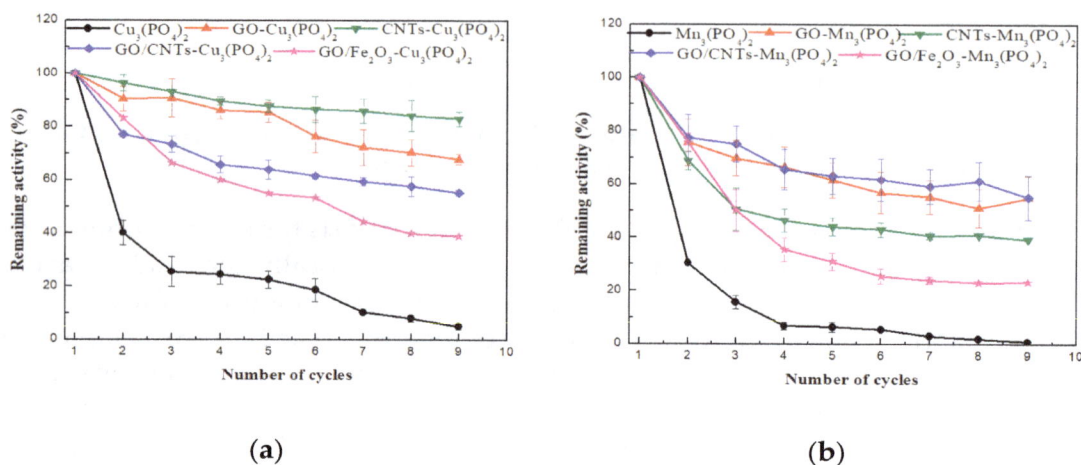

(a)　　　　　　　　　　　　　　　　　　　　(b)

Figure 5. Operational stability of: (a) $Cu_3(PO_4)_2$-based CaLB-HNFs and (b) $Mn_3(PO_4)_2$-based CaLB-HNFs. The 100% percentage corresponds to the lipase hydrolytic activity at the first catalytic cycle.

3.3. Transesterification of Tyrosol by CaLB Nanoflowers in Non-Aqueous Media

The prepared nanomaterials-modified CaLB-HNFs were used for the synthesis of tyrosol esters in non-aqueous media. Tyrosol is a natural phenolic antioxidant derived from various plants, such as olive and green tea. This abundant product has been associated with many health-related benefits as well as plenty of industrial applications [54,55]. An increase of tyrosol lipophilicity is suggested to improve its antioxidant activity [56,57]. Thus, the enzymatic lipophilization of tyrosol may be of great interest. For this reason, GO/Fe_2O_3-$Cu_3(PO_4)_2$ and GO/Fe_2O_3-$Mn_3(PO_4)_2$ CaLB-HNFs were used as biocatalysts for the transesterification of tyrosol with vinyl butyrate (Figure S4) in a variety of organic solvents, as well as in environmentally friendly ionic and deep eutectic solvents; the results are presented in Table 3. GO/CNTs-based HNFs were also used for the transesterification of tyrosol; the results are presented in Table S2.

Table 3. Conversion yields for the enzymatic transesterification of tyrosol with vinyl butyrate in non-aqueous media catalyzed by GO/Fe_2O_3-$Cu_3(PO_4)_2$ and GO/Fe_2O_3-$Mn_3(PO_4)_2$ CaLB HNFs.

Reaction Medium	Conversion Yield (%)	
	GO/Fe_2O_3-$Cu_3(PO_4)_2$ CaLB-HNFs	GO/Fe_2O_3-$Mn_3(PO_4)_2$ CaLB-HNFs
n-Hexane	99.6 ± 0.4	100.0 ± 0.3
Acetonitrile	80.3 ± 0.3	80.7 ± 0.8
2-Methyl-2-butanol	30.2 ± 1.1	52.6 ± 1.2
tert-Butyl-methylether	98.9 ± 0.5	99.7 ± 0.6
tert-Butanol	23.2 ± 0.2	22.5 ± 0.4
[BMIM][PF$_6$]	13.6 ± 1.6	20.0 ± 4.7
ChCl:U	33.2 ± 2.8	26.7 ± 4.6

As seen in Table 3, both CaLB-HNFs were able to catalyze the transesterification of tyrosol, achieving high conversion yields in most of the non-aqueous solvents. It has been recently proposed that the hydrophobic surface of the hybrid nanoflowers benefits synthetic reactions in non-aqueous solvents by promoting the oriented delivery of the substrates near the hydrophobic surface of nanoflowers and, thus, to the active sites of the enzyme [49]. Conversion yields of transesterification seem to strongly depend on the nature of the organic solvent, namely its polarity and viscosity. More specifically, the nanoflower-catalyzed reactions in non-polar solvents, e.g., n-hexane and *tert*-butyl-methylether, exhibited high conversion yields up to 100%. Moreover, the reaction rate of the transesterification reaction catalyzed by GO/Fe$_2$O$_3$–Mn$_3$(PO$_4$)$_2$ CaLB-HNFs in hexane and *tert*-butyl-methylether was up to 73-fold higher in comparison with that in other media (Table S3). Solvents with low polarity enable enzymes to preserve the essential water molecules bound on their surface in order to maintain their natural conformation and be fully functional [58,59]. On the other hand, the use of more hydrophilic solvents with higher affinity to interact with water molecules [60], such as 2-methyl-2-butanol and *tert*-butanol, led to a decrease of the conversion yield of the transesterification reaction.

Both HNFs were able to catalyze the transesterification of tyrosol in eco-friendly alternatives of organic solvents, such as ionic liquids and deep eutectic solvents ([BMIM][BF$_6$] and ChCl:U, respectively). Those green solvents have been widely employed for enzymatic biotransformations, as they present high chemical and thermal stability, low vapour pressure, low toxicity, and the ability to enhance the catalytic performance of the enzymes [17,55,61]. As seen in Table 3, the conversion yield is decreased in the ionic liquid compared to organic solvents, which could be attributed to the low dispersability of the nanoflowers in these media. Moreover, the high viscosity of [BMIM][BF$_6$] (381 cP at 25 °C)[62] and ChCl:U (1200 mPa s at 25 °C)[63] could lead to mass-transfer limitations, restricting the biocatalytic activity of immobilized CaLB [64,65].

The use of magnetic nanobiocatalysts could facilitate the separation from the reaction solution through the application of an external magnetic field and, thus, the reuse of the biocatalyst [52]. Considering this aspect, magnetic CaLB-HNFs were applied in consecutive reaction cycles for tyrosol transesterification. Figure 6 presents the remaining catalytic activity of the GO/Fe$_2$O$_3$-Mn$_3$(PO$_4$)$_2$ CaLB-HNFs for successive catalytic cycles. As seen, these magnetic nanoflowers presented excellent operational stability after eight consecutive reaction cycles (576 h of total operation) without any loss of biocatalytic activity, making this hybrid nanobiocatalyst one of the most robust nanobiocatalysts reported until now for similar reaction processes. This enhanced operational stability, in comparison with the one described above for the hydrolysis of *p*-NPB (Figure 5), could be attributed to the fact that non-polar organic solvents do not remove protein-bound water that is crucial for maintaining protein structure and function, leading to a more rigid conformation of the immobilized biocatalyst [66,67].

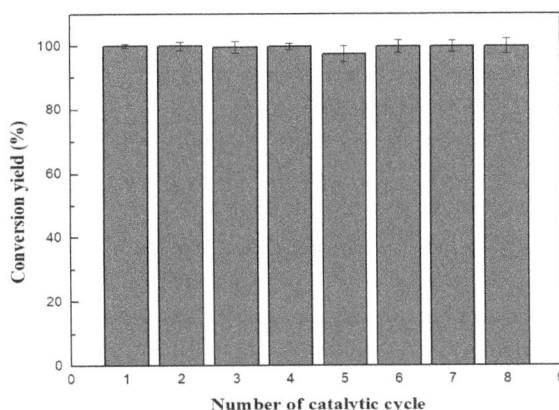

Figure 6. Operational stability of GO/Fe$_2$O$_3$-Mn$_3$(PO$_4$)$_2$ CaLB-HNFs after eight reaction cycles for the enzymatic transesterification of tyrosol with vinyl butyrate in *tert*-butyl-methylether. Each reaction was carried out for 72 h at 50 °C.

4. Conclusions

Herein, we report the preparation and characterization of novel hybrid nanoflowers comprised of copper (II) or manganese (II) ions combined with magnetic nanoparticles and carbon-based nanomaterials. These nanoflowers can be effectively used as versatile host platforms for the immobilization of an industrially relevant enzyme (CaLB) through biomimetic mineralization. The metal ion and the nature of the nanomaterials affect the structural and catalytic characteristics of the immobilized lipase in different manners. The nanomaterials-modified hybrid nanoflowers presented an excellent catalytic performance in the production of tyrosol esters in different organic solvents and environmental-friendly ionic solvents. Furthermore, CaLB-magnetic HNFs (combining GO and maghemite nanoparticles) exhibited remarkable operational stability for the tyrosol transesterification reaction, as the nanobiocatalyst retained almost its entire catalytic activity even after eight successive reaction cycles, indicating that these bio-nanoconjugates could potentially be used as efficient tools for heterogeneous biocatalytic transformations in large-scale applications.

Supplementary Materials:
Figure S1: SEM images of: (A1, A2) $Cu_3(PO_4)_2$ CaLB-HNFs; (B1-B3) GO-$Cu_3(PO_4)_2$ CaLB-HNFs; (C1-C3) CNTs-$Cu_3(PO_4)_2$ CaLB-HNFs; (D1-D3) GO/CNTs-$Cu_3(PO_4)_2$ CaLB-HNFs; (E1, E2) GO/Fe_2O_3-$Cu_3(PO_4)_2$ CaLB-HNFs; (F1-F3) $Mn_3(PO_4)_2$ CaLB-HNFs; (G1-G3) GO-$Mn_3(PO_4)_2$ CaLB-HNFs; (H1,H2) CNTs-$Mn_3(PO_4)_2$ CaLB-HNFs; (I1-I3) GO/CNTs-$Mn_3(PO_4)_2$ CaLB-HNFs; (J1, J2) GO/Fe_2O_3-$Mn_3(PO_4)_2$ CaLB-HNFs, Figure S2: EDS spectra of: (a) GO/Fe_2O_3-$Cu_3(PO_4)_2$ CaLB-HNFs; (b) GO/Fe_2O_3-$Mn_3(PO_4)_2$ CaLB-HNFs, Table S1: Elemental analysis of GO/Fe_2O_3-based CaLB-HNFs by EDS, Figure S3: XRD patterns of: (a) $Cu_3(PO_4)_2$-based CaLB-HNFs; (b) $Mn_3(PO_4)_2$-CaLB-HNFs, Figure S4: Transesterification of tyrosol with vinyl butyrate catalyzed by CaLB, Table S2: Conversion yields for the enzymatic transesterification of tyrosol with vinyl butyrate in non aqueous media, by GO/CNTs-$Cu_3(PO_4)_2$ CaLB-HNFs, Table S3: Reaction rates (mM h^{-1}) of tyrosol transesterification catalyzed by GO/Fe_2O_3-$Mn_3(PO_4)_2$ CaLB-HNFs in non-aqueous media.

Author Contributions: H.S. contributed to the overall design of the experiments, results interpretation, and manuscript writing. D.G., A.A., A.P., and E.P.G. contributed to results interpretation and reviewed the manuscript. R.F. performed the synthesis and experiments with nanoflowers, results interpretation, and manuscript writing. M.P. contributed to results interpretation and manuscript writing. M.A.H., A.E., K.S., and D.M. performed the synthesis of nanomaterials. S.E.M., E.D.S., and X.R.D. contributed to the characterization of nanoflowers and the results analysis. K.T. performed the Raman characterization of nanoflowers, contributed to results analysis, and reviewed the manuscript.

Acknowledgments: This work made use of the Cornell Center for Materials Research Shared Facilities which are supported through the NSF MRSEC program (DMR-1719875).

References

1.   Ansari, S.A.; Husain, Q. Potential applications of enzymes immobilized on/in nano materials: A review. *Biotechnol. Adv.* **2012**, *30*, 512–523. [CrossRef] [PubMed]

2.   Cipolatti, E.P.; Valério, A.; Henriques, R.O.; Moritz, D.E.; Ninow, J.L.; Freire, D.M.G.; Manoel, E.A.; Fernandez-Lafuente, R.; De Oliveira, D. Nanomaterials for biocatalyst immobilization – state of the art and future trends. *RSC Adv.* **2016**, *6*, 104675–104692. [CrossRef]

3.   Di Credico, B.; Redaelli, M.; Bellardita, M.; Calamante, M.; Cepek, C.; Cobani, E.; D'Arienzo, M.; Evangelisti, C.; Marelli, M.; Moret, M.; et al. Step-by-Step Growth of HKUST-1 on Functionalized TiO_2 Surface: An Efficient Material for CO_2 Capture and Solar Photoreduction. *Catalysts* **2018**, *8*, 353. [CrossRef]

4.   Di Credico, B.; Cobani, E.; Callone, E.; Conzatti, L.; Cristofori, D.; D'Arienzo, M.; Dirè, S.; Giannini, L.; Hanel, T.; Scotti, R.; et al. Size-controlled self-assembly of anisotropic sepiolite fibers in rubber nanocomposites. *Appl. Clay Sci.* **2018**, *152*, 51–64. [CrossRef]

5.   Pavlidis, I.V.; Patila, M.; Bornscheuer, U.T.; Gournis, D.; Stamatis, H. Graphene-based nanobiocatalytic systems: recent advances and future prospects. *Trends Biotechnol.* **2014**, *32*, 312–320. [CrossRef] [PubMed]

6. Shende, P.; Kasture, P.; Gaud, R. Nanoflowers: the future trend of nanotechnology for multi-applications. *Artif. Cells Nanomed. Biotechnol.* **2018**, *46*, 413–422. [CrossRef]

7. Cui, J.; Jia, S. Organic–inorganic hybrid nanoflowers: A novel host platform for immobilizing biomolecules. *Co-ord. Chem. Rev.* **2017**, *352*, 249–263. [CrossRef]

8. Ge, J.; Lei, J.; Zare, R.N. Protein–inorganic hybrid nanoflowers. *Nat. Nanotechnol.* **2012**, *7*, 428–432. [CrossRef]

9. Batule, B.S.; Park, K.S.; Gautam, S.; Cheon, H.J.; Kim, M.I.; Park, H.G. Intrinsic peroxidase-like activity of sonochemically synthesized protein copper nanoflowers and its application for the sensitive detection of glucose. *Sens. Actuators B Chem.* **2018**, *283*, 749–754. [CrossRef]

10. Chen, X.; Xu, L.; Wang, A.; Li, H.; Wang, C.; Pei, X.; Zhang, P.; Wu, S.G. Efficient synthesis of the key chiral alcohol intermediate of Crizotinib using dual-enzyme@CaHPO4 hybrid nanoflowers assembled by mimetic biomineralization. *J. Chem. Technol. Biotechnol.* **2019**, *94*, 236–243. [CrossRef]

11. Altinkaynak, C.; Tavlasoglu, S.; Kalin, R.; Sadeghian, N.; Ozdemir, H.; Ocsoy, I.; Özdemir, N. A hierarchical assembly of flower-like hybrid Turkish black radish peroxidase-Cu^{2+} nanobiocatalyst and its effective use in dye decolorization. *Chemosphere* **2017**, *182*, 122–128. [CrossRef]

12. Li, P.; Zhang, B.; Fan, L.; Wang, H.; Tian, L.; Ali, N. Papain/Zn$_3$(PO$_4$)$_2$ hybrid nanoflower: preparation, characterization and its enhanced catalytic activity as an immobilized enzyme. *RSC Adv.* **2016**, *6*, 46702–46710.

13. López-Gallego, F.; Yate, L. Selective biomineralization of Co$_3$ (PO$_4$)$_2$-sponges triggered by His-tagged proteins: efficient heterogeneous biocatalysts for redox processes. *Chem. Commun.* **2015**, *51*, 8753–8756. [CrossRef] [PubMed]

14. Ocsoy, I.; Dogru, E.; Usta, S. A new generation of flowerlike horseradish peroxides as a nanobiocatalyst for superior enzymatic activity. *Enzym. Microb. Technol.* **2015**, *75*, 25–29. [CrossRef]

15. Escobar, S.; Velasco-Lozano, S.; Lu, C.-H.; Lin, Y.-F.; Mesa, M.; Bernal, C.; López-Gallego, F. Understanding the functional properties of bio-inorganic nanoflowers as biocatalysts by deciphering the metal-binding sites of enzymes. *J. Mater. Chem. B* **2017**, *5*, 4478–4486. [CrossRef]

16. Patel, S.K.; Otari, S.V.; Li, J.; Kim, D.R.; Kim, S.C.; Cho, B.-K.; Kalia, V.C.; Kang, Y.C.; Lee, J.-K. Synthesis of cross-linked protein-metal hybrid nanoflowers and its application in repeated batch decolorization of synthetic dyes. *J. Hazard. Mater.* **2018**, *347*, 442–450. [CrossRef] [PubMed]

17. Papadopoulou, A.A.; Tzani, A.; Polydera, A.C.; Katapodis, P.; Voutsas, E.; Detsi, A.; Stamatis, H. Green biotransformations catalysed by enzyme-inorganic hybrid nanoflowers in environmentally friendly ionic solvents. *Environ. Sci. Pollut. Res.* **2018**, *25*, 26707–26714. [CrossRef]

18. Zhu, L.; Gong, L.; Zhang, Y.; Wang, R.; Ge, J.; Liu, Z.; Zare, R.N. Rapid Detection of Phenol Using a Membrane Containing Laccase Nanoflowers. *Chem. Asian J.* **2013**, *8*, 2358–2360. [CrossRef] [PubMed]

19. Fu, M.; Xing, J.; Ge, Z. Preparation of laccase-loaded magnetic nanoflowers and their recycling for efficient degradation of bisphenol A. *Sci. Total Environ.* **2019**, *651*, 2857–2865. [CrossRef] [PubMed]

20. Bao, J.; Huang, T.; Wang, Z.; Yang, H.; Geng, X.; Xu, G.; Samalo, M.; Sakinati, M.; Huo, D.; Hou, C. 3D graphene/copper oxide nano-flowers based acetylcholinesterase biosensor for sensitive detection of organophosphate pesticides. *Sens. Actuators B Chem.* **2019**, *279*, 95–101. [CrossRef]

21. Zhu, X.; Huang, J.; Liu, J.; Zhang, H.; Jiang, J.; Yu, R. A dual enzyme–inorganic hybrid nanoflower incorporated microfluidic paper-based analytic device (µPAD) biosensor for sensitive visualized detection of glucose. *Nanoscale* **2017**, *9*, 5658–5663. [CrossRef]

22. Sharma, N.; Parhizkar, M.; Cong, W.; Mateti, S.; Kirkland, M.A.; Puri, M.; Sutti, A. Metal ion type significantly affects the morphology but not the activity of lipase–metal–phosphate nanoflowers. *RSC Adv.* **2017**, *7*, 25437–25443. [CrossRef]

23. Zeng, J.; Xia, Y. Not just a pretty flower. *Nat. Nanotechnol.* **2012**, *7*, 415–416. [CrossRef] [PubMed]

24. Lin, Z.; Xiao, Y.; Wang, L.; Yin, Y.; Zheng, J.; Yang, H.; Chen, G. Facile synthesis of enzyme–inorganic hybrid nanoflowers and their application as an immobilized trypsin reactor for highly efficient protein digestion. *RSC Adv.* **2014**, *4*, 13888–13891. [CrossRef]

25. Cui, J.; Zhao, Y.; Liu, R.; Zhong, C.; Jia, S. Surfactant-activated lipase hybrid nanoflowers with enhanced enzymatic performance. *Sci. Rep.* **2016**, *6*, 27928. [CrossRef]

26. Wang, X.; Shi, J.; Li, Z.; Zhang, S.; Wu, H.; Jiang, Z.; Yang, C.; Tian, C. Facile One-Pot Preparation of Chitosan/Calcium Pyrophosphate Hybrid Microflowers. *ACS Appl. Mater. Interfaces* **2014**, *6*, 14522–14532. [CrossRef]

27. Li, H.; Hou, J.; Duan, L.; Ji, C.; Zhang, Y.; Chen, V. Graphene oxide-enzyme hybrid nanoflowers for efficient water soluble dye removal. *J. Hazard. Mater.* **2017**, *338*, 93–101. [CrossRef] [PubMed]

28. Li, K.; Wang, J.; He, Y.; Abdulrazaq, M.A.; Yan, Y. Carbon nanotube-lipase hybrid nanoflowers with enhanced enzyme activity and enantioselectivity. *J. Biotechnol.* **2018**, *281*, 87–98. [CrossRef]

29. Stergiou, D.V.; Diamanti, E.K.; Gournis, D.; Prodromidis, M.I.; Prodromidis, M. (Mamas) Comparative study of different types of graphenes as electrocatalysts for ascorbic acid. *Electrochem. Commun.* **2010**, *12*, 1307–1309. [CrossRef]

30. Tsoufis, T.; Tomou, A.; Gournis, D.; Douvalis, A.P.; Panagiotopoulos, I.; Kooi, B.; Georgakilas, V.; Arfaoui, I.; Bakas, T. Novel Nanohybrids Derived from the Attachment of FePt Nanoparticles on Carbon Nanotubes. *J. Nanosci. Nanotechnol.* **2008**, *8*, 5942–5951. [CrossRef]

31. Tzitzios, V.K.; Bakandritsos, A.; Georgakilas, V.; Basina, G.; Boukos, N.; Bourlinos, A.B.; Niarchos, D.; Petridis, D. Large-Scale Synthesis, Size Control, and Anisotropic Growth of &gamma-Fe2O3 Nanoparticles: Organosols and Hydrosols. *J. Nanosci. Nanotechnol.* **2007**, *7*, 2753–2757. [PubMed]

32. Prestrelski, S.; Tedeschi, N.; Arakawa, T.; Carpenter, J. Dehydration-induced conformational transitions in proteins and their inhibition by stabilizers. *Biophys. J.* **1993**, *65*, 661–671. [CrossRef]

33. Bradford, M.M. A rapid and sensitive method for the quantitation of microgram quantities of protein utilizing the principle of protein-dye binding. *Anal. Biochem.* **1976**, *72*, 248–254. [CrossRef]

34. Huang, Y.; Fang, J.; Omenya, F.; O'Shea, M.; Chernova, N.A.; Zhang, R.; Wang, Q.; Quackenbush, N.F.; Piper, L.F.J.; Scanlon, D.; et al. Understanding the stability of $MnPO_4$. *J. Mater. Chem.* A **2014**, *2*, 12827. [CrossRef]

35. Kharbish, S.; Andráš, P.; Luptáková, J.; Milovská, S. Raman spectra of oriented and non-oriented Cu hydroxy-phosphate minerals: Libethenite, cornetite, pseudomalachite, reichenbachite and ludjibaite. *Spectrochim. Acta - Part A Mol. Biomol. Spectrosc.* **2014**, *130*, 152–163. [CrossRef] [PubMed]

36. Sahana, M.B.; Vasu, S.; Sasikala, N.; Anandan, S.; Sepehri-Amin, H.; Sudakar, C.; Gopalan, R. Raman spectral signature of Mn-rich nanoscale phase segregations in carbon free LiFe1-xMnxPO4 prepared by hydrothermal technique. *RSC Adv.* **2014**, *4*, 64429–64437. [CrossRef]

37. Tsirka, K.; Katsiki, A.; Chalmpes, N.; Gournis, D.; Paipetis, A.S. Mapping of Graphene Oxide and Single Layer Graphene Flakes—Defects Annealing and Healing. *Front. Mater.* **2018**, *5*, 1–11. [CrossRef]

38. Kaniyoor, A.; Ramaprabhu, S. A Raman spectroscopic investigation of graphite oxide derived graphene. *AIP Adv.* **2012**, *2*, 32183. [CrossRef]

39. Rong, J.; Zhang, T.; Qiu, F.; Zhu, Y. Preparation of Efficient, Stable, and Reusable Laccase–$Cu_3(PO_4)_2$ Hybrid Microspheres Based on Copper Foil for Decoloration of Congo Red. *ACS Sustain. Chem. Eng.* **2017**, *5*, 4468–4477. [CrossRef]

40. Lee, H.R.; Chung, M.; Kim, M.I.; Ha, S.H. Preparation of glutaraldehyde-treated lipase-inorganic hybrid nanoflowers and their catalytic performance as immobilized enzymes. *Enzym. Microb. Technol.* **2017**, *105*, 24–29. [CrossRef]

41. Chatzikonstantinou, A.V.; Gkantzou, E.; Gournis, D.; Patila, M.; Stamatis, H. Stabilization of Laccase Through Immobilization on Functionalized GO-Derivatives. In *Enzyme Nanoarchitectures: Enzymes Armored with Graphene*, 1st ed.; Kumar, C.V., Ed.; Elsevier Inc.: Amsterdam, The Netherlands, 2018; Volume 609, pp. 47–81. ISBN 9780128152409.

42. Patila, M.; Diamanti, E.K.; Bergouni, D.; Polydera, A.C.; Gournis, D.; Stamatis, H. Preparation and biochemical characterisation of nanoconjugates of functionalized carbon nanotubes and cytochrome c. *Nanomed. Res. J.* **2018**, *3*, 10–18.

43. Tzialla, A.A.; Pavlidis, I.V.; Felicissimo, M.P.; Rudolf, P.; Gournis, D.; Stamatis, H. Lipase immobilization on smectite nanoclays: Characterization and application to the epoxidation of α-pinene. *Bioresour. Technol.* **2010**, *101*, 1587–1594. [CrossRef] [PubMed]

44. Secundo, F.; Carrea, G. Mono- and disaccharides enhance the activity and enantioselectivity ofBurkholderia cepacia lipase in organic solvent but do not significantly affect its conformation. *Biotechnol. Bioeng.* **2005**, *92*, 438–446. [CrossRef] [PubMed]

45. Secundo, F.; Barletta, G.L.; Dumitriu, E. Carre Can an Inactivating Agent Increase Enzyme Activity in Organic Solvent? Effects of 18-Crown-6 on Lipase Activity, Enantioselectivity, and Conformation. *Biotechnol. Bioeng.* **2007**, *97*, 12–18. [CrossRef] [PubMed]

46. Hao, M.; Fan, G.; Zhang, Y.; Xin, Y.; Zhang, L. Preparation and characterization of copper-Brevibacterium cholesterol oxidase hybrid nanoflowers. *Int. J. Boil. Macromol.* **2019**, *126*, 539–548. [CrossRef] [PubMed]

47. Zhang, H.; Fei, X.; Tian, J.; Li, Y.; Zhi, H.; Wang, K.; Xu, L.; Wang, Y. Synthesis and continuous catalytic application of alkaline protease nanoflowers–PVA composite hydrogel. *Catal. Commun.* **2018**, *116*, 5–9. [CrossRef]

48. Garcia-Galan, C.; Berenguer-Murcia, Á.; Fernandez-Lafuente, R.; Rodrigues, R.C. Potential of Different Enzyme Immobilization Strategies to Improve Enzyme Performance. *Adv. Synth. Catal.* **2011**, *353*, 2885–2904. [CrossRef]

49. Gao, J.; Kong, W.; Zhou, L.; He, Y.; Ma, L.; Wang, Y.; Yin, L.; Jiang, Y. Monodisperse core-shell magnetic organosilica nanoflowers with radial wrinkle for lipase immobilization. *Chem. Eng. J.* **2017**, *309*, 70–79. [CrossRef]

50. Jiang, W.; Wang, X.; Yang, J.; Han, H.; Li, Q.; Tang, J. Lipase-inorganic hybrid nanoflower constructed through biomimetic mineralization: A new support for biodiesel synthesis. *J. Colloid Interface Sci.* **2018**, *514*, 102–107. [CrossRef]

51. Patila, M.; Pavlidis, I.V.; Kouloumpis, A.; Dimos, K.; Spyrou, K.; Katapodis, P.; Gournis, D.; Stamatis, H. Graphene oxide derivatives with variable alkyl chain length and terminal functional groups as supports for stabilization of cytochrome c. *Int. J. Boil. Macromol.* **2016**, *84*, 227–235. [CrossRef]

52. Orfanakis, G.; Patila, M.; Catzikonstantinou, A.V.; Lyra, K.-M.; Kouloumpis, A.; Spyrou, K.; Katapodis, P.; Paipetis, A.; Rudolf, P.; Gournis, D.; et al. Hybrid Nanomaterials of Magnetic Iron Nanoparticles and Graphene Oxide as Matrices for the Immobilization of β-Glucosidase: Synthesis, Characterization, and Biocatalytic Properties. *Front. Mater.* **2018**, *5*, 1–11. [CrossRef]

53. Bilal, M.; Asgher, M.; Iqbal, M.; Hu, H.; Zhang, X. Chitosan beads immobilized manganese peroxidase catalytic potential for detoxification and decolorization of textile effluent. *Int. J. Boil. Macromol.* **2016**, *89*, 181–189. [CrossRef] [PubMed]

54. Rodríguez-Morató, J.; Boronat, A.; Kotronoulas, A.; Pujadas, M.; Pastor, A.; Olesti, E.; Pérez-Mañá, C.; Khymenets, O.; Fitó, M.; Farré, M.; et al. Metabolic disposition and biological significance of simple phenols of dietary origin: hydroxytyrosol and tyrosol. *Drug Metab. Rev.* **2016**, *48*, 1–19. [CrossRef]

55. Papadopoulou, A.A.; Katsoura, M.H.; Chatzikonstantinou, A.; Kyriakou, E.; Polydera, A.C.; Tzakos, A.G.; Stamatis, H. Enzymatic hybridization of α-lipoic acid with bioactive compounds in ionic solvents. *Bioresour. Technol.* **2013**, *136*, 41–48. [CrossRef]

56. Aissa, I.; Sghair, R.M.; Bouaziz, M.; Laouini, D.; Sayadi, S.; Gargouri, Y. Synthesis of lipophilic tyrosyl esters derivatives and assessment of their antimicrobial and antileishmania activities. *Lipids Heal.* **2012**, *11*, 13. [CrossRef]

57. Zhou, D.-Y.; Sun, Y.-X.; Shahidi, F. Preparation and antioxidant activity of tyrosol and hydroxytyrosol esters. *J. Funct. Foods* **2017**, *37*, 66–73. [CrossRef]

58. Zaks, A.; Klibanov, A.M. The effect of water on enzyme action in organic media. *J. Boil. Chem.* **1988**, *263*, 8017–8021.

59. Klibanov, A.M.; Klibanov, A.M.; Klibanov, A.M. Improving enzymes by using them in organic solvents. *Nat. Cell Boil.* **2001**, *409*, 241–246. [CrossRef] [PubMed]

60. Sharma, S.; Kanwar, S.S. Organic Solvent Tolerant Lipases and Applications. *Sci. World J.* **2014**, *2014*, 1–15.

61. Sheldon, R.A. Biocatalysis and Biomass Conversion in Alternative Reaction Media. *Chem. A Eur. J.* **2016**, *22*, 12984–12999. [CrossRef]

62. Dyson, P.J.; Laurenczy, G. Determination of the Viscosity of the Ionic Liquids [bmim][PF6] and [bmim][TF2N] Under High CO2 Gas Pressure Using Sapphire NMR Tubes. *Zeitschrift für Naturforschung B* **2008**, *63*, 681–684. [CrossRef]

63. Stefanovic, R.; Ludwig, M.; Webber, G.B.; Atkin, R.; Page, A.J. Nanostructure, Hydrogen Bonding and Rheology in Choline Chloride Deep Eutectic Solvents as a Function of the Hydrogen Bond Donor. *Phys. Chem. Chem. Phys.* **2017**, *19*, 3297–3306. [CrossRef] [PubMed]

64. Qin, J.; Zou, X.; Lv, S.; Jin, Q.; Wang, X. Influence of ionic liquids on lipase activity and stability in alcoholysis reactions. *RSC Adv.* **2016**, *6*, 87703–87709. [CrossRef]

65. Zeuner, B.; Ståhlberg, T.; Van Buu, O.N.; Kunov-Kruse, A.J.; Riisager, A.; Meyer, A.S. Dependency of the hydrogen bonding capacity of the solvent anion on the thermal stability of feruloyl esterases in ionic liquid systems. *Green Chem.* **2011**, *13*, 1550–1557. [CrossRef]

66. Wang, S.; Meng, X.; Zhou, H.; Liu, Y.; Secundo, F.; Liu, Y. Enzyme Stability and Activity in Non-Aqueous Reaction Systems: A Mini Review. *Catalysts* **2016**, *6*, 32. [CrossRef]
67. Stepankova, V.; Bidmanova, S.; Koudelakova, T.; Prokop, Z.; Chaloupkova, R.; Damborsky, J. Strategies for Stabilization of Enzymes in Organic Solvents. *ACS Catal.* **2013**, *3*, 2823–2836. [CrossRef]

7

Ionic Nanocomplexes of Hyaluronic Acid and Polyarginine to form Solid Materials

María Gabriela Villamizar-Sarmiento [1,2], Ignacio Moreno-Villoslada [3], Samuel Martínez [1,4], Annesi Giacaman [5,6], Victor Miranda [2], Alejandra Vidal [5], Sandra L. Orellana [3], Miguel Concha [5], Francisca Pavicic [5], Judit G. Lisoni [7], Lisette Leyton [1,4,*] and Felipe A. Oyarzun-Ampuero [1,2,*]

[1] Advanced Center of Chronic Diseases (ACCDiS), Universidad de Chile, Santos Dumont 964, Independencia, Santiago 8380494, Chile

[2] Departamento de Ciencias y Tecnología Farmacéuticas, Facultad de Ciencias Químicas y Farmacéuticas, Universidad de Chile, Santos Dumont 964, Independencia, Santiago 8380494, Chile

[3] Instituto de Ciencias Químicas, Facultad de Ciencias, Universidad Austral de Chile, Isla Teja, Casilla 567, Valdivia 5090000, Chile

[4] Laboratory of Cellular Communication, Program of Cell and Molecular Biology, Institute of Biomedical Sciences (ICBM), Faculty of Medicine, University of Chile, Av. Independencia 1027, Santiago 8380453, Chile

[5] Instituto de Anatomía, Histología y Patología, Facultad de Medicina, Universidad Austral de Chile, Valdivia 5090000, Chile

[6] Jeffrey Modell Center of Diagnosis and Research in Primary Immunodeficiencies. Faculty of Medicine, University of La Frontera, Temuco 4780000, Chile

[7] NM MultiMat, Instituto de Ciencias Físicas y Matemáticas, Facultad de Ciencias, Universidad Austral de Chile, Valdivia 5090000, Chile

[*] Correspondence: lleyton@med.uchile.cl (L.L.); foyarzuna@ciq.uchile.cl (F.A.O.-A.)

Abstract: We report on the design, development, characterization, and a preliminary cellular evaluation of a novel solid material. This material is composed of low-molecular-weight hyaluronic acid (LMWHA) and polyarginine (PArg), which generate aqueous ionic nanocomplexes (INC) that are then freeze-dried to create the final product. Different ratios of LMWHA/PArg were selected to elaborate INC, the size and zeta potential of which ranged from 100 to 200 nm and +25 to −43 mV, respectively. Turbidimetry and nanoparticle concentration analyses demonstrated the high capacity of the INC to interact with increasing concentrations of LMWHA, improving the yield of production of the nanostructures. Interestingly, once the selected formulations of INC were freeze-dried, only those comprising a larger excess of LMWHA could form reproducible sponge formulations, as seen with the naked eye. This optical behavior was consistent with the scanning transmission electron microscopy (STEM) images, which showed a tendency of the particles to agglomerate when an excess of LMWHA was present. Mechanical characterization evidenced low stiffness in the materials, attributed to the low density and high porosity. A preliminary cellular evaluation in a fibroblast cell line (RMF-EG) evidenced the concentration range where swollen formulations did not affect cell proliferation (93–464 µM) at 24, 48, or 72 h. Considering that the reproducible sponge formulations were elaborated following inexpensive and non-contaminant methods and comprised bioactive components, we postulate them with potential for biomedical purposes. Additionally, this systematic study provides important information to design reproducible porous solid materials using ionic nanocomplexes.

Keywords: self-assembly; sponges; ionic nanocomplexes; polyarginine; hyaluronic acid; cell proliferation

1. Introduction

Aqueous ionic nanocomplexes (INC) are structures formed in water by the association of high-molecular-weight molecules, such as polymers, with complementary charged low- or high-molecular-weight polyions, such as ionic cross-linkers (tripolyphosphate), dyes, oligomers, or polyelectrolytes [1–10]. INC can be formed following the very simple procedure of mixing aqueous solutions of oppositely charged components at room temperature. This procedure avoids the use of organic/toxic solvents and the application of high mechanical energies, thus being ideal for biological uses. In addition, the procedure is low cost, both economically and environmentally, which facilitates the adoption of INC at the industrial level. The main nanostructures that are obtained as INC are nanogels, massive nanoprecipitates, or swollen aggregates [8,11–14]. Opposite to gelation (where only the larger component is able to be allocated in the surface, determining the surfacial charge), ionic complexation between two polymeric species allows the net charge of the INC to be selected by simply varying the ratio between oppositely charged species, thus allowing the production of formulations with tuneable electrostatic characteristics [13]. Depending on the binding forces between the ionic reactants and the ratio between their corresponding apparent charge concentrations, colloidal suspensions of the INC can be achieved just by tuning the absolute amount of matter in the mixture. At high electroneutralization regimes, the system tends to produce macroprecipitates, so the colloidal suspensions need to be highly diluted to keep colloidal stability [8]. However, an excess of one of the components allows stability of the colloidal suspension at a more concentrated regime, producing colloidal particles charged enough to ensure stability through electrostatic repulsions. Although the molecule to add in excess is normally of high molecular weight [8], examples are also found in which a low-molecular-weight component is added in excess, which, instead of diffusing out of the colloidal particles, keeps associated to them, determining the net charge of the particle and thus being responsible for the mixture stability [4].

Interestingly, upon removal of water from colloidal suspensions of INC, a solid material may arise [7,8]. In this sense, freeze-drying appears as a suitable technique to obtain micro- and nanoporous materials from suspensions of INC. Sponges are solid porous structures that can easily be manipulated and applied to selected biological tissues [15–17]. Polymeric materials prepared in a spongy form can be very useful in tissue engineering as scaffolds, which can reinforce, replace, and support some organs of the body, and also as non-scaffold materials that are able to promote cell growth [18]. If they are enriched with drugs, they are also usable as active drug delivery systems. In any case, such formulations must possess several essential properties, such as biocompatibility, biodegradability (if necessary), and absence of cytotoxicity, which primarily depend on the composition and the elaboration method of the material. Nowadays, different techniques have been developed to prepare sponge-like structures from polymers, such as phase separation, electrospinning, freeze-drying, etc.

Glycosaminoglycans, such as hyaluronic acid (HA), keratan sulfate, and chondroitin sulfate, are negatively charged biopolymers and good candidates to provide biological and functional properties as they are part of the extracellular matrix in a variety of tissues. HA is composed of repeating units of disaccharides, which include D-glucuronic acid and N-acetylglucosamine molecules linked by –(1–4) and –(1–3) glycoside bonds. This compound is involved in numerous processes occurring in the body, such as wound healing, ovulation, fertilization, signal transduction, and tumor physiology [19]. The biocompatibility and related negligible side effects make HA one of the more readily available compounds used in many fields of medicine as a biologically active molecule and as excipient in drug delivery systems [19,20].

Polyaminoacids are promising macromolecules for the development of biological active compounds and drug delivery systems. This is based on the fact that these molecules are structurally similar to polypeptides and are thus degraded by human enzymes; their accumulation within the organism is minimal. Interestingly, the cationic polyaminoacid polyarginine (PArg) shows interesting biological properties, such as being able to translocate through cell membranes, thereby promoting the uptake of molecules associated with it [21,22]. This interesting feature of PArg has been exploited to develop drug delivery systems used for gene therapy [23], protein/vaccine delivery [24], and cancer treatment [25]. Furthermore, PArg enhances the absorption of drugs across epithelia [26], a property that may be utilized for mucosal drug delivery. Interestingly, the toxicity of the bare polycationic PArg may be minimized by its complexation with polyanionic species, such as HA [27].

The aim of this work was to study, for the first time, the formation of reproducible sponges using INC comprising LMWHA and PArg prepared at different LMWHA/PArg as the input. The methodology to prepare the INC, which was further used to test sponge formation, was similar to the one previously developed by us [13] but explored new combinations. The colloidal suspensions were studied in terms of turbidity, nanoparticle concentration, apparent hydrodynamic diameter, zeta potential, and shape. The solid materials were prepared by freeze-drying (using standard conditions: 0.02 mbar, −54 °C, and 24 h) the obtained INC suspensions, a strategy that is significantly different from others focused at obtaining solid materials from solubilized components [28–31]. Optical and scanning electronic microscopy (SEM) images of the solid material (sponge) were used to support the analysis of the final product. Mechanical characterization evidenced low stiffness in the materials, attributed to the low density and high porosity. Finally, we conducted a preliminary cellular evaluation in fibroblast (RMF-EG cell line). This evidenced the concentration range where swollen formulations did not affect cell proliferation at 24, 48, or 72 h, thus projecting potential doses to be administered in further in vitro or preclinical/clinical studies.

2. Materials and Methods

2.1. Chemicals

Low-molecular-weight hyaluronic acid (LMWHA, Mw ~29 kDa) was purchased from Inquiaroma (Barcelona, Spain). The equivalent weight of the LMWHA considering the number of possible charges was 403 g/mol of ionizable groups. Polyarginine (PArg, Mw ~5–15 kDa) was purchased from Sigma Aldrich (St. Louis, MO, USA). The equivalent weight of the PArg considering the number of possible charges was 192.5 g/mol of ionizable groups. For cell culture studies (RMF-EG cells), we used Dulbecco's Modified Eagle Medium (DMEM-HG, Gibco, Paisley, UK), fetal bovine serum (FBS, Gibco, Paisley, UK), and 1% penicillin–streptomycin (Gibco, Paisley, UK). 3-(4,5-dimethylthiazol-2-yl)-5-(3-carboxymethoxyphenyl)-2-(4-sulfophen-yl)-2H-tetrazolium inner salt (MTS) and phenazine methosulfate (PMS) were purchased from Promega (Madison, WI, USA). All other reagents were of the highest analytical grade. Milli-Q water was used for experimentation.

2.2. Solid Material Preparation and Characterization

The method comprised three steps: (1) prepare suspensions of polymeric INC containing LMWHA and PArg; (2) freeze-dry the obtained INC suspensions to form the solid material; and (3) sterilize the solid material resulting from freeze-drying the INC suspensions.

INC suspensions were prepared following a procedure similar to the one described by Oyarzun-Ampuero et al. [13]. Different mass of LMWHA (5, 10, 12, 13, 15, 20, 25, 30, 35, and 50 µmol) were dissolved in 4.5 mL of Milli-Q water and added to a solution prepared by dissolving 12 µmol of PArg in 4.5 mL of Milli-Q water. The mixing was done in 50-mL cylindrical plastic containers with 38 mm diameter while stirring at room temperature. Magnetic stirring was maintained for 10 min to enable complete stabilization of the systems. The formed INC suspensions were then characterized. Size and zeta potential were determined by photon correlation spectroscopy and laser Doppler anemometry using a Zetasizer Nano-ZS (Malvern Instruments, Malvern, UK). Each batch was analyzed in triplicate. Turbidimetry studies were done using a UV–Vis spectrophotometer (Perkin Elmer, Lambda 25, Waltham, MA, USA), choosing a wavelength where the individual components (LMWHA and PArg) did not present absorption bands. In this research, the wavelength of 540 nm was selected. The nanoparticle concentration was determined by nanoparticle tracking analysis (NTA) using a NanoSight NS300 (Malvern Instruments, Malvern, UK). Each batch was diluted from 10 to 1000 times with Milli-Q water to achieve an optimum concentration range of 10^7–10^9 particles/mL. A minimum of five videos (one minute each one) of the particles moving under Brownian motion were captured by the NanoSight. The videos were then analyzed for size distribution and particle concentration using the built-in NTA v 3.0 software (Malvern, UK). The morphology of the INC suspensions was determined by scanning transmission electron microscopy (STEM), model Inspect F-50 (FEI, Hillsboro, OR, USA). STEM images were obtained by sticking a droplet (20 µL) of the formulation to a cooper grid (200 mesh, covered with Formvar) for 2 min, then removing the droplet with filter paper (avoiding the paper from touching the grid), then washing the grid twice with a droplet of Milli-Q water for 1 min, and then removing the droplet with filter paper. Subsequently, the sample was stained with a solution of 1% (w/v) phosphotungstic acid by adding a droplet of this solution to the grid for 2 min and then removing with filter paper. Finally, the grid was dried at room temperature for at least 1 h before being analyzed.

In order to prepare the solid materials, 9 mL of the prepared INC suspensions were frozen at −20 °C for 24 h in the cylindrical plastic container they were produced in and then transferred to a freeze-dryer (Christ, Alpha Plus 1-2 LD, Osterode am Harz, Germany). The sublimation proceeded at 0.02 mbar for 24 h (condenser temperature of −54 °C). The morphology and porosity of the solid materials were examined with naked eye and by SEM (LEO 420, Cambrigde, England). For SEM analysis, the samples were cut with a razor blade and coated with a gold layer. Porosity threshold of the sponges was analyzed theoretically after their non-floatability in cyclohexane, a low-density organic solvent, was corroborated so that the maximum volume limit value for the solid part of the sponges could be easily calculated from their mass. The porosity threshold was then calculated as follows:

$$Porosity = \frac{volume\ of\ the\ sponges - maximum\ volume\ limit\ of\ the\ solid\ part\ of\ the\ sponges}{volume\ of\ the\ sponges}$$

For mechanical characterizations of the sponges, a series of 200-µL samples were transferred in 96-well plates, frozen (−20 °C, 24 h), and then freeze-dried (using standard conditions: 0.02 mbar, −54°C, and 24 h). The analyses were performed in the compression mode, and hardness and apparent Young's modulus (E_{app}) were obtained. The materials were placed on a fixture base table (TA-BTKIT, Brookfield) and compressed, carefully centered, with a cylindrical TA-39 probe of 2 mm diameter. The resolution of the texture analysis system was 0.1 g and 0.1 mm. The test speed was set at 0.7 mm s^{-1}, the load trigger value ranged from 0.7 to 1.0 g, and the maximum load was set at 2 g. For the final E_{app} analysis, data for 25 sponges were considered. Finally, the stability of the sponges in water and PBS was evaluated by optical microscopy (Olympus CKX41, Arquimed, Tokyo, Japan) using a digital camera (Digital Sight DS-Fi2, Nikon, Tokyo, Japan) with the Micrometrics SE Premium® software. The samples were placed at 37 °C (room temperature) on a slide and under the 4× objective. Subsequently, 10 µL of Milli-Q water or PBS was added, and the behavior of each sponge was recorded using the Open Broadcaster software (v.23.0.2, OBS Studios Contributors). To sterilize the solid material resulting from freeze-drying the INC suspensions, the sponges were sterilized under ultraviolet 25 W light for 4 min (UV lamp, Biolight, Santiago, Chile), sealed in plastic bags in laminar flow hood, and stored in desiccators containing dried silica gel in order to avoid moisture before their use. This method was previously used by our group, and the resulting sponges were demonstrated to maintain their biological potential in in vitro [32,33] and in vivo [34] studies.

2.3. Cellular Studies

2.3.1. Material Preparation and Administration to Cells

A formulation of INC comprising 12 mg of LMWHA and 2.4 mg and PArg (charge ratio LMWHA/PArg = 2.4), was prepared following the procedure described in Section 2.2. Aliquots of the above preparation were diluted in Milli-Q water in order to obtain 93, 186, 464, and 1856 µM in 100 µL; transferred in 96-well plates; and freeze-dried as described in Section 2.2. For the transfer of the material to the cells, 100 µL of culture medium was added to the sponges in the plates, and swollen formulations were then pipetted to the 96-well plates containing the cells.

2.3.2. Proliferation Assay

Five thousand RMF-EG cells were seeded in 96-well plates. After the cells adhered to the plates (2 h), the culture medium (100 µL) was extracted, and the selected formulation (charge ratio LMWHA/PArg = 2.4, prepared and transferred as described in Section 2.3.1) was added in order to achieve different concentrations (93, 185, 464, and 1856 µM). After the evaluation in terms of cellular proliferation at different time intervals (24, 48, and 72 h), the medium was replaced with 80 µL of serum-free medium plus 20 µL of MTS:PMS (20:1), mixed, and then incubated for 2 h at 37 °C. We used proliferation kit MTS (Promega, Madison, WI, USA). The reduction of MTS to formazan was determined by measuring the absorbance of this solution at 490 nm by spectrophotometry. This methodology is in accordance with the ISO 10993-5 guidelines.

2.4. Statistical Analysis

The results are shown as the mean ± standard error of the mean for $n = 3$. The results were analyzed using one-way ANOVA tests and Dunn posttests. Statistical significance was set at $p < 0.05$.

3. Results and Discussion

3.1. LMWHA/PArg INC Suspensions

Table 1 shows the results of apparent size and zeta potential of the prepared LMWHA/PArg INC. The methodology followed involved fixing the amount of PArg and varying the amount of LMWHA in order to achieve different charge ratios. It can be seen that most of the mixtures showed apparent size in the range between 100 and 200 nm, with low polydispersity indexes. Interestingly, macroprecipitates were formed at a LMWHA/PArg ratio of 1.1, expressed in relative number of equivalents, evidencing the highest electroneutralization between negatively and positively charged polymers. In this respect, it can also be seen that, at lower ratios, stable INC with positive zeta potential were obtained due to the excess of the positively charged PArg, while at higher ratios, the INC showed negative zeta potential due to the excess of LMWHA in the particles. In addition, there were appreciable differences in size, zeta potential, and polydispersity. These characteristics were influenced by the specific properties of each component (i.e., rigidity, linear charge, and molecular weight) and by the total mass of the polymeric formulations. In fact, formulations developed under the same strategy but using high-molecular-weight polymers showed higher size and higher polydispersity [13,35,36], indicating the role of the polymeric molecular weight on the homogeneity of the mixtures. The high correlation between the charge ratio of the components and the physicochemical properties of the formulations could be attributed to the low molecular weight of the polymers and also to their low molecular mass polydispersion. These selected parameters (concentration, charge ratio, and molecular weight of polymers) are ideal for designing specific nanoformulations in terms of size, low size polydispersity, and net charge.

Table 1. Physicochemical properties of formulations prepared with different ratios of low-molecular-weight hyaluronic acid/polyarginine (LMWHA/PArg) and evaluated in Milli-Q water (mean ± SD, $n = 3$).

Mass Ratio LMWHA/PArg	Charge Ratio [LMWHA]/[PArg]	Size (nm)	Polydispersity Index	Zeta Potential (mV)
2.0/2.4	0.4	126 ± 29	0.3–0.4	21.6 ± 5
4.0/2.4	0.8	128 ± 2	0.1–0.2	24.35 ± 3
4.7/2.4	0.9	138 ± 23	0.1–0.2	23.3 ± 3
5.4/2.4	1.1	Precipitation	—	—
6.0/2.4	1.2	141 ± 10	0.1–0.2	−17.9 ± 3
8.0/2.4	1.6	145 ± 9	0.2–0.3	−33.1 ± 3
10/2.4	2.0	149 ± 32	0.2–0.3	−35.0 ± 4
12/2.4	2.4	146 ± 18	0.2–0.3	−36.2 ± 3
14/2.4	2.8	186 ± 71	0.1–0.2	−39.7 ± 0.4
20/2.4	4.0	166 ± 13	0.1–0.2	−42.6 ± 0.1

With the aim of more in-depth characterization of the colloidal suspensions, turbidity and nanoparticle tracking analyses of the formulations were studied. Due to the fact that stable colloidal suspensions maintain INC homogeneously dispersed in the aqueous phase, turbidity may give information about the stability of the suspensions as well as a qualitative idea regarding the interplay of size and amount of INC formed [37]. Figure 1 shows the values of the apparent absorbance of the colloidal suspensions at 540 nm, where functional groups of LMWHA and PArg did not show absorption bands. It can be seen that there was almost a linear increase in turbidity as the total mass of LMWHA increased (not considering the precipitation zone) due to an increase in the mass of the suspensions. The increase in the size of the INC as more LMWHA was added was moderate, as can be seen in Table 1. Therefore, the increase in turbidity was presumably mainly caused by the increase in the number of formed nanoparticles.

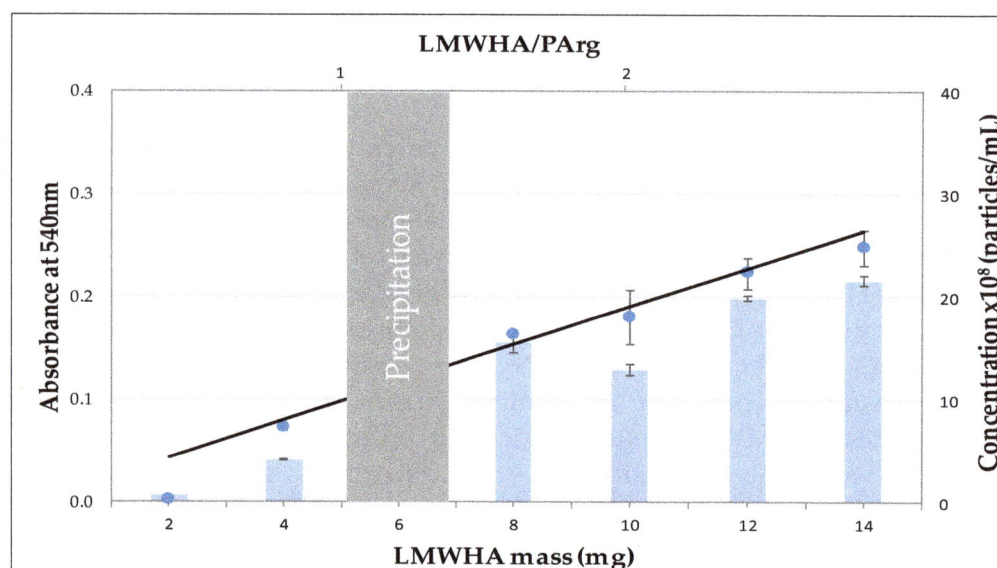

Figure 1. Turbidity (dots) and nanoparticle concentration (bars) as a function of the LMWHA mass and ratio of the components (LMWHA/PArg) (mean ± SD, $n = 3$).

The above presumption was corroborated by nanoparticle tracking analysis, which showed that the concentration of nanoparticles in the colloidal suspensions increased similarly to the turbidity. In fact, a linear tendency related to the increases in LMWHA mass and LMWHA/PArg ratio was also observed. Because the amount of PArg was fixed in these experiments, the increase in the number and size of particles indicate that the formation mechanism of the INC allowed the incorporation of more LMWHA to preexisting particles, making them generally bigger and showing higher zeta potential (in absolute value), and that the low-molecular-weight polymers were subjected to interaction equilibrium, allowing a higher number of contacts as the absolute concentration of the reactants increased. Importantly, as evidenced above, turbidity (analyzed by UV–Vis spectrophotometer) represents a simple and inexpensive methodology to preliminarily study (in terms of colloidal behavior) nanoparticle concentration and stability to design new nanoparticle formulations. It could also be useful to analyze batch-to-batch reproducibility for routinary analyses.

STEM microscopy experiments were developed in order to visualize selected formulations (LMWHA/PArg: 0.4, 0.8, 1.2, 1.6, 2.0, 2.4). As evidenced in Figure 2, the nanoparticles showed a spheroidal shape with an apparent size between 100 and 200 nm. In addition, an aggregation/agglomeration pattern could be seen as the LMWHA/PArg ratio increased. The samples at charge ratio of ≤1.6 showed more homogenous distribution with lower aggregation between the particles, while an agglomeration pattern between the nanoparticles was observed at higher ratios. It can be expected that the observed agglomeration pattern at higher mass of LMWHA could influence the formation of the solid material from the LMWHA/PArg nanocomplexes.

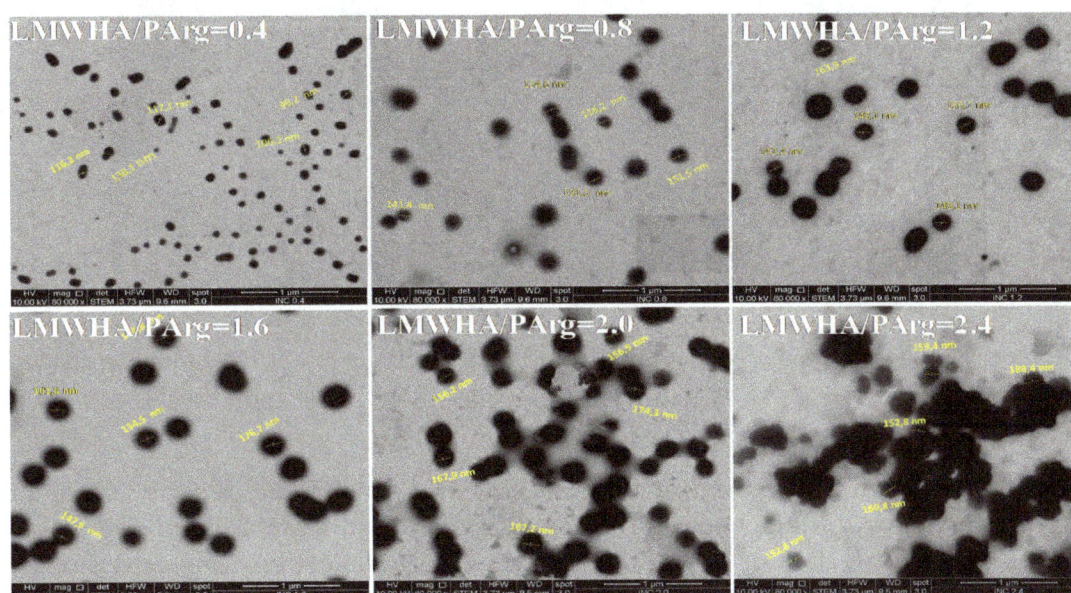

Figure 2. Scanning transmission electron microscopy (STEM) images of ionic nanocomplexes (INC) containing LMWHA and PArg at a ratio of 0.4, 0.8, 1.2, 1.6, 2.0, and 2.4 (scale bar of 1 μm).

3.2. Solid Materials

Figure 3 shows the obtained materials after freeze-drying 9 mL of the colloidal suspensions, whose characteristics are shown in Table 1 and Figures 1 and 2, in 50-mL cylindrical containers with 38 mm diameter. The selected LMWHA/PArg ratios were 0.4, 0.8, 1.2, 1.6, 2.0, and 2.4. It can be noticed that the formation of well-structured sponges failed in the case of the two compositions bearing excess of PArg as well as when freeze-drying the highly-neutralized mixture showing a LMWHA/PArg ratio of 1.2. In contrast, the suspensions showing an excess of LMWHA presented better characteristics as sponges, showing quasi cylindrical shape with diameter of around 30 mm, which was slightly lower than that of the container they were produced in, and height of 3–6 mm, as a result of shrinking on their z axis. The formed sponges were low-density, highly porous materials with density lower than 10^{-3} g/cm^3 and porosity higher than 99%. Interestingly, we observed by compression analyses that the sponges had low stiffness. Their stress–strain curves showed a nonlinear behavior with viscoelastic characteristics (Figure S1, Supplementary Material). Hardness of around 8 g was obtained between 7–15% deformation, and the E_{app} values were around 253 ± 77 kPa (n = 25), a fact related to the low density and high porosity of the materials. In addition, water and PBS were added to the sponges at 37 °C, and the behavior was observed by microscopy. As shown in the videos (video S1, video S2), the microfibers were rapidly hydrated, and the material were dispersed in small pieces into the medium. Materials with similar behavior when exposed to biorelevant media have been proposed for in vitro/in vivo testing for wound healing purposes [32,34,38] and/or to be enriched with active molecules for therapeutic purposes [39–41]. Furthermore, other interesting characterizations regarding the interplay between cells and solid materials and obtaining the stiffness have been published. In this sense, the proposal from Liverani et. al. (2017) could represent an approach to be considered in the future to characterize this and other solid materials [42].

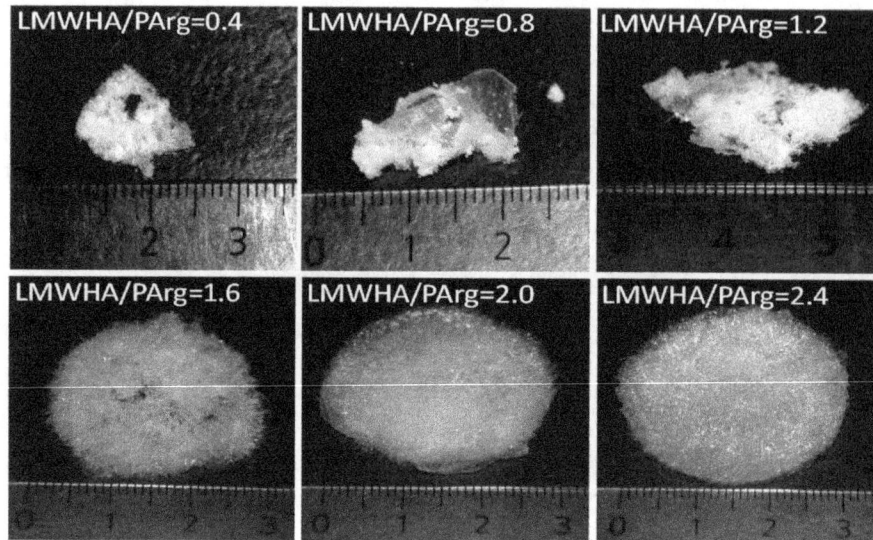

Figure 3. Optical images of materials obtained after freeze-drying 9 mL of INC aqueous suspensions containing LMWHA and PArg at a ratio of 0.4, 0.8, 1.2, 1.6, 2.0, and 2.4 in cylindrical plastic containers with 38 mm diameter.

The production of well-formed sponges from a colloidal suspension depends on several factors, among which we can name the shape and dimensions of the container compared to the volume of the colloidal suspension, the amount of components of the colloidal suspensions, and the nature of the material components. Upon freezing colloidal suspensions, the colloidal particles are concentrated during ice formation at the boundary of ice crystals, submitted to the out-of-equilibrium process called "ice-segregation-induced self-assembly" (ISISA) [8,43,44]. The migration of the solutes during freezing is determined by their hydrophilicity and mutual interactions at increasing local concentrations and lower temperatures. In this sense, as freezing at temperatures around −20 °C is produced from outside to inside the limits of the suspension volume, small ions may migrate with liquid water and concentrate at the inner part of the frozen cylinder. Molecules and particles with lower diffusion coefficients, interacting with complementary charged polyions, and amphiphilic or hydrophobic components may migrate more slowly, producing structured deposits at the boundary of ice crystals. Molecular rigidity may enhance the cohesive forces between particles and molecules subjected to electrostatic interactions. Sublimation of ice crystals after freezing during freeze-drying once the solutes are rearranged at the boundary of ice crystals then allows well-structured porous materials to be obtained. In this sense, several facts may explain the results found in this investigation. PArg is a more flexible, more amphiphilic polymer compared with the more rigid, more hydrophilic HA. In addition, the compositions with a lower content of LMWHA present less total mass of solutes. All this favors migration of the components, a weak structure at the boundary of ice crystals during freezing, and a tendency of the system to collapse during sublimation due to attractive interactions and gravity. In contrast, the higher total mass of the LMWHA-rich compositions and molecular rigidity of the polysaccharide favor the reinforcement of the structures around ice crystals and achieve the necessary tensile strength to keep the porous structure during and after sublimation.

The obtained materials presented a porous structure made of a combination of micrometric morphologies, such as microfibers and microsheets, showing a high surface area, as can be seen from the SEM images in Figure 4. As evidenced, materials with LMWHA/PArg ratios of ≤1.2 showed a more entangled and holey network of microstructures and lower pore size. As the LMWHA content increased in the materials, higher pores were observed, which was related to the formation of more extended, non-collapsed microstructures, in accordance with the facts revealed by optical observation. In addition, it can be seen that, at ratios of ≥1.6, larger fibers and microsheets were increasingly formed, which is in agreement with the tendency described for the nanoparticles evidenced by STEM (Figure 2) and for the sponges evidenced by optical images showing more structured materials (Figure 3). Interestingly, the images did not show texturing on the surface of the laminar structures. This fact suggests that there were no remaining molecules (such as uncomplexed polymers). The above can be explained by the tendency of the solutes to migrate in the freezing process, i.e., once ice is produced upon freezing, uncomplexed chains (if any) will migrate and bind the aggregates as the local concentration increases, making them part of the final product.

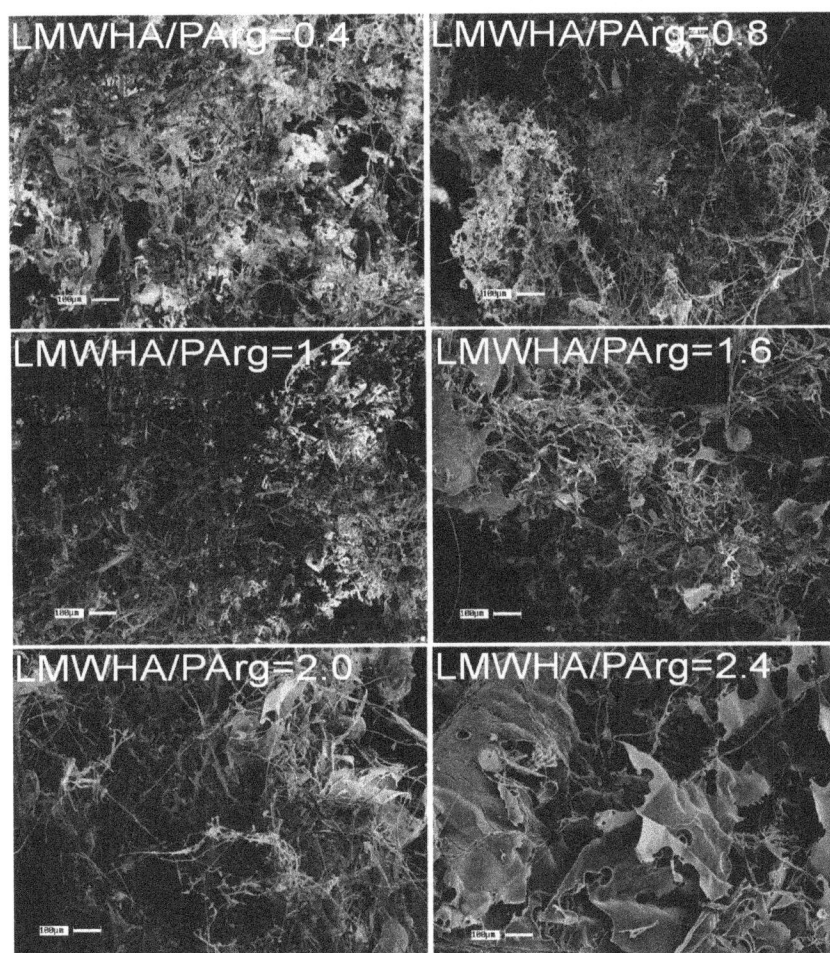

Figure 4. SEM images of materials obtained after freeze-drying 9 mL of INC containing LMWHA and PArg at different ratios of 0.4, 0.8, 1.2, 1.6, 2.0, and 2.4 in cylindrical plastic containers with 38 mm diameter. Scale bar of 100 mm.

3.3. Cellular Studies

Fibroblasts represent an adequate model cell to test the safety of new materials proposed to be applied in the skin for different therapeutic purposes [45–47]. Sponges based on a LMWHA/PArg ratio of 2.4 were selected due to their more structured characteristics (Figures 3 and 4) and also because they show very fast swellability when exposed to aqueous media. The abovementioned characteristics are important because they favor a better manipulation of the solid formulations to be applied or when swollen formulations are administered (as in the present study). Different doses of the sponges treated with culture medium were transferred to cells in order to achieve concentrations of 93, 185, 464, and 1856 μM (see Section 2.3). As evidenced in Figure 5, the concentrations of 93, 185, and 464 μM did not significantly affect the fibroblast cell proliferation at 24, 48, or 72 h. The absence of toxicity for those doses could be used as a reference for further in vitro or preclinical/clinical studies. In contrast, the highest concentration (1856 μM) decreased the cell proliferation from 48 h. Additionally, cell morphology could be clearly seen in phase-contrast microphotographs (Figure S2). The results were similar to those observed by MTS, demonstrating that a considerable amount of dead fibroblast was observed at the highest concentration (1586 μM). In fact, at 932 μM, cells started to accumulate, showing a less healthy appearance. This behavior could be reasonably explained due to the larger content of polymeric species in the culture, possibly affecting the homeostatic equilibrium between the cells, the culture medium, and the transfer of O_2/CO_2 to the environment.

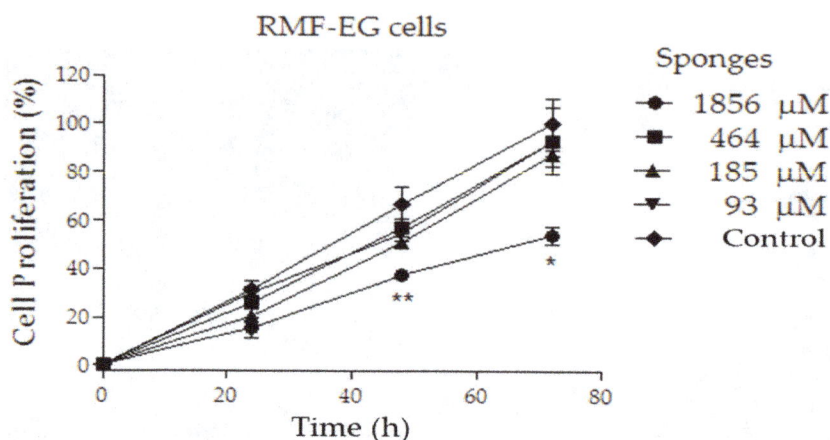

Figure 5. Proliferation curve of fibroblast (RMF-EG cell line) exposed to different doses of sponges of LMWHA/PArg = 2.4. Significant differences were obtained when comparing the values from each time with respect to the respective control value (mean ± SD, $n = 3$) and are indicated (** $p \leq 0.01$, * $p \leq 0.05$).

3.4. Final Remarks

Low-density sponge-like materials, as those presented in this paper, are solid pharmaceutical forms to be administered in tissues and can easily be obtained from INC. The formation of INC between biocompatible ionic polymers therefore emerges as a useful tool to achieve solid, highly porous materials with biomedical potential. Thus, the strategy for this type of formulations is (i) to formulate INC that provides good balance of charges and intimate mixing of oppositely charged polymeric species and (ii) to fabricate low-density sponges by freeze-drying (avoiding cryoprotectants). This has several advantages for better manipulation, such as high stability, easy storage and transportation [38], together with high therapeutic potential [7,8,32,34,38,39,41] despite their low stiffness and high lability toward hydration (see video S1, video S2). Importantly, the hydrated material stands as a gel-like mucus (if intending to dry INC by freeze-drying and further reconstitute them through hydration, cryoprotectants are required [36,48,49]), and cell cultures studies (done following the ISO 10993-5 guidelines) correspond to the cellular response to this hydrated hydrogel. The physical and mechanical characteristics of the sponges facilitate their administration to patients as they are able to be directly applied into selected tissues [34,50–54]. Due to the need to keep stability of the colloidal suspensions of INC to be freeze-dried, the total maximum concentration of reactants is normally low, furnishing low density to the final solid materials and avoiding excessive metabolic stress when applied, which is adequate for therapeutic purposes. Another advantage of these materials regarding possible commercial purposes is that they are elaborated under mild conditions in aqueous medium, avoiding toxic excipients or covalent cross-linkers. In addition, these materials can also serve as drug carriers due to the countless number of active molecules that can be incorporated in the colloidal suspensions, giving rise to the solid materials. The incorporation of the extra active molecules must be studied case-by-case for their possible interactions with the polymeric reactants, their influence on INC formation and stability [7,41], and their behavior in the freeze-drying process.

Here, we have shown the formation of low-density sponge-like materials made from INC of low-molecular-weight polymers, among which one was an anionic polysaccharide (LMWHA) and the other was a cationic polyaminoacid (PArg). Both higher stability as a solid material and good swellability in culture media were furnished by an excess of LMWHA over PArg (LMWHA/PArg ratio of 2.4). In vitro experiments showed a limit of cytotoxic concentration of this selected sponge for RMF-EG fibroblasts in the range of 464–1856 μM. This affords potential safe doses for further studies on the application of this material for medical purposes. The critical role of PArg in these materials as a biodegradable and biocompatible polycationic substrate able to interact with LMWHA, allowing the formation of the corresponding INC, could also be further extended to specific uses where its cell-penetrating capacities could promote the intracellular access of selected drugs and genes attached to the solid material [55–57].

4. Conclusions

In the present work, we demonstrated, for the first time, that aqueous INC formed by LMWHA and PArg were able to generate sponges after freeze-drying of the nanosuspensions. Interestingly, only those INC comprising an excess of HA were able to form sponges. NTA showed the formation of an increasing number of INC as the excess of LMWHA increased up to 2.4 over PArg. STEM experiments showed an increasing tendency of the particles to agglomerate. This phenomenon may be attributable to the higher total mass of these formulations, together with the higher rigidity afforded by HA. Mechanical characterization evidenced materials with low stiffness, attributed to the low density and high porosity. Finally, we provided a preliminary cellular evaluation in fibroblasts, evidencing the concentration range where a selected formulation did not affect cell proliferation up to 72 h, thus projecting potential doses to be administered in further in vitro or preclinical/clinical studies. Considering that the generated materials are composed of biodegradable and biocompatible compounds, we postulate them as candidates with potential for biomedical purposes. Additionally, this systematic study provides important information for researchers to design reproducible porous solid materials using INC of selected compositions as input.

Author Contributions: M.G.V.-S., A.G., and V.M. developed LMWHA/PArg INC suspensions and sponges; they also contributed to the characterization using several techniques. S.M. and F.P. performed cellular assays. J.G.L. contributed to the mechanical characterization. I.M.-V., S.L.O., and F.A.O.-A. guided the INC and sponge formation. A.V., M.C., and L.L. guided the biological evaluation. All authors contributed with the writing of the manuscript.

References

1. Faul, C.F.J.; Antonietti, M. Ionic Self-Assembly: Facile Synthesis of Supramolecular Materials. *Adv. Mater.* **2003**, *15*, 673–683. [CrossRef]

2. Flores, M.E.; Sano, N.; Araya-Hermosilla, R.; Shibue, T.; Olea, A.F.; Nishide, H.; Moreno-Villoslada, I. Self-association of 5, 10, 15, 20-tetrakis-(4-sulfonatophenyl)-porphyrin tuned by poly(decylviologen) and sulfobutylether-β-cyclodextrin. *Dye. Pigment.* **2015**, *112*, 262–273. [CrossRef]

3. Fuenzalida, J.P.; Nareddy, P.K.; Moreno-Villoslada, I.; Moerschbacher, B.M.; Swamy, M.J.; Pan, S.; Goycoolea, F.M. On the role of alginate structure in complexing with lysozyme and application for enzyme delivery. *Food Hydrocoll.* **2016**, *53*, 239–248.

4. Gómez-Tardajos, M.; Pino-Pinto, J.P.; Díaz-Soto, C.; Flores, M.E.; Gallardo, A.; Elvira, C.; Reinecke, H.; Nishide, H.; Moreno-Villoslada, I. Confinement of 5,10,15,20-tetrakis-(4-sulfonatophenyl)-porphyrin in novel poly(vinylpyrrolidone)s modified with aromatic amines. *Dye. Pigment.* **2013**, *99*, 759–770. [CrossRef]

5. Gröhn, F.; Klein, K.; Brand, S. Facile route to supramolecular structures: Self-assembly of dendrimers and naphthalene dicarboxylic acids. *Chem. A Eur. J.* **2008**, *14*, 6866–6869. [CrossRef] [PubMed]

6. Pichon, A. Ionic self-assembly: Porphyrins pair up. *Nat. Chem.* **2010**, *2*, 611. [CrossRef] [PubMed]

7. Pino-Pinto, J.P.; Oyarzun-Ampuero, F.; Orellana, S.L.; Flores, M.E.; Nishide, H.; Moreno-Villoslada, I. Aerogels containing 5,10,15,20-tetrakis-(4-sulfonatophenyl)-porphyrin with controlled state of aggregation. *Dye. Pigment.* **2017**, *139*, 193–200. [CrossRef]

8. Sanhueza, L.; Castro, J.; Urzúa, E.; Barrientos, L.; Oyarzun-Ampuero, F.; Pesenti, H.; Shibue, T.; Sugimura, N.; Tomita, W.; Nishide, H.; et al. Photochromic Solid Materials Based on Poly(decylviologen) Complexed with Alginate and Poly (sodium 4-styrenesulfonate). *J. Phys. Chem. B* **2015**, *119*, 13208–13217. [CrossRef]

9. Willerich, I.; Ritter, H.; Grohn, F. Structure and thermodynamics of ionic dendrimer-dye assemblies. *J. Phys. Chem. B* **2009**, *113*, 3339–3354. [CrossRef]

10. Zhang, T.; Brown, J.; Oakley, R.J.; Faul, C.F.J. Towards functional nanostructures: Ionic self-assembly of polyoxometalates and surfactants. *Curr. Opin. Colloid Interface Sci.* **2009**, *14*, 62–70. [CrossRef]

11. Janes, K.A.; Fresneau, M.P.; Marazuela, A.; Fabra, A.; Alonso, M.J. Chitosan nanoparticles as delivery systems for doxorubicin. *J. Control. Release* **2001**, *73*, 255–267. [CrossRef]

12. Kleine-Brueggeney, H.; Zorzi, G.K.; Fecker, T.; El Gueddari, N.E.; Moerschbacher, B.M.; Goycoolea, F.M. A rational approach towards the design of chitosan-based nanoparticles obtained by ionotropic gelation. *Colloids Surf. B Biointerfaces* **2015**, *135*, 99–108. [CrossRef] [PubMed]

13. Oyarzun-Ampuero, F.A.; Goycoolea, F.M.; Torres, D.; Alonso, M.J. A new drug nanocarrier consisting of polyarginine and hyaluronic acid. *Eur. J. Pharm. Biopharm.* **2011**, *79*, 54–57. [CrossRef] [PubMed]

14. Luo, Y.; Wang, Q. Recent development of chitosan-based polyelectrolyte complexes with natural polysaccharides for drug delivery. *Int. J.Biol. Macromol.* **2014**, *64*, 353–367. [CrossRef] [PubMed]

15. Madihally, S.V.; Matthew, H.W.T. Porous chitosan scaffolds for tissue engineering. *Biomaterials* **1999**, *20*, 1133–1142. [CrossRef]

16. Ho, M.-H.; Kuo, P.-Y.; Hsieh, H.-J.; Hsien, T.-Y.; Hou, L.-T.; Lai, J.-Y.; Wang, D.-M. Preparation of porous scaffolds by using freeze-extraction and freeze-gelation methods. *Biomaterials* **2004**, *25*, 129–138. [CrossRef]

17. Katoh, K.; Tanabe, T.; Yamauchi, K. Novel approach to fabricate keratin sponge scaffolds with controlled pore size and porosity. *Biomaterials* **2004**, *25*, 4255–4262. [CrossRef] [PubMed]

18. Ivan'kova, E.M.; Dobrovolskaya, I.P.; Popryadukhin, P.V.; Kryukov, A.; Yudin, V.E.; Morganti, P. In-situ cryo-SEM investigation of porous structure formation of chitosan sponges. *Polym. Test.* **2016**, *52*, 41–45. [CrossRef]

19. Salwowska, N.M.; Bebenek, K.A.; Żądło, D.A.; Wcisło-Dziadecka, D.L. Physiochemical properties and application of hyaluronic acid: A systematic review. *J. Cosmet. Dermatol.* **2016**, *15*, 520–526. [CrossRef]

20. Highley, C.B.; Prestwich, G.D.; Burdick, J.A. Recent advances in hyaluronic acid hydrogels for biomedical applications. *Curr. Opin. Biotechnol.* **2016**, *40*, 35–40. [CrossRef]

21. Lundberg, M.; Wikström, S.; Johansson, M. Cell surface adherence and endocytosis of protein transduction domains. *Mol. Ther.* **2003**, *8*, 143–150. [CrossRef]

22. Kanwar, J.R.; Gibbons, J.; Verma, A.K.; Kanwar, R.K. Cell-penetrating properties of the transactivator of transcription and polyarginine (R9) peptides, their conjugative effect on nanoparticles and the prospect of conjugation with arsenic trioxide. *Anti Cancer Drugs* **2012**, *23*, 471–482. [CrossRef] [PubMed]

23. Torchilin, V.P. Tat peptide-mediated intracellular delivery of pharmaceutical nanocarriers. *Adv. Drug Deliv. Rev.* **2008**, *60*, 548–558. [CrossRef] [PubMed]

24. Lingnau, K.; Riedl, K.; von Gabain, A. IC31® and IC30, novel types of vaccine adjuvant based on peptide delivery systems. *Expert Rev. Vaccines* **2007**, *6*, 741–746. [CrossRef] [PubMed]

25. Miklán, Z.; Orbán, E.; Csík, G.; Schlosser, G.; Magyar, A.; Hudecz, F. New daunomycin–oligoarginine conjugates: Synthesis, characterization, and effect on human leukemia and human hepatoma cells. *Pept. Sci.* **2009**, *92*, 489–501. [CrossRef] [PubMed]

26. Miyamoto, M.; Natsume, H.; Iwata, S.; Ohtake, K.; Yamaguchi, M.; Kobayashi, D.; Sugibayashi, K.; Yamashina, M.; Morimoto, Y. Improved nasal absorption of drugs using poly-l-arginine: Effects of concentration

and molecular weight of poly-l-arginine on the nasal absorption of fluorescein isothiocyanate–dextran in rats. *Eur. J. Pharm. Biopharm.* **2001**, *52*, 21–30. [CrossRef]

27. Pensado, A.; Fernandez-Piñeiro, I.; Seijo, B.; Sanchez, A. Anionic nanoparticles based on Span 80 as low-cost, simple and efficient non-viral gene-transfection systems. *Int. J. Pharm.* **2014**, *476*, 23–30. [CrossRef] [PubMed]

28. Isago, Y.; Suzuki, R.; Isono, E.; Noguchi, Y.; Kuroyanagi, Y. Development of a Freeze-Dried Skin Care Product Composed of Hyaluronic Acid and Poly (γ-Glutamic Acid) Containing Bioactive Components for Application after Chemical Peels. *Open J. Regen. Med.* **2014**, *3*, 45–53. [CrossRef]

29. Niiyama, H.; Kuroyanagi, Y. Development of novel wound dressing composed of hyaluronic acid and collagen sponge containing epidermal growth factor and vitamin C derivative. *J. Artif. Organs* **2014**, *17*, 81–87. [CrossRef]

30. Kang, H.J.; Jang, Y.J.; Park, I.K.; Kim, H.J. An Evaluation of the Efficacy and Characterization of a Cross-Linked Hyaluronic Acid/Collagen/Poloxamer Sheet for Use in Chronic Wounds. *J. Wound Manag. Res.* **2018**, *14*, 26–36. [CrossRef]

31. Park, S.-N.; Lee, H.J.; Lee, K.H.; Suh, H. Biological characterization of EDC-crosslinked collagen–hyaluronic acid matrix in dermal tissue restoration. *Biomaterials* **2003**, *24*, 1631–1641. [CrossRef]

32. Concha, M.; Vidal, A.; Giacaman, A.; Ojeda, J.; Pavicic, F.; Oyarzun-Ampuero, F.A.; Torres, C.; Cabrera, M.; Moreno-Villoslada, I.; Orellana, S.L. Aerogels made of chitosan and chondroitin sulfate at high degree of neutralization: Biological properties toward wound healing. *J. Biomed. Mater. Res. Part B Appl. Biomater.* **2018**, *106*, 2464–2471. [CrossRef] [PubMed]

33. Orellana, S.L.; Giacaman, A.; Pavicic, F.; Vidal, A.; Moreno-Villoslada, I.; Concha, M. Relevance of charge balance and hyaluronic acid on alginate-chitosan sponge microstructure and its influence on fibroblast growth. *J. Biomed. Mater. Res. A* **2016**, *104*, 2537–2543. [CrossRef] [PubMed]

34. Vidal, A.; Giacaman, A.; Oyarzun-Ampuero, F.A.; Orellana, S.; Aburto, I.; Pavicic, M.F.; Sanchez, A.; Lopez, C.; Morales, C.; Caro, M.; et al. Therapeutic potential of a low-cost device for wound healing: A study of three cases of healing after lower-extremity amputation in patients with diabetes. *Am. J.* **2013**, *20*, 394–398. [CrossRef] [PubMed]

35. Almalik, A.; Donno, R.; Cadman, C.J.; Cellesi, F.; Day, P.J.; Tirelli, N. Hyaluronic acid-coated chitosan nanoparticles: Molecular weight-dependent effects on morphology and hyaluronic acid presentation. *J. Control. Release* **2013**, *172*, 1142–1150. [CrossRef] [PubMed]

36. Oyarzun-Ampuero, F.A.; Rivera-Rodríguez, G.R.; Alonso, M.J.; Torres, D. Hyaluronan nanocapsules as a new vehicle for intracellular drug delivery. *Eur. J. Pharm. Sci.* **2013**, *49*, 483–490. [CrossRef] [PubMed]

37. Fuenzalida, J.P.; Flores, M.E.; Móniz, I.; Feijoo, M.; Goycoolea, F.; Nishide, H.; Moreno-Villoslada, I. Immobilization of Hydrophilic Low Molecular-Weight Molecules in Nanoparticles of Chitosan/Poly (sodium 4-styrenesulfonate) Assisted by Aromatic–Aromatic Interactions. *J. Phys. Chem. B* **2014**, *118*, 9782–9791. [CrossRef] [PubMed]

38. Orellana, S.; Giacaman, A.; Vidal, A.; Morales, C.; Oyarzun-Ampuero, F.; Lisoni, J.; Henríquez-Báez, C.; Morán-Trujillo, L.; Miguel, C.; Moreno-Villoslada, I. Chitosan/chondroitin sulfate aerogels with high polymeric electroneutralization degree: Formation and mechanical properties. *Pure Appl.Chemi.* **2018**, *90*, 901–911. [CrossRef]

39. Díaz, C.; Catalán-Toledo, J.; Flores, M.E.; Orellana, S.L.; Pesenti, H.; Lisoni, J.; Moreno-Villoslada, I. Dispersion of the Photosensitizer 5,10,15,20-Tetrakis(4-Sulfonatophenyl)-porphyrin by the Amphiphilic Polymer Poly (vinylpirrolidone) in Highly Porous Solid Materials Designed for Photodynamic Therapy. *J. Phys. Chem. B* **2017**, *121*, 7373–7381. [CrossRef]

40. Coronel, A.; Catalán-Toledo, J.; Fernández-Jaramillo, H.; Godoy-Martínez, P.; Flores, M.E.; Moreno-Villoslada, I. Photodynamic action of methylene blue subjected to aromatic-aromatic interactions with poly (sodium 4-styrenesulfonate) in solution and supported in solid, highly porous alginate sponges. *Dye. Pigment.* **2017**, *147*, 455–464. [CrossRef]

41. Gomez, A.; Marcela Bonilla, J.; Alejandra Coronel, M.; Martínez, J.; Morán-Trujillo, L.; Orellana, S.; Vidal, A.; Giacaman, A.; Morales, C.; Torres-Gallegos, C.; et al. Antibacterial activity against Staphylococcus aureus of chitosan/chondroitin sulfate nanocomplex aerogels alone and enriched with erythromycin and elephant garlic (*Allium ampeloprasum L. var. ampeloprasum*) extract. *Pure Appl. Chem.* **2018**, *90*, 885–900. [CrossRef]

42. Liverani, C.; Mercatali, L.; Cristofolini, L.; Giordano, E.; Minardi, S.; Della Porta, G.; De Vita, A.; Miserocchi, G.; Spadazzi, C.; Tasciotti, E.; et al. Investigating the Mechanobiology of Cancer Cell–ECM Interaction through Collagen-Based 3D Scaffolds. *Cell. Mol. Bioeng.* **2017**, *10*, 223–234. [CrossRef]

43. Gutiérrez, M.C.; Ferrer, M.L.; del Monte, F. Ice-Templated Materials: Sophisticated Structures Exhibiting Enhanced Functionalities Obtained after Unidirectional Freezing and Ice-Segregation-Induced Self-Assembly. *Chem. Mater.* **2008**, *20*, 634–648. [CrossRef]

44. Zhang, X.; Li, C.; Yunjun, L. Aligned/Unaligned Conducting Polymer Cryogels with Three-Dimensional Macroporous Architectures from Ice-Segregation-Induced Self-Assembly of PEDOT-PSS. *Langmuir* **2011**, *27*, 1915–1923. [CrossRef] [PubMed]

45. Pakkaner, E.; Yalcin, D.; Uysal, B.; Top, A. Self-assembly behavior of the keratose proteins extracted from oxidized Ovis aries wool fibers. *Int. J. Biol. Macromol.* **2019**, *125*, 1008–1015. [CrossRef]

46. Liu, L.; Cai, R.; Wang, Y.; Tao, G.; Ai, L.; Wang, P.; Yang, M.; Zuo, H. Polydopamine-Assisted Silver Nanoparticle Self-Assembly on Sericin/Agar Film for Potential Wound Dressing Application. *Int. J. Mol. Sci.* **2018**, *19*, 2875. [CrossRef]

47. Zashikhina, N.N.; Volokitina, M.V.; Korzhikov-Vlakh, V.A.; Tarasenko, I.I.; Lavrentieva, A.; Scheper, T.; Ruhl, E.; Orlova, R.V.; Tennikova, T.B.; Korzhikova-Vlakh, E.G. Self-assembled polypeptide nanoparticles for intracellular irinotecan delivery. *Eur. J. Pharm. Sci.* **2017**, *109*, 1–12. [CrossRef]

48. Villamizar-Sarmiento, M.G.; Molina-Soto, E.F.; Guerrero, J.; Shibue, T.; Nishide, H.; Moreno-Villoslada, I.; Oyarzun-Ampuero, F.A. A New Methodology to Create Polymeric Nanocarriers Containing Hydrophilic Low Molecular-Weight Drugs: A Green Strategy Providing a Very High Drug Loading. *Mol. Pharm.* **2019**. [CrossRef]

49. Guerrero, S.; Inostroza-Riquelme, M.; Contreras-Orellana, P.; Diaz, V.; Lara, P.; Vivanco-Palma, A.; cárdenas, A.; Miranda, V.; Robert, P.; Leyton, L.; et al. Curcumin-loaded nanoemulsion: A new safe and effective formulation to prevent tumor reincidence and metastasis. *Nanoscale* **2018**, *10*, 22612–22622. [CrossRef]

50. Li, Y.; Yao, M.; Wang, X.; Zhao, Y. Effects of gelatin sponge combined with moist wound-healing nursing intervention in the treatment of phase III bedsore. *Exp. Ther. Med.* **2016**, *11*, 2213–2216. [CrossRef]

51. Joshi, V.; Vaja, R.; Richens, D. Cost analysis of gentamicin-impregnated collagen sponges in preventing sternal wound infection post cardiac surgery. *J. Wound Care* **2016**, *25*, 22–25. [CrossRef] [PubMed]

52. Femminella, B.; Iaconi, M.C.; Di Tullio, M.; Romano, L.; Sinjari, B.; D'Arcangelo, C.; De Ninis, P.; Paolantonio, M. Clinical Comparison of Platelet-Rich Fibrin and a Gelatin Sponge in the Management of Palatal Wounds After Epithelialized Free Gingival Graft Harvest: A Randomized Clinical Trial. *J Periodontol.* **2016**, *87*, 103–113. [CrossRef] [PubMed]

53. Singh, M.; Bhate, K.; Kulkarni, D.; Santhosh Kumar, S.N.; Kathariya, R. The effect of alloplastic bone graft and absorbable gelatin sponge in prevention of periodontal defects on the distal aspect of mandibular second molars, after surgical removal of impacted mandibular third molar: A comparative prospective study. *J. Maxillofac. Oral. Surg.* **2015**, *14*, 101–106. [CrossRef] [PubMed]

54. Ozbalci, G.S.; Tuncal, S.; Bayraktar, K.; Tasova, V.; Ali Akkus, M. Is gentamicin-impregnated collagen sponge to be recommended in pilonidal sinus patient treated with marsupialization? A prospective randomized study. *Ann. Ital. Chir.* **2014**, *85*, 576–582. [PubMed]

55. Cui, Y.; Sui, J.; He, M.; Xu, Z.; Sun, Y.; Liang, J.; Fan, Y.; Zhang, X. Reduction-Degradable Polymeric Micelles Decorated with PArg for Improving Anticancer Drug Delivery Efficacy. *ACS Appl. Mater. Interfaces* **2016**, *8*, 2193–2203. [CrossRef] [PubMed]

56. Alhakamy, N.A.; Dhar, P.; Berkland, C.J. Charge Type, Charge Spacing, and Hydrophobicity of Arginine-Rich Cell-Penetrating Peptides Dictate Gene Transfection. *Mol. Pharm.* **2016**, *13*, 1047–1057. [CrossRef] [PubMed]

57. Reimondez-Troitino, S.; Alcalde, I.; Csaba, N.; Inigo-Portugues, A.; de la Fuente, M.; Bech, F.; Riestra, A.C.; Merayo-Lloves, J.; Alonso, M.J. Polymeric nanocapsules: A potential new therapy for corneal wound healing. *Drug Deliv. Transl. Res.* **2016**, *6*, 708–721. [CrossRef] [PubMed]

8

"Chocolate" Gold Nanoparticles—One Pot Synthesis and Biocompatibility

Neelika Roy Chowdhury [1], Allison J. Cowin [2], Peter Zilm [3] and Krasimir Vasilev [1,2,*]

[1]   School of Engineering, University of South Australia, Mawson Lakes SA 5095, Australia;
      neelika.roy_chowdhury@mymail.unisa.edu.au
[2]   Future Industries Institute, University of South Australia, Mawson Lakes SA 5095, Australia;
      allison.cowin@unisa.edu.au
[3]   Microbiology Laboratory, Adelaide Dental School, The University of Adelaide, Adelaide SA 5005, Australia;
      peter.zilm@adelaide.edu.au
*   Correspondence: krasimir.vasilev@unisa.edu.au

Abstract: The chemical synthesis of nanoparticles can involve and generate toxic materials. Here, we present for the first time, a one pot direct route to synthesize gold nanoparticles (AuNPs) using natural cacao extract as both a reducing and stabilizing agent. The nanoparticles were characterized by UV-visible spectroscopy (UV-VIS), dynamic light scattering (DLS), and transmission electron microscopy (TEM); and have excellent biocompatibility with human primary dermal fibroblasts.

Keywords: green synthesis; cacao; non-cytotoxic

1. Introduction

For decades, nanoparticles (NP) of noble metals such as gold, silver, and platinum have captivated the researchers and the general public with their remarkable physical and chemical properties, as well as for their potent therapeutic power [1]. Gold is one of the least reactive among the noble metals [2], but its nanoparticulate forms possess unique chemical, electrical, and optical properties [3]. These properties, which are size and shape dependent, can be tuned by a variety of means, such as the synthesis route, reactants, and experimental conditions [4]. Numerous applications have benefited from the special properties of gold nanoparticles, including optics, imaging, sensing, catalysis [5,6] and biomedicine [7,8] (in particular dentistry, cancer diagnostics, and photothermal and photodynamic therapies) [9–11].

As a result of the growing interest in gold nanoparticles, numerous chemical and physical synthesis routes have been proposed during the last decades [12,13]. However, major drawbacks with some of the conventional synthesis methods include toxic, hazardous chemicals and challenging reaction parameters [14–16]. Modern synthetic trends are shifting to alternative synthetic routes to minimize the use of harmful chemicals. Several studies have reported the benefits of the biosynthesis approaches using plant extracts, and unicellular and multicellular organisms [14,17]. Although chemically complex, phytochemicals have major advantages over other biosynthesis methods, as they are generally non-toxic to mammalian cell types and to the environment [14,18]. The use of plant derivatives also reduces the possibility for the absorption of toxic chemicals on the surface of the AuNPs [19]. Various studies have demonstrated the power of phytochemicals in gold nanoparticles (AuNPs) synthesis as well as the biocompatibility of the generated AuNPs to different cell types [14,17].

Here, we report for the first time, on the potential of cacao extract as a reducing and stabilizing agent in the synthesis of AuNPs. In addition to being a popular constituent in various foods and beverages, cacao has been speculated to alleviate health disparities such as aging, inflammation, depression, cancer, and stress [20–24]. The hypothesis behind this work is that oxalic acid, which is a constituent of cacao, can reduce Au^{3+} in $HAuCl_4$ to metallic gold and stabilize the resultant nanoparticle colloidal solution. This hypotheses is also further substantiated by our previous work, where we reported the synthesis of silver nanoparticles (AgNPs) facilitated by cacao extract [25]. Herein, we extend this synthesis approach to prepare biocompatible 'green' gold nanoparticles and explore their properties. This easy single-step synthesis route was optimized and the prepared samples were characterized using UV-visible spectroscopy (UV-VIS), dynamic light scattering (DLS), and transmission electron microscopy (TEM). Finally, primary human dermal fibroblast (HDFs) cells were used to evaluate the biocompatibility of the gold nanoparticles.

2. Materials and Methods

2.1. Reagents and Chemicals

Hydrogen tetrachloroaurate ($HAuCl_4$), penicillin, streptomycin, and L-glutamine were bought from Pro Sci Tech, Kirwan, Australia. Cold pressed cacao powder was obtained from Forest Super Foods, Melbourne, Australia, and stored in an air-tight container (Goodguys, Adelaide, Australia). NaOH pellets, phosphate buffer saline (PBS) tablets, foetal bovine serum (FBS), nitric acid (70%), and Dulbecco's Modified Eagle Medium (DMEM) were purchased from Sigma-Aldrich, Sydney, Australia. Hydrochloric acid (36%) was procured from Ajax Finechem Pty. Ltd., Sydney, Australia. All of the reagents were used as received. Ultra-pure MilliQ water (resistivity 18.2 Ω, Sigma-Aldrich, Sydney, Australia) was used for all of the experimental and cleaning procedures.

2.2. Synthesis of Gold Nanoparticles

The aqueous extracts of cacao were prepared by mixing a varying amount of cacao powder (Table 1) in 10 mL of ultrapure water (MilliQ system, Millipore Corp., Burlington, MA, USA) at room temperature. The extract obtained after the filtration (0.45 μm—sterile EO, Sartorius Stedim Australia Pty. Ltd., Dandenong South, Australia) of the suspension was stored for the synthesis of AuNPs. Then, the cacao extracts were mixed with aqueous solution of gold chloride (0.1 mg/mL in MilliQ water) (Table 1). The reaction mixtures were stirred continuously for 30 min at 100 °C under reflux. After 30 min, the heating source was removed, the reaction mixtures were cooled down to room temperature (25 °C), and stirring continued for 24 h. The periodic (30 min, 1 h, 2 h, 3 h, 4 h, and 24 h) monitoring of the prepared AuNPs was carried out using a UV-VIS spectrophotometer. The samples, S1, S2, S3, S4, and S5, refer to the AuNPs suspensions synthesized with 0.1, 1, 2.5, 10, and 50 mg/mL of cacao extract, respectively. The pH of the nanoparticle solutions was six.

Table 1. Concentrations of reactants used for the synthesis of different cacao-gold nanoparticles (AuNPs).

Sample	Gold Chloride (mg/mL)	Cacao (mg/mL)
S1	0.1	0.5
S2	0.1	1
S3	0.1	2.5
S4	0.1	10
S5	0.1	50

2.3. Characterization

The progress of the reaction was periodically monitored using a Cary 5 UV-VIS spectrophotometer (Varian Australia Pty. Ltd., Mulgrave, Australia) at room temperature in the wavelength range of 400–800 nm. All of AuNPs' suspensions were diluted 2X (*v/v*) with MilliQ water prior to UV-VIS spectral characterization. MilliQ water was used as a blank throughout the experiment. Quartz cuvettes were used for all of the measurements.

All of the samples were diluted to a suitable concentration using MilliQ water prior to DLS analysis to determine the hydrodynamic diameter of the nanoparticles. A Nicomp 380 particle size analyzer (Nicomp Particle Sizing Systems, Port Richey, FL, USA) operating at 25 °C was used for all of the DLS and zeta potential measurements. The mean hydrodynamic diameters reported are the average of the three measurements taken of the three independent nanoparticle batches (separate syntheses). Disposable plastic cuvettes were used for all of the measurements. All of the analyses were carried out at pH-6.

A 'JEOL 2100F' (Tokyo, Japan) transmission electron microscope (TEM), operated at an acceleration voltage of 200 kV, was used to determine the size and morphology of the synthesized cacao-AuNPs. Samples for the TEM analysis were prepared by depositing a small volume (10 µL) of the AuNPs solution on a carbon coated copper grid (ProSciTech, Kirwan, Australia). The grid was left to dry overnight at room temperature prior to TEM analysis. The crystal structure of the AuNPs were determined with the selected area electron diffraction (SAED) pattern obtained from TEM images.

2.4. Fibroblasts Study

Primary derived HDFs were gifted from Dr. Louise Smith, the University of South Australia. The HDFs were harvested and grown, as described elsewhere [26]. Briefly, cells were grown from frozen stocks and maintained in DMEM at 37 °C in 95% humidity and 5% CO_2. The DMEM was changed every 3–4 days. Ethics approval was approved by the Ethical Committee at the Queen Elizabeth Hospital and the University of South Australia Human Ethics Committee, described elsewhere [26].

The viability of the cacao-AuNPs treated primary human dermal fibroblast (HDFs) cells was tested using a resazurin assay based on the reduction of non-fluorescent resazurin by metabolically active living cells to form fluorescent resorufin which was quantified using a microplate reader. Cells (1×10^4 cells per well in DMEM) were seeded in 24 well plates on air plasma cleaned (5 min, 40 W, 2×10^{-1} mbar) thermanox coverslips and incubated (24 h in 95% air, 5% CO_2 at 37 °C) until they reached 50 and 80% confluency. The DMEM was supplemented with FBS, penicillin (100 IU), and streptomycin (100 lg) (Invitrogen). After incubation, the DMEM was removed and the cells (50 and 80% confluent) were briefly washed with PBS. The cells were then treated with different concentrations of AuNPs (containing 500, 250, and 125 µg/mL of Au) prepared in warm DMEM and incubated for 24 and 72 h. After each incubation period, the media was aspirated again and the cells were rinsed with PBS. A stock solution of 110 mg/mL resazurin was prepared in phosphate buffered saline and filter sterilized using a 0.2 mm filter. The stock was then diluted 1:10 in fresh warm DMEM and 600 µL of the diluted solution was added to each well. After 1 h, 200 µL of the reduced solution was transferred into a 96 well plate and the fluorescent intensity was recorded using a plate reader ($\lambda_{ex} = 544$ nm and $\lambda_{em} = 590$ nm). Fresh DMEM without any AuNPs served as the control. The media from the wells were replaced with fresh DMEM every 2–3 days during the course of the assay. The percentage of cell viability was calculated with the following equation.

$$\text{Viability (\%)} = 100 \times \text{Absorption}_{test} / \text{Absorption}_{control}$$

2.5. Statistical Analysis

All of the statistical analyses were performed using graph pad prism 6 software. All of the data were expressed as mean ± standard error mean (SEM). Statistical significance was determined using one-way ANOVA with a Dunnett's post-test. All of the experiments were performed in biological and technical triplicates on three separate days.

3. Results and Discussion

3.1. Synthesis and Characterization of AuNPs

The procedure for synthesizing gold nanoparticles with cacao extract is outlined in Figure 1. Briefly, under rigorous stirring, cacao extract was mixed with an aqueous solution of gold chloride (HAuCl$_4$) at 100 °C (boiling temperature). Upon the addition of the reactants, the solution became increasingly darker and changed from transparent light yellow to purple-red within 5 min. The color change was a visual indication of the formation of AuNPs (Figure 2A–E insets). The samples, S1, S2, S3, S4, and S5, refer to the AuNPs suspensions synthesized with increasing the concentration of the cacao extract by 0.5, 1, 2.5, 10, and 50 mg/mL, respectively (see Table 1).

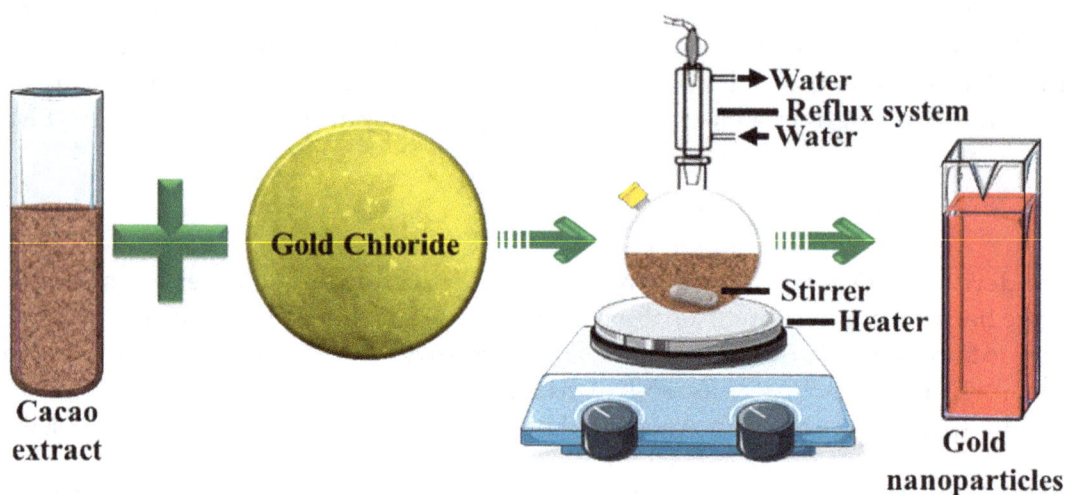

Figure 1. Schematic illustration of gold nanoparticles (AuNPs) synthesis from cacao extract.

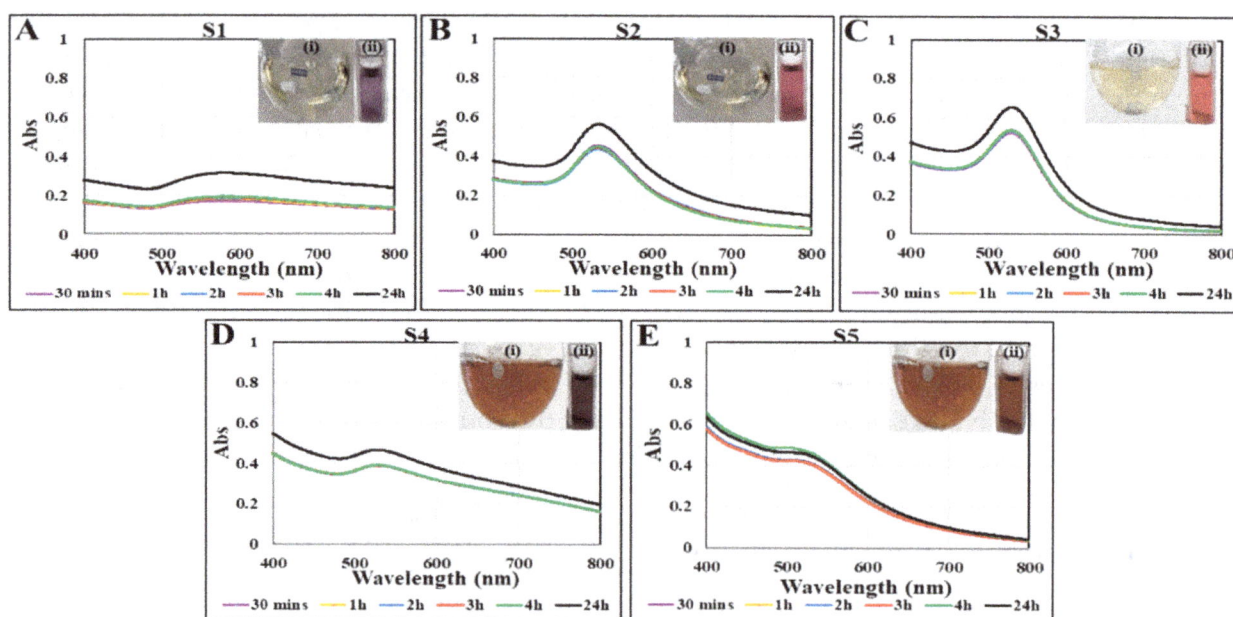

Figure 2. UV-Vis spectra of gold nanoparticles S1, (**A**); S2 (**B**); S3 (**C**); S4 (**D**); and S5 (**E**) obtained using cacao extract. Absorption spectra of samples recorded at 30 min, 1, 2, 3, 4, and 24 h from the initiation of AuNPs synthesis at 100 °C. Insets displaying the AuNPs solution before (i) and after (ii) synthesis.

The intense purple-red color of the reaction mixture is due to the well-known phenomenon of plasmon resonance (PR), which is the result of resonant oscillations of the semiconfined electrons in the nanoparticles with the incident photons [27]. Several distinct parameters, including amount, size, and shape of the nanoparticles, interparticle electronic interactions, and the surrounding media have an influence on the position, shape, and intensity of the PR band [28,29]. UV-visible spectrophotometry was employed to evaluate the time course of the reaction kinetics by taking measurements at intervals of 30 min, 1, 2, 3, 4, and 24 h. UV-visible profiles of the reaction mixture at these time points are shown in Figure 2. The spectra of all of these samples exhibited a maximum of absorption (λ_{max}) at around 530 nm, consistent with the plasmon resonance absorption band of AuNPs at ~510–560 nm [30], which confirmed the formation of gold nanoparticles. As an overall trend, the reaction had a fast-initial phase, much of the reduction being completed within 30 min. Clearly, there is no significant change in the absorption intensity within time points of 30 min, 1, 2, 3, and 4 h. However, a noticeable increase was observed after 24 h, which demonstrates that the reaction was slowly continuing, but was completed within this period as there was no further changes in the adsorption intensity beyond 24 h.

The UV-visible spectra pointed to some interesting trends. Initially, when the cacao concentration increased to 2.5 mg/mL (S3) (Figure 2A–C), the intensity of the plasmon resonance absorption increased, which suggests the formation of more gold nanoparticles. In the same time, the maximum of the peak shifted to the left, indicating a decrease in nanoparticles size. A further increase in the cacao concentration to 10 and 50 mg/mL (S4 and S5, Figure 2D,E) led to a decrease in the plasmon resonance absorption and a broadening of the spectra, pointing to aggregations in the system.

Oxalic acid in cacao is a natural reducing agent, which we showed to reduce silver ions into nanoparticles [25,31]. Oxalic acid exists as oxalate ions ($C_2O_4^{2-}$) in the experimental conditions (pH > 4.3) used for this study. The standard potential of the $C_2O_4^{2-}/CO_2$ oxido-reduction couple is $E^0_{red} = -0.49$ V, while the one of $AuCl_4^-/Au(s)$ is $E^0_{red} = 1.002$ V. The oxido-reduction reaction scheme resulting from the reduction of gold ions by oxalate is as follows:

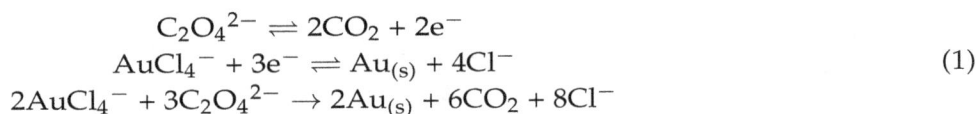

$$C_2O_4^{2-} \rightleftharpoons 2CO_2 + 2e^-$$
$$AuCl_4^- + 3e^- \rightleftharpoons Au_{(s)} + 4Cl^-$$
$$2AuCl_4^- + 3C_2O_4^{2-} \rightarrow 2Au_{(s)} + 6CO_2 + 8Cl^- \qquad (1)$$

The synthesized AuNPs were characterized for their hydrodynamic diameter and zeta potential. The results are summarized in Table 2. The hydrodynamic diameter (as determined by the DLS) of the samples S1, S2, and S3 was 54, 29, and 18 nm, respectively. These results are in good agreement with the UV-vis absorption spectra, where the PR maximum shifted to shorter wavelengths, indicating a smaller particle size. The hydrodynamic diameter of samples S4 and S5 could not be reliably determined because of aggregates, also suggested by the UV-vis spectra.

The zeta potential (ζ) provides important cues about the stabilization mechanisms in the colloidal suspension of nanoparticles. The zeta potential (Table 2) of samples S1, S2, and S3 was between -11 mV to -17 mV, and this range is known to confer incipient stability for colloids [32]. The negative charge on the surface of the nanoparticles appears to play a significant role by ensuring repulsion between the particles in the suspension. The samples were very stable and even after a month, no visible particle agglomeration was observed.

Table 2. Hydrodynamic diameter and zeta potential of green AuNPs synthesized using cacao extract.

Sample	Hydrodynamic Size (nm)	Zeta Potential (mV)
S1	54.4 ± 9.1	−11.65
S2	28.7 ± 3.4	−14.10
S3	17.9 ± 1.5	−17.52

The TEM images in Figure 3 shows that the morphology of the as-synthesized AuNPs was mostly spherical, and there was no particle aggregation when the cacao concentration was below 10 mg/mL (Figure 3A–C). The particle size analysis of samples S1, S2, S3, and S4 are shown in Figure 3E–H, respectively. The AuNPs had an average particle size of 35 ± 10 nm (S1), 20 ± 9.1 nm (S2), 10 ± 11.6 nm (S3), and 7 ± 4.2 nm (S4). The particle sizes of samples S1, S2, and S3, determined from the TEM images, are smaller than the hydrodynamic diameter measured by DLS. These variations in the particles sizes are the result of the different measurement principles used by these two methods [33]. The size distribution analysis of sample S5 could not be performed reliably because of the presence of particle aggregation. However, both methods suggest an overall trend of particle size reduction when increasing the concentration of cacao was observed.

The selected area electron diffraction (SAED) analysis confirmed that the synthesized AuNPs are crystalline in nature (Figure 3I–K). As a result of random orientation of crystal planes, concentric diffraction rings were observed in the SAED patterns of the samples S1, S2, and S3, and the reflection rings were indexed to (111), (200), and (220) planes of the face centered cubic (fcc) crystalline lattice of gold.

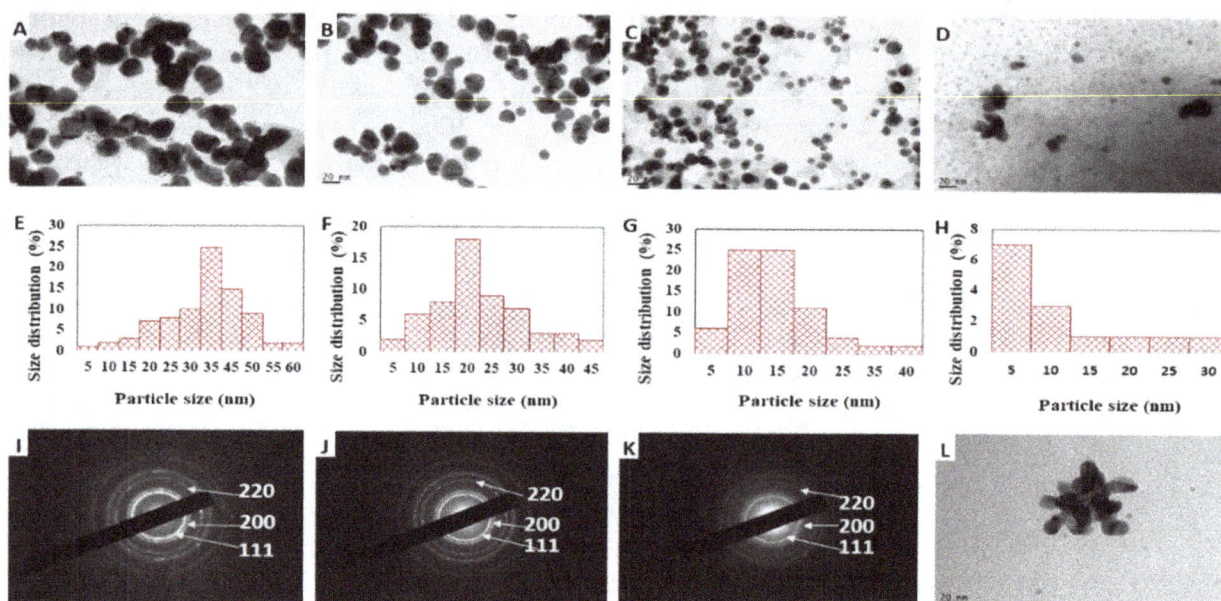

Figure 3. TEM images of the cacao-AuNPs S1 (**A**), S2 (**B**), S3 (**C**), S4 (**D**), and S5 (**L**). The corresponding histograms showing the particle size distribution (**E**, **F**, **G**, and **H** for S1, S2, S3, and S4, respectively). Representative SAED patterns of S1 (**I**), S2 (**J**), and S3 (**K**).

3.2. Viability of Human Dermal Fibroblasts after Exposure to AuNPs

Gold nanoparticles have found numerous applications in advanced medical therapies ranging from the sensing to treatment of cancers. It is thus important to evaluate the cytotoxicity of this new nanomaterial to human cells. Human dermal fibroblasts were selected for this experiment because these cells play important role in connective tissue. We tested AuNPs resulting from S1, S2, and S3 only since these preparations were free of aggregations, which is important for potential applications.

Cells having two different levels of confluence (50% and 80%) were assessed for their viability after 24 and 72 h of exposure to concentrations of cacao-AuNPs that contained 125, 250, and 500 μg/mL of Au. Representative microphotographs of the untreated and HDFs treated with AuNPs are shown in Figure S1. After 24 h of exposure, 50% confluent HDFs showed no morphological changes in any treatment groups compared with the control (Figure S1A). At a treatment time of 72 h (Figure S1B), the cells morphology is similar to the control for all three of the AuNPs exposure concentrations. No cell shrinkage or floating cells were observed in the AuNPs treated HDFs at both time points. For the 80%

confluent cells, at 24 h of treatment, the cells are well spread out (Figure S1C) and exhibited a typical fibroblast morphology. When the treatment time was increased to 72 h, the area of cell spreading decreased when the HDFs were exposed to sample S3 (Figure S1D). As all of the AuNPs samples contained the same amount of gold, this effect could be due to the smaller size of the AuNPs in S3.

The viability of the HDFs in the culture conditions was determined using the resazurin assay. Figure 4A,B represents the influence of AuNPs on a 50% confluent HDFs after 24 and 72 h of treatment. The cells showed a greater than 90% viability for all of the AuNPs-HDFs treatments compared to the control. Interestingly, at both time points, the number of viable HDFs treated with S3 were significantly higher than the control, suggesting that S3 may have contributed in the proliferation of the HDFs as well.

When 80% confluent HDFs were treated with AuNPs for 24 and 72 h, the results were different (Figure 4C,D respectively). At 24 h, there was a significant increase in the viable cell number for S2 and S3 at all of the tested concentrations of Au, but S1 remained non-significant compared to the control. However, when the exposure time was increased to 72 h, a non-significant reduction in the cell viability was observed for S3. The degrees of freedom (DF) and probability (P) values determined from the viability assay for samples S1, S2, and S3 are provided in the Supporting Information (Table S1).

It is important to note that a variable degree of cytotoxicity against a wide range of cells has been reported for gold nanoparticles [34]. However, our data indicate that none of the cacao extract derived AuNPs samples caused any acute toxicity to HDFs. In general, cacao and its phytochemical constituents are known to be beneficial for humans [22] and to promote wound healing [21,35]. In this respect, HDFs are extremely important for controlling the wound healing process [36]. The fact that there was not an adverse cytotoxic effect observed on these cells indicates that the new cacao-AuNPs have a good biocompatibility and may be useful in the field of biomedicine. The synthesized AuNPs also have potentials as nano drug carriers. The negative surface charge and the carboxyl acid groups of the oxalic acid in these AuNPs can be used to bind and deliver other antibiotics or medically relevant drugs [37].

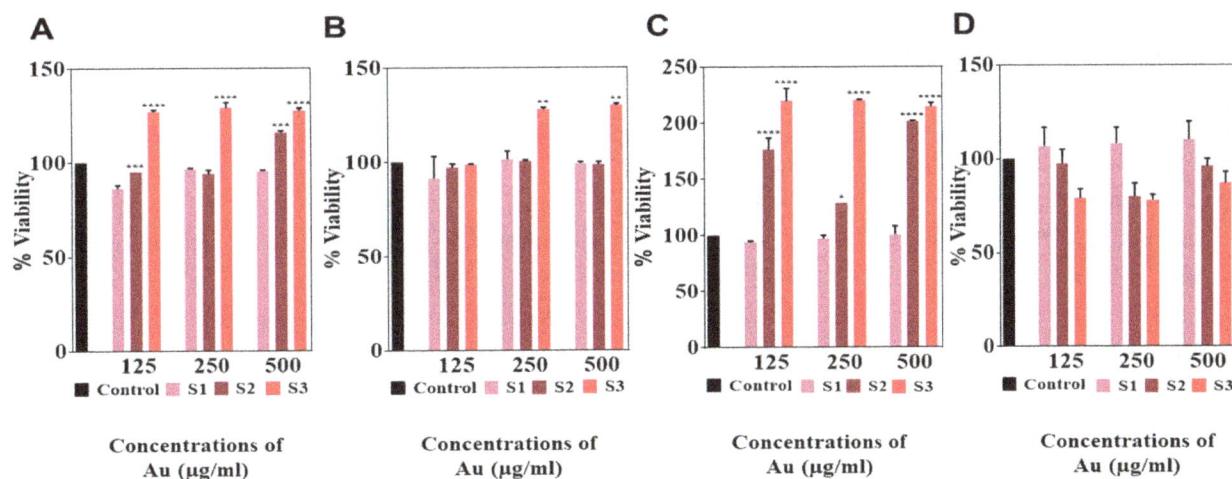

Figure 4. Determination of AuNPs generated cytotoxicity towards primary human dermal fibroblasts (HDFs). Cells were treated with different concentrations of AuNPs for 24 and 72 h. Cell viability of 50% confluent HDFs after 24 (**A**) and 72 h (**B**) exposure; and 80% confluent cells after 24 (**C**) and 72 h (**D**). The untreated cells served as controls. The results are represented as ±standard error mean (SEM) ($n = 3$). * $p < 0.05$, ** $p < 0.01$, *** $p < 0.001$, **** $p < 0.0001$. Asterisks indicate statistical significance compared to the control.

4. Conclusions

Collectively, we developed a fast, single-step, and reproducible method for the synthesis of gold nanoparticles using the extract of cacao as a reducing and stabilizing agent. The resultant AuNPs were mostly spherical, had a crystalline structure, and were negatively charged. We determined the experimental conditions that lead to stable colloidal suspensions, which are important for future applications. Furthermore, the size of the nanoparticles could be tuned by adjusting the concentration of the reactants. In vitro studies suggested that the cacao derived AuNPs are biocompatible, as none of the tested formulations exhibited cytotoxicity towards 50% and 80% confluent HDFs. This is important as gold nanoparticles have gained significant attention for application in fields of medical diagnostics and therapies. The toxic chemical free method for gold nanoparticles preparation developed in this work presents also opportunities in other fields, such as sensing. The surface of the nanoparticles can potentially be functionalised with desired ligands, which will provide opportunities for surface immobilization to surfaces for various applications. Another possibility, reinforced by the tunability of nanoparticles sizes, would be attachment of drugs and biomolecules to provide vehicles for delivery of cargo inside biological cells. Overall, this exciting, simple, green, and single-step new procedure for AuNPs preparation provides endless opportunities in numerous fields of research and practical application.

Author Contributions: Conceptualization, K.V. and N.R.C.; methodology, N.R.C.; validation, N.R.C., P.Z., and K.V.; formal analysis, N.R.C. and K.V.; investigation, N.R.C.; resources, A.J.C., P.Z., and K.V.; data curation, N.R.C.; writing (original draft preparation), N.R.C.; writing (review and editing), N.R.C., A.J.C., P.Z., and K.V.; supervision, A.J.C., P.Z., and K.V.; project administration, A.J.C., P.Z., and K.V.; and funding acquisition, K.V.

Acknowledgments: The authors acknowledge Michael Roberts at the School of Pharmacy, University of South Australia for sourcing and collection of skin.

References

1.   Lansdown, A.B. Silver and Gold. In *Patty's Toxicology*, 6th ed.; Wiley-Blackwell: Hoboken, NJ, USA, 2012; Volume 1, pp. 75–112.

2.   Hammer, B.; Norskov, J.K. Why gold is the noblest of all the metals. *Nature* **1995**, *376*, 238–240. [CrossRef]

3.   Kim, E.Y.; Kumar, D.; Khang, G.; Lim, D.-K. Recent advances in gold nanoparticle-based bioengineering applications. *J. Mater. Chem. B* **2015**, *3*, 8433–8444. [CrossRef]

4.   Saha, K.; Agasti, S.S.; Kim, C.; Li, X.; Rotello, V.M. Gold nanoparticles in chemical and biological sensing. *Chem. Rev.* **2012**, *112*, 2739–2779. [CrossRef] [PubMed]

5.   Teimouri, M.; Khosravi-Nejad, F.; Attar, F.; Saboury, A.A.; Kostova, I.; Benelli, G.; Falahati, M. Gold nanoparticles fabrication by plant extracts: Synthesis, characterization, degradation of 4-nitrophenol from industrial wastewater, and insecticidal activity—A review. *J. Clean. Prod.* **2018**, *184*, 740–753. [CrossRef]

6.   Villa, A.; Dimitratos, N.; Chan-Thaw, C.E.; Hammond, C.; Veith, G.M.; Wang, D.; Manzoli, M.; Prati, L.; Hutchings, G.J. Characterisation of gold catalysts. *Chem. Soc. Rev.* **2016**, *45*, 4953–4994. [CrossRef] [PubMed]

7.   Benelli, G. Gold nanoparticles—Against parasites and insect vectors. *Acta Trop.* **2018**, *178*, 73–80. [CrossRef] [PubMed]

8.   Kumari, M.; Mishra, A.; Pandey, S.; Singh, S.P.; Chaudhry, V.; Mudiam, M.K.R.; Shukla, S.; Kakkar, P.; Nautiyal, C.S. Physico-Chemical Condition Optimization during Biosynthesis lead to development of Improved and Catalytically Efficient Gold Nano Particles. *Sci. Rep.* **2016**, *6*, 27575. [CrossRef] [PubMed]

9.   Thakor, A.S.; Jokerst, J.; Zavaleta, C.; Massoud, T.F.; Gambhir, S.S. Gold Nanoparticles: A Revival in Precious Metal Administration to Patients. *Nano Lett.* **2011**, *11*, 4029–4236. [CrossRef] [PubMed]

10.   Dykman, L.; Khlebtsov, N. Gold nanoparticles in biomedical applications: Recent advances and perspectives. *Chem. Soc. Rev.* **2012**, *41*, 2256–2282. [CrossRef] [PubMed]

11.   Giljohann, D.A.; Seferos, D.S.; Daniel, W.L.; Massich, M.D.; Patel, P.C.; Mirkin, C.A. Gold nanoparticles for biology and medicine. *Angew. Chem. Int. Ed.* **2010**, *49*, 3280–3294. [CrossRef] [PubMed]

12. Daniel, M.-C.; Astruc, D. Gold nanoparticles: Assembly, supramolecular chemistry, quantum-size-related properties, and applications toward biology, catalysis, and nanotechnology. *Chem. Rev.* **2004**, *104*, 293–346. [CrossRef] [PubMed]

13. Zhao, P.; Li, N.; Astruc, D. State of the art in gold nanoparticle synthesis. *Coord. Chem. Rev.* **2013**, *257*, 638–665. [CrossRef]

14. Akhtar, M.S.; Panwar, J.; Yun, Y.-S. Biogenic synthesis of metallic nanoparticles by plant extracts. *ACS Sustain. Chem. Eng.* **2013**, *1*, 591–602. [CrossRef]

15. Dreaden, E.C.; Alkilany, A.M.; Huang, X.; Murphy, C.J.; El-Sayed, M.A. The golden age: Gold nanoparticles for biomedicine. *Chem. Soc. Rev.* **2012**, *41*, 2740–2779. [CrossRef] [PubMed]

16. Maurer-Jones, M.A.; Gunsolus, I.L.; Murphy, C.J.; Haynes, C.L. Toxicity of engineered nanoparticles in the environment. *Anal. Chem.* **2013**, *85*, 3036–3049. [CrossRef] [PubMed]

17. Kumar, V.; Yadav, S.K. Plant-mediated synthesis of silver and gold nanoparticles and their applications. *J. Chem. Technol. Biotechnol.* **2009**, *84*, 151–157. [CrossRef]

18. Lee, J.; Kim, H.Y.; Zhou, H.; Hwang, S.; Koh, K.; Han, D.-W.; Lee, J. Green synthesis of phytochemical-stabilized Au nanoparticles under ambient conditions and their biocompatibility and antioxidative activity. *J. Mater. Chem.* **2011**, *21*, 13316–13326. [CrossRef]

19. Smitha, S.; Philip, D.; Gopchandran, K. Green synthesis of gold nanoparticles using Cinnamomum zeylanicum leaf broth. *Spectrochim. Acta Part A Mol. Biomol. Spectrosc.* **2009**, *74*, 735–739. [CrossRef] [PubMed]

20. Calderón-Garcidueñas, L.; Mora-Tiscareño, A.; Franco-Lira, M.; Cross, J.V.; Engle, R.; Aragón-Flores, M.; Gómez-Garza, G.; Jewells, V.; Medina-Cortina, H.; Solorio, E. Flavonol-rich dark cocoa significantly decreases plasma endothelin-1 and improves cognition in urban children. *Front. Pharmacol.* **2013**, *4*, 104. [CrossRef] [PubMed]

21. Davis, S.C.; Perez, R. Cosmeceuticals and natural products: Wound healing. *Clin. Dermatol.* **2009**, *27*, 502–506. [CrossRef] [PubMed]

22. Dillinger, T.L.; Barriga, P.; Escárcega, S.; Jimenez, M.; Lowe, D.S.; Grivetti, L.E. Food of the gods: Cure for humanity? A cultural history of the medicinal and ritual use of chocolate. *J. Nutr.* **2000**, *130*, 2057S–2072S. [CrossRef] [PubMed]

23. Neukam, K.; Stahl, W.; Tronnier, H.; Sies, H.; Heinrich, U. Consumption of flavanol-rich cocoa acutely increases microcirculation in human skin. *Eur. J. Nutr.* **2007**, *46*, 53–56. [CrossRef] [PubMed]

24. Scholey, A.B.; French, S.J.; Morris, P.J.; Kennedy, D.O.; Milne, A.L.; Haskell, C.F. Consumption of cocoa flavanols results in acute improvements in mood and cognitive performance during sustained mental effort. *J. Psychopharmacol.* **2010**, *24*, 1505–1514. [CrossRef] [PubMed]

25. Chowdhury, N.R.; MacGregor-Ramiasa, M.; Zilm, P.; Majewski, P.; Vasilev, K. 'Chocolate' silver nanoparticles: Synthesis, antibacterial activity and cytotoxicity. *J. Colloid Interface Sci.* **2016**, *482*, 151–158. [CrossRef] [PubMed]

26. MacNeil, S.; Shepherd, J.; Smith, L. Production of tissue-engineered skin and oral mucosa for clinical and experimental use. In *3D Cell Culture: Methods and Protocols*; Humana Press: New York, NY, USA, 2011; Volume 695, pp. 129–153.

27. Hutter, E.; Fendler, J.H.; Roy, D. Surface Plasmon Resonance Studies of Gold and Silver Nanoparticles Linked to Gold and Silver Substrates by 2-Aminoethanethiol and 1,6-Hexanedithiol. *J. Phys. Chem. B* **2001**, *105*, 11159–11168. [CrossRef]

28. Moores, A.; Goettmann, F. The plasmon band in noble metal nanoparticles: An introduction to theory and applications. *New J. Chem.* **2006**, *30*, 1121–1132. [CrossRef]

29. Tsuda, T.; Sakamoto, T.; Nishimura, Y.; Seino, S.; Imanishi, A.; Kuwabata, S. Various metal nanoparticles produced by accelerated electron beam irradiation of room-temperature ionic liquid. *Chem. Commun.* **2012**, *48*, 1925–1927. [CrossRef] [PubMed]

30. Jain, P.K.; Lee, K.S.; El-Sayed, I.H.; El-Sayed, M.A. Calculated absorption and scattering properties of gold nanoparticles of different size, shape, and composition: Applications in biological imaging and biomedicine. *J. Phys. Chem. B* **2006**, *110*, 7238–7248. [CrossRef] [PubMed]

31. Khan, Z.; Hussain, J.I.; Kumar, S.; Hashmi, A.A.; Malik, M.A. Silver Nanoparticles: Green Route, Stability and Effect of Additives. *J. Biomater. Nanobiotechnol.* **2011**, *2*, 390–399. [CrossRef]

32. Salopek, B.; Krasic, D.; Filipovic, S. Measurement and application of zeta-potential. *Rudarsko-Geolosko-Naftni Zbornik* **1992**, *4*, 147–151.

33. Ito, T.; Sun, L.; Bevan, M.A.; Crooks, R.M. Comparison of nanoparticle size and electrophoretic mobility measurements using a carbon-nanotube-based coulter counter, dynamic light scattering, transmission electron microscopy, and phase analysis light scattering. *Langmuir* **2004**, *20*, 6940–6945. [CrossRef] [PubMed]

34. Boisselier, E.; Astruc, D. Gold nanoparticles in nanomedicine: Preparations, imaging, diagnostics, therapies and toxicity. *Chem. Soc. Rev.* **2009**, *38*, 1759–1782. [CrossRef] [PubMed]

35. Davis, S.; Mertz, P.; Eaglstein, W. Second-degree burn healing: The effect of occlusive dressings and a cream. *J. Surg. Res.* **1990**, *48*, 245–248. [CrossRef]

36. Pan, Z.; Lee, W.; Slutsky, L.; Pernodet, N.; Rafailovich, M.H.; Clark, R.A.F. Adverse effects of titanium dioxide nanoparticles on human dermal fibroblasts and how to protect cells. *Small* **2009**, *5*, 511–520. [CrossRef] [PubMed]

37. Karthika, V.; Kaleeswarran, P.; Gopinath, K.; Arumugam, A.; Govindarajan, M.; Alharbi, N.S.; Khaled, J.M.; Al-anbr, M.N.; Benelli, G. Biocompatible properties of nano-drug carriers using TiO_2-Au embedded on multiwall carbon nanotubes for targeted drug delivery. *Mater. Sci. Eng. C* **2018**, *90*, 589–601. [CrossRef] [PubMed]

9

Effects of Sample Preparation on Particle Size Distributions of Different Types of Silica in Suspensions

Rodrigo R. Retamal Marín [1,*], **Frank Babick** [1], **Gottlieb-Georg Lindner** [2], **Martin Wiemann** [3] and **Michael Stintz** [1]

[1] Research Group Mechanical Process Engineering, Institute of Process Engineering and Environmental Technology, Technische Universität Dresden, Münchner Platz 3, D-01062 Dresden, Germany; frank.babick@tu-dresden.de (F.B.); michael.stintz@tu-dresden.de (M.S.)

[2] Evonik Resource Efficiency GmbH, Brühler Straße 2, 50389 Wesseling, Germany; gottlieb-georg.lindner@evonik.com

[3] IBE R&D Institute for Lung Health gGmbH, Mendelstr 11, D-48149 Münster, Germany; martin.wiemann@ibe-ms.de

* Correspondence: rodrigo.retamal@tu-dresden.de

Abstract: The granulometric characterization of synthetic amorphous silica (SAS) nanomaterials (NMs) still demands harmonized standard operation procedures. SAS is produced as either precipitated, fumed (pyrogenic), gel and colloidal SAS and these qualities differ, among others, with respect to their state of aggregation and aggregate strength. The reproducible production of suspensions from SAS, e.g., for biological testing purposes, demands a reasonable amount of dispersing energy. Using materials representative for each of the types of SAS, we employed ultrasonic dispersing (USD) at energy densities of 8–1440 J/mL and measured resulting particle sizes by dynamic light scattering and laser diffraction. In this energy range, USD had no significant impact on particle size distributions of colloidal and gel SAS, but clearly decreased the particle size of precipitated and fumed SAS. For high energy densities, we observed a considerable contamination of SAS suspensions with metal particles caused by abrasion of the sonotrode's tip. To avoid this problem, the energy density was limited to 270 J/mL and remaining coarse particles were removed with size-selective filtration. The ultrasonic dispersion of SAS at medium levels of energy density is suggested as a reasonable compromise to produce SAS suspensions for toxicological in vitro testing.

Keywords: nanomaterials (NMs); nanostructured; synthetic amorphous silica (SAS); ultrasonic dispersing (USD); energy density; sample preparation; in vitro testing

1. Introduction

Modification of physico-chemical properties of nanomaterials (NMs) or nanostructured materials allows the control and variation of design, development and improvement of new products. Synthetic amorphous silica (SAS) comprise an important group of NMs, which are added to industrial as well as consumer products such as cosmetic or foods [1–7] within which they serve e.g., as stabilizers, thickeners, pigments, flow enhancing agents, or UV absorbers [3,8–10]. Based on some concern regarding possible health impacts and safety risks of NMs, legal authorities request the toxicological analysis of SAS NMs by means of in vivo and in vitro studies [11–15].

Generally, ultrasonic dispersing or separation has been used for sample preparation of nanomaterials for safety assessment [16]. These studies can support the optimization of nanosynthesis or nano-applications for the sake of a "Green Synthesis of Nanomaterials". One example for such green nano-application of amorphous silica in entomology and parasitology as a nanopesticide has been considered safe for humans because of the specific mechanisms of action [17].

An important aspect of exposure and toxicological analyses is the characterization of NMs with respect to particle size. Most SAS occur in an aggregated state with particle sizes ranging from nanometer-sized primary particles to micrometer-sized aggregates or agglomerates [18–20]. However, the sample preparation for a specific in vitro test should consider the particle size-distribution which is of relevance for a given exposure pathway [21–24]. For example, inhalation of particles into the human respiratory tract leads to a fractionation of particles: larger agglomerates are deposited in the nasopharyngeal region (5–30 µm), small agglomerates are partially deposited in the tracheobronchial region (1–5 µm), and only small (<1 µm) and nano-sized materials (1–100 nm) may penetrate into the alveolar region of the lung [25–30]. Thus, toxicity testing of NMs using in vitro lung models demands the preparation of properly suspended samples under defined conditions.

In general, studies on environmental and health risk assessment focus on transport and deposition of NMs in real-life exposure scenarios. Both processes are governed by the mobility of aggregates and agglomerates, for which reason the size of aggregates and agglomerates needs to be measured. This is different from the nanomaterial definition recommendation of the European Commission, which is based on number-weighted distribution of the minimum size of isolated particles or constituent particles within aggregates and agglomerates [31,32].

The analysis of nanomaterials (NMs) and nanostructured materials requires standard operation procedures (SOPs) for the preparation of suspension samples to ensure defined granulometric states [33–35]. Therefore, the preparation and analysis of nanomaterial suspensions needs a high degree of standardization with respect to primary sample preparation (stock suspension), secondary sample preparation, conditioning (e.g., adjusting suspension composition or concentration), sample splitting and finally measurement/interpretation. All these steps need to be considered for the characterization of liquid-suspended powders and for the comparison of different SOPs in view of their reproducibility. Although wetting and low energy dispersion of SAS powders in the suspension are substantial components of (primary) sample preparation, further ultrasonic dispersing (USD) is needed as it is the most versatile method to disintegrate large particle agglomerates into small particle aggregates or primary particles. At the same time, stabilization and homogenization of the dispersed particles are necessary.

USD is a rather intense type of dispersing, which relies on the hydrodynamic stress caused by collapsing cavitation bubbles. As USD can be performed with different types of equipment (e.g., ultrasonic bath, high-power probe sonicator, or cup-horn sonication) methods ensuring its reproducibility are mandatory. Several studies have shown that the energy density (measured in J/mL) serves as a well-suited parameter for obtaining a largely identical degree of USD of nanomaterials among different laboratories [23,36,37]. Of note, the application of different ultrasonic dispersing methods (e.g., variation of the sonotrode geometry) and parameter settings (e.g., sonication time and vibration amplitude) requires the application of the energy density concept, applicable to ultrasonication, rotor-stator systems [37,38] or high-pressure dispersing [39–41]. In all these cases, the ultrasonic dispersing energy density can be used as a main parameter for comparing available and new sample preparation protocols or SOPs.

To achieve a reproducible particle size analysis, it is necessary to use adequate sample preparation techniques. The USD facilitates the disintegration of submicron agglomerates, which are therefore of special relevance for the preparation of stable and homogeneous distribution of particles in the suspension and contribute to achieve a stable dispersion. The particles are under interaction of different dispersion forces which control their random dispersion in the sample volume. The energy density has been used in Table 1 to compare several studies, which are developed for the application to nanostructured materials).

Table 1. Published protocols of ultrasonic dispersing (USD) specifically designed for application to nanostructured materials; characteristic parameters including the (range of) inserted energy density.

Protocol	Sample Volume	Dispersing Time	(Calorimetric) Energy Density
Tantra 2016 [42]; Pradhan 2016 [43]	6 mL	16 min	1176 J/mL
Rasmussen et al., 2013 [35]	15 mL	10 min	500–400 J/mL
	10 mL	16 min	2500 J/mL
Taurozzi et al., 2012 [44]	50 mL	5 min	300 J/mL
Jensen et al., 2011 [33]	6 mL	16 min	3140 J/mL
Bihari et al., 2008 [45]	1 mL	1 min	420 J/mL
Mandzy et al., 2005 [46]	-	Time frames (2 h)	5700 J/mL
Pohl et al., 2005 [37]	10–42 mL	17–630 s	400–30,000 J/mL
Pohl et al., 2004 [47]	3–6 mL	-	100–2000 J/mL

Table 1 shows that the energy densities used in several recent studies to disperse nanomaterials differ by more than two orders of magnitude. This raises the question, as to which extent they influence the results of particle size measurement. Previous studies on different grades of NMs such as SiO_2, Al_2O_3, or TiO_2 have shown that even with a comparatively high energy density of up to 5 kJ/mL a maximum dispersion cannot be achieved for all materials [20]. However, as administration of such energy densities requires extensive cooling of the samples and prolonged periods of ultrasonic treatment, we were seeking for a reasonable compromise to achieve an acceptable dispersion of nanomaterials.

In this paper, we examine the effect of USD energy on the dispersion of SAS and characterize resulting particle size distributions (PSD). Despite their identical chemical composition, SAS products show considerable variations with respect to the synthetic routes, particle morphology, and product properties. The synthesis of silica is realized either in aqueous solution based on sodium silicate solution or in gaseous phase from $SiCl_4$ [48,49]. The types of silica originating from silica synthesis processes in aqueous solution are silica gel (SG), precipitated silica (PS) and colloidal silica (CS). Fumed silica (FS), also referred to as pyrogenic silica, is synthesized from gaseous phase. SAS products are nanostructured NMs [46] as they consist of aggregates and agglomerates of nanosized constituent particles (FS, PS and SG, cf. [50]) or well-dispersed nano-objects (CS). Accordingly, the preparation of suspensions of FS, PS and SG requires defined dispersion procedures for their use, e.g., in toxicity studies, and characterization, whereas this is not really necessary for colloidal silica [19,20]. Furthermore, the different types of silica have a characteristic morphology due to their varied synthetic processes. This is an important issue that needs to be considered for comparison and data interpretation (e.g., reproducibility, effectiveness).

2. Materials and Methods

2.1. Materials

This study analyzed SAS products, representative for FS, PS, SG and CS, respectively. While FS, PS, SG were provided as untreated hydrophilic powders, CS was provided as an aqueous suspension. Important physico-chemical properties are summarized in Table 2.

Table 2. SAS properties.

SAS Type Internal Code	Fumed Silica F-3	Precipitated Silica P-2	Silica Gel G-1	Colloidal Silica C-1
BET [1] (m^2/g)	300	440	700	200 [2]
solid content for suspensions (wt.-%)	-	-	-	40
pH [3]	5	6.5	4.4	9.7
electric conductivity (μS/cm) at 25 °C	4	160	55	4771.6

[1] BET: Surface measured according to Brunauer, Emmet and Teller [51,52]. [2] measured from freeze dried material. [3] suspended in ultra-pure water (1 wt.-%, 25 °C).

Figure 1a–c show SEM (JEOL Ltd, Tokyo, Japan) images of typical aggregates of FS, PS and SG [50]. In contrast, CS (Figure 1d) contains isolated spherical nanoparticles, which have gathered into an agglomerate-like structure upon drying on the TEM grid.

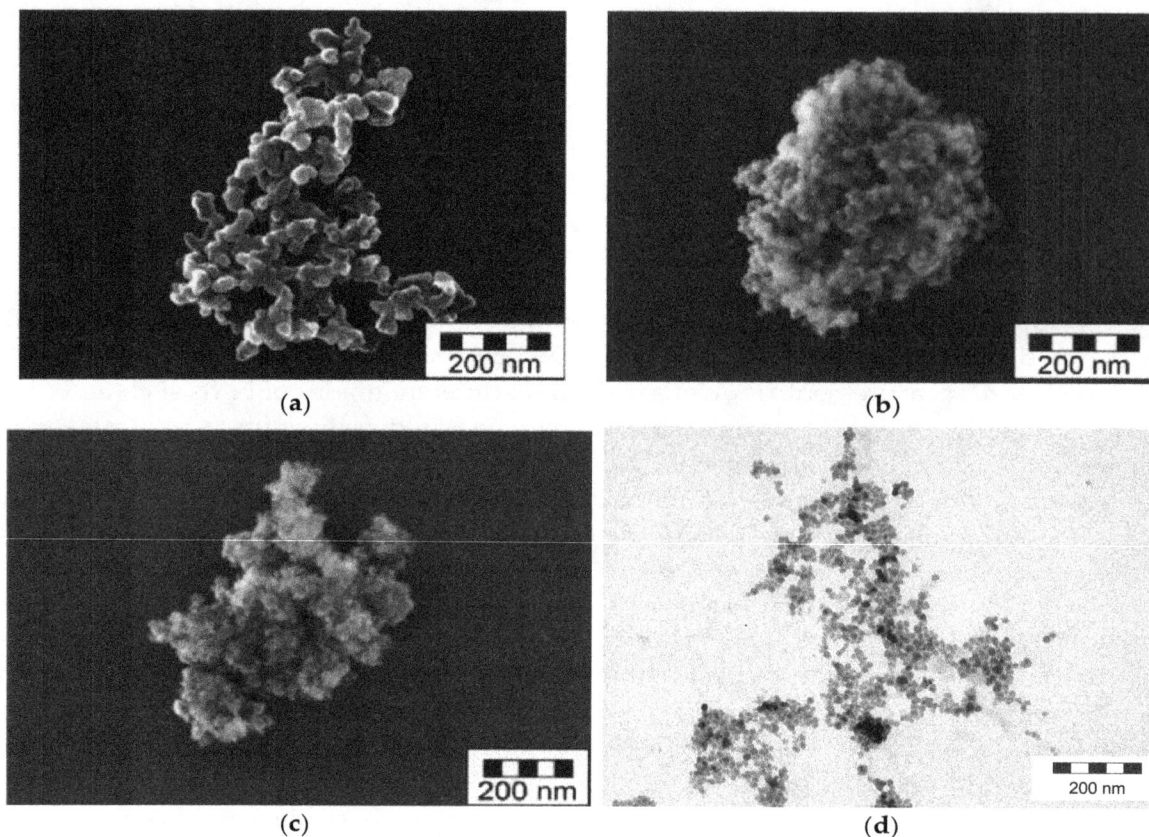

(a)

(b)

(c)

(d)

Figure 1. SEM and TEM images of different silica types. (**a**) fumed (pyrogenic) silica (opened fractal-like aggregates), (**b**) precipitated silica (compact fractal-like aggregates), (**c**) silica gel (compact and microporous fractal-like aggregates) and (**d**) colloidal silica (isolated spherical nanoparticles or small aggregates, here dried on TEM grid to opened agglomerates).

2.2. Instruments and Procedures for Sample Preparation

To prepare stock suspensions of SAS 1 wt.-% of silica powders (FS, PS, SG) were dispersed in 100 mL de-ionized water (18.3 MΩcm, 0.2 μm filtered). To avoid re-agglomeration, the particles were placed in a liquid environment that ensured high surface charges. The pH value of prepared silica suspension is far from the isoelectric point of silica (e.g., pH 1.8–2.5) and it has a low electric conductivity (see Table 2) [53–55]. Powders were dispersed by different treatments: Firstly, by means of a paddle stirrer (PS) (model RW 11 basic, IKA, Staufen, Germany), which administered the lowest input of mechanical energy into the suspension and which was used for the homogenization of SAS suspensions. Of note, the geometric size of the paddle stirrer and the sample beaker as well as stirring velocity ensured hydrodynamic equivalence to the "paddle apparatus" specified in Ph. Eur. 5.7. (2006) [56]. Secondly, by means of a turbulent shear rotor stator (RS) (Ultra-Turrax T25, IKA) which was used to achieve advanced dispersion and to investigate changes of PSD of different SAS types upon progressive dispersion energy. The RS provides shear forces, which cause shear stress on particle agglomerates. Thirdly, by means of immersion horns (three different instruments, see Table 3) which were used in most experiments [33–35]. The sonotrode or horn is in direct contact with the suspension and the dispersion effect is associated to cavitation, which occurs in highly intensive sound fields. Furthermore, the cavitation causes the formation of vapor cavities in a liquid (bubbles) which steadily grow to a critical size, at which they turn instable and implode [20]. This implosion produces high temperature and rapid micro-jets, which exert mechanical stress to the particles close to the formed bubble [57]. This mechanical stress leads to the fragmentation or at the least to the erosion upon the direct contact with the imploding cavities [58,59].

The USD equipment consisted of three different instruments equipped with different sonotrodes (a few mm up to a few cm) (see Table 3), which were operated at frequencies in the range of 20 kHz to 100 kHz. Instruments had a nominal power consumption of a few Watt to approximately 1 kW [18]. The calorimetric energy input was measured at different dispersing instruments. Most of them work at frequencies in the range of 20 kHz. In pilot experiments, various tip diameters of the sonotrodes were selected according to the geometry of the beaker and the sample volume required by the sample preparation protocols.

Table 3. Technical data of ultrasonic dispersion instruments operated at approximately 20 kHz.

Model	Vibra-Cell 72412 [1]	UDS751 [2]	SONIFIER 450D [3]
Code	**V**	**T**	**B**
company	Sonics and Materials	Topas GmbH	Branson Ultrasonics
normal capacity (W)	600	200	400
tip diameter (Ø, mm)	13 19	3 7 14	5 13
amplitude (%)	0–100	0–100	10–100

[1] Vibra-Cell 72412 (Sonics & Materials, Newtown, CT, United States). [2] UDS751 (Topas GmbH, Dresden, SN, Germany).
[3] SONIFIER 450D (Branson Ultrasonic Corporation, Danbury, CT, United States).

For the particle size analysis of all silica types in this study one USD equipment was selected, the generator Vibra-Cell 72412 (Sonics and Materials; 20 kHz, nominal power: 600 W together with a 19 mm tip diameter solid probe. The tip of the probe was replaced for each series of dispersing experiments. USD was performed at maximum amplitude (100%) in a pulsed mode (2s:2s) with the probe being uniformly immersed in the sample. The same type of glass beaker was employed for all samples; the beakers were placed in cooled water during the USD. Even though, samples were steadily heated-up with ongoing USD, for which reason the USD was interrupted after a maximum of 4 min to cool down the complete the sample and the ultrasonic probe. This procedure ensured that the sample temperature stayed below 33 °C [20].

2.3. Instruments for Particle Size Analysis

USD leads to deagglomeration and disintegration of aggregates and the corresponding change in the granulometric state was quantified by laser diffraction (LD) and dynamic light scattering (DLS). However, these standard analytical techniques are based on mathematical models, which are not perfectly applicable to the examined particle systems, e.g., because they assume spherical particles (e.g., Stokes-Einstein relation and Mie theory for DLS), or because they do not cover the whole size range of broad distributions (e.g., Fraunhofer diffraction theory for LD) [18–20,60,61]. To avoid these technical limitations and for a better interpretation of measured data it is advantageous to apply both techniques to well-adapted dilutions of the same sample.

LD measurements were carried out with a HELOS KR (Sympatec, Clausthal-Zellerfeld, Germany) for angular ranges below 35° (i.e., forward scattering). Within this study an angular range of 0.1° to 9° (measurement range R3) was used, which is sensitive for particles of 0.5 µm to 175 µm, but which is insensitive to nanoparticles ($x \leq 100$ nm). These measurements were rather insensitive to small, weakly scattering particles << 1 µm.

To quantify particle sizes in sub-micrometer range, DLS measurements were conducted. The employed instrumentation, HPPS (Malvern, UK), bases on backscattered (173°) and sideward scattered light, respectively. Measured DLS signals (i.e., autocorrelation functions) were analyzed with inversion procedures, which compute complete size distributions, and cumulant analysis. The latter yields a polydispersity index, PDI, and a characteristic mean particle size, x_{cum}, which is the harmonic mean of the intensity-weighted size distribution. The samples are filled in closed cuvettes (4 mL), which are placed in the temperature-controlled sample holder at least 15 min before the measurements.

2.4. Estimation of the Calorimetric Energy Input

The effectivity of the ultrasonication, in comparison with other dispersion procedures regarding size reduction, demands the necessity to evaluate the applied acoustic energy per unit suspension volume (E_V) [18,20,23,37,46]. The calculation of ultrasonic dispersion energy cannot be calculated directly. There are two ways to estimate the inserted acoustic energy. One way is the estimation from electrical energy consumption ($E_{V,el}$) and the other possibility is through the generated heat after implosion of bubbles ($E_{V,cal}$). The electrical energy consumption depends on many factors: transformation of energy of ultrasonic dispersions instruments (e.g., normal capacity, range of frequencies, types and probe diameter) that must be considered for comparison and validation. Furthermore, another important point to consider is the acoustic reflections inside the probe depending on, for example, sample volume, density and the propagation velocity of sound of liquid (acoustic impedance) [62].

This study uses the calorimetric energy input as decisive parameter for the effect of USD. Therefore, the USD devices needed to be calibrated regarding the calorimetric power input for the setup and settings (sample volume, horn diameter; nominal ultrasonic amplitude) employed in dispersing the SAS samples. The calibration comprises the evaluation of temperature increase by ultrasonication of a defined volume of de-ionized water. In this study, USD was conducted at 100 mL suspension sample placed in a 150 mL cylindrical borosilicate beaker. Please note that other studies worked with higher volumes (e.g., 500 mL, [23,34]) or smaller ones (e.g., 6 mL [33,42,43]). A further difference to other studies is that the beaker was placed in insulating foam, to minimize heat exchange with environment. The whole setup, including the ultrasonic probe as well as a thermometer with short response-time, was allowed to thermally equilibrate. Then ultrasonication was started and the temperature within the beaker was recorded as function of time.

The calorimetric energy input into a suspension sample by ultrasonication can be calculated from the heat production rate P_{cal} which is a function of the dispersion time t_{disp}. The former parameter is valid for defined conditions of ultrasonication, i.e., for defined sample properties and USD settings. It can be determined by specifically designed experiments that measure the temperature increase when sonicating the particle-free dispersion medium (e.g., water). Furthermore, it is necessary to consider the mass and specific heat of the beaker to have a correct determination of heat production and to evaluate the initial slope of temperature increase, which needs to be less than 4 Kelvin ($\Delta T \leq 4$ K). Furthermore, the following proposed equation to estimate the calorimetric energy input assumes that the sonotrode has a zero heat capacity and that there is no heat exchange with the environment:

$$P_{cal} = (m_w c_{p,w} + m_b c_{p,b}) \cdot \Delta T / t_{disp} \tag{1}$$

where m_w denotes the mass of liquid, m_b denotes the mass of beaker, $c_{p,w}$ specific heat of water, $c_{p,b}$ specific heat of beaker, ΔT the temperature increase and t the dispersion time [63,64]. The calorimetric energy density can be determined as follows:

$$E_{V,cal} = \frac{P_{cal} \cdot t_{disp}}{V} = (m_W c_{p,W} + m_B c_{p,B}) \cdot \Delta T / V \tag{2}$$

with V being the suspension volume.

The energy density is considered to be a most important process parameter when dispersing suspensions and emulsions. Frequently, a power-law relationship can be established between average particle size x and energy density $E_{V,cal}$ (e.g., [47]):

$$\overline{x} \propto E_{V,cal}^{-b} \tag{3}$$

where "average particle size" can be any characteristic distribution parameter (e.g., median size or arithmetic mean) referring to a defined type of quantity, in which the size distribution is weighted (e.g., number, volume or scattering intensity). The exponent b describes the material's dispersibility under the specified conditions.

Finally, there are some assumptions to estimate values for calorimetric calibration of ultrasonic instruments. The first assumption is the uniform temperature of water and beaker, second—zero heat capacity of ultrasonic probe and third—no heat transfers out of the system beaker-water.

3. Results and Discussion

3.1. Calorimetric Calibration of Probe Sonication

While the ultrasonic wave propagates through the dispersion medium, its energy is absorbed and converted into heat [65,66]. Figure 2 exemplifies this progressive heating during USD for a non-insulated beaker. The images give an impression of how the heat generated upon ultrasonication is transferred from the sonotrode into the surrounding medium and associated masses. Therefore, it is important to consider some issues such as sample volume, or isolating foam to achieve a correct calorimetric calibration of probe sonication, as explained in the proposed protocol (see Section 2.4).

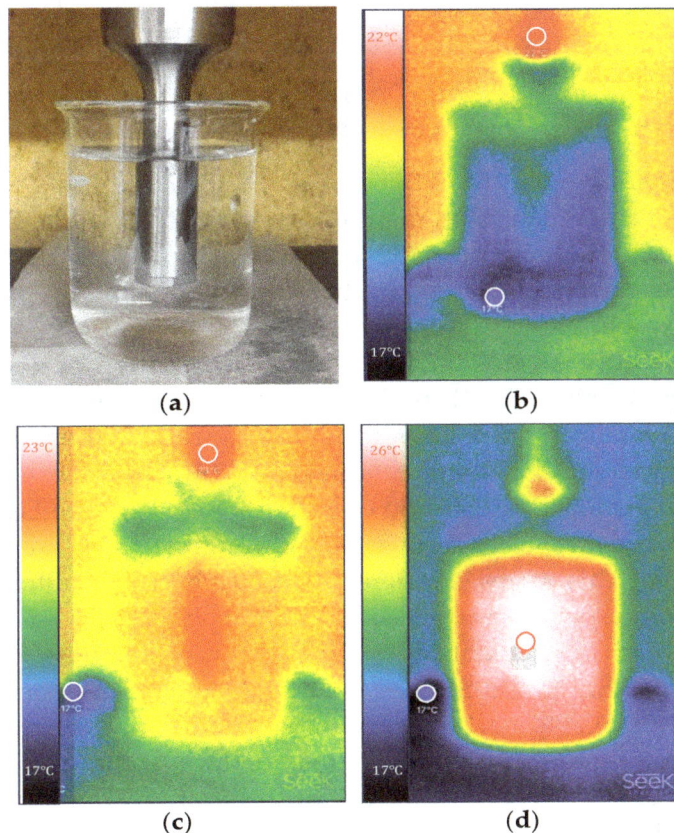

Figure 2. Progressive heating of the sonotrode, sample fluid, and beaker in the course of USD. Images were captured with a thermal imaging camera (Seek Thermal) after different dispersing periods; (**a,b**): 0 s, (**c**): 30 s, (**d**): 60 s. Temperature (in degree Celsius) is shown on a pseudo-color scale whose range was automatically adjusted to the temperature peak (T_{min}/T_{max}) of the measurement (i.e., (**a**) 17 °C/22 °C, (**b**) 17 °C/23 °C, (**c**) 17 °C/26 °C).

The calorimetric calibration was calculated for different ultrasonic dispersing instruments (according to the protocol in Section 2.4) to compare their heat production after setting of different parameters (see Section 2.2). The calorimetric measurements delivered the results in Figure 3.

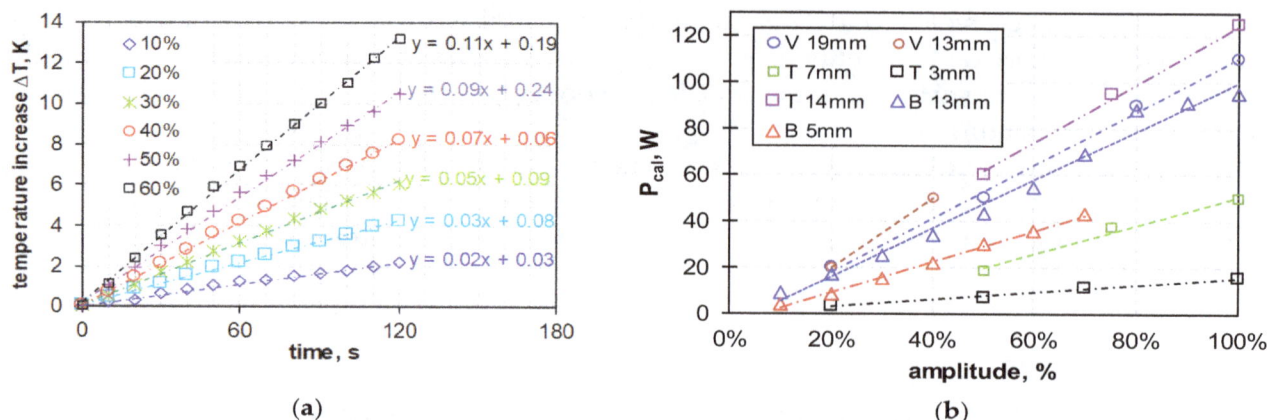

Figure 3. Calorimetric calibration curves of ultrasonic dispersing instruments. (**a**) temperature increases (ΔT) in Kelvin (K) over time, obtained with a 13 mm tip diameter sonotrode (Branson 450D) and increasing amplitudes (10–60% from maximum, as indicated in the graph). (**b**) Heat production (P_{cal}, W) of the different dispersing instruments and sonotrode geometries Vibra-Cell 72412 (V), Topas UDS751 (T) and Branson SONIFIER 450F (B) (outlined in Table 3) as a function of increasing amplitude (in % from maximum).

3.2. Sample Preparation by Probe Sonication

3.2.1. Impact of USD on Particle Size Distribution of SAS

Previous studies have shown that the PSD of, e.g., FS may be highly polydisperse [20,54]) and covers a wide range from of a few nanometers up to several micrometers [18–20]. As outlined above (Section 2.3), the appropriate granulometric analysis of such samples requires a combination of LD and DLS. This section addresses the effectiveness of ultrasonic dispersing on particle size of silica types measured with both techniques.

Suspensions of the SAS types were prepared by a combination of dispersing procedures and defined calorimetric energy densities (EV, J/mL). Depending on the instrumentation (see Section 2.3) dispersing energies were weak from propeller stirrer (PS), moderate from rotor-stator (RS), or intense from ultrasonication (US, different dispersing energies) and the selected calorimetric energy densities' values are based on the calorimetric calibration of probe sonication (see Figure 3). The calorimetric energy density of the US treatment ranged from 8 J/mL through 18 J/mL and 270 J/mL to a maximum energy input of 1440 J/mL. This stepwise increase of energy input allows for a comprehensive characterization of SAS with respect to particle size and morphology.

Figures 4–6 show the LD measurements of the PSD results of the transformed distribution density of upon increasing $E_{V,cal}$ for PS, FS, and SG (see Section 2.1). The transformed distribution density represents in accordance with ISO 9276-1:1998 provide the differential size distribution on a log scaled abscissa. Areas under the curve represent the volume portions of the size classes [67]. The result of a long-term sedimentation of silica suspensions (after five months) is shown in parallel.

Precipitated silica (PS, 400 m²/g, 1 wt.-%) shows a clear tendency of deagglomeration by increasing dispersion energy (Figure 4a). Especially the presence of coarse, micrometer-sized agglomerates was diminished, and this effect started mostly at 18 J/mL. In line with these results the sedimentation profiles of ultrasonicated PS suspensions show an increasing degree of opacity upon 270 J/mL and 1440 J/mL (Figure 4b), suggesting the presence of slowly settling, light scattering particles in the sub-micrometer range. However, although the zone of opacity was wider upon 1440 J/mL, the volume of the white matter at the bottom was similar.

Largely similar to PS was fumed (pyrogenic) silica (FS, 300 m²/g, 1 wt.-%) deagglomerated by increasing USD energy (Figure 5a). The deagglomeration of large particles was achieved already with a moderate dispersion (RS) and increased further upon administration of USD energy. Interestingly, the mono-modal PSD of FS became bimodal upon 18, 270, and 1440 J/mL, such that a fine and a large fraction could be distinguished (Figure 5a). Of note, the large fraction generated by 270 J/mL and 1440 J/mL comprised larger particles as compared to the fraction induced by 18 J/mL. This suggests that elevated energy levels at least in part can provoke an agglomeration of FS. The effect was most obvious at an USD energy of 270 J/mL. These results confirm previous studies, where the coarser particles appeared at energy density levels above 171 J/mL [20]. The sedimentation profiles of FS at higher dispersion energies showed a similar degree of opacity in the supernatant, but an increased volume of the white sediment. This suggests that high USD energy leads to the formation of larger particles from the same mass of FS (see Figure 5b).

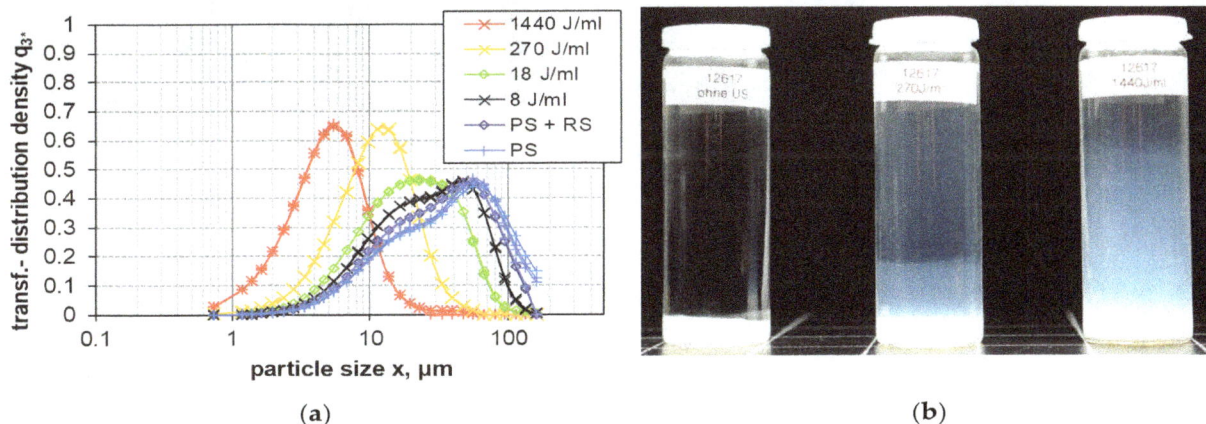

(a)

(b)

Figure 4. Particle size distribution of PS (440 m^2/g) dispersed by different procedures. (**a**) Size distribution measured by laser diffraction spectroscopy plotted as the transformed distribution density (q^{3*}). Weak (PS), moderate (RS) and intense (US) dispersion with increasing ultrasonic energies (indicated in the diagram) were employed. (**b**) Silica suspensions after five months (left 1 J/mL, middle 270 J/mL, and right 1440 J/mL).

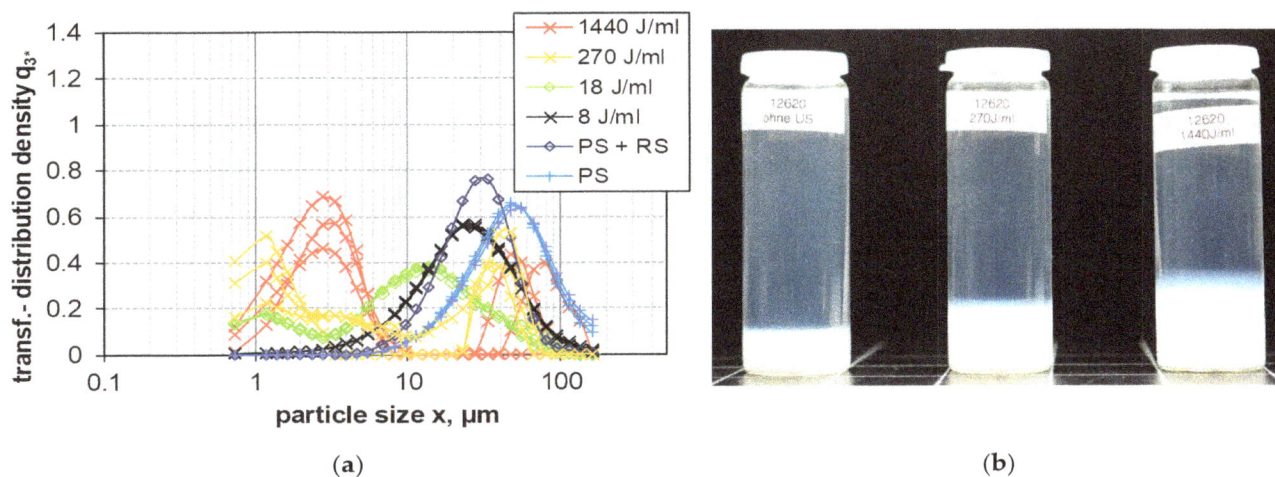

(a)

(b)

Figure 5. Particle size distribution of FS (300 m^2/g) dispersed by different procedures. (**a**) Size distribution measured by laser diffraction spectroscopy plotted as the transformed distribution density (q^{3*}). Weak (PS), moderate (RS) and intense (US) dispersion with increasing ultrasonic energies (indicated in the diagram) were employed. (**b**) Silica suspensions after five months (left 1 J/mL, middle 270 J/mL, and right 1440 J/mL).

As can be seen in Figure 6a, the PSD of silica gel (SG, 700 m^2/g, 1 wt.-%) remained unchanged upon increasing energy density. SG consists of compact (dense) and microporous fractal-like aggregates (see Figure 1), which appear to be insensitive to high USD energies and undergo a rapid sedimentation of particles (by median $x_{50,3} = 6$ µm), irrespective of USD treatment (see Figure 6b), although both ultrasonic treatments led to a similar degree of opacity of the supernatant which may be indicative of smaller particles (not measurable by LD).

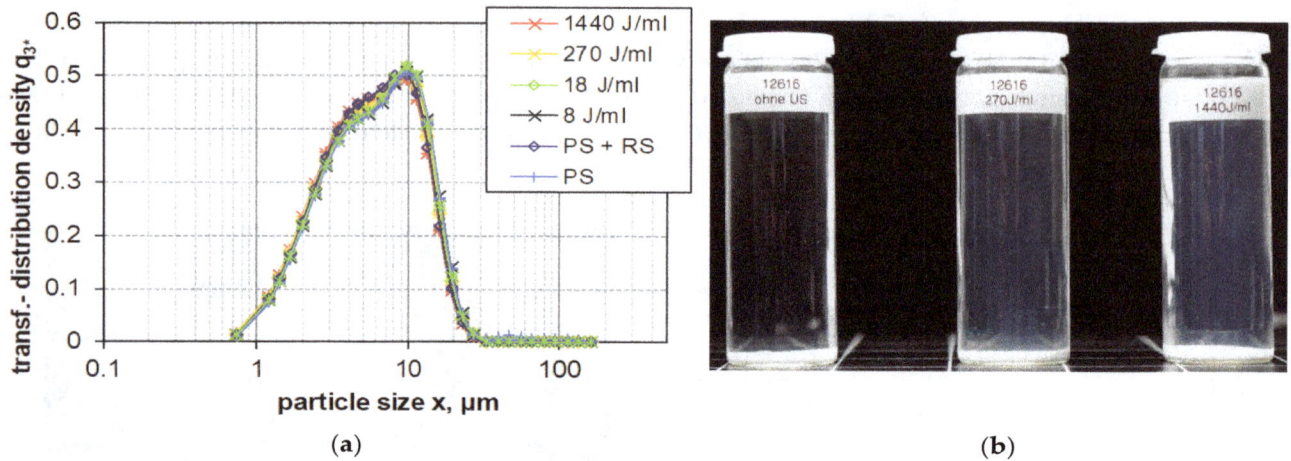

(a)   (b)

Figure 6. Particle size distribution of SG (BET: 700 m^2/g) dispersed by different procedures. (**a**) Size distribution measured by laser diffraction spectroscopy plotted as the transformed distribution density (q^3*). Weak (PS), moderate (RS) and intense (US) dispersion with increasing ultrasonic energies (indicated in the diagram) were employed. (**b**) Silica suspensions after five months (left 1 J/mL, middle 270 J/mL, and right 1440 J/mL).

The effect of increasing dispersion energy on particle size of PS, SG and FS, as measured by LD, is compared in Figure 7, which shows the trend analysis for the $x_{50,3}$ and $x_{99,3}$ quantiles of the volume weighted size distribution. Whereas the particle size of SG is not altered by increased dispersion energy, particle size of PS is constantly lowered. In the case of FS, only the $x_{50,3}$ value reflects the decrease in particle size, whereas the $x_{99,3}$ is inconsistent and shows an upwards trend demonstrating the coarsening or reagglomeration upon high USD energy. Figure 8 shows a comparison the stability of intensely dispersed silica types at 1440 J/mL after long-term sedimentation (5 months). The different sediment degrees of silica types support the LD results and provide a subjective information about sedimentation velocity in the gravitational field of the earth.

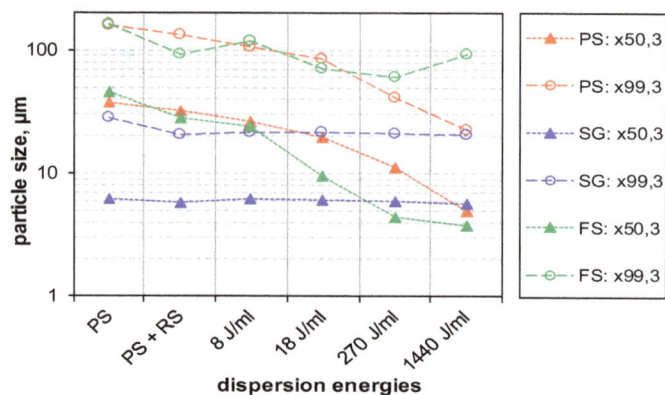

Figure 7. Particle size distribution of silica suspension after administration of increasing dispersion energy as measured by laser diffraction. Curves show the trends for the $x_{50,3}$ and $x_{99,3}$ quantiles of the volume weighted size distribution. USD energy density of ultrasonic treatment is indicated in J/mL.

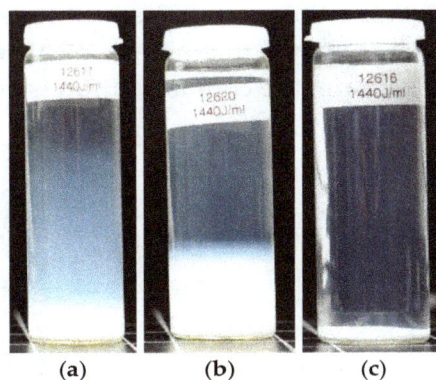

Figure 8. Intensely dispersed SAS samples (USD, $E_{V,cal}$: 1440 J/mL) after long-term sedimentation (5 months): (**a**) PS; (**b**) FS; (**c**) SG.

Since LD is not sensitive for silica particles smaller than 1 µm [68], the characterization of sub-micrometer particles (1 nm–10 µm) was carried out with DLS. Data of three silica types (FS, SG, PS) was expressed as intensity-weighted size distribution, using the characteristic values mean size (xcum) and the polydispersity index (PDI) obtained by cumulant analysis. Figure 9a compares the as result calculated logarithmic normal distribution (LND) and shows the impact of 270 J/mL and 1440 J/mL on particle size. In Figure 9b the Intensity-weighted transformed distribution density functions are shown. While the particle size of SG and FS remains nearly unchanged in the lower size range, the long-term sedimentation e cumulative size distribution curve of PS is shifted leftwards, indicating that particle size had shifted to submicron and nano range (<1 µm). Fumed silica (FS) shows a minimal tendency to increase the size of submicron particles upon high energy density.

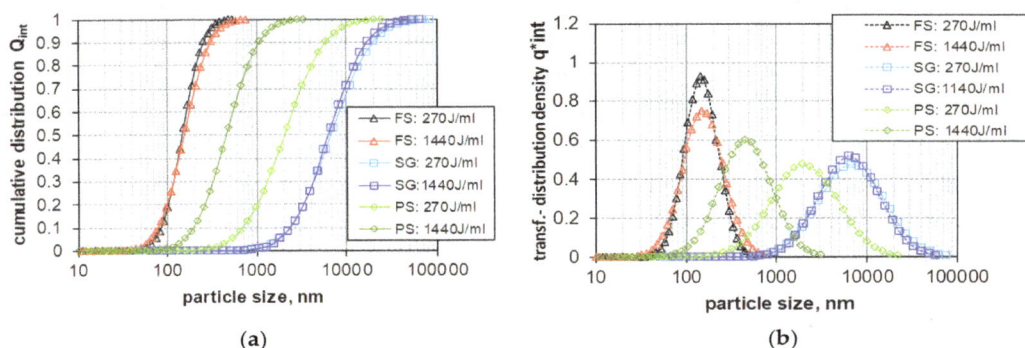

Figure 9. Particle size distribution (PSD) of amorphous silica suspensions dispersed by two different ultrasonic dispersion energy densities (270 and 1440 J/mL). PSD was measured by dynamic light scattering (DLS). FS: fumed silica, SG: gel silica, PS: precipitated silica. (**a**) Intensity-weighted sum functions for different energy density values. (**b**) Intensity-weighted transformed distribution density functions for different energy density values. Logarithmic normal distribution.

Figure 10a shows the granulometric state of colloidal silica (CS) over the full range of dispersion energies as used in Figure 7. While stirring (PS, PS + RS) had no effect on particle size, ultrasonic treatment surprisingly led to a larger and broader PSD indicated by an increase in hydrodynamic size and PDI. The effect started at a low energy density of 8 J/mL and was found to be strongly augmented upon higher USD energies. We found that wear particles from the sonotrode's tip, the larger of which appeared as a sediment at the bottom of the vial (Figure 10b), made a major contribution to this effect. Due to the strong light scattering properties of such metal particles (compared to the small and weakly scattering CS particles), even low amounts of wear particles contaminate the light optic measurements. If this increase of the mean particles size would be caused by agglomerated colloidal silica particles, they would be visible as a sediment layer after 5 months. In the case of silica gel scattering, intensity of the micrometer particles hides the contamination signals during measurement whereas, after settling, the contamination is embedded in the silica sediment.

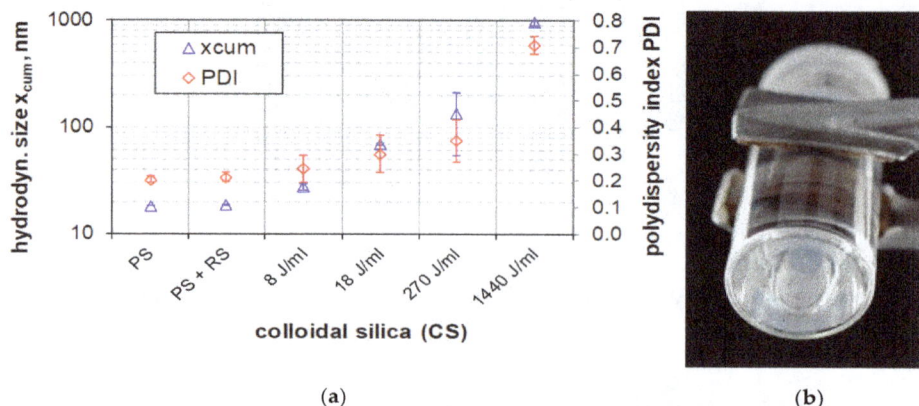

(a)   (b)

Figure 10. Particle size distribution of colloidal silica dispersed by different procedures measured by DLS. (**a**) Mean particle size (xcum) and corresponding polydispersity index (PDI) as determined with cumulant analysis. (**b**) Bottom view of a vial with an intensely dispersed colloidal silica sample ($E_{V,cal}$: 1440 J/mL) taken after 5 months of gravitational settling.

3.2.2. Sample Contamination with Probe Sonication

Figure 11a,b show the effect of high dispersion energy on the sonotrode's tip if delivered over prolonged period. Wear particles ablated from the tip contaminate suspension and can observed as a black sediment (i.e., coarse titanium particles) and/or as a well as a grey discoloration of the suspension (Figure 11c). Figure 12 shows a SEM picture of sonotrode abrasion particles collected from the bottom of silica suspension. As shown in a previous study, the abrasion of the ultrasonic probe and sample contamination occurs in the moment of ultrasonication; the number of particles increases linearly with time [20]. Furthermore, it was shown that in suspension of pyrogenic silica (PS, 1 wt.-%) at a dispersion energy >171 J/mL sonotrode wear particles contribute to the PSD and interfere with the sample analysis by LD [20]. Nevertheless, this widespread sonicator type is superior to other indirect sonicator types (e.g., ultrasonic bath, cup horn) [36,69] due to its high effectiveness of USD and with regard to the best possible disintegration of agglomerates and aggregates in a short time [20]. Therefore, a restriction of the USD energy appears reasonable. To remove larger particles at lower dispersion energy, e.g., from PS suspensions, we developed a dispersion protocol combining stirring, USD and filtration steps (see Section 3.3).

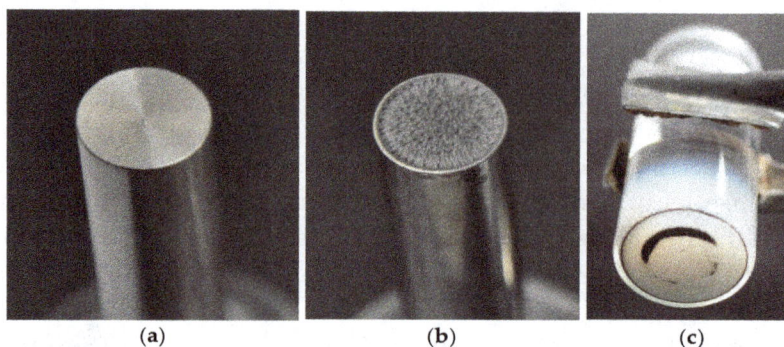

(a)   (b)   (c)

Figure 11. Comparison of new and used sonotrode with sonotrode abrasion as a consequence of long ultrasonic dispersion time (**a,b**); sonotrode abrasion sediment on the bottom of a precipitated silica suspension sample after high USD energy (**c**) (i.e., $E_{V,cal}$: 1440 J/mL).

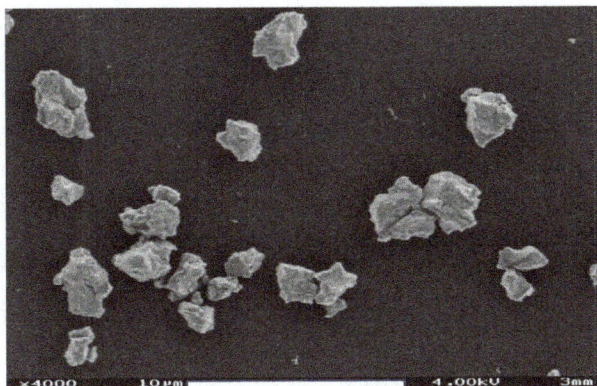

Figure 12. Scanning electron microscope image of wear particles from the sonotrode tip. Abrasion particles were collected from the sediment of a silica suspension sample after high USD energy (i.e., $E_{V,\text{cal}}$: 1440 J/mL).

3.3. Sample Preparation with Size-Selective Filtration

Testing the in vitro toxicity of nanomaterials requires that the size distribution of particles in cell culture media is well defined. With respect to inhalation exposure, which may be tested by the alveolar macrophage assay [24], larger non-respirable coarse particles need to be removed so that a mass-per-volume- or surface-per-volume dose metrics can be applied [15]. Ideally, particle size distribution should reflect inhalable fractions with aerodynamic diameters smaller than 4 µm. However, as outlined above for paddle stirring (PS), this would require high ultrasonic energy and bears the risk of metal particle contamination (see Figures 11 and 12). To circumvent this risk, a size classification by controlled filtration (with 100% fines penetration) was developed, using a commercially available nylon gaze with a pore size of nominally 5 µm (Bückmann, Germany). Figure 13 shows the grade efficiency function $T(x)$ demonstrating that glass spheres below 7 µm can freely permeate the filter, whereas spheres larger than 15 µm were retained.

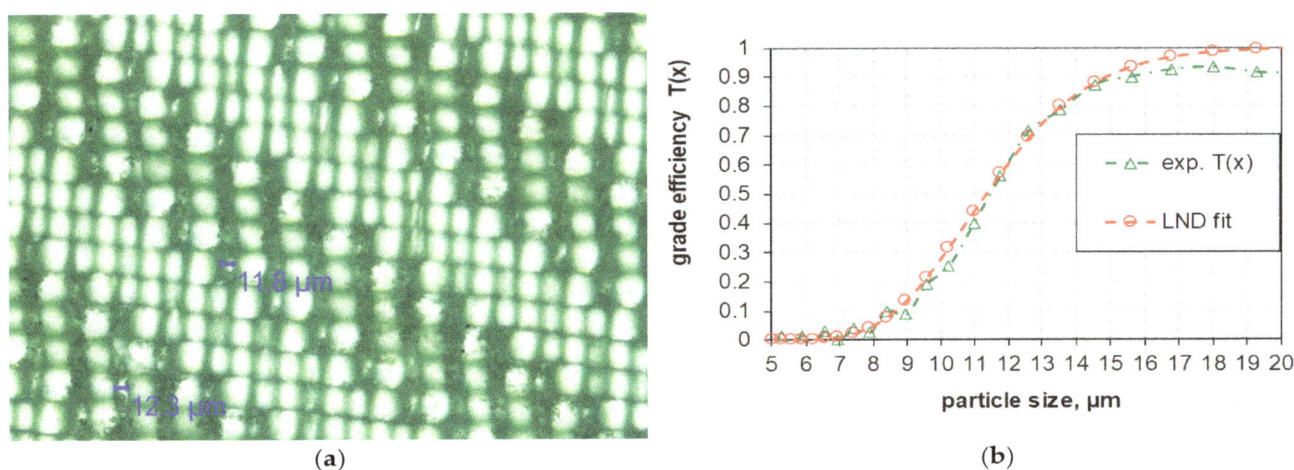

(a)   (b)

Figure 13. (a) Light microscope image of polymer gaze for size-selective suspension filtration and (b) $T(x)$: grade efficiency function after gaze test filtration with glass spheres.

The theoretical volume weighted cumulative distribution function of the filtrate $Q_{3,f}(x)$ was derived from the experimentally determined grade efficiency function $T(x)$ and the feed distribution $Q_{3,i}(x)$:

$$\Delta Q_{3,\,f}(x) = (1 - T(x)) \cdot \left(\frac{\Delta Q_{3,i}(x)}{1 - \sum T(x) \cdot \Delta Q_{3,i}(x)} \right) \tag{4}$$

In the next step ultrasonication and filtration where combined to prepare a suspension suitable for in vitro testing. The method is shown exemplarily for the PS used in this study: the powder was suspended (1 mg/mL) in de-ionized water by means of a magnet stirrer (700 rpm, 10 min). Thereafter the PS sample was filtrated by gaze filter with a grade efficiency curve shown in Figure 13b and a cut-off size of 11.5 μm. The filtrate of the silica sample was then dispersed with $E_{V,cal}$: 18 J/mL and $E_{V,cal}$: 270 J/mL. Results were expressed in Figure 14a as cumulative particle volume curves (Q3, green curve) and compared to the effects of progressive dispersion energy on the size distribution of non-filtered PS (red curve).

Figure 14a shows the advantage of combined moderate ultrasonication by 270 J/mL with filtration (green line) to remove particles larger than 11 μm in comparison with the high ultrasonic energy result (red line), including possible sample pollution. LD results (Figure 14b) show that low energies (i.e., weak dispersion (PS), moderate dispersion (RS) without ultrasonication) leave a considerable amount of micrometer-sized agglomerates in the suspension, whereas ultrasonic dispersion with 8, 18, 270 and 1440 J/mL progressively reduced micrometer-sized agglomerates.

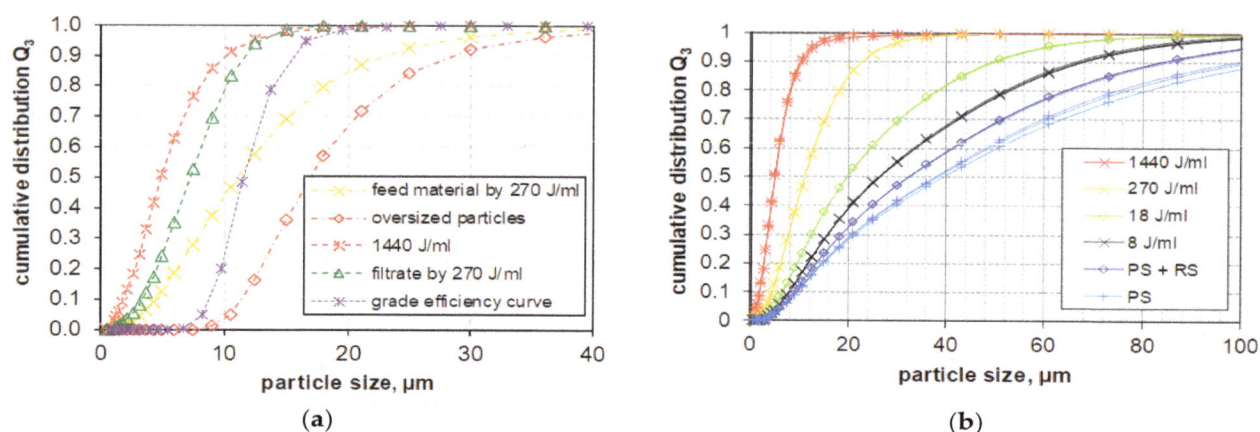

Figure 14. (**a**) filter grade efficiency curve and calculated results for penetrated (filtrate) and retained particle size distributions from feed size distribution (yellow) in comparison to not filtered but with 1440 J/mL dispersed sample (**b**) Evolution of PSD during ultrasonication of precipitated silica (440 m^2/g); cumulative distribution functions measured by LD.

4. Conclusions

The effect of dispersion energy on particle size distribution of nanomaterial suspensions depends not only on a defined dispersion procedure (e.g., dispersion time, sample volume) but also on the silica types (e.g., morphology).

Ultrasonic dispersion energy density is a main parameter for comparability of sample preparation protocols. Sonication is limited by sample pollution with wear particles from the probe. Therefore, upper limit dispersion energy density values must be determined. In the case of silica it is recommended to apply dispersion energy density only up to 300 J/mL.

The resulting particle size distributions strongly depend on the type of silica. Fumed SAS reach PSDs in the submicron range even at low values of ultrasonic energy density; continued sonication leads to a steady, yet slight size reduction. Gel and colloidal SAS are hardly or even adversely affected by increasing ultrasonic dispersion energies. The PSDs of precipitated SAS strongly depends on the increasing ultrasonic dispersion energy, changing constantly to smaller sizes.

Additional size-selective filtration can remove the large and settling particles without the risk of sample contamination by too high ultrasonic energy dispersion. A combination with the above-described ultrasonic dispersion provides a general SOP for the preparation of well-defined suspensions of SAS nanoparticles for in vitro toxicological tests.

Author Contributions: R.R.R.M. is the main author who largely conducted the experimental investigations (including design of experiments and data analysis) and contributed most to the manuscript. F.B. contributed to the development of the sample preparation procedures and the interpretation of the measurement data. He also critically reviewed the original draft of the manuscript. G.G.L. was involved as an expert for the material system, who helped to develop the methodology of sample preparation and measurement. M.W. developed the filtration method and wrote parts of the manuscript. M.S. supervised this study and prepared its basic design. In addition, he performed the characterization of the size-selective filter. He also wrote parts of the manuscript.

Acknowledgments: The authors thank Evonik Resource Efficiency GmbH for providing silica samples and, phys.-chem. data.

References

1. Borm, P.J.A.; Robbins, D.; Haubold, S.; Kuhlbusch, T.; Fissan, H.; Donaldson, K.; Schins, R.; Stone, V.; Kreyling, W.; Lademann, J.; et al. The potential risks of nanomaterials: A review carried out for ECETOC. *Part. Fibre Toxicol.* **2006**, *3*, 11. [CrossRef] [PubMed]

2. Mogharabi, M.; Abdollahi, M.; Faramarzi, M.A. Toxicity of nanomaterials; an undermined issue. *DARU J. Pharm. Sci.* **2014**, *22*, 59. [CrossRef] [PubMed]

3. Fruijtier-Pölloth, C. The safety of nanostructured synthetic amorphous silica (SAS) as a food additive (E 551). *Arch. Toxicol.* **2016**, *90*, 2885–2916. [CrossRef] [PubMed]

4. Lorenz, C.; von Goetz, N.; Scheringer, M.; Wormuth, M.; Hungerbühler, K. Potential exposure of German consumers to engineered nanoparticles in cosmetics and personal care products. *Nanotoxicology* **2011**, *5*, 12–29. [CrossRef] [PubMed]

5. He, X.; Hwang, H.-M. Nanotechnology in food science: Functionality, applicability, and safety assessment. *J. Food Drug Anal.* **2016**, *24*, 671–681. [CrossRef] [PubMed]

6. Winkler, H.C.; Suter, M.; Naegeli, H. Critical review of the safety assessment of nano-structured silica additives in food. *J. Nanobiotechnol.* **2016**, *14*, 44. [CrossRef] [PubMed]

7. Retamal Marín, R.R.; Babick, F.; Stintz, M. Physico-chemical separation process of nanoparticles in cosmetic formulations. *J. Phys. Conf. Ser.* **2017**, *838*, 012004. [CrossRef]

8. Froggett, S.J.; Clancy, S.F.; Boverhof, D.R.; Canady, R.A. A review and perspective of existing research on the release of nanomaterials from solid nanocomposites. *Part. Fibre Toxicol.* **2014**, *11*, 17. [CrossRef] [PubMed]

9. Epstein, H.A.; Kielbassa, A. Nanotechnology in Cosmetic Products, Bio-Nanotechnology: A Revolution in Food. *Biomed. Health Sci.* **2013**, 414–423. [CrossRef]

10. Wu, M.S.; Sun, D.S.; Lin, Y.C.; Cheng, C.L.; Hung, S.C.; Chen, P.K.; Yang, J.H.; Chang, H.H. Nanodiamonds protect skin from ultraviolet B-induced damage in mice. *J. Nanobiotechnol.* **2015**, *13*, 35. [CrossRef] [PubMed]

11. Lin, W.; Huang, Y.W.; Zhou, X.D.; Ma, Y. In vitro toxicity of silica nanoparticles in human lung cancer cells. *Toxicol. Appl. Pharmacol.* **2006**, *217*, 252–259. [CrossRef] [PubMed]

12. Foged, C.; Brodin, B.; Frokjaer, S.; Sundblad, A. Particle size and surface charge affect particle uptake by human dendritic cells in an in vitro model. *Int. J. Pharm.* **2005**, *298*, 315–322. [CrossRef] [PubMed]

13. Noël, A.; Maghni, K.; Cloutier, Y.; Dion, C.; Wilkinson, K.J.; Hallé, S.; Tardif, R.; Truchon, G. Effects of inhaled nano-TiO$_2$ aerosols showing two distinct agglomeration states on rat lungs. *Toxicol. Lett.* **2012**, *214*, 109–119. [CrossRef] [PubMed]

14. Yang, H.; Wu, Q.Y.; Li, M.Y.; Lao, C.S.; Zhang, Y.J. Pulmonary Toxicity in Rats Caused by Exposure to Intratracheal Instillation of SiO$_2$ Nanoparticles. *Biomed. Environ. Sci.* **2017**, *30*, 264–279. [CrossRef] [PubMed]

15. Wiemann, M.; Vennemann, A.; Sauer, U.G.; Wiench, K.; Ma-Hock, L.; Landsiedel, R. An in vitro alveolar macrophage assay for predicting the short-term inhalation toxicity of nanomaterials. *J. Nanobiotechnol.* **2016**, *14*, 16. [CrossRef] [PubMed]

16. Marvanová, S.; Kulich, P.; Skoupý, R.; Hubatka, F.; Ciganek, M. Size-segregated urban aerosol characterization by electron microscopy and dynamic light scattering and influence of sample preparation. *Atmos. Environ.* **2018**. [CrossRef]

17. Benelli, G. Mode of action of nanoparticles against insects. *Environ. Sci. Pollut. Res.* **2018**, *25*, 12329–12341. [CrossRef] [PubMed]

18. Babick, F. Suspensions of colloidal particles and aggregates. In *Particle Technology Series*; Valverde Millán, J.M., Ed.; Springer: Berlin/Heidelberg, Germany, 2016; Volume 20, ISBN 978-3-319-30661-2.

19. Babick, F.; Schieß, K.; Stintz, M. Characterization of Pyrogenic Powders with Conventional Particle Sizing Technique: I. Prediction of Measured Size Distributions. *Part. Part. Syst. Charact.* **2012**, *29*, 104–115. [CrossRef]

20. Retamal Marín, R.R.; Babick, F.; Stintz, M. Ultrasonic dispersion of nanostructured materials with probe sonication—Practical aspects of sample preparation. *Powder Technol.* **2017**, *318*, 451–458. [CrossRef]

21. Oberdörster, G.; Maynard, A.; Donaldson, K.; Castranova, V.; Fitzpatrick, J.; Ausman, K.; Carter, J.; Karn, B.; Kreyling, W.; Lai, D.; et al. Principles for characterizing the potential human health effects from exposure to nanomaterials: Elements of a screening strategy. *Part. Fibre Toxicol.* **2005**, *2*, 8. [CrossRef] [PubMed]

22. Passagne, I.; Morille, M.; Rousset, M.; Pujalté, I.; L'Azou, B. Implication of oxidative stress in size-dependent toxicity of silica nanoparticles in kidney cells. *Toxicology* **2012**, *299*, 112–124. [CrossRef] [PubMed]

23. Taurozzi, J.; Hackley, V.; Wiesner, M. Ultrasonic dispersion of nanoparticles for environmental, health and safety assessment—Issues and recommendations. *Nanotoxicology* **2011**, *5*, 711–729. [CrossRef] [PubMed]

24. Veith, L.; Vennemann, A.; Breitenstein, D.; Engelhard, C.; Wiemann, M.; Hagenhoff, B. Detection of SiO_2 nanoparticles in lung tissue by ToF-SIMS imaging and fluorescence microscopy. *Analyst* **2017**, *142*, 2631. [CrossRef] [PubMed]

25. Maier, M.; Hannebauer, B.; Holldorff, H.; Albers, P. Does Lung Surfactant Promote Disaggregation of Nanostructured Titanium Dioxide? *J. Occup. Environ. Med.* **2006**, *48*, 1314–1320. [CrossRef] [PubMed]

26. Bakand, S.; Hayes, A.; Dechsakulthorn, F. Nanoparticles: A review of particle toxicology following inhalation exposure. *Inhal. Toxicol.* **2012**, *24*, 125–135. [CrossRef] [PubMed]

27. Clippinger, A.J.; Ahluwalia, A.; Allen, D.; Bonner, J.C.; Casey, W.; Castranova, V.; David, R.M.; Halappanavar, S.; Hotchkiss, J.A.; Jarabek, A.M.; et al. Expert consensus on an in vitro approach to assess pulmonary fibrogenic potential of aerosolized nanomaterials. *Arch. Toxicol.* **2016**, *90*, 1769–1783. [CrossRef] [PubMed]

28. Nemmar, A.; Hoylaerts, M.F.; Hoet, P.H.M.; Vermylen, J.; Nemery, B. Size effect of intratracheally instilled particles on pulmonary inflammation and vascular thrombosis. *Toxicol. Appl. Pharmacol.* **2003**, *186*, 38–45. [CrossRef]

29. Arick, D.Q.; Choi, Y.H.; Kim, H.C.; Won, Y. Effects of nanoparticles on the mechanical functioning of the lung. *Adv. Colloid Interface Sci.* **2015**, *225*, 218–228. [CrossRef] [PubMed]

30. ISO. *Nanotechnologies—Health and Safety Practices in Occupational Settings Relevant to Nanotechnologies*; TC229, ISO/TR 12885; ISO: Geneva, Switzerland, 2008.

31. European Commission. Commission Recommendation of 18 October 2011 on the Definition of Nanomaterial (2011/696/EU). *Off. J. Eur. Union* **2011**, *54*, 38–40.

32. Linsinger, T.P.J.; Roebben, G.; Gilliland, D.; Calzolai, L.; Rossi, F.; Gibson, N.; Klein, C. *Requirements on Measurements for the Implementation of the European Commission Definition of the Term 'Nanomaterial'*; Publications Office of the European Union: Luxembourg, 2012. [CrossRef]

33. Jensen, K.; Kembouche, Y.; Christiansen, E.; Jacobsen, N.; Wallin, H.; Guiot, C.; Spalla, O.; Witschger, O. The Generic NANOGENOTOX Dispersion Protocol: Final Protocol for Producing Suitable Manufactured Nanomaterial Exposure Media. NANOGENOTOX Joint Action, European Commission. 2011. Available online: www.nanogenotox.eu/files/PDF/web%20nanogenotox%20dispersion%20protocol.pdf (accessed on 20 June 2018).

34. OECD. *Guidelines for the Testing of Chemicals, Section 3 Test No. 318: Dispersion Stability of Nanomaterials in Simulated Environmental Media*; OECD Publishing: Washington, DC, USA, 2017; ISBN 9789264284142.

35. Rasmussen, K.; Mech, A.; Mast, J.; de Temmerman, P.-J.; Waegeneers, N.; van Steen, F.; Pizzolon, J.C.; de Temmerman, L.; van Doren, E.; Jensen, K.A.; et al. *Synthetic Amorphous Silicon Dioxide (NM-200, NM-201, NM-202, NM-203, NM-204): Characterisation and Physico-Chemical Properties*; Report EUR 26046; European Commission: Brussels, Belgium, 2013.

36. Hartmann, N.; Jensen, K.; Baun, A.; Rasmussen, K.; Rauscher, H.; Tantra, R.; Cupi, D.; Gilliland, D.; Pianella, F.; Sintes, J.R. Techniques and Protocols for Dispersing Nanoparticle Powders in Aqueous Media—Is there a Rationale for Harmonization? *J. Toxicol. Environ. Health B* **2015**, *18*, 299–326. [CrossRef] [PubMed]

37. Pohl, M.; Schubert, H.; Schuchmann, H. Herstellung stabiler Dispersionen aus pyrogener Kieselsäure. *Chem. Ing. Tech.* **2005**, *77*, 258–262. [CrossRef]

38. Wengeler, R.; Ruslim, F.; Nirschl, H.; Merkel, T. Dispergierung feindisperser Agglomerate mit Mikro-Dispergierelementen. *Chem. Ing. Tech.* **2004**, *76*, 659–662. [CrossRef]

39. Wengeler, R.; Teleki, A.; Vetter, M.; Pratsinis, S.; Nirschl, H. High-pressure liquid dispersion and fragmentation of flame-made silica agglomerates. *Langmuir* **2006**, *22*, 4928–4935. [CrossRef] [PubMed]

40. Sauter, C.; Schuchmann, H. High pressure for dispersing and deagglomerating nanoparticles in aqueous solutions. *Chem. Eng. Technol.* **2007**, *30*, 1401–1405. [CrossRef]

41. Bałdyga, J.; Makowski, Ł.; Orciuch, W.; Sauter, C.; Schuchmann, H. Agglomerate dispersion in cavitating flows. *Chem. Eng. Res. Des.* **2009**, *87*, 474–484. [CrossRef]

42. Tantra, R. *Nanomaterial Characterization: An Introduction*; Wiley: Hoboken, NJ, USA, 2016; ISBN 9781118753460.

43. Pradhan, S.; Hedberg, J.; Wold, E.B.S.; Wallinder, I.O. Effect of sonication on particle dispersion, administered dose and metal release of non-functionalized, non-inert metal nanoparticles. *J. Nanopart Res.* **2016**, *18*, 285. [CrossRef] [PubMed]

44. Taurozzi, J.; Hackley, V.; Wiesner, M. Preparation of Nanoparticle Dispersions from Powdered Material using Ultrasonic Disruption. *NIST Spec. Publ.* **2012**. [CrossRef]

45. Bihari, P.; Vippola, M.; Schultes, S.; Praetner, M.; Khandoga, A.G.; Reichel1, C.A.; Coester, C.; Tuomi, T.; Rehberg, M.; Krombach, F. Optimized dispersion of nanoparticles for biological in vitro and in vivo studies. *Part. Fibre Toxicol.* **2008**, *5*, 14. [CrossRef] [PubMed]

46. Mandzy, N.; Grulke, E.; Druffel, T. Breakage of TiO$_2$ agglomerates in electrostatically stabilized aqueous dispersions. *Powder Technol.* **2005**, *160*, 121–126. [CrossRef]

47. Pohl, M.; Hogekamp, S.; Hoffmann, N.; Schuchmann, H. Dispergieren und Desagglomerieren von Nanopartikeln mit Ultraschall. *Chem. Ing. Tech.* **2004**, *76*, 392–396. [CrossRef]

48. Napierska, D.; Thomassen, L.C.J.; Lison, D.; Martens, J.A.; Hoet, P.H. The nanosilica hazard: Another variable entity. *Part. Fibre Toxicol.* **2010**, *7*, 39. [CrossRef] [PubMed]

49. Brinker, C.F.; Schrerer, G.W. *Sol-Gel Science. The Physics and Chemistry of Sol-Gel Processing*, 2nd ed.; Academic Press: London, UK, 1990; ISBN 9780121349707.

50. Albers, P.; Maier, M.; Reisinger, M.; Hannebauer, B.; Weinand, R. Physical boundaries within aggregates—differences between amorphous, para-crystalline, and crystalline Structures. *Cryst. Res. Technol.* **2015**, *50*, 846–865. [CrossRef]

51. Brunauer, S.; Emmett, P.H.; Teller, E. Adsorption of Gases in Multimolecular Layers. *J. Am. Chem. Soc.* **1938**, 309–319. [CrossRef]

52. ISO 9277:2010. *Determination of the Specific Surface Area of Solids by Gas Adsorption—BET Method*; ISO: Geneva, Switzerland, 2010.

53. Hosokawa, M.; Nogi, K.; Naito, M.; Yokoyama, T. *Nanoparticle Technology Handbook*; Elsevier Science: New York, NY, USA, 2007; ISBN 978-0-444-53122-3.

54. Retamal Marín, R.R.; Babick, F.; Hillemann, L. Zeta potential measurements for non-spherical colloidal particles—Practical issues of characterisation of interfacial properties of nanoparticles. *Colloids Surf. A* **2017**, *532*, 516–521. [CrossRef]

55. Karina Maria Paciejewska, Untersuchung des Stabilitätsverhaltens von binären kolloidalen Suspensionen, Ph.D. Thesis, Technische Universität Dresden, Dresden, Germany, December 2010. Available online: http://d-nb.info/1019001267/34 (accessed on 20 June 2018).

56. *European Pharmacopoeia: Supplement 5.7, Band 5 von European Pharmacopoeia: Supplement*, 5th ed.; Convention on the Elaboration of a European Pharmacopoeia, 4805–4806; Council of Europe: Strasbourg, France, 2006.

57. Hauptmann, P.; Sorge, G. *Ultraschall in Wissenschaft und Technik*; Durchschnittliche Kundenbewertung: Leipzig, Germany, 1985; ISBN 0-906674-38-7.

58. Kusters, K.; Pratsinis, S.; Thoma, S.; Smith, D. Ultrasonic fragmentation of agglomerate powders. *Chem. Eng. Sci.* **1993**, *48*, 4119–4127. [CrossRef]

59. Aoki, M.; Ring, T.; Haggerty, J. Analysis and modeling of the ultrasonic dispersion technique. *Adv. Ceram. Mater.* **1987**, *2*, 209–212. [CrossRef]

60. Etzler, F.M.; Deanne, R. Particle Size Analysis: A Comparison of Various Methods II. *Part. Part. Syst. Charact.* **1997**, *14*, 278–282. [CrossRef]

61. Bayat, H.; Rastgo, M.; Zadeh, M.M.; Vereecken, H. Particle size distribution models, their characteristics and fitting capability. *J. Hydrol.* **2015**, *529*, 872–889. [CrossRef]

62. Marton, L.; Marton, C. *Methods of Experimental Physics: Ultrasonics*; Academic Press: Cambridge, MA, USA, 1981; ISBN 0-12-475961-0.

63. Raman, V.; Abbas, A. Experimental investigations on ultrasound mediated particle breakage. *Ultrason. Sonochem.* **2008**, *15*, 55–64. [CrossRef] [PubMed]

64. Raso, J.; Mañas, P.; Pagán, R.; Sala, F. Influence of different factors on the output power transferred into medium by ultrasound. *Ultrason. Sonochem.* **1999**, *5*, 157–162. [CrossRef]

65. Edmond, P.D. *Methods in Experimental Physics*; Elsevier: New York, NY, USA, 1981; Volume 19, ISBN 978-0-12-475961-9.

66. Bhatia, A.B. *Ultrasonic Absorption. An Introduction to the Theory of Sound Absorption and Dispersion in Gases, Liquids, and Solids*; Oxford University Press: New York, NY, USA, 1967; ISBN 9780486649177.

67. ISO 9276-1:1998. *Representation of Results of Particle size Analysis—Part 1: Graphical Representation*; ISO: Geneva, Switzerland, 1998.

68. Kuchenbecker, P.; Gemeinert, M.; Rabe, T. Interlaboratory study of particle size distribution measurements by laser diffraction. *Part. Part. Syst. Charact.* **2012**, *29*, 304–310. [CrossRef]

69. Mawson, R.; Rout, M.; Ripoll, G.; Swiergon, P.; Singh, T.; Koerzer, K.; Juliano, P. Production of particulates from transducer erosion: Implications on food safety. *Ultrason. Sonochem.* **2014**, *21*, 2122–2130. [CrossRef] [PubMed]

Green Micro- and Nanoemulsions for Managing Parasites, Vectors and Pests

Lucia Pavoni [1]**, Roman Pavela** [2]**, Marco Cespi** [1]**, Giulia Bonacucina** [1]**, Filippo Maggi** [1]**,
Valeria Zeni** [3]**, Angelo Canale** [3]**, Andrea Lucchi** [3]**, Fabrizio Bruschi** [4] **and Giovanni Benelli** [3,*]

[1]   School of Pharmacy, University of Camerino, via Sant'Agostino, 62032 Camerino, Italy
[2]   Crop Research Institute, Drnovska 507, 161 06 Prague 6, Ruzyne, Czech Republic
[3]   Department of Agriculture, Food and Environment, University of Pisa, via del Borghetto 80, 56124 Pisa, Italy
[4]   Department of Translational Research, N.T.M.S., University of Pisa, 56124 Pisa, Italy
*   Correspondence: giovanni.benelli@unipi.it

Abstract: The management of parasites, insect pests and vectors requests development of novel, effective and eco-friendly tools. The development of resistance towards many drugs and pesticides pushed scientists to look for novel bioactive compounds endowed with multiple modes of action, and with no risk to human health and environment. Several natural products are used as alternative/complementary approaches to manage parasites, insect pests and vectors due to their high efficacy and often limited non-target toxicity. Their encapsulation into nanosystems helps overcome some hurdles related to their physicochemical properties, for instance limited stability and handling, enhancing the overall efficacy. Among different nanosystems, micro- and nanoemulsions are easy-to-use systems in terms of preparation and industrial scale-up. Different reports support their efficacy against parasites of medical importance, including *Leishmania, Plasmodium* and *Trypanosoma* as well as agricultural and stored product insect pests and vectors of human diseases, such as *Aedes* and *Culex* mosquitoes. Overall, micro- and nanoemulsions are valid options for developing promising eco-friendly tools in pest and vector management, pending proper field validation. Future research on the improvement of technical aspects as well as chronic toxicity experiments on non-target species is needed.

Keywords: agricultural pests; dengue; filariasis; insecticides; larvicides; mosquito control; stored product insects

1. Introduction

1.1. Micro- and Nanoemulsions

Over the past decades, pharmaceutical, food and agricultural research has focused the attention on the development of delivery systems able to encapsulate, protect and deliver lots of different compounds. One of the most versatile tools is represented by colloidal dispersions, which are heterogeneous systems in which the inner phase is dispersed into a continuous medium. Micro- and nanoemulsions (MEs and NEs respectively) are self-emulsifying colloidal systems, having the internal phase usually smaller than 100 nm, dispersed in a liquid medium [1]. This characteristic enhances some physicochemical properties, i.e., stability and bioavailability. In fact, the small size of the internal phase allows the system to bypass the problems related to the gravity force, avoiding phenomena as creaming or sedimentation. Moreover, the low surface and interfacial tensions promote suitable spreading and penetration of the active compounds [2].

MEs and NEs are generally composed of an aqueous phase, an oily phase, a surfactant agent and a possible cosurfactant. For this reason, they are able to incorporate both hydrophilic and lipophilic compounds [3]. The choice of MEs and NEs components are strictly related to their application. For example, it is possible to select several oily phases between synthetic oils, (ethyl oleate, squalene and triglycerides), mineral oils and vegetable oil (e.g., olive, sunflower and soybean oil). Generally, the oily phase is used to solubilise and carry lipophilic molecules, but sometimes the oily fraction, as in the case of plant essential oils (EOs), can also be the active ingredient. EOs have been widely used in traditional medicine around the world since the Middle Ages, mainly for their antimicrobial and antioxidant properties.

A fundamental aspect about the formulation of EOs is the selection of suitable surfactant agents. The amphiphilic properties of a surfactant are represented by the hydrophilic–lipophilic balance (HLB) value. The choice of the suitable HLB value depends on the nature of the continuous phase. However, it should be desirable to select a surfactant with an intermediate value because it will partition between the aqueous and the oily phase, lowering the interfacial tension and conferring the optimal curvature of the layer, to guarantee the formation and stabilisation of the droplets. Depending on the chemical properties, surfactants can be divided into different classes: anionic, cationic, non-ionic and zwitterionic. The most diffused are polisorbates (anionic), such as Tween 80 (HLB 16.7) and Span 80 (HLB 8.6). In recent years there has been a growing interest in exploiting the surfactant properties of natural products such as polysaccharides, proteins (lectin) and sugar esters, which are desirables for the development of eco-friendly formulations. MEs and NEs have been deeply investigated, since they possess some practical advantages: easiness of formulation, industrial scale-up and high potential for use in several applications.

Apart from the terminology, these two systems present some substantial differences that it is necessary to highlight to better understand the mechanisms of their formation: (i) physicochemical behaviour, (ii) properties and (iii) applications. A summary of the main features of MEs and NEs is reported in Figure 1.

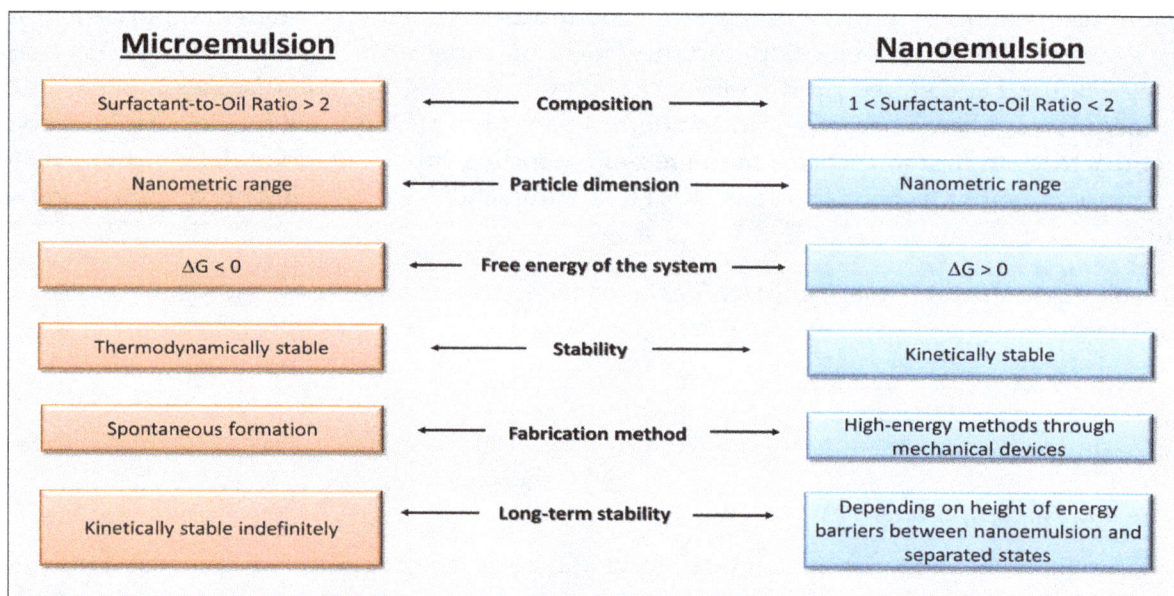

Figure 1. Comparison of the main physicochemical properties between micro- and nanoemulsion.

First, it is important to highlight that, despite the prefixes 'micro' and 'nano' define two different orders of magnitude, i.e., 10^{-6} and 10^{-9}, respectively, the size of the dispersed phase (generally oily droplets) for both of these two systems fall in the nanometric range. According to the literature it is not possible to exactly define a range of particle size distribution, since different authors report different results within the nanometric order of magnitude [3,4]. In any case, it has been reported that MEs are characterised by a smaller size of the dispersed phase respect to NEs [5,6].

ME has been defined as "a system of water, oil and amphiphile, which is a single optically isotropic and thermodynamically stable liquid solution" [7]. Introduced for the first time in 1944 by Hoar and Schulman, MEs were initially investigated for oil recovery from underground reservoirs [8,9]. Furthermore, the interest around them spread into several application fields. MEs were studied in detail in the pharmaceutical field as promising drug delivery systems for lipophilic compounds. As previously mentioned, they show several advantages such as solubilization of lipophilic compounds, enhancement of physicochemical stability respect to the related macro-systems (emulsions), improvement of the active ingredients bioavailability, achievement of a controlled drug delivery system, easiness of preparation and scale-up [10]. However, their real use is limited by the high amount of surfactant requested for the formation of such system, being these agents irritant against mucous membranes and potentially hazardous for the environment [11,12].

On the contrary, one of the most important advantages of NEs is the presence of low amounts of surfactant, generally less than 10%, compared to almost 15% in MEs, and a low surfactant-to-oil ratio (SOR) necessary for their formation, that is, >2 in MEs and comprised between 1 and 2 in NEs [2,5,6]. Briefly, NE is defined as "a thermodynamically unstable colloidal dispersion consisting of two immiscible liquids, with one of the liquids being dispersed as small spherical droplets (r < 100 nm) in the other liquid" [5]. It can be considered as a conventional emulsion, with the only difference of a smaller size of the dispersed phase. However, the most influential parameter varying in these two nanostructured colloidal dispersions is their free energy, conferring them different features in terms of preparation, formulation and stability.

As reported in the previous definitions, MEs are thermodynamically stable while NEs are kinetically stable. This is due to the free energy possessed by the separate state (oil + water) respect to the colloidal systems. MEs are energetically favoured, with ΔG values lower than the respective separate phases. On the contrary, NEs (oily droplet in water) possess higher free energy than those of the separate phases, water and oil.

The preparation methods of MEs and NEs are a direct consequence of this aspect. In fact, being the formation of MEs favourable, they can be obtained spontaneously by mixing oil, water and surfactant, without any external energy input. However, the application of magnetic stirring or heating could be convenient to expedite the process in order to overcome the kinetic barriers.

The energetic process that drives the MEs formation is based on the following formula [13].

$$\Delta G = \gamma \, \Delta A - T \, \Delta S \qquad (1)$$

where ΔG is the free energy of the final system (ME), γ is the interfacial tension oil–water, ΔA is the variation of the interfacial area, T is the temperature and ΔS is the variation of the system entropy.

Briefly, ΔG must be negative so that a process occurs spontaneously. Since ΔA is very high in a ME (because of the formation of lots of small oily droplets that increases the interfacial area), this process is promoted by a very slow interfacial tension (γ) and by the entropy of the system that rises for the transition of the separate phases into only one containing a large number of particles; this allows obtaining a negative ΔG value.

The formation of MEs is strictly dependent on the sensitive SOR and, to determine the optimal one, is used to build a pseudoternary phase diagram. This kind of system, in fact, needs a very low interfacial tension and a favourable packaging of surfactant molecules, given by the relative interaction between their hydrophobic tails and the oil phase. This allows the formation of a fluid film at the oil–water interface [14]. Usually, the addition of cosurfactant agents is required, generally alcohols, to facilitate this phenomenon, useful to reduce the amount of surfactant as well [15]. Being MEs dynamic systems, we have to take into account that the interface is continuously subjected to a rearrangement of its structure and to the Brownian motion of the internal phase, with a possible variation of its radius [16].

Since NEs are thermodynamically unstable, the free energy of the systems, ΔG (Formula (1)), will be always positive. Thus, to exceed this value, an external energy input results to be necessary.

Depending on the physicochemical mechanisms, the methods used for NE preparation can be divided into high-energy and low-energy methods. The first ones use mechanical devices able to provide the force needed for the disruption of the dispersed phase into very small droplets, in the range of nanometres (r < 200 nm). Generally, NE formation follows a two steps procedure. In the first phase there is the formation of a macroemulsion through a mechanic stirrer. In the second one the macroemulsion is converted into a NE.

The most common devices used for this process are microfluidizer, sonicator, and high-pressure homogenizer. This last device uses high pressure value to pump the macroemulsion in a very narrow orifice that promotes the breaking of big droplets into smallest ones. The same result is achieved through ultrasound waves that lead to the dispersion process by means of cavitation phenomenon.

Although these approaches seem to be robust, they show some limitations concerning costs, process implementation and industrial scale-up [17].

On the contrary, low-energy methods are simpler, cheaper and more effective in producing smaller droplets. However, they require an accurate knowledge of the process parameters, showing some limitations in the ingredients and conditions [6,18].

Generally, low-energy methods are based on the phase inversion, transforming a W/O macroemulsion into an O/W NE through the variation in composition (emulsion inversion point (EIP)) or temperature (phase inversion temperature (PIT)). At the inversion point, the interfacial tension is so low that very fine droplets can be obtained, only with the support of low energy input.

Briefly, the phase inversion due to the PIT method is linked to the presence of surfactants that, based on a temperature change, modify their affinity for the hydrophilic or lipophilic phase. With EIP method there is a modification in the composition (water, surfactants, electrolytes) of the final system, which leads to a variation of the lipophilic-hydrophilic balance, with a consequent change in the curvature of surfactant layer. The free energy of the system influences the long-term stability behaviour as well. MEs should remain stable indefinitely, if the initial conditions about the chemical composition and storage will keep unchanged.

NEs, instead, will remain in a metastable state that will guarantee the stability of the systems if the energy barrier between the two different energy states remains high enough to avoid the reversion of the system and the phase separation. This occurs because of such instability phenomena such as coalescence, flocculation and Ostwald ripening, which, bringing growth of droplets, lead to creaming. It represents the migration of the dispersed phase influenced by buoyancy.

Coalescence is due to the merger of small droplets into bigger ones, while, in flocculation, droplets become very closer to move as a unique phase. These phenomena are related to the surfactant layer on the droplets surface that guarantees the steric stabilisation as much as the thickness of the layer is comparable with the droplets size. For this reason, NEs are not particularly affected by coalescence and flocculation, as compared to a traditional emulsion.

On the contrary, NEs are more prone to Ostwald ripening. This phenomenon can be defined as: "the process of disappearance of small particles or droplets by dissolution and deposition on the larger particles or droplets. The driving force for Ostwald ripening is the difference in solubility between the small and the large particles. The smaller particles (with higher radius of curvature) are more soluble than the larger ones (with lower radius of curvature). With time, the smaller particles or droplets dissolve, and their molecules diffuse in the bulk and deposit on the larger ones. This results in a shift of the particle or droplet size distribution to larger values" [19]. It is a thermodynamic process, being larger particles energetically favoured over the smaller ones.

Since the aqueous solubility of the oily droplets strongly influences the occurrence of this phenomenon, a suitable solution could be the addition of non-polar compounds that condition positively the distribution of the droplets in the oily phase. Some of the most used "ripening inhibitors" are medium-chain triglycerides (MCT), corn oil and sunflower oil [20–22].

Concluding, some of the most influential parameters on the NE stability are:

(i)   The SOR and relative concentrations; they influence the interfacial tension. It is not possible to stabilize a fixed relationship between these parameters because they are strictly related to the nature of the compounds that confer unique properties to the systems, which, in turn, differ from each other.

(ii)   The ionic strength of the dispersion medium; it affects the repulsive forces between the droplets of the dispersed phase. As the ionic strength increases, the repulsive forces decrease and the systems will be prone to instability.

(iii)   The solubility of the dispersed phase; it allows droplets to move towards the continuous phase with the appearance of Ostwald ripening.

(iv)   The temperature; it affects the solubility with the above-mentioned consequences. Moreover, it influences the energy balance of the system as well.

1.2. Applications

Thanks to the previously mentioned advantages, such nanosystems have been widely exploited in different fields as a tool for oil recovery, fuel and reaction medium in chemical applications [23–25]. However, in this section we are going to focus the attention on their applications in food, agrochemical, cosmetic and pharmaceutical fields.

About the food area, they have been developed to improve and extend the use of low water-soluble compounds or food-derived bioactive compounds with poor bioavailability. Such delivery nanosystems seem to be a suitable tool to solve this kind of problems. A significant example has been reported by Yu and Huang [26]. They demonstrated that curcumin showed a 9-fold increase in oral bioavailability when encapsulated into NEs. Moreover, it was faster digested as well, through lipolysis, respect to the unformulated compound. In the last years, NEs have been considered as a fundamental tool for the delivery of functional substances in functional foods or fortified beverages such as fatty acids, polyphenols, vitamins, micronutrients, antioxidants and others [27]. For example, O/W NE was exploited in order to encapsulate and deliver Omega-3 fatty acids in yoghurts [28].

Being extremely stable in a wide range of pH, MEs and NEs are very useful for encapsulating nutrients and protecting them from environmental conditions such as temperature or light-mediated oxidation and from possible transformation by means of enzymatic reactions and hydrolysis [29]. They formulated a valid solution to maintain suitable organoleptic properties of foods and beverages. In fact, MEs and NEs can encapsulate volatile molecules and control the release of flavours. Moreover, they can be used to prevent contamination of products and to prolong their shelf-life, both directly, for example by adding a preservative NE inside food, or indirectly, by functionalizing the packaging system in the same way [30,31]. Besides these advantages, MEs and NEs in food chain show some limitations, due to the nature of the components. For examples, in a food product the oily phase should be a triglyceride. Since the solubilization of a long chain triglyceride (LCT) is hard to obtain, it should be preferable to choose between a medium and short chain triglyceride [32].

Actually, the real limiting step in food grade nanosystem formulation is related to surfactant, because many of them are not allowed for human consumption or just at very low concentrations. Some of the admitted ones are sugar esters, monoglycerides, lecithins, glycolipids, fatty alcohols and fatty acids [33]. This issue is, nowadays, a great object of study. A large number of authors in fact, through the building of pseudoternary phase diagram of food grade components, tried to find suitable and stable formulations based only on food-grade compounds [32].

Regarding the pharmaceutical field, modern technology is progressing toward developing efficient drug delivery tools, with particular attention to an increase of bioavailability, a controlled release of the drug, a targeted biological effect and good storage stability over time. All these goals could be pursued by the exploitation of MEs and NEs. Being composed of hydrophilic and lipophilic domains, they are versatile systems able to incorporate and solubilise drugs of both natures. Araya et al. proved that MEs enhanced the oral bioavailability of poor water soluble drugs, as Ibuprofen and Ketoprofen, increasing their solubility and their plasma concentration from 60 to 20,000 times [34].

Since MEs/NEs can raise the bioavailability, the administered dose of drugs could be reduced minimizing possible side effects. These formulations behave as controlled release tools, both in O/W and W/O systems. In fact, in the first case, the oily phase acts as reservoir of active compounds, while when the oil is the external phase limits the diffusion [35]. However, the rate of drug release is influenced by the composition of the environment, such as pH and ionic strength, and features of the nanosystems, i.e., droplet dimension, type of MEs or NEs, nature of the drug and route of administration. Moreover, a limiting step is represented by the ability of the drug to cross the biological barrier, such as mucosa cells or skin [36].

Oral delivery of such nanosystems should be very useful to carry on poor water-soluble drugs, since they allow to overcome the dissolution issue on gastric fluids, which generally is strictly related to bioavailability. Moreover, they reduce the hepatic first-pass metabolism favouring the passage of the drugs in the bloodstream [35]. The small size of the internal phase and the presence of surfactants improve the drug absorption in the gastrointestinal (GI) tract, in the first case, enhancing the permeability of biological barriers, and, in the second case, promoting a wide and deep distribution [3,37].

Yin et al. showed how a ME, composed by Capryol 90 (oil), Cremophor EL (surfactant) and Transcutol (cosurfactant), increased the bioavailability of docetaxel, as compared to the related commercial product, after oral administration in rats. This result was obtained through the cumulative effect of enhanced drug solubility, improved permeability and inhibition of P-glycoprotein (P-gp) efflux [38]. Thanks to their low viscosity and possibility to be sterilised by filtration, MEs and NEs are very favourable in parenteral administration as well [4]. Moreover, they showed an appreciable physical stability in plasma [39].

Both O/W and W/O systems are suitable for parenteral formulations. Generally, O/W systems are used to deliver lipophilic compounds in order to obtain a controlled release of the drugs. Thus, they are administered by the intravenous, intramuscular or subcutaneous routes. On the contrary, W/O systems, applied as subcutaneous or intramuscular administration, are suitable to encapsulate hydrophilic drugs in order to obtain prolonged release delivery systems [4].

Dordevic et al. optimised a risperidone-based NE and monitored the pharmacokinetic parameters of the active ingredient. After intraperitoneal administration in rats, they obtained a 1.2–1.5-fold increase of bioavailability, 1.1–1.8-fold decrease in liver distribution, and 1.3-fold increase of brain uptake of risperidone as compared to the drug solution [40]. MEs and NEs are widely studied and used for topical, ocular and nasal administration as well [13,41]. The topical route has been investigated mainly in the cosmetic field, exploiting these systems in order to obtain a better penetration of the active molecules through the skin barrier [42].

Intranasal route should be exploited to deliver active molecules directly on the brain. Vyas et al. developed a mucoadhesive clonazepam-based ME for the epilepsy treatment. The concentration of this molecule in the brain was found to be 2-fold higher when compared with intravenous administration, indicating an enhanced distribution and bioavailability of the active ingredient in the site of action [43].

As in other nanosystems, functionalisation of MEs and NEs allows to build up targeted drug delivery systems, which are able to address the activity mainly in a desired target site.

Shiokawa et al. reported the formulation of aclainomycin A, a lipophilic antitumour-antibiotic drug, through a ME linked to folate molecules. They showed that, the use of folate, helpfully modified with PEG molecules, can be considered as an effective strategy to target MEs on tumour cells [44].

Another interesting field of application of nanosystems is the agricultural one. In particular, nanotechnology is starting to revolutionise the pest management, providing innovative tools, i.e., nanoemulsions, nanoparticles and nanocapsules for the delivery of pesticide compounds (Figure 2).

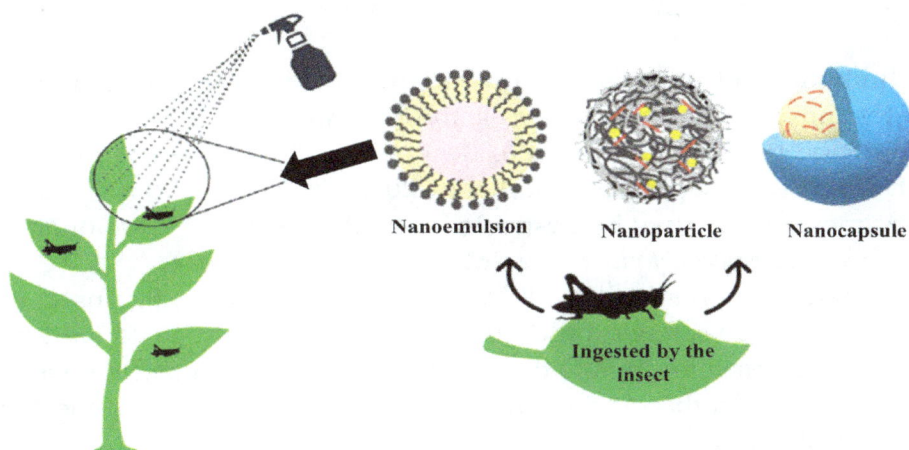

Figure 2. The most used nanosystems in insect pest control (adapted from Medina-Pérez et al. [45], with permission of Elsevier, 2019).

Among several nanodelivery systems, MEs and NEs are the easiest ones to handling and formulate. In particular, they are necessary in the presence of compounds with low water solubility that require a delivery system for their application in the field [46]. Du et al. carried out a systematic study about the formation of O/W NE based on methyl laurate as oil phase and alkyl polyglycoside and polyoxyethylene 3-lauryl ether as surfactants [47]. Moreover, they evaluated the effect of β-cypermethrin on the stability and physicochemical properties of the system.

The encapsulation process improves the physicochemical stability of pesticides and prevents the degradation of active agents [48]. Song et al. (2009) proved that the encapsulation of triazophos—an organophosphorus insecticide—is able to prevent the hydrolysis of the active compound [49]. In terms of bioactivity, these compounds result to be more effective. Nanosystems are able to ensuring their release to the target site, also providing a controlled release of the molecules at the site of action and thus reducing the required concentration of applied pesticides [2,49]. Moreover, thanks to the small size of the dispersed phase, the active compounds could improve their spreading, deposition and permeation on the target site.

2. Green Micro- and Nanoemulsions

In the last years, the growing interest of the global community on the planet fate is leading towards a more responsible and sustainable exploitation of natural resources. In particular, the worth of plants, as primary sources of ingredients for the realisation of a great variety of products, has been revaluated. In fact, some plant-based materials offer superior performance characteristics as compared to the synthetic ones. Nowadays, they have started to be applied in several fields such as pharmaceuticals, nutraceuticals, cosmetics and agrochemicals. Relying also on longstanding uses in the traditional medicine systems, they are generally employed as essential oils (EOs) and extracts, acting as flavouring agents, dyes, fortifying agents in functional foods or actual active ingredients [50].

EOs are mixtures of volatile and lipophilic molecules (mainly terpenoids and phenylpropanoids), produced in secretory structures of aromatic plants, in particular those belonging to angiosperms, such as Apiaceae, Asteraceae, Geraniaceae, Lamiaceae, Lauraceae, Myrtaceae and Verbenaceae, as products of their secondary metabolism [51].

EOs have been widely employed in the flavour and fragrance industry. They also find industrial application in foodstuffs (e.g., soft drinks, food and packaging) and cosmetics (e.g., perfumes, skin and hair care products). Regarding their medical properties, EOs are mainly used as antimicrobial agents.

Recent studies have attested pesticide properties of several EOs, natural pure compounds and extracts. The use of plant sources in crop protection dates back to 2000 years ago [52]. However, in the 20th century a wide spread of synthetic pesticides started to take hold. They were favourable thanks to a high and long-lasting efficacy. If, on the one hand, they increased crop yield, on the other hand, their overuse led to toxic effects on humans and the environment with occurrence of resistance in pests [53–55].

The current limitations of their use are pushing discovery and development of less harmful products. One of the most promising solutions is the exploitation of plant-based pesticides. In fact, if the synthetic pesticide market is expected to decline by 1.5% per year, biopesticides have been estimated to reach the 20% of the pesticide market by 2025 [56,57].

The oldest and most widely used biopesticide is pyrethrum, a pure compound derived from the dried flowers of *Tanacetum cinerariifolium* (Trevir.) Sch.Bip. (Asteraceae) [58]. Actually, it has taken around 80% of the biopesticide market [59]. By virtue of its low toxicity against both mammals and environment, it presents a high safety profile [60]. However, its synthetic derivatives, also known as pyrethroids, have been designed to emulate the activity of the natural molecule. Despite their efficacy, they showed to be hazardous for the environment because of their long-lasting effects and high toxicity against non-target organisms [61].

Nicotine and the other alkaloids of tobacco represent another class of botanical pesticides. They act on the nervous system of pest, mimicking the neurotransmitter acetylcholine. Their use is now declined for their proved toxicity on human beings. The same problem has been observed for rotenone, isolated from *Derris elliptica* (Wall.) Benth. roots. Even though it is one of the most effective biopesticides, its high toxicity towards aquatic organisms and mammals deeply limited its use [62].

Neem (*Azadirachta indica* A. Juss.) is source of a very interesting compound, azadirachtin, a limonoid with considerable pesticide activity. It has shown bactericidal, fungicidal, and insecticidal properties, acting as a feeding and oviposition deterrent and as a growth inhibitor [63]. A fundamental aspect is its safety profile: no persistence in soil, no adverse effects on water or groundwater organisms, no toxicity to mammals [64,65].

Eco-friendly alternatives in biopesticides include the wide group of EOs. One of the most promising aspects in the exploitation of EOs is their lack of toxicity on mammals; they are generally harmless for the environment when compared with synthetic pesticides [66]. Their safety profile is guaranteed by the fact that most of EOs have been recognised as Generally Recognised As Safe (GRAS) substances by the Food and Drug Administration (FDA) and by the Environmental Protection Agency (EPA) of the United States [67]. For these reasons, a possible residue of EO-based pesticides on crop does not constitute a risk for human health.

It has been reported that EOs, such as thymol-containing EOs or EOs compounds, such as eugenol or α-terpineol, showed LC_{90} values two or three order of magnitude higher as compared to synthetic commercial products, such as endosulfan, against Juvenile Rainbow Trouts [68]. Pavela et al. reported that Apiaceae EOs have no toxicity against non-target organisms, as adult microcrustaceans *D. magna* and adult earthworms *E. fetida*, unlike α-cypermethrin that, even in much lower concentrations, caused almost 100% mortality [69].

Beyond the proofs about their safety, in the last years several studies have been carried out on the pesticide efficacy of EOs. Results showed that such substances exert a marked activity against pests, both in direct and indirect way. They act as chemosterilant, fumigant, ovicidal and repellent agents, altering growth, development and feeding behaviour [70–73]. In a recent review, Pavela collected the results published about the pesticide activity of EOs deriving from around 122 different species. Their efficacy could be expressed by an exciting data: 77 EOs showed $LC_{50} < 50$ ppm [74]. Their bioactivity is strictly linked to the presence of different compounds present in the mixture of each EO, monoterpene and sesquiterpene hydrocarbons, phenolic monoterpenes, oxygen containing mono- and sesquiterpenes and phenylpropanoids [75].

The main mechanism of action is linked to the ability of EOs to interfere with the cell membrane. Their accumulation leads to the disruption of the cell wall, leakage of the cellular contents and perturbation of homeostasis [76,77]. All these alterations lead to cell death. It has been reported that several EO constituents act in this manner [78,79]. Nevertheless, EOs, as well as plant extracts, are able to interfere with the nervous system of pests and vectors, inducing even death [80].

For example plant extracts, in particular alkaloids, can act at different levels of the pest nervous system [81]. They can function as competitive inhibitors of the acetylcholinesterase (AChE) enzyme, with consequent accumulation of the neurotransmitter in the synapses, followed by a state of permanent stimulation of the postsynaptic membrane [82]. Moreover they could be antagonist of GABA receptors as well, causing hyperexcitation, convulsion and death of the pest due to reduction of neuronal inhibition [83]. However, the most important target site of EOs is the octopaminergic system [80,84]. Octopamine is a neuromodulator and the absence of octopamine receptors in mammals is the factor that determines the distinction between target and non-target organisms. Acting on the octopaminergic system, the active compounds will be harmless for non-target organisms [72,85].

In addition to the above-mentioned advantages on the exploitation of EOs as biopesticides, a fundamental aspect is their synergistic effect. Synergism occurs in EOs since they are a mixture of 20–60 compounds, where all the components cooperate to enhance the bioactivity [86,87]. This results in a high efficacy since they act with different and complementary mechanisms of action and the combined effect is usually higher than those of the single components, allowing the reduction of the effective dose. Moreover, the mutual synergism represents a suitable tool to fight the development of resistance phenomenon, which is common with synthetic pesticides, that normally have only a target site [75].

Since EOs showed to be among the best candidates as botanical pesticides, we can ask why its commercial spread is still limited. The reason is strictly linked to their physicochemical properties, such as lipophilic nature and thus poor water solubility, scarce stability, high volatility, thermal decomposition and oxidative degradation [88]. These aspects translate into reduced efficacy and handling difficulties [72,85]. Moreover, being volatile compounds, EOs show low persistence in the environment and a scarce accumulation in soil and water [89].

All these reasons are encouraging researchers to find out suitable solutions to protect and deliver EOs. Currently, the selected strategy is the encapsulation method. Encapsulation is a process through which an active compound is coated or entrapped into a matrix. In this way, the bioactive molecule is isolated and protected by the matrix from the surrounding environment and its release depends on the external conditions and the matrix nature as well [88].

In this respect, in the last years nanotechnology revealed to be the best approach for the exploitation of EOs, allowing to overcome the limitations related to their use [48,90–95]. Although nanotechnology represents an innovative tool able to revolutionise pest management science, it remains a big, but exciting, challenge. An example of EOs stabilisation has been reported by Cespi et al. [96]. They found a suitable solution allowing the use of *Smyrnium olusatrum* L. EO, an oil difficult to handle for stability problems related to the high concentrations of its main constituent, isofuranodiene, which easily undergoes crystallisation. After a systematic study based on an experimental design, they found the best ME capable of encapsulating and protecting EO thanks to the presence of ethyl oleate that avoids the crystallisation issue. Moreover, this formulation proved to be stable over one year and maintained unchanged the bioactivity of EO. Pavela et al. used the same strategy to vehiculate isofuranodiene, the main active compound of *S. olusatrum* EOs [97]. Isofuranodiene-based ME (0.75%) has been tested against *Culex quinquefasciatus* Say showing potent larvicidal effects, with LC_{50} value of 17.7 mL·L^{-1}.

The advantage of MEs and NEs to deliver EOs is not only related to the enhancement of the physicochemical stability but also to the improvement of bioavailability [2,49]. For this reason, the bioactivity of EO-based nanosystems is often higher than those of free EOs. Osman Mohamed Ali et al. carried out a study on the encapsulation of neem and citronella EOs in O/W NEs, to exploit their pest control properties. Stunning *in vivo* results were obtained towards phytopathogenic fungi *Rhizoctonia solani* (Cooke) Wint. and *Athelia rolfsii* (Curzi) C.C. Tu & Kimbr.; EO-based NEs showed exceptional effectiveness, which was higher than those of free EOs [98]. The higher activity of EO-based MEs compared to free EOs (*Trachyspermum ammi* (L.) Sprague ex Turrill, *Pimpinella anisum* L. and *Crithmum maritimum* L.) has been also demonstrated by Pavoni et al. on different species of bacteria and fungi [99]. Moreover, Liang et al. tested the antibacterial activity of peppermint EO NE and the relative free EO on *Listeria monocytogenes* and *Staphyloccoccus aureus* [100]. Although they showed comparable MIC values, the surprising difference was related to the long-term inhibition growth given by NE. Such formulation, by increasing the stability and solubility of EO, was capable of establishing a sustained release. The dispersed phase in fact acts as a nanotank releasing active ingredient over time [21].

Furthermore, the small size of the internal phase improves mobility and penetration with an increase of the activity, and the high surface area of the oily drops enhances the efficacy [101]. Salvia-Trujillo et al. demonstrated the advantageous bioactivity of EO-based nanosystem as compared to the related coarse emulsion [102]. In this case, the difference has been made by the size of the oily droplets, highlighting once again the great advantages generated by such nanosystems.

More explicative examples of EO-based MEs/NEs as biopesticides will be reported in-depth in the following sections.

3. Green Micro- and Nanoemulsions as Insecticides

3.1. Hemiptera

Hemiptera is an order of insects comprising ~68,000 species. Some of them, including many aphids, are important agricultural pests, damaging crops by the direct action of sucking sap, but also harming them indirectly by being the vectors of bacteria, phytoplasmas, spiroplasmas and viruses. They often produce copious amounts of honeydew which encourages the growth of sooty mould. Significant pests include the cottony cushion scale, a pest of citrus fruit trees, the green peach aphid and other aphids which attack crops worldwide and transmit plant diseases. Although several studies have been reported on the activity of EOs against Hemiptera species, only few authors investigated their effectiveness on the same target when encapsulated into MEs or NEs [103–106].

Among the few examples available, Fernandes et al. developed an insecticidal NE based on *Manilkara subsericea* (Mart.) Dubard extract [107]. The efficacy of hexane-soluble fraction from ethanolic extract of *M. subsericea* on *Dysdercus peruvianus* has been previously reported by the same authors [108]. *D. peruvianus* is an Hemiptera species (Pyrrhocoridae) that acts on cotton crops causing huge harvest losses [109]. Since the apolar fraction of the extract is water insoluble, the exploitation of NE technology seemed to be a favourable strategy. After a wide screening on the suitable HLB value of surfactants and the mean droplet size, the following NE composition has been chosen: 5% of *M. subsericea* extract solubilised in 5% octyldodecyl myristate (oil phase) and 5% of surfactants (sorbitan monooleate/polysorbate 80). This NE, characterised by mean droplet size of 155 nm and PDI value of 0.15, proved to be a good insecticide. In fact, it showed its activity since the first day of treatment (12% of mortality), that was sustained over time, with a mortality index of 66% of the insect population after 30 days. Moreover, the safety of this NE was confirmed noting the lack of effects against acetylcholinesterase as well as no acute toxicity on mice.

As said before, aphids represent ones of the world's major insect pests, causing serious economic damage to a range of temperate and tropical crops. This ranges from grain crops and brassicas to potato, cotton, vegetable and fruit crops. For this reason, the investigation on botanical remedies to manage these pests gained great importance and generated several studies on a wide number of EOs and aphids species [110]. Santana et al. (2012) tested the activity of *Thymus vulgaris* L. and *Lavandula latifolia* Medik. on different aphid species, namely *Rhopalosiphum padi* (L.) and *Myzus persicae* Sulzer [111]. Isman (2000) evaluated the fumigant toxicity of four EOs on *Aphis gossypii* Glover, the pest that affects mainly cotton crops, as well as a variety of plants such as citrus, coffee, cocoa, pepper, potato and many ornamental plants [85,112]. On the same target Kalaitzaki et al. tested a formulation of natural pyrethrins, a combination of six esters extracted from the flowers of *T. cinerariifolium* [113]. They solubilised pyrethrins in lemon oil obtaining, initially a W/O ME that was suddenly diluted in water, leading to the formation of an O/W NE. Results about insecticidal activity showed lower LC_{50} and LC_{90} values of pyrethrin-based NE as compared to those of pyrethrum commercial products (761.8 vs. 965.5 mg/mL and 4011.2 vs. 5224.0 mg/mL, respectively).

Pascual-Villalobos et al. performed a wide screening of the repellence activity of 10 EOs and 18 pure compounds against *R. padi*, the major pest of cereal crops on a world scale [114,115]. To face the volatility issue related to the nature of EOs, authors encapsulated the most active ones in NEs, in particular aniseed and peppermint EOs, as well as geraniol, *cis*-jasmone and farnesol. The effectiveness of NEs were evaluated in terms of repellence (RD_{50} and RD_{90}) and mortality after 24 h. Interestingly, some results showed that the smaller were the oil droplets the higher was the repellence activity. In particular, citral-based NE at 2%, having a particle size of 99 nm, showed a repellence index of 66, while the same formulation with larger particles (816 nm) exerted low activity.

3.2. Mosquitoes

Mosquitoes are the vectors of pathogens and parasites of medical and veterinary importance leading to the spread of diseases such as malaria, filariasis, dengue, yellow fever, Japanese encephalitis and Zika virus, just to cite the most important, some of them are lethal, especially in developing countries [116]. Thus, the effective management of these vector populations is a worthy challenge. At the moment, the main approaches to control their spread are: (1) killing adult species through the use of insecticides, (2) reduction of adults population interfering with their fecundity and oviposition or (3) killing mosquito young instars [74].

Although several pesticide products are available on the market, their dangerous effects on the environment along with the development of resistance bring to the need of new sustainable and eco-friendly tools. In the last years, research focused the attention on those EOs suitable as active ingredients in botanical larvicides. Pavela reported, from the literature, the activity of 122 EOs as mosquito larvicides [74]. Interestingly, 77 of them showed LC_{50} value < 50 ppm. Moreover, Pavela assessed the acute toxicity of 30 aromatic compounds of EOs against *C. quinquefasciatus* [87], which is the main vector of the lymphatic filariasis and has been investigated as a vector of Zika virus as well [117,118]. For this reason, several authors investigated the effect of different EOs encapsulated into MEs/NEs against this target.

Oliveira et al. improved the water solubility of *Pterodon emarginatus* Vogel oleoresin through its dispersion in a polisorbate 80/sorbitan monooleate NE, at 1:1 oil–surfactant ratio [119]. This formulation caused the death of around 100% of *C. quinquefasciatus* larvae after an exposure time of 48 h at the concentrations of 100 and 200 mg/L, probably due to morphological alterations on the final abdomen segment of the larvae. Since the *P. emarginatus*-based NE did not exert any toxicity on the green algae *Chlorella vulgaris* Beijerinck, it can be considered an eco-friendly botanical product. The effect of EOs formulations on non-target organisms have been investigated in depth by Pavela et al. on the microcrustacean *Daphnia magna* Straus, the aquatic worm *Tubifex tubifex* (Müller) as well as the earthworm *Eisenia fetida* (Savigny) [69,97]. Moreover, they proved the larvicidal activity of MEs based on Apiaceae EOs, as those of *T. ammi*, *C. maritimum* and *P. anisum*, and on isofuranodiene, the major volatile compound of *S. olusatrum* EO, evaluating the chronic and acute toxicity on *C. quinquefasciatus*. These formulations showed remarkable efficacy, with LC_{50} values of 1.57, 2.23, 4.01 and 17.7 mL/L, respectively.

Several studies have been conducted on the effectiveness of OEs-based MEs and NEs against *Aedes aegypti* L. larvae, the major vector of dengue and yellow fever. In particular, *Rosmarinus officinalis* L. and *Ocimum basilicum* L.-based NEs showed evident efficacy on larval mortality, in a time and dose-dependent manner [120,121]. Interestingly, several authors reported how the exploitation of nanotechnology in pest management could be useful to enhance, not only the stability of EOs, but also their efficacy as pesticide agents.

Balasubramani et al. [122] reported a study based on the larvicidal activity of *Vitex negundo* L. EO on *A. aegypti*. The encapsulated EO showed higher toxicity as compared to the free one, with lower LC_{50} and LC_{90} values. MEs and NEs, in fact, providing a higher dispersion of the lipophilic phase into an aqueous one, could increase the concentration of active ingredients dispersed at the interface leading to direct improvement of the interaction with the target [123].

An important parameter related to the EOs activity is the size of the oily droplets. In fact, Anjali et al. [124] observed that the smaller was the droplets size, the higher was the formulation efficacy. In particular, neem oil NE with a medium diameter of 31 nm caused the mortality of 86% of *C. quinquefasciatus* larvae after 24 h, while NEs of 93 and 251 nm showed a percentage of mortality of 73% and 48%, respectively.

Sugumar et al. [125] compared the activity of *Eucalyptus globulus* Labill. EO encapsulated both in NE and bulk emulsion against *C. quinquefasciatus*. It was observed that, at the concentration of 250 ppm, NE caused 100% of mortality after only 4 h, while the bulk emulsion obtained the same result after 24 h. It is possible to suppose that the size reduction of oil droplets, and thus the increment of the surface area, lead to a better interaction and penetration of the active ingredients into the target organisms [126].

3.3. Stored Product Beetles

Cereal crops can be still considered a main food source for mankind [127]. However, their yield could be compromised by pest infestations during storage. This leads to an extensive loss of crops in term of quality and quantities. In fact pests, not only reduce the amount of grains, but also create suitable environmental conditions for the growth of moulds [128]. The most widespread insect of stored products is *Tribolium castaneum* Herbst, also known as the red flour beetle, which is able to release carcinogenic substances [129].

Botanical research found out several EOs able to fight stored product pests, in particular *T. castaneum*, acting through contact, fumigant, growth inhibitory, antifeedant and repellent actions [130]. Starting from this knowledge, several authors worked on the development of suitable formulation of EOs for their real application. Hashem et al. encapsulated *P. anisum* EO, known to be effective against *T. castaneum*, into a NE, in order to enhance its physicochemical properties [131]. 10% EO-based NE showed a mortality index of 81.33% after 12 days of exposure. Moreover, such system was able to significantly affect the development of progeny and reduce the grain weight loss (%). Morphological and histological evaluations showed that the EO-based NE adhered to several body parts and penetrated through the cuticle, causing cellular necrosis. On the same target, other authors tested EOs obtained from three species of *Achillea*, *A. biebersteinii* Afan., *A. santolina* Falk and *A. millefolium* E.Mey. [130]. They showed how the EO bioactivity depends on the kind of exposure and thus, the mechanism of action. In fact, fumigant toxicity proved to be more effective respect to the topical and contact ones. In particular, the EO-based NE showed significant higher fumigant toxicity as compared to the free EOs, with almost one order of magnitude lower LD_{95} values. Moreover, authors proved that these nanosystems were more effective, in terms of mortality, on adults as compared to larvae, although they strongly affected their growth and development.

Interestingly, Pant et al. added a new ingredient to EO-based NEs that was proven to enhance the effectiveness of the system [132]. They formulated 10% eucalyptus EO NE to test against *T. castaneum*, using karanja and jatropha aqueous filtrates (at increasing concentration from 20% to 60%) in place of water. Such filtrates, obtained from the de-oiled seed cakes, showed to possess insecticidal properties [133,134]. This study reported how the presence of aqueous filtrates improves the physicochemical properties of the formulations, reducing the medium size of the dispersed phase and the PDI value. Moreover, they enhanced the shelf-life of EO for long periods of time reducing its volatility. In fact, after two months, in presence of filtrates, the concentration of EO active ingredients remained unchanged, while in presence of water it decreased to 5%.

Eucalyptus globulus-based NE has been investigated against the species *Sitophilus granarius* L., as well [135]. This formulation showed higher efficacy on this pest when compared with free EO. In addition, such NE showed to be safe, since it did not show mortality and did not cause biochemical alterations in rats.

Choupanian et al. investigated the activity of neem oil NEs against *T. castaneum* and *Sitophilus oryzae* L., also known as the rice weevil [136]. Authors underlined as the effectiveness of a system could depend, not only on the presence and amount of active ingredients, but rather on the formulation parameters. In this case, the choice of the surfactant was carefully evaluated. In fact, polysorbate and alkylpolyglucoside have been compared. NEs obtained with polysorbate showed smaller droplets size and enhanced stability as compared to those containing the other surfactant. Moreover, by their reduced size, they showed higher activity since the active ingredient could penetrate the insect cuticle and come in contact with the target. Moreover, the study reported higher pest mortality of NEs as compared to commercial products and the crude oil extract. These results could be ascribed again to the reduced droplets size of the NEs that caused 100% of mortality in both species after 48 h. Although the previous mentioned species are the most common pests that affect stored products, researchers investigated EO-based NEs against other species as well, obtaining encouraging results about the effectiveness of such nanosystems on the preservation of cereal crops from the infestation of several different pests species [130].

4. Green Micro- and Nanoemulsions as Insect and Tick Repellents

As detailed in the paragraph above, hematophagous insects act as main vectors of several diseases, such as Zika virus, dengue, malaria and yellow fever, causing more than one million deaths per year [137,138]. There is need of new specific drugs or vaccines to treat or prevent such diseases; however, one possible approach to control them is represented by reliable vector control tools, with proven epidemiological impact. One of the simpler ways to deal with this is the employ of repellent products. Repellents are chemical molecules able to prevent the arthropod landing on the skin and the consequent bite [139]. They act through a topical action forming a vapour layer having an intolerable odour for a given arthropod species, preventing its contact with human skin. It is desirable that such molecules do not penetrate in the bloodstream but, rather remain in the stratum corneum [140].

The ideal arthropod repellent should possess some key features: (i) broad spectrum of activity, (ii) long-lasting effect (>8 h), (iii) no toxicity for human being and environment, no skin irritation and low penetration, (iv) odourless to humans and unbearable to arthropods [139]. Generally, repellents are lipophilic volatile molecules, thus they need a suitable vehicle or formulation to be administered.

Now only five/six compounds have been recognised and approved by the Environment Protection Agency (EPA) and the Center for Disease Control and Prevention (CDC) as active repellent ingredients. They have been admitted for skin products thanks to their low toxicity [138]. Three of them are synthetic compounds. The most known and used, since 1957, is N,N-diethyl-3-methylbenzamide (DEET). Despite its high efficacy and long-lasting effect, several studies proved its toxicity due to high skin absorption [141,142]. Its overuse may cause encephalopathy, dermal toxicity, cardiovascular diseases and psychosis and hence, its use has been now restricted and forbidden for pregnant women and children [143]. Other recognised synthetic compounds are ethyl butylacetylaminopropionate (IR3535) and picaridin. The first one is not harmful if ingested, inhaled, or used onto the skin and thus, it can be accepted for human use. Picaridin can be compared to DEET in terms of efficacy and long-lasting effect but it showed only slow toxicity [138,139].

Given the toxicity and resistance issues related to synthetic repellents, one of the biggest challenges for the scientific community is the identification of new efficient and safe compounds [142]. Since ancient times human being has used plants as means to protect himself from insects and pests, by burning or bruising them or by applying their extracts directly on the skin [144,145]. In fact, plants can produce some by-products properly to defend themselves against bloodsucking arthropods. Generally they act binding the odorant-binding proteins in the arthropod's antennae for cuing, preventing their approach [145].

At present, research is focused on the exploitation of EOs to find out new effective natural repellents [146]. Their activity seems to be related to the presence of isoprenoid molecules. In particular, the combination of monoterpenes and sesquiterpenes in the mixture of EOs is considered to be responsible for their repellent activity [147]. Several studies reported that monoterpenes as citronellol, limonene, camphor and thymol showed effective repellent activity [148–150]. Citronellal and eucalyptus EOs have been recognised as skin treatments by EPA while PMD (*p*-menthane-3,8-diol), a compound of *Corymbia citriodora* (Hook.) K.D.Hill & L.A.S.Johnson. EOs, is the only natural repellent recommended by CDC, showing no adverse effects on human health [146]. Although EOs efficacy and safety have been widely proved, their use is still restricted due to some drawbacks related to their physicochemical properties. In fact, they showed rapid evaporation and a short action. Moreover, the application of pure EOs on the human skin could cause irritation [139].

To overcome these limitations the best strategy could be the encapsulation of such active ingredients to develop suitable formulations able to protect and control the release of EOs. The main systems developed for the formulation of repellent EOs are micro-/nanocapsules, MEs/NEs, liposomes, solid lipid nanoparticles and polymeric micelles [139]. Containing oily and water insoluble substances, MEs and NEs could be considered among the best choices as EOs vehicle.

Nowadays the classical repellent formulations on the market are spray solutions and lotions. The first ones require a high amount of alcohol to solubilise the active ingredients while the second ones are emulsions with low stability. On the contrary, NEs and MEs are able to overcome these issues. In fact, they are highly stable, low viscous to be easily spread on the skin and physiologically acceptable in terms of composition [139].

Nuchuchua et al. carried out a study on NEs based on citronella (*Cymbopogon citratus* (DC.) Stapf), hairy basil (*Ocimum americanum* L.) and vetiver (*Vetiveria zizanioides* (L.) Nash) EOs [151]. They evaluated their physicochemical properties, the *in vitro* release, the *in vivo* efficacy on *Ae. aegypti* and the toxicity against normal human foreskin fibroblast (NHF) cells. They compared the different formulations before and after high-pressure homogenisation. After this high-energy process, smaller oily droplets, in the range of 150 to 160 nm, were obtained. They resulted to have a better stability, expressed as zeta potential values, after 2 months. Moreover, the small size of the oily droplets showed to play an important role in the formulation efficacy. In fact, NEs showed a higher release rate, based on a diffusion mechanism, and longer repellent activity. Authors supposed that formulations having smaller size should be able to form a whole film on the skin to prolong the activity. The best formulation was the NE composed of 10% citronella, 5% hairy basil and 5% vetiver EOs, in terms of size, stability and efficacy (4.7 h of protection). Also, Sakuluku et al. investigated the effects of high pressure homogenisation, concentration of surfactant and presence of glycerol on the physicochemical properties and mosquito repellent activity of 20% citronella EO NEs [152]. The best conditions to obtain effective NEs were as follows: concentration of surfactant at 2.5% and water:glycerol at 0:100 ratio. In fact, they demonstrated to influence the kinetic release and the activity against *Ae. aegypti*, as well as the droplet size and the long-term stability. The high amount of glycerol, and thus the high viscosity of the system, delayed the release of EOs, resulting in a prolonged repellent activity on time.

Drapeau et al. formulated PMD based-MEs to evaluate against *Ae. aegypti* [138]. They compared a "surfactantless" ME, composed of water, propanol and PMD and a classical ME, obtained through the construction of a ternary phase diagram. The presence of surfactants led to a prolonged activity, that increased from 315 min of the "surfactantless" ME to 385 min of the classical ME, as well as the reduction of the amount of propanol. The selected formulation was composed of: 46% of H_2O, 20% (*w/w*) of PMD, 25% of PrOH, 2% of Cremophor RH40 (surfactant), 3% of Texapon N70 (surfactant), 1% of 2-ethylhexane-1,3-diol (cosurfactant) and 3% of ethyl (−)-(*S*)-lactate (cosolvent). The addition of these two additives seemed to increase the activity of PMD. The cosurfactant has been selected for its repellent properties, while ethyl (−)-(*S*)-lactate could act as lactic acid competitor on human skin, a good attractant for mosquitos [153–155].

Lastly, Navayan et al. showed how MEs could be a suitable tool to prolong the repellent activity of EOs [156]. In fact, 5%, 10% and 15% eucalyptus EO-based MEs showed a protection time against Culicidae of 82, 135 and 170 min, respectively, while free EO at the same concentrations showed lower time of activity, i.e., 34, 47 and 59 min, respectively. The results obtained through the encapsulation of EO were similar to those of DEET at the same concentrations. Notably, this work outlined how nanosystems could be a desirable tool to increase EOs protection, reduce their volatility, promote their release and prolong the activity on time.

5. Green Micro- and Nanoemulsions as Acaricides

Mite control is economically important for assuring the survival of several vegetables and ornamental plants in greenhouses. For this purpose, conventional pesticides have been widely applied. They include organotin compounds, mitochondrial electron transport inhibitor-acaricides (fenazaquin, fenpyroximate, pyridaben and tebufenpyrad) and pyrethroids. Although they resulted to be very effective, their use has been limited due to the development of pest resistance and the non-target, environmental and human toxicity. These issues have highlighted the need to find out new alternatives for pest management. Botanical pesticides seem to be a valid alternative to the synthetic ones, and are in the field of acaricides products as well. In particular, EOs showed to be the most important natural sources of compounds with acaricidal activity [157–160].

Choi et al. tested the activity of fifty-three EOs against eggs and adults of *Tetranychus urticae* Koch as well as adults of the biocontrol agent *Phytoseiulus persimilis* Athias-Henriot [161]. This study revealed that the most active EOs were: caraway (*Carum carvi* L.) seed, citronella java (*Cymbopogon winterianus* Jowitt), lemon eucalyptus (*C. citriodora*), pennyroyal (*Mentha pulegium* L.), and peppermint (*M. x piperita* L.) EOs showed >90% of toxicity against adults of both mite species. From the obtained results, authors supposed that EOs were delivered and acted on the vapour phase, affecting the respiratory system of mites.

Although their safety and effectiveness, EOs showed a short lasting effect related to their rapid volatilisation and/or degradation [125]. Thus, their encapsulation in liquid sprayable MEs and NEs could be a suitable solution.

Concerning mite species of public health importance, Xu et al. investigated the acaricidal activity of neem oil against *Sarcoptes scabiei* expressed as the speed of kill (min) [162]. Authors compared the effectiveness of pure EO, the EO-based emulsion and the EO-based ME. Neem EO-ME demonstrated the highest acaricidal activity with a lethal time of 192 min followed by 212 min of EO-emulsion and 337 min of pure EO. As expected, the encapsulation process and the small size of the dispersed phase enhanced the activity of EOs and the interaction with target organisms. Moreover, the study reported that ME without active ingredient showed the ability to kill mites. It has been supposed that it could be due to the presence of sodium dodecyl benzene sulfonate (SDBS) in the mixture of surfactants. In fact, given its activity, it has been used to enhance the efficacy of the active ingredients [162].

Research aimed to the effective management of tick species has also been carried out. Chaisri et al. tested the activity of citronella EO on *Rhipicephalus microplus* (Canestrini) [163]. In this study, results have been expressed as larval and adult mortality. ME showed higher acaricidal efficacy compared with the pure citronella EO. In particular, larval mortality after 24 h occurred at the concentration of 0.78% EO-based ME in respect to the concentration of 3.125% of free EO. Also in this case, it could be supposed that the small size of oily droplets, <50 nm, and the presence of surfactants, Tween 20/propylene glycol 3:1, gave a synergistic effect. In particular, surfactants could interfere with the lipids of mites epicuticle, favouring the penetration of active ingredients [153,164].

dos Santos et al. proved the use of cinnamon (*Cinnamomum verum* J. Presl) EO as efficient tool to control ticks on cattle [165]. Indeed, this EO was evaluated against *R. microplus* through both *in vitro* and *in vivo* tests, the latter performed on infested dairy cows. Authors also formulated nanocapsules and NEs. They resulted to be very useful for the exploitation of cinnamon EO acaricidal activity. In fact, nanoencapsulated EO showed to be effective at low concentration (0.5%), ten times lower than that of pure EO (5%). Thus, such nanosystems at 0.5% were able to reduce infestation, oviposition and fertility of *R. microplus*. In fact, the encapsulation of EOs produced an improvement of the active ingredient stability and of its protection and guaranteed sustained release over time.

Nevertheless, the advantages of nanotechnology cannot be ever observed. Galli et al. investigated the activity of *E. globulus* EO [166]. For the purpose, they used the same formulation, concentrations, target and procedures of those previously reported. In this case, pure EO showed to be effective decreasing the reproduction of ticks. On the contrary EO-based nanocapsules and NEs exerted low efficacy. However, it is possible to find an explanation of this result on the short exposure time (30 s) of the pests to nanosystems. This time should be not sufficient for the release of EO [167].

Mossa et al. recently investigated the acaricidal activity of emulsion and NE based on garlic (*Allium sativum* L.) EO on two eriophyid olive mites: *Aceria oleae* (Nalepa) and *Tegolophus hassani* (Keifer) [168]. After several stability studies, they found out a suitable and stable formulation, respect to the classical emulsion giving phase separation after two days. It was composed of 5% of garlic EOs, oil/Tween20 at 1:1.2 ratio and it was obtained through a sonication process for 35 min. Beyond the stability issue, garlic EO-based NE was demonstrated to be more effective than the respective emulsion. In fact, NE showed LC_{50} values of 298.22 and 309.634 µg/mL on *A. oleae* and *T. hassani*, respectively, over to 584.878 and 677.830 µg/mL of the emulsion. Moreover, they proved to be safe for mammal administration as they did not produce toxicity in rats.

Badawy et al. formulated four different NEs based on two EOs—*Callistemon viminalis* (Sol. ex Gaertn.) G.Don and *Origanum vulgare* L.—and two monoterpenes—R-limonene and pulegone [169]. They investigated the activity of 10% concentrated NEs on *T. urticae* in terms of contact toxicity, fumigant toxicity and on bean plants under greenhouse conditions. Although all the formulations showed high efficacy, the monoterpene-based NEs proved to be more toxic against the target organism and with a more rapid outbreak of the activity. Moreover, the fumigant toxicity was more pronounced than contact toxicity. As mentioned above, this could be explained by the fact that such compounds are delivered on vapour phase and act mainly on the respiratory system [161].

6. Green Micro- and Nanoemulsions for Developing Antiparasitic Drugs

Micro- and nanoemulsions can also be useful tools to boost the bioactivity and increase the stability of antiparasitic drugs [170]. In the following paragraphs, we will outline the major achievements in the development of green micro- and nanoemulsions targeting both protozoan and helminth parasites.

6.1. Parasitic Protozoa

6.1.1. *Toxoplasma gondii*

The apicomplexan *Toxoplasma gondii* (Nicolle & Manceaux) infects approximately two billion people worldwide [171]; however, seroprevalence is declining in Western Countries [172].

New drugs are needed for the treatment of toxoplasmosis, particularly in immunocompromised patients or in congenitally infected subjects [173]. Among the new possible drugs, atovaquone is under evaluation for its ability to suppress protozoan parasites with a broad-spectrum activity. However, the use of this drug is limited by its extremely low water solubility and bioavailability. NEs prepared with atovaquone, based on grape seed oil using spontaneous emulsification method, showed increased bioavailability and efficacy for treatment of toxoplasmosis. In fact, *in vitro* this NE resulted active against *T. gondii*, using both RH and another strain (namely, the so-called Tehran strains), cultured on HeLa cells. Such results were confirmed in *in vivo* studies in mice treated orally; these resulted with a lower number of tissue cysts compared to animals treated with the standard preparation, by virtue of better bioavailability [174].

6.1.2. *Leishmania* spp.

They are vector-borne parasites belonging to *Leishmania* genus, order Trypanosomatida. They cause diseases with different clinical pictures: cutaneous (CL), mucocutaneous (MCL) and visceral (VL) [175].

Studies have been carried out on the effects of aromatic/heterocyclic sulphonamides, in the low nanomolar range, on the β-carbonic anhydrase (CA, EC 4.2.1.1) of *Leishmania* spp., which resulted effectively inhibited, without, however, any effect on parasite viability. The same drugs, formulated as

NEs in clove oil, inhibited the growth of either *Leishmania infantum* Nicolle or *Leishmania amazonensis* Lainson & Shaw, being less cytotoxic than the widely used antifungal amphotericin B, as revealed by haemolytic assay [176].

NEs as a delivery system for copaiba (*Copaifera* sp. Linnaeu) and andiroba (*Carapa guianensis* Aublet) oils (nanocopa and nanoandiroba with an average particle size of 76.1 and 88.1 nm, respectively) were tested on *L. infantum* (VL) and *L. amazonensis* (CL). Nanocopa and nanoandiroba resulted toxic to promastigotes of both *Leishmania* species. In particular, ultrastructural analyses by scanning electron microscopy showed a shift of the parasite to oval shape and the retraction of flagella, as early as 1 h after treatment, with concentrations near the IC_{50} values. Furthermore, the treatment with such NEs reduced infectivity of the two species in macrophage cultures. Beneficial results were obtained also in mice experimentally infected with *L. amazonensis* or *L. infantum* (i.e., reduction in lesion size, parasite burden and inflammation). Animals affected by CL treated for eight weeks with NEs showed delay in lesion development. In VL model, around 50% reduction in parasite burden in liver and spleen of mice treated with nanocopa and nanoandiroba was found as compared with control untreated animals [177].

Nanotechnology has allowed the advancement of photodynamic therapy (PDT). In fact, many photosensitisers (PS), insoluble in water, need a nanocarrier as a physiologically acceptable carrier. NEs are efficient in solubilising liposoluble drugs, like the PS, in water. A zinc phthalocyanine (PS) oil-in-water NE, essential clove oil and polymeric surfactant (Pluronic® F127) for the formulation of a topical delivery system for use in PDT was used against *L. amazonensis* and *L. infantum*. The toxicity in the dark and the photobiological activity of the formulations were evaluated *in vitro* on *Leishmania* and macrophages. The zinc phthalocyanine NE was effective in PDT against *Leishmania* spp. with several advantages compared to other topical treatments like paromomycin and amphotericin B. These drugs have many disadvantages like local side effects and a very high cost, often limiting their use [178].

The antiparasitic activity of nanoemulsionated EO of a *Lavanudula* species was tested against *Leishmania major*, a species responsible for CL. In particular, NE with EO of *L. angustifolia* Mill. (where 1,8-cineol and linalool were the major components), as well as of *Rosmarinus officinalis* L., induced significant mortality of the parasite [179]. The NE of *L. angustifolia* and *R. officinalis* EOs showed antiparasitic effects that were much more significant than those obtained with the nonemulsioned EO of *R. officinalis* [180].

A taxonomically related parasite to *Leishmania* is *Trypanosoma evansi* Steel, the etiological agent of the disease known as "Surra" and "Mal das Cadeiras" which affects horses in Brazil, and sometimes also humans. The *in vitro* trypanocidal activity of the nanoemulsified *Schinus molle* L. EO was tested; this NE reduced the number of living parasites even totally, when the highest concentration was used (1%) contrary to the non-emulsified EO, which gave only 68% of mortality as a maximum [181].

6.1.3. *Plasmodium* spp.

Plasmodium parasites cause malaria, a disease which represents one of the major public health problem at global level with 219 million cases of malaria and 435,000 deaths estimated in 2017, particularly concentrated in Africa [182].

NEs loaded with arteether (ART), a semisynthetic derivative of artemisinin, by virtue of their solubility and consequently bioavailability, enhanced efficacy against *Plasmodium yoelii nigeriensis*, in a mouse model of experimental malaria. The *in vitro* release profile of the ART-NEs showed 62% drug release within 12 h; no significant effect on cell viability was observed. The authors focused the attention on a particular NE, loaded with ART (ART-NE), ART-NE-V, which showed a significantly enhanced bioavailability. This NE was well tolerated in the experimentally infected mice with no abnormality in behaviour, food/water consumption and general activity of the animals throughout the treatment and post treatment period. ART-NE-V, administered orally, had an 80% curative rate in comparison to the 100% cure rate achieved by intramuscular route at the same dose and to the 30% curative rate obtained in mice treated with ART in ground nut oil [183].

6.2. Helminths

Echinococcus granulosus

This parasite is the aetiological agent of cystic echinococcosis (CE), a zoonotic infection with economic and public health importance worldwide distributed. CE can result in a substantial human disease burden and have a relevant economic impact on animal productivity [184,185].

EOs from *Zataria multiflora* Boiss. were tested on the cestode *Echinococcus granulosus sensu lato* [186]. The effect was tested on the protoscoleces, isolated in liver hydatid cysts collected from naturally infected sheep. NEs at different concentrations (1–2 mg/mL) induced mortality levels up to 100% after 20 and 10 min, respectively, a scolicidal activity significantly higher than that obtained with nonemulsified oil [187]. *In vivo* studies in infected mice showed that the largest cysts were significantly reduced in size, as well as their total number, in animals treated with NE, compared to those treated with nonemulsified oil [186].

The *in vitro* and *ex vivo* activity of *Melaleuca alternifolia* (Maiden & Betche) Cheel oil (tea tree oil (TTO)), its NE formulation (NE-TTO) and its major component (terpinen-4-ol) were evaluated for their effects against *Echinococcus ortleppi* (another *Echinococcus* species, also known as G5 and clearly closely related to the genotypes of *E. canadensis*). This *Echinococcus* species infects cattle, which represents the principal intermediate host, mainly distributed in Europe, Africa, some areas of Asia and South America [188]. In *ex-vivo* studies the TTO, NE-TTO and the terpinen-4-ol were directly injected in the cysts isolated from cattle. The protoscolicidal action of the TTO major compound, terpinen-4-ol, resulted very promising. In fact, just after 5 min of exposure, non-viable *E. ortleppi* protoscoleces were obtained, at the concentration of 2 mg/mL. The results obtained in this study showed protoscolicidal effect at all tested formulations and concentrations. However, the effects of TTO were higher than those of NE-TTO but this latter had the ability to reduce the volatilisation of the compound and consequently to increase the protoscolicidal effect at the action site [189].

7. Green Formulations against Nematodes Attacking Plants

Meloidogyne spp.

The root-knot nematodes (*Meloidogyne* spp.) are key pests threating several crops of economic importance. Their control is mainly based on the use of chemical nematicides. However, following the withdrawal of several synthetic nematicides because of their detrimental effects on soil biodiversity, natural products of botanical origin have been investigated for their possible use against these agricultural pests. Indeed, besides effectiveness for nematode control, botanicals assure beneficial effects on structure and residual life (e.g., microorganisms) of the soil. Among the most promising natural substances with nematicidal activity, glucosinolates, isothiocyanates, aliphatic acids (e.g., acetic, butyric, hexanoic and decanoic acids), alkaloids, piperamides, flavonoids (e.g., quercetin-7-glucoside), limonoids (azadirachtin, meliacins), quassinoids (e.g., chaparrinone, glaucarubolone, klaineanone, samaderines B and E), saponins and triterpene acids (e.g., 11-oxo triterpenic, pomolic, lantanolic, lantoic, camarin, lantacin, camarinin and ursolic acids), cyanogenic glycosides, polyacetylenes, phenolic acids (e.g., salicylic, gallic, *p*-hydroxybenzoic, vanillic, caffeic, and ferulic acids), fatty acids (e.g., linoleic and oleic acids) and volatile compounds (e.g., ascaridole, 2-undecanone, furfural, benzaldehyde, thymol, geraniol, eugenol, linalool, decenal and decadienal) are the most important ones [190–193]. Among them, isothiocyanates and neem azadirachtin have been encapsulated in marketed formulations effective against the growth and development of *Meloidogyne* spp. with limited effects on soil biodiversity [193,194]. Also, the EOs from *Foeniculum vulgare* Mill., *Pimpinella anisum* L., *Eucalyptus melliodora* A Cunn ex Schauer, *Origanum vulgare* L., *O. dictamnus* L., *Mentha pulegium* L. and *Melissa officinalis* L. were effective against *M. incognita* (Kof. & White) Chitwood showing EC_{50} values of 0.2, 0.3, 0.8, 1.6, 1.7, 3.2 and 6.2 $\mu L \cdot mL^{-1}$, respectively [194,195]. Among their main constituents, benzaldehyde, γ–eudesmol, methyl chavicol, carvone, pulegone and (*E*)-anethole were ideal candidate ingredients for nematicidal formulations [194,195]. On the other hand, efforts about formulating these botanical active ingredients in micro- and nanoemulsions remain limited, outlining the urgent need of future research.

8. Green Micro- and Nanoemulsions in the Real World

As reported above, researchers in entomology and parasitology are making great efforts for the improvement of pest control in terms of efficacy and safety for environment and human being. The potential of EOs and plant extracts as biopesticides and their exploitation through nanoencapsulation opened new challenging strategies for Integrated Pest/Vector Management (IPM/IVM). From the literature analysis (Scopus database, 27 June 2019), it can be observed that in the last 20 years approximately 100 documents were published concerning the employment of MEs and NEs for the vehiculation of pesticides (Figure 3). Interestingly, MEs were firstly studied and the maximum interest was reached around 2010. On the contrary, the use of NEs as pesticide formulations was more recent, reaching the highest attention in the last 2–3 years. Another aspect to be highlighted is represented by the nature of active ingredients employed as pesticide. Regarding MEs, the use of botanical and synthetic pesticides is almost the same along the years, while for NEs there is always a stronger prevalence (~70%) of studies on natural pesticides. These results seem to highlight a temporal correlation between the diffusion of biopesticides and the development of NEs for their application.

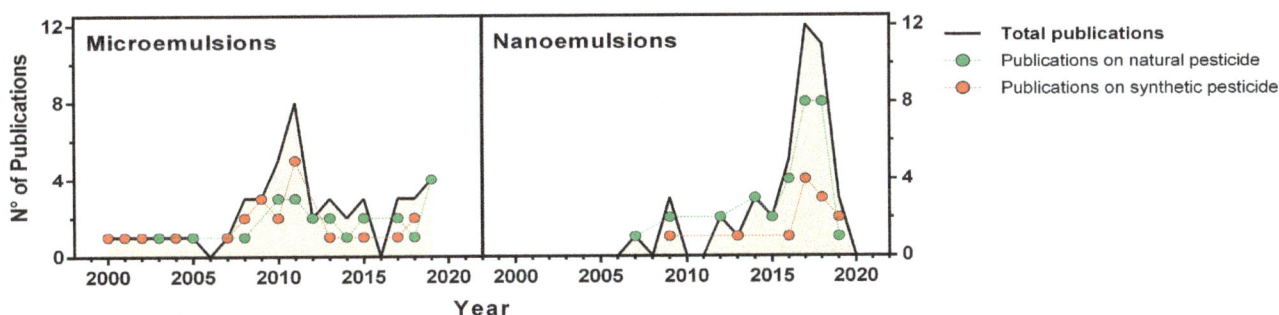

Figure 3. Number of publications on micro-(left) and nanoemulsion (right) vehicles for natural and synthetic pesticides per year.

Even though literature reported several studies with effective results, in the real world the exploitation of EO-based MEs and NEs is still limited. Currently, the pronounced effectiveness of chemical pesticides is still predominant respect to the eco-friendly advantages of the botanical ones. However, the common awareness about the need of a more sustainable world will likely lead towards a radical change in favour of the exploitation of green solutions in the near future.

Although some patents reported the nanoformulation of chemical pesticides or the nanoencapsulation of EOs [196,197], only few of them describe EO-based MEs or NEs as biopesticides. Enan et al. patented MEs as tool for the encapsulation and delivery of two or more EOs for pest control [198]. In particular, they used unsaturated C12-C26 fatty acids and/or salts and saturated C6-C14 fatty acids and/or salts as surfactants to enhance the activity of the ingredients, resulting in an improvement of the pesticide efficacy. According to the authors, this approach brings to a reduction of the active ingredient amount required to obtain an effective pest control.

Since scientific studies showed promising results, in the last years some botanical pesticides started to be available on the market. For example, Prev-Am® Plus is a fungicide and insecticide, based on orange (*Citrus x aurantium* L.) EO, that acts for direct contact. Since Prev-Am® Plus biodegrades rapidly and does not have a high environmental persistence, it is an excellent product for the Integrated Pest Management (IPM) programs, helping in the management of resistance and ensuring a minimal impact on beneficial insects. It can be used on a wide range of crops such as olive trees, vines and citrus fruits and it is allowed in organic agriculture.

Given the well-known repellent activity of EOs, a personal repellent based on EOs has been commercialised. Repel® is a spray containing 30% of lemon eucalyptus EO. It was proven to be able to repel mosquitoes, in particular the vectors of Zika, West Nile, Dengue and Chikungunya viruses, for up to six hours.

Also, Bayer® launched on the market Requiem® EC, an emulsifiable concentrated formulation based on terpenes originally discovered in an insecticidal plant—*Dysphamia ambrosioides* (L.) Mosyakin & Clemants. It is a contact insecticide/acaricide for use in the control or suppression of many foliar-feeding species, including aphids, thrips, plant-feeding mites, whiteflies, mealy bugs, leafminers, Lygus bugs, leafhoppers and moths attacking crops such as citrus, grapes, potatoes and others. Its low toxicity on mammalian and non-target organisms makes it a reduced-risk insecticide.

9. Regulatory Remarks

The EU regulates the botanical products used for the control of parasites, arthropod pests and vectors through two different regulations, the EC No. 1107/2009 and the EU No. 528/2012. The first one regards the plant protection products, addressing their risk evaluation and regulating the authorisation of commercialisation in the crop protection field. The second one, named Biocidal Products Regulation (BPR), takes into account "any substance or mixture exerting a controlling effect on any harmful organism by any means other than mere physical or mechanical action".

Interestingly, while EC No. 1107/2009 does not mention nanomaterials at all, the BPR poses specific issue, stating that "where nanomaterials are used in that product, the risk to human health, animal health and the environment has been assessed separately". For this reason, BPR excludes the possibility of "simplified authorisation procedure" followed for "low-risk" products, in the case of biocide containing nanomaterials. Moreover, the BPR highlights the necessity of a proper methodology for the risk evaluation for nanomaterials.

Although nanotechnology showed to be a great opportunity to achieve a more rational Integrated Pest Management (IPM), the lack of knowledge on the fate and effects on humans and environment of nanomaterials represents, nowadays, an important limitation on their widespread exploitation. It is needed an increased regulatory oversight to ensure their appropriate identification and risk assessment evaluation. In this direction, the European Community is addressing innovative methodologies able to evaluate the risk of nanopesticides and nanomaterials in general. In particular, the European Chemicals Agency (ECHA) is starting to define the guidelines for the monitoring and the evaluation of nanomaterials in the environment, and for the support about their registration procedure (four appendices for nanomaterials applicable to Chapters R.6, R.7a, R.7b and R.7c of the IR&CSA guidance) [199].

Among the different risk assessment procedures, the Quantitative Structure-Activity Relationship/Quantitative Structure-Property Relationship (QSAR/QSPR) appears one of the most promising tools for chemicals. In this regard, the scientific community is moving towards an innovative tool, nano-QSAR/QSPR, introducing the computational approach in the risk assessment of nanomaterials. Several studies focused on how nano-QSAR/QSPR should be supported by the development of new interpretative descriptors for the nanosystems. Moreover, they highlighted the need to model different classes of nanomaterials, given their wide variability in the molecular structure and mechanism of toxicity [200,201].

Currently, the most studied nanomaterials through nano-QSAR/QSPR for risk evaluation are metal oxide and carbon nanoparticles [202–204].

Although nano-QSAR/QSPR is showing to be a useful approach on the risk assessment on nanomaterials, it should be improved by increasing the experimental data on the toxicity of all the different nanomaterials classes, that are still restricted, allowing nano-QSAR/QSPR to be a real tool for the prediction of nanomaterials fate.

Even though much progress has been made, the efforts that are underway to improve the risk assessment procedures of nanomaterials should continue. A pragmatic and internationally accepted nanomaterial decision framework is necessary in order to clarify all the potential toxicological issues, opening to a large-scale diffusion of all the nano-based products.

10. Conclusions and Key Challenges for Future Research

Control of pests and vectors is a highly current issue since they are known to affect the health of the planet. Acting as vectors of devastating pathogens, many pests constitute a threat for the health and survival of living beings, as plants, animals and, above all, human beings. Although in the last decades chemical pesticides have been considered the solution to this problem, nowadays we are becoming aware that they are nothing more than a palliative. In fact, their efficacy has been overshadowed by two main drawbacks, the environmental hazards and the resistance development, linked to their overuse.

Nowadays, a possible solution has been found on the exploitation of botanical compounds, in particular EOs, which showed to possess antiparasitic, insecticidal, larvicidal, acaricidal, ovicidal, fumigant, repellent and chemosterilant effects among other biological properties. They could ensure a sustainable and eco-friendly way to control parasite and pest spreading. In this direction, several efforts have been done in the scientific research fields. For example, several botanical species have been deeply investigated to find out a high number of new active compounds. Anyway, suitable and innovative solutions could be reached only through a multidisciplinary approach. In fact, the physicochemical limits of biological compounds could be overcome only thanks to the development of suitable formulations. For this reason, technological research could offer the real solution to exploit the great advantages and the effectiveness of botanical compounds. Besides insecticides and acaricides, this is also true also for the development of new nematicides, as well as to develop drugs against parasites of public health importance.

In this scenario, nanotechnologies represent the tool of choice. Since they can encapsulate the active compound in a suitable way to protect them and, at the same time, to exalt their efficacy, botanical compound-based nanosystems could represent the turning point in the pest management. Among the different nanosystems available, the MEs and NEs proved to be the most suitable as vehicles for botanicals when those are characterised by high lipophilicity.

Although promising results have been reported in the literature, a strong gap between the theoretical research and the practical application still persists. In this direction, in the near future it is necessary to improve and examine in-depth different aspects of green nanotechnologies; in particular, (i) industrialisation of botanical species plantation in order to increase the amount and the yield of active ingredients, (ii) standardisation of products in terms of quali-quantitative composition, (iii) optimisation of the formulation process to enhance the stability and efficacy of nanosystems, (iv) reduction of the costs of production, (v) evaluation of the real long-term effects of the new products on the environment and non-target organisms and (vi) definition of a clear normative framework able to facilitate the commercial authorisation of botanical compound-based nanosystems.

Author Contributions: L.P., G.B. (Giulia Bonacucina), F.B. and G.B. (Giovanni Benelli) conceived and designed the manuscript. L.P., F.M., M.C. and G.B. (Giulia Bonacucina) drafted the paragraphs dedicated to nanoemulsion preparation. L.P., R.P., V.Z., A.C., A.L. and G.B. (Giovanni Benelli) drafted paragraphs dedicated to insects and mites. F.B. drafted the paragraphs dedicated to parasites. All authors critically reviewed and approved the final version of the manuscript.

Acknowledgments: Two anonymous reviewers kindly improved an earlier version of our manuscript. The authors are grateful to Tracy Jin, Cassie Zhang and the editorial staff of *Nanomaterials* for their support during the development and drafting of this Invited Review.

References

1.   Anton, N.; Vandamme, T.F. Nano-emulsions and micro-emulsions: Clarifications of the critical differences. *Pharm. Res.* **2011**, *28*, 978–985. [CrossRef] [PubMed]

2.   Tadros, T.; Izquierdo, P.; Esquena, J.; Solans, C. Formation and stability of nano-emulsions. *Adv. Colloid Interface Sci.* **2004**, *108*, 303–318. [CrossRef] [PubMed]

3.   Talegaonkar, S.; Azeem, A.; Ahmad, F.; Khar, R.; Pathan, S.; Khan, Z. Microemulsions: A Novel Approach to Enhanced Drug Delivery. *Recent Pat. Drug Deliv.* **2008**, *2*, 238–257. [CrossRef]

4.   Gasco, M.R. Microemulsions in the pharmaceutical field: Perspectives and applications. *Surfactant Sci. Ser.* **1997**, *66*, 97–122.

5.   McClements, D.J. Nanoemulsions versus microemulsions: Terminology, differences, and similarities. *Soft Matter* **2012**, *8*, 1719–1729. [CrossRef]

6.   Rao, J.; McClements, D.J. Formation of flavor oil microemulsions, nanoemulsions and emulsions: Influence of composition and preparation method. *J. Agric. Food Chem.* **2011**, *59*, 5026–5035. [CrossRef]

7.   Danielsson, I.; Lindman, B. The definition of microemulsion. *Colloids Surf.* **1981**, *3*, 391–392. [CrossRef]

8.   Hoar, T.P.; Schulman, J.H. Transparent Water-in-Oil Dispersions: The Oleopathic Hydro-Micelle. *Nature* **1943**, *152*, 102–103. [CrossRef]

9.   Bera, A.; Mandal, A. Microemulsions: A novel approach to enhanced oil recovery: A review. *J. Pet. Explor. Prod. Technol.* **2015**, *5*, 255–268. [CrossRef]

10. Pavoni, L.; Benelli, G.; Maggi, F.; Bonacucina, G. Green nanoemulsion interventions for biopesticide formulations. In *Nano-Biopesticides Today and Future Perspectives*; Academic Press: Cambridge, MA, USA, 2019; pp. 133–160, ISBN 978-0-12-815829-6.

11. Venhuis, S.H.; Mehrvar, M. Health effects, environmental impacts, and photochemical degradation of selected surfactants in water. *Int. J. Photoenergy* **2004**, *6*, 115–125. [CrossRef]

12. Wilhelm, K.P.; Cua, A.B.; Wolff, H.H.; Maibach, H.I. Surfactant-induced stratum corneum hydration *in vivo*: Prediction of the irritation potential of anionic surfactants. *J. Investig. Dermatol.* **1993**, *101*, 310–315. [CrossRef] [PubMed]

13. Lawrence, M.J.; Rees, G.D. Microemulsion-based media as novel drug delivery systems. *Adv. Drug Deliv. Rev.* **2012**, *64*, 175–193. [CrossRef]

14. Schulman, J.H.; Stoeckenius, W.; Prince, L.M. Mechanism of Formation and Structure of Micro Emulsions by Electron Microscopy. *J. Phys. Chem* **1959**, *63*, 1677–1680. [CrossRef]

15. Alany, R.G.; Rades, T.; Agatonovic-Kustrin, S.; Davies, N.M.; Tucker, I.G. Effects of alcohols and diols on the phase behaviour of quaternary systems. *Int. J. Pharm.* **2000**, *196*, 141–145. [CrossRef]

16. Lam, A.C.; Schechter, R.S. The theory of diffusion in microemulsion. *J. Colloid Interface Sci.* **1987**, *120*, 56–63. [CrossRef]

17. Salvia-Trujillo, L.; Rojas-Graü, M.A.; Soliva-Fortuny, R.; Martín-Belloso, O. Effect of processing parameters on physicochemical characteristics of microfluidized lemongrass essential oil-alginate nanoemulsions. *Food Hydrocoll.* **2013**, *30*, 401–407. [CrossRef]

18. Chang, Y.; McClements, D.J. Optimization of orange oil nanoemulsion formation by isothermal low-energy methods: Influence of the oil phase, surfactant, and temperature. *J. Agric. Food Chem.* **2014**, *62*, 2306–2312. [CrossRef]

19. Tadros, T. *Ostwald Ripening BT Encyclopedia of Colloid and Interface Science*; Tadros, T., Ed.; Springer: Berlin/Heidelberg, Germany, 2013; p. 820, ISBN 978-3-642-20665-8.

20. Chang, Y.; McLandsborough, L.; McClements, D.J. Physical properties and antimicrobial efficacy of thyme oil nanoemulsions: Influence of ripening inhibitors. *J. Agric. Food Chem.* **2012**, *60*, 12056–12063. [CrossRef]

21. Donsì, F.; Annunziata, M.; Vincensi, M.; Ferrari, G. Design of nanoemulsion-based delivery systems of natural antimicrobials: Effect of the emulsifier. *J. Biotechnol.* **2012**, *159*, 342–350. [CrossRef]

22. Terjung, N.; Löffler, M.; Gibis, M.; Hinrichs, J.; Weiss, J. Influence of droplet size on the efficacy of oil-in-water emulsions loaded with phenolic antimicrobials. *Food Funct.* **2012**, *3*, 290–301. [CrossRef]

23. Shah, D.; Micelles, D.O. *Microemulsions and Monolayers: Science and Technology*; CRC Press: New York, NY, USA, 1998.

24. Holmberg, K. Organic and bioorganic reactions in microemulsions. *Adv. Colloid Interface Sci.* **1994**, *51*, 137–174. [CrossRef]

25. Lopez-Quintela, M.A. Synthesis of nanomaterials in microemulsions: Formation mechanisms and growth control. *Curr. Opin. Colloid Interface Sci.* **2003**, *8*, 137–144. [CrossRef]

26. Yu, H.; Huang, Q. Improving the oral bioavailability of curcumin using novel organogel-based nanoemulsions. *J. Agric. Food Chem.* **2012**, *60*, 5373–5379. [CrossRef] [PubMed]

27. Saifullah, M.; Ahsan, A.; Shishir, M.R.I. Production, Stability and Application of Micro and Nanoemulsion in Food Production and the food Processing Industry. *Emulsions* **2016**, *3*, 405–442. [CrossRef]

28. Chee, C.P.; Gallaher, J.J.; Djordjevic, D.; Faraji, H.; McClements, D.J.; Decker, E.A.; Hollender, R.; Peterson, D.G.; Roberts, R.F.; Coupland, J.N. Chemical and sensory analysis of strawberry flavoured yogurt supplemented with an algae oil emulsion. *J. Dairy Res.* **2005**, *72*, 311–316. [CrossRef]

29. Silva, H.D.; Cerqueira, M.Â.; Vicente, A.A. Nanoemulsions for Food Applications: Development and Characterization. *Food Bioprocess Technol.* **2012**, *5*, 854–867. [CrossRef]

30. Donsi, F.; Ferrari, G. Essential oil nanoemulsions as antimicrobial agents in food. *J. Biotechnol.* **2016**, *233*, 106–120. [CrossRef]

31. Alexandre, E.M.C.; Lourenço, R.V.; Bittante, A.M.Q.B.; Moraes, I.C.F.; do Amaral Sobral, P.J. Gelatin-based films reinforced with montmorillonite and activated with nanoemulsion of ginger essential oil for food packaging applications. *Food Packag. Shelf Life* **2016**, *10*, 87–96. [CrossRef]

32. Flanagan, J.; Singh, H. Microemulsions: A potential delivery system for bioactives in food. *Crit. Rev. Food Sci. Nutr.* **2006**, *46*, 221–237. [CrossRef]

33. Kralova, I.; Sjöblom, J. Surfactants used in food industry: A review. *J. Dispers. Sci. Technol.* **2009**, *30*, 1363–1383. [CrossRef]

34. Araya, H.; Tomita, M.; Hayashi, M. The novel formulation design of O/W microemulsion for improving the gastrointestinal absorption of poorly water soluble compounds. *Int. J. Pharm.* **2005**, *305*, 61–74. [CrossRef] [PubMed]

35. Bonacucina, G.; Cespi, M.; Misici-falzi, M.; Palmieri, G.F. Colloidal Soft Matter as Drug Delivery System. *J. Pharm. Sci.* **2009**, *98*, 1–42. [CrossRef] [PubMed]

36. Paul, B.K.; Moulik, S.P. Uses and applications of microemulsions. *Curr. Sci. Assoc.* **2001**, *80*, 990–1001.

37. Kim, C.K.; Cho, Y.J.; Gao, Z.G. Preparation and evaluation of biphenyl dimethyl dicarboxylate microemulsions for oral delivery. *J. Control. Release* **2001**, *70*, 149–155. [CrossRef]

38. Yin, Y.M.; Cui, F.D.; Mu, C.F.; Choi, M.K.; Kim, J.S.; Chung, S.J.; Shim, C.K.; Kim, D.D. Docetaxel microemulsion for enhanced oral bioavailability: Preparation and *in vitro* and *in vivo* evaluation. *J. Control. Release* **2009**, *140*, 86–94. [CrossRef] [PubMed]

39. Von Corswant, C.; Thorén, P.; Engström, S. Triglyceride-based microemulsion for intravenous administration of sparingly soluble substances. *J. Pharm. Sci.* **1998**, *87*, 200–208. [CrossRef] [PubMed]

40. Ðorđević, S.M.; Santrač, A.; Cekić, N.D.; Marković, B.D.; Divović, B.; Ilić, T.M.; Savić, M.M.; Savić, S.D. Parenteral nanoemulsions of risperidone for enhanced brain delivery in acute psychosis: Physicochemical and *in vivo* performances. *Int. J. Pharm.* **2017**, *533*, 421–430. [CrossRef] [PubMed]

41. Gupta, S.; Moulik, S.P. Biocompatible microemulsions and their prospective uses in drug delivery. *J. Pharm. Sci.* **2008**, *97*, 22–45. [CrossRef] [PubMed]

42. Majeed, A.; Bashir, R.; Farooq, S.; Maqbool, M. Preparation, Characterization and Applications of Nanoemulsions: An Insight. *J. Drug Deliv.* **2019**, *9*, 520–527. [CrossRef]

43. Vyas, T.K.; Babbar, A.K.; Sharma, R.K.; Singh, S.; Misra, A. Intranasal Mucoadhesive Microemulsions of Clonazepam: Preliminary Studies on Brain Targeting. *J. Pharm. Sci.* **2006**, *95*, 570–580. [CrossRef]

44. Shiokawa, T.; Hattori, Y.; Kawano, K.; Ohguchi, Y.; Kawakami, H.; Toma, K.; Maitani, Y. Effect of polyethylene glycol linker chain length of folate-linked microemulsions loading aclacinomycln A on targeting ability and antitumor effect *in vitro* and *in vivo*. *Clin. Cancer Res.* **2005**, *11*, 2018–2025. [CrossRef] [PubMed]

45. Medina-Pérez, G.; Fernández-Luqueño, F.; Campos-Montiel, R.G.; Sánchez-López, K.B.; Afanador-Barajas, L.N.; Prince, L. Nanotechnology in crop protection: Status and future trends. In *Nano-Biopesticides Today and Future Perspectives*; Academic Press: Cambridge, MA, USA, 2019; pp. 17–45, ISBN 978-0-12-815829-6.

46. Khater, H.; Govindarajan, M.; Benelli, G. *Natural Remedies in the Fight Against Parasites*; InTech, BoD–Books on Demand: London, UK, 2017; ISBN 953513289X.

47. Du, Z.; Wang, C.; Tai, X.; Wang, G.; Liu, X. Optimization and Characterization of Biocompatible Oil-in-Water Nanoemulsion for Pesticide Delivery. *ACS Sustain. Chem. Eng.* **2016**, *4*, 983–991. [CrossRef]

48. Perlatti, B.; de Souza Bergo, P.L.; Fernandes, J.B.; Forim, M.R. Polymeric nanoparticle-based insecticides: A controlled release purpose for agrochemicals. In *Insecticides-Development of Safer and More Effective Technologies*; IntechOpen: London, UK, 2013.

49. Song, S.; Liu, X.; Jiang, J.; Qian, Y.; Zhang, N.; Wu, Q. Stability of triazophos in self-nanoemulsifying pesticide delivery system. *Colloids Surf. A Physicochem. Eng. Asp.* **2009**, *350*, 57–62. [CrossRef]

50. Lubbe, A.; Verpoorte, R. Cultivation of medicinal and aromatic plants for specialty industrial materials. *Ind. Crops Prod.* **2011**, *34*, 785–801. [CrossRef]

51. Fahn, A. Structure and function of secretory cells. *Adv. Bot. Res.* **2000**, *31*, 37–75.

52. Isman, M.B. Botanical insecticides, deterrents, and repellents in modern agriculture and an increasingly regulated world. *Annu. Rev. Entomol.* **2006**, *51*, 45–66. [CrossRef] [PubMed]

53. Chen, M.; Chang, C.H.; Tao, L.; Lu, C. Residential exposure to pesticide during childhood and childhood cancers: A meta-analysis. *Pediatrics* **2015**, *136*, 719–729. [CrossRef]

54. Goulson, D. An overview of the environmental risks posed by neonicotinoid insecticides. *J. Appl. Ecol.* **2013**, *50*, 977–987. [CrossRef]

55. McCaffery, A.; Nauen, R. The insecticide resistance action committee (IRAC): Public responsibility and enlightened industrial self-interest. *Outlooks Pest Manag.* **2006**, *17*, 11–14.

56. Thakore, Y. The biopesticide market for global agricultural use. *Ind. Biotechnol.* **2006**, *2*, 194–208. [CrossRef]

57. Isman, M.B. A renaissance for botanical insecticides? *Pest Manag. Sci.* **2015**, *71*, 1587–1590. [CrossRef] [PubMed]

58. Pavela, R. History, presence and perspective of using plant extracts as commercial botanical insecticides and farm products for protection against insects—A review. *Plant Prot. Sci.* **2016**, *52*, 229–241.

59. Isman, M.B. Problems and opportunities for the commercialization of botanical insecticides. In *Biopesticides of Plant Origin*; Regnault-Roger, C., Philogene, B.J.R., Vincent, C., Eds.; Lavoisier: Paris, France, 2005; pp. 283–291.

60. Collins, D.A. A review of alternatives to organophosphorus compounds for the control of storage mites. *J. Stored Prod. Res.* **2006**, *42*, 395–426. [CrossRef]

61. Singh, A.; Srivastava, V.K. Toxic effect of synthetic pyrethroid permethrin on the enzyme system of the freshwater fish Channa striatus. *Chemosphere* **1999**, *39*, 1951–1956. [CrossRef]

62. Guleria, S.; Jammu, T. *Integrated Pest Management: Innovation-Development Process*; Springer: Dordrecht, The Netherlands; Heidelberg, Germany, 2009.

63. Benelli, G.; Canale, A.; Toniolo, C.; Higuchi, A.; Murugan, K.; Pavela, R.; Nicoletti, M. Neem (*Azadirachta indica*): Towards the ideal insecticide? *Nat. Prod. Res.* **2017**, *31*, 369–386. [CrossRef] [PubMed]

64. Raizada, R.B.; Srivastava, M.K.; Kaushal, R.A.; Singh, R.P. Azadirachtin, a neem biopesticide: Subchronic toxicity assessment in rats. *Food Chem. Toxicol.* **2001**, *39*, 477–483. [CrossRef]

65. Mehlhorn, H.; Al-Rasheid, K.A.S.; Abdel-Ghaffar, F. The Neem tree story: Extracts that really work. In *Nature Helps*; Springer: Heidelberg, Germany, 2011; pp. 77–108.

66. Pavela, R.; Benelli, G. Essential Oils as Ecofriendly Biopesticides? Challenges and Constraints. *Trends Plant Sci.* **2016**, *21*, 1000–1007. [CrossRef] [PubMed]

67. Burt, S. Essential oils: Their antibacterial properties and potential applications in foods—A review. *Int. J. Food Microbiol.* **2004**, *94*, 223–253. [CrossRef] [PubMed]

68. Stroh, J.; Wan, M.T.; Isman, M.B.; Moul, D.J. Evaluation of the acute toxicity to juvenile Pacific coho salmon and rainbow trout of some plant essential oils, a formulated product, and the carrier. *Bull. Environ. Contam. Toxicol.* **1998**, *60*, 923–930. [CrossRef]

69. Pavela, R.; Benelli, G.; Pavoni, L.; Bonacucina, G.; Cespi, M.; Cianfaglione, K.; Bajalan, I.; Morshedloo, M.R.; Lupidi, G.; Romano, D.; et al. Microemulsions for delivery of Apiaceae essential oils—Towards highly effective and eco-friendly mosquito larvicides? *Ind. Crops Prod.* **2019**, *129*, 631–640. [CrossRef]

70. Dubey, N.K. *Natural Products in Plant Pest Management*; CABI: Wallingford, UK, 2011; ISBN 184593671X.

71. Isman, M.B. *Botanical Insecticides, Deterrents, Repellents and Oils*; CABI: Oxfordsh, UK, 2010; pp. 433–445.

72. Koul, O.; Walia, S.; Dhaliwal, G.S. Essential oils as green pesticides: Potential and constraints. *Biopestic. Int.* **2008**, *4*, 63–84.

73. Nerio, L.S.; Olivero-Verbel, J.; Stashenko, E. Repellent activity of essential oils: A review. *Bioresour. Technol.* **2010**, *101*, 372–378. [CrossRef] [PubMed]

74. Pavela, R. Essential oils for the development of eco-friendly mosquito larvicides: A review. *Ind. Crops Prod.* **2015**, *76*, 174–187. [CrossRef]

75. Rattan, R.S. Mechanism of action of insecticidal secondary metabolites of plant origin. *Crop Prot.* **2010**, *29*, 913–920. [CrossRef]

76. Lambert, R.J.W.; Skandamis, P.N.; Coote, P.J.; Nychas, G. A study of the minimum inhibitory concentration and mode of action of oregano essential oil, thymol and carvacrol. *J. Appl. Microbiol.* **2001**, *91*, 453–462. [CrossRef] [PubMed]

77. Tian, J.; Ban, X.; Zeng, H.; He, J.; Chen, Y.; Wang, Y. The mechanism of antifungal action of essential oil from dill (*Anethum graveolens* L.) on Aspergillus flavus. *PLoS ONE* **2012**, *7*, e30147. [CrossRef] [PubMed]

78. Ceylan, E.; Fung, D.Y.C. Antimicrobial activity of spices 1. *J. Rapid Methods Autom. Microbiol.* **2004**, *12*, 1–55. [CrossRef]

79. Di Pasqua, R.; Hoskins, N.; Betts, G.; Mauriello, G. Changes in membrane fatty acids composition of microbial cells induced by addiction of thymol, carvacrol, limonene, cinnamaldehyde, and eugenol in the growing media. *J. Agric. Food Chem.* **2006**, *54*, 2745–2749. [CrossRef]

80. Enan, E. Insecticidal activity of essential oils: Octopaminergic sites of action. *Comp. Biochem. Physiol. Part C Toxicol. Pharm.* **2001**, *130*, 325–337. [CrossRef]

81. Jankowska, M.; Rogalska, J.; Wyszkowska, J.; Stankiewicz, M. Molecular targets for components of essential oils in the insect nervous system—A review. *Molecules* **2018**, *23*, 34. [CrossRef]

82. Mills, C.; Cleary, B.V.; Walsh, J.J.; Gilmer, J.F. Inhibition of acetylcholinesterase by tea tree oil. *J. Pharm. Pharm.* **2004**, *56*, 375–379. [CrossRef]

83. Priestley, C.M.; Williamson, E.M.; Wafford, K.A.; Sattelle, D.B. Thymol, a constituent of thyme essential oil, is a positive allosteric modulator of human GABAA receptors and a homo–oligomeric GABA receptor from Drosophila melanogaster. *Br. J. Pharm.* **2003**, *140*, 1363–1372. [CrossRef]

84. Enan, E.E. Molecular response of Drosophila Melanogaster Tyramine Receptor Cascade to Plant Essential Oils. *Insect Biochem. Mol. Biol.* **2005**, *35*, 309–321. [CrossRef] [PubMed]

85. Isman, M.B. Plant essential oils for pest and disease management. *Crop Prot.* **2000**, *19*, 603–608. [CrossRef]

86. Benelli, G.; Pavela, R.; Canale, A.; Cianfaglione, K.; Ciaschetti, G.; Conti, F.; Nicoletti, M.; Senthil-Nathan, S.; Mehlhorn, H.; Maggi, F. Acute larvicidal toxicity of five essential oils (*Pinus nigra, Hyssopus officinalis, Satureja montana, Aloysia citrodora* and *Pelargonium graveolens*) against the filariasis vector *Culex quinquefasciatus*: Synergistic and antagonistic effects. *Parasitol. Int.* **2017**, *66*, 166–171. [CrossRef] [PubMed]

87. Pavela, R. Acute toxicity and synergistic and antagonistic effects of the aromatic compounds of some essential oils against *Culex quinquefasciatus* Say larvae. *Parasitol. Res.* **2015**, *114*, 3835–3853. [CrossRef] [PubMed]

88. Turek, C.; Stintzing, F.C. Stability of essential oils: A review. *Compr. Rev. Food Sci. Food Saf.* **2013**, *12*, 40–53. [CrossRef]

89. Isman, M.B.; Miresmailli, S.; Machial, C. Commercial opportunities for pesticides based on plant essential oils in agriculture, industry and consumer products. *Phytochem. Rev.* **2011**, *10*, 197–204. [CrossRef]

90. Benelli, G. Plant-mediated biosynthesis of nanoparticles as an emerging tool against mosquitoes of medical and veterinary importance: A review. *Parasitol. Res.* **2016**, *115*, 23–34. [CrossRef]

91. Haldar, K.M.; Haldar, B.; Chandra, G. Fabrication, characterization and mosquito larvicidal bioassay of silver nanoparticles synthesized from aqueous fruit extract of putranjiva, *Drypetes roxburghii* (Wall.). *Parasitol. Res.* **2013**, *112*, 1451–1459. [CrossRef]

92. Benelli, G. Gold nanoparticles–against parasites and insect vectors. *Acta Trop.* **2018**, *178*, 73–80. [CrossRef] [PubMed]

93. Arjunan, N.K.; Murugan, K.; Rejeeth, C.; Madhiyazhagan, P.; Barnard, D.R. Green synthesis of silver nanoparticles for the control of mosquito vectors of malaria, filariasis, and dengue. *Vector-Borne Zoonotic Dis.* **2012**, *12*, 262–268. [CrossRef] [PubMed]

94. Ghormade, V.; Deshpande, M.V.; Paknikar, K.M. Perspectives for nano-biotechnology enabled protection and nutrition of plants. *Biotechnol. Adv.* **2011**, *29*, 792–803. [CrossRef] [PubMed]

95. Pavela, R.; Murugan, K.; Canale, A.; Benelli, G. *Saponaria officinalis*-synthesized silver nanocrystals as effective biopesticides and oviposition inhibitors against *Tetranychus urticae* Koch. *Ind. Crops Prod.* **2017**, *97*, 338–344. [CrossRef]

96. Cespi, M.; Quassinti, L.; Perinelli, D.R.; Bramucci, M.; Iannarelli, R.; Papa, F.; Ricciutelli, M.; Bonacucina, G.; Palmieri, G.F.; Maggi, F. Microemulsions enhance the shelf-life and processability of *Smyrnium olusatrum* L. essential oil. *Flavour Fragr. J.* **2017**, *32*, 159–164. [CrossRef]

97. Pavela, R.; Pavoni, L.; Bonacucina, G.; Cespi, M.; Kavallieratos, N.G.; Cappellacci, L.; Petrelli, R.; Maggi, F.; Benelli, G. Rationale for developing novel mosquito larvicides based on isofuranodiene microemulsions. *J. Pest Sci.* **2019**, *92*, 909–921. [CrossRef]

98. Osman Mohamed Ali, E.; Shakil, N.A.; Rana, V.S.; Sarkar, D.J.; Majumder, S.; Kaushik, P.; Singh, B.B.; Kumar, J. Antifungal activity of nano emulsions of neem and citronella oils against phytopathogenic fungi, *Rhizoctonia solani* and *Sclerotium rolfsii*. *Ind. Crops Prod.* **2017**, *108*, 379–387. [CrossRef]

99. Pavoni, L.; Maggi, F.; Mancianti, F.; Nardoni, S.; Ebani, V.V.; Cespi, M.; Bonacucina, G.; Palmieri, G.F. Microemulsions: An effective encapsulation tool to enhance the antimicrobial activity of selected EOs. *J. Drug Deliv. Sci. Technol.* **2019**. [CrossRef]

100. Liang, R.; Xu, S.; Shoemaker, C.F.; Li, Y.; Zhong, F.; Huang, Q. Physical and antimicrobial properties of peppermint oil nanoemulsions. *J. Agric. Food Chem.* **2012**, *60*, 7548–7555. [CrossRef]

101. Sasson, Y.; Levy-Ruso, G.; Toledano, O.; Ishaaya, I. Nanosuspensions: Emerging novel agrochemical formulations. In *Insecticides Design Using Advanced Technologies*; Springer: Berlin, Germany, 2007; pp. 1–39.

102. Salvia-Trujillo, L.; Rojas-Graü, A.; Soliva-Fortuny, R.; Martín-Belloso, O. Physicochemical characterization and antimicrobial activity of food-grade emulsions and nanoemulsions incorporating essential oils. *Food Hydrocoll.* **2015**, *43*, 547–556. [CrossRef]

103. Zhao, N.N.; Zhang, H.; Zhang, X.C.; Luan, X.B.; Zhou, C.; Liu, Q.Z.; Shi, W.P.; Liu, Z.L. Evaluation of acute toxicity of essential oil of garlic (*Allium sativum*) and its selected major constituent compounds against overwintering *Cacopsylla chinensis* (Hemiptera: Psyllidae). *J. Econ. Entomol.* **2013**, *106*, 1349–1354. [CrossRef] [PubMed]

104. Mann, R.S.; Tiwari, S.; Smoot, J.M.; Rouseff, R.L.; Stelinski, L.L. Repellency and toxicity of plant-based essential oils and their constituents against *Diaphorina citri* Kuwayama (Hemiptera: Psyllidae). *J. Appl. Entomol.* **2012**, *136*, 87–96. [CrossRef]

105. González, W.J.O.; Gutiérrez, M.M.; Murray, A.P.; Ferrero, A.A. Composition and biological activity of essential oils from Labiatae against *Nezara viridula* (Hemiptera: Pentatomidae) soybean pest. *Pest Manag. Sci.* **2011**, *67*, 948–955. [CrossRef] [PubMed]

106. Tian, B.L.; Liu, Q.Z.; Liu, Z.L.; Li, P.; Wang, J.W. Insecticidal Potential of Clove Essential Oil and Its Constituents on *Cacopsylla chinensis* (Hemiptera: Psyllidae) in Laboratory and Field. *J. Econ. Entomol.* **2015**, *108*, 957–961. [CrossRef] [PubMed]

107. Fernandes, C.P.; de Almeida, F.B.; Silveira, A.N.; Gonzalez, M.S.; Mello, C.B.; Feder, D.; Apolinário, R.; Santos, M.G.; Carvalho, J.C.T.; Tietbohl, L.A.C.; et al. Development of an insecticidal nanoemulsion with *Manilkara subsericea* (Sapotaceae) extract. *J. Nanobiotechnol.* **2014**, *12*, 1–9. [CrossRef] [PubMed]

108. Fernandes, C.P.; Xavier, A.; Pacheco, J.P.F.; Santos, M.G.; Mexas, R.; Ratcliffe, N.A.; Gonzalez, M.S.; Mello, C.B.; Rocha, L.; Feder, D. Laboratory evaluation of the effects of *Manilkara subsericea* (Mart.) Dubard extracts and triterpenes on the development of *Dysdercus peruvianus* and *Oncopeltus fasciatus*. *Pest Manag. Sci.* **2013**, *69*, 292–301. [CrossRef] [PubMed]

109. Stanisçuaski, F.; Ferreira-DaSilva, C.T.; Mulinari, F.; Pires-Alves, M.; Carlini, C.R. Insecticidal effects of canatoxin on the cotton stainer bug *Dysdercus peruvianus* (Hemiptera: Pyrrhocoridae). *Toxicon* **2005**, *45*, 753–760. [CrossRef] [PubMed]

110. Gutiérrez, C.; Fereres, A.; Reina, M.; Cabrera, R.; González-Coloma, A. Behavioral and Sublethal Effects of Structurally Related Lower Terpenes on *Myzus persicae*. *J. Chem. Ecol.* **1997**, *23*, 1641–1650. [CrossRef]

111. Santana, O.; Cabrera, R.; Gimenez, C.; González-Coloma, A.; Sánchez-Vioque, R.; De los Mozos-Pascual, M.; Rodríguez-Conde, M.F.; Laserna-Ruiz, I.; Usano-Alemany, J.; Herraiz, D. Perfil químico y biológico de aceites esenciales de plantas aromáticas de interés agro-industrial en Castilla-La Mancha (España). *Grasas Y Aceites* **2012**, *63*.

112. Blackman, R.L.; Eastop, V.F. *Aphids on the World's Crops: An Identification and Information Guide*; John Wiley & Sons Ltd.: Hoboken, NJ, USA, 2000; ISBN 0471851914.

113. Kalaitzaki, A.; Papanikolaou, N.E.; Karamaouna, F.; Dourtoglou, V.; Xenakis, A.; Papadimitriou, V. Biocompatible colloidal dispersions as potential formulations of natural pyrethrins: A structural and efficacy study. *Langmuir* **2015**, *31*, 5722–5730. [CrossRef] [PubMed]

114. Pascual-Villalobos, M.J.; Cantó-Tejero, M.; Vallejo, R.; Guirao, P.; Rodríguez-Rojo, S.; Cocero, M.J. Use of nanoemulsions of plant essential oils as aphid repellents. *Ind. Crops Prod.* **2017**, *110*, 45–57. [CrossRef]

115. Blackman, R.L.; Eastop, V.F. Taxonomic issues. *Aphids Crop Pests* **2007**, 1–29.

116. James, A.A. Mosquito molecular genetics: The hands that feed bite back. *Science* **1992**, *257*, 37–39. [CrossRef] [PubMed]

117. Jambulingam, P.; Subramanian, S.; de Vlas, S.J.; Vinubala, C.; Stolk, W.A. Mathematical modelling of lymphatic filariasis elimination programmes in India: Required duration of mass drug administration and post-treatment level of infection indicators. *Parasit. Vectors* **2016**, *9*, 501. [CrossRef] [PubMed]

118. Benelli, G.; Romano, D. Mosquito vectors of Zika virus. *Entomol. Gen.* **2017**, *36*, 309–318. [CrossRef]

119. Oliveira, A.E.M.F.M.; Duarte, J.L.; Cruz, R.A.S.; Souto, R.N.P.; Ferreira, R.M.A.; Peniche, T.; Conceição, E.C.; Oliveira, L.A.R.; Faustino, S.M.M.; Florentino, A.C.; et al. *Pterodon emarginatus* oleoresin-based nanoemulsion as a promising tool for *Culex quinquefasciatus* (Diptera: Culicidae) control. *J. Nanobiotechnol.* **2017**, *15*, 1–11. [CrossRef] [PubMed]

120. Duarte, J.L.; Amado, J.R.R.; Oliveira, A.E.M.F.M.; Cruz, R.A.S.; Ferreira, A.M.; Souto, R.N.P.; Falcão, D.Q.; Carvalho, J.C.T.; Fernandesa, C.P. Evaluation of larvicidal activity of a nanoemulsion of *Rosmarinus officinalis* essential oil. *Braz. J. Pharm.* **2015**, *25*, 189–192. [CrossRef]

121. Ghosh, V.; Mukherjee, A.; Chandrasekaran, N. Formulation and characterization of plant essential oil based nanoemulsion: Evaluation of its larvicidal activity against *Aedes aegypti*. *Asian J. Chem.* **2013**, *25*, S321.

122. Balasubramani, S.; Rajendhiran, T.; Moola, A.K.; Kumari, R.; Diana, B. Development of nanoemulsion from *Vitex negundo* Lessential oil and their efficacy of antioxidant antimicrobial and larvicidal activities (*Aedes aegypti* L.).) *Environ. Sci. Pollut. Res.* **2017**, *24*, 15125–15133. [CrossRef]

123. Gaysinsky, S.; Taylor, T.M.; Davidson, P.M.; Bruce, B.D. Antimicrobial Efficacy of Eugenol Microemulsions in Milk against *Listeria monocytogenes* and *Escherichia coli* O157:H7. *J. Food Prot.* **2007**, *70*, 2631–2637. [CrossRef]

124. Anjali, C.; Sharma, Y.; Mukherjee, A.; Chandrasekaran, N. Neem oil (*Azadirachta indica*) nanoemulsion-a potent larvicidal agent against *Culex quinquefasciatus*. *Pest Manag. Sci.* **2012**, *68*, 158–163. [CrossRef] [PubMed]

125. Sugumar, S.; Clarke, S.K.; Nirmala, M.J.; Tyagi, B.K.; Mukherjee, A.; Chandrasekaran, N. Nanoemulsion of eucalyptus oil and its larvicidal activity against *Culex quinquefasciatus*. *Bull. Entomol. Res.* **2014**, *104*, 393–402. [CrossRef]

126. Dwivedy, A.K.; Singh, V.K.; Prakash, B.; Dubey, N.K. Nanoencapsulated *Illicium verum* Hook. f. essential oil as an effective novel plant-based preservative against aflatoxin B1 production and free radical generation. *Food Chem. Toxicol.* **2018**, *111*, 102–113. [CrossRef] [PubMed]

127. Alonso-Amelot, M.E.; Avila-Núñez, J.L. Comparison of seven methods for stored cereal losses to insects for their application in rural conditions. *J. Stored Prod. Res.* **2011**, *47*, 82–87. [CrossRef]

128. Magan, N.; Hope, R.; Cairns, V.; Aldred, D. Post-harvest fungal ecology: Impact of fungal growth and mycotoxin accumulation in stored grain. In *Epidemiology of Mycotoxin Producing Fungi*; Springer: Berlin, Germany, 2003; pp. 723–730.

129. Hodges, R.J.; Robinson, R.; Hall, D.R. Quinone contamination of dehusked rice by *Tribolium castaneum* (Herbst) (Coleoptera: Tenebrionidae). *J. Stored Prod. Res.* **1996**, *32*, 31–37. [CrossRef]

130. Nenaah, G.E. Chemical composition, toxicity and growth inhibitory activities of essential oils of three *Achillea* species and their nano-emulsions against *Tribolium castaneum* (Herbst). *Ind. Crops Prod.* **2014**, *53*, 252–260. [CrossRef]

131. Hashem, A.S.; Awadalla, S.S.; Zayed, G.M.; Maggi, F.; Benelli, G. *Pimpinella anisum* essential oil nanoemulsions against *Tribolium castaneum*—Insecticidal activity and mode of action. *Environ. Sci. Pollut. Res.* **2018**, *25*, 18802–18812. [CrossRef] [PubMed]

132. Pant, M.; Dubey, S.; Patanjali, P.K.; Naik, S.N.; Sharma, S. Insecticidal activity of eucalyptus oil nanoemulsion with karanja and jatropha aqueous filtrates. *Int. Biodeterior. Biodegrad.* **2014**, *91*, 119–127. [CrossRef]

133. Kesari, V.; Das, A.; Rangan, L. Physico-chemical characterization and antimicrobial activity from seed oil of *Pongamia pinnata*, a potential biofuel crop. *Biomass Bioenergy* **2010**, *34*, 108–115. [CrossRef]

134. Sharma, S.; Verma, M.; Prasad, R.; Yadav, D. Efficacy of non-edible oil seedcakes against termite (*Odontotermes obesus*). *J. Sci. Ind. Res.* **2011**, *70*, 1037–1041.

135. Mossa, A.T.H.; Abdelfattah, N.A.H.; Mohafrash, S.M.M. Nanoemulsion of camphor (*Eucalyptus globulus*) essential oil, formulation, characterization and insecticidal activity against wheat weevil, *Sitophilus granarius*. *Asian J. Crop Sci.* **2017**, *9*, 50–62. [CrossRef]

136. Choupanian, M.; Omar, D.; Basri, M.; Asib, N. Preparation and characterization of neem oil nanoemulsion formulations against *Sitophilus oryzae* and *Tribolium castaneum* adults. *J. Pestic. Sci.* **2017**, *42*, 158–165. [CrossRef] [PubMed]

137. van der Goes van Naters, W.; Carlson, J.R. Insects as chemosensors of humans and crops. *Nature* **2006**, *444*, 302. [CrossRef] [PubMed]

138. Drapeau, J.; Verdier, M.; Touraud, D.; Kröckel, U.; Geier, M.; Rose, A.; Kunz, W. Effective insect repellent formulation in both surfactantless and classical microemulsions with a long-lasting protection for human beings. *Chem. Biodivers.* **2009**, *6*, 934–947. [CrossRef] [PubMed]

139. Tavares, M.; da Silva, M.R.M.; de Oliveira de Siqueira, L.B.; Rodrigues, R.A.S.; Bodjolle-d'Almeira, L.; dos Santos, E.P.; Ricci-Júnior, E. Trends in insect repellent formulations: A review. *Int. J. Pharm.* **2018**, *539*, 190–209. [CrossRef] [PubMed]

140. Pinto, I.C.; Cerqueira-Coutinho, C.S.; Santos, E.P.; Carmo, F.A.; Ricci-Junior, E. Development and characterization of repellent formulations based on nanostructured hydrogels. *Drug Dev. Ind. Pharm.* **2017**, *43*, 67–73. [CrossRef]

141. Rowland, M.; Freeman, T.; Downey, G.; Hadi, A.; Saeed, M. DEET mosquito repellent sold through social marketing provides personal protection against malaria in an area of all–night mosquito biting and partial coverage of insecticide–treated nets: A case–control study of effectiveness. *Trop. Med. Int. Heal.* **2004**, *9*, 343–350. [CrossRef]

142. Abou-Donia, M.B. Neurotoxicity resulting from coexposure to pyridostigmine bromide, DEET, and permethrin: Implications of Gulf War chemical exposures. *J. Toxicol. Environ. Heal. Part A* **1996**, *48*, 35–56. [CrossRef]

143. Qiu, H.; McCall, J.W.; Jun, H.W. Formulation of topical insect repellent N, N-diethyl-m-toluamide (DEET): Vehicle effects on DEET *in vitro* skin permeation. *Int. J. Pharm.* **1998**, *163*, 167–176. [CrossRef]

144. Moore, S.J.; Lenglet, A.; Hill, N. Plant-based insect repellents. In *Insect Repellents: Principles Methods, and Use*; CRC Press: Boca Raton, FL, USA, 2006.

145. Seyoum, A.; Pålsson, K.; Kung'a, S.; Kabiru, E.W.; Lwande, W.; Killeen, G.F.; Hassanali, A.; Knots, B.G.J. Traditional use of mosquito-repellent plants in western Kenya and their evaluation in semi-field experimental huts against *Anopheles gambiae*: Ethnobotanical studies and application by thermal expulsion and direct burning. *Trans. R. Soc. Trop. Med. Hyg.* **2002**, *96*, 225–231. [CrossRef]

146. Rehman, J.U.; Ali, A.; Khan, I.A. Plant based products: Use and development as repellents against mosquitoes: A review. *Fitoterapia* **2014**, *95*, 65–74. [CrossRef] [PubMed]

147. Jaenson, T.G.T.; Pålsson, K.; Borg-Karlson, A.K. Evaluation of extracts and oils of mosquito (Diptera: Culicidae) repellent plants from Sweden and Guinea-Bissau. *J. Med. Entomol.* **2006**, *43*, 113–119. [CrossRef] [PubMed]

148. Sukumar, K.; Perich, M.J.; Boobar, L.R. Botanical derivatives in mosquito control: A review. *J. Am. Mosq. Control Assoc.* **1991**, *7*, 210–237. [PubMed]

149. Jantan, I.; Zaki, Z.M. Development of environment-friendly insect repellents from the leaf oils of selected Malaysian plants. *Asean Rev. Biodivers. Environ. Conserv.* **1998**, *6*, 1–7.

150. Yang, Y.C.; Lee, E.H.; Lee, H.S.; Lee, D.K.; Ahn, Y.J. Repellency of aromatic medicinal plant extracts and a steam distillate to Aedes aegypti. *J. Am. Mosq. Control Assoc.* **2004**, *20*, 146–149. [PubMed]

151. Nuchuchua, O.; Sakulku, U.; Uawongyart, N.; Puttipipatkhachorn, S.; Soottitantawat, A.; Ruktanonchai, U. *In Vitro* Characterization and Mosquito (*Aedes aegypti*) Repellent Activity of Essential-Oils-Loaded Nanoemulsions. *AAPS PharmSciTech* **2009**, *10*, 1234–1242. [CrossRef] [PubMed]

152. Sakulku, U.; Nuchuchua, O.; Uawongyart, N.; Puttipipatkhachorn, S.; Soottitantawat, A.; Ruktanonchai, U. Characterization and mosquito repellent activity of citronella oil nanoemulsion. *Int. J. Pharm.* **2009**, *372*, 105–111. [CrossRef] [PubMed]

153. Kogan, A.; Garti, N. Microemulsions as transdermal drug delivery vehicles. *Adv. Colloid Interface Sci.* **2006**, *123–126*, 369–385. [CrossRef] [PubMed]

154. Steib, B.M. The Effect of Lactic Acid on Odour-Related Host Preference of Yellow Fever Mosquitoes. *Chem. Senses* **2001**, *26*, 523–528. [CrossRef]

155. Bernier, U.R.; Kline, D.L.; Posey, K.H.; Booth, M.M.; Yost, R.A.; Barnard, D.R. Synergistic Attraction of Aedes aegypti (L.) to Binary Blends of L-Lactic Acid and Acetone, Dichloromethane, or Dimethyl Disulfide. *J. Med. Entomol.* **2009**, *40*, 653–656. [CrossRef] [PubMed]

156. Navayan, A.; Moghimipour, E.; Khodayar, M.J.; Vazirianzadeh, B.; Siahpoosh, A.; Valizadeh, M.; Mansourzadeh, Z. Evaluation of the Mosquito Repellent Activity of Nano-sized Microemulsion of *Eucalyptus globulus* Essential Oil Against Culicinae. *Jundishapur J. Nat. Pharm. Prod.* **2017**, *12*. [CrossRef]

157. Miresmailli, S.; Isman, M.B. Efficacy and persistence of rosemary oil as an acaricide against twospotted spider mite (Acari: Tetranychidae) on greenhouse tomato. *J. Econ. Entomol.* **2006**, *99*, 2015–2023. [CrossRef] [PubMed]

158. Çalmaşur, Ö.; Aslan, İ.; Şahin, F. Insecticidal and acaricidal effect of three Lamiaceae plant essential oils against *Tetranychus urticae* Koch and *Bemisia tabaci* Genn. *Ind. Crops Prod.* **2006**, *23*, 140–146. [CrossRef]

159. Laborda, R.; Manzano, I.; Gamón, M.; Gavidia, I.; Pérez-Bermúdez, P.; Boluda, R. Effects of *Rosmarinus officinalis* and *Salvia officinalis* essential oils on *Tetranychus urticae* Koch (Acari: Tetranychidae). *Ind. Crops Prod.* **2013**, *48*, 106–110. [CrossRef]

160. Han, J.; Kim, S.; Choi, B.; Lee, S.; Ahn, Y. Fumigant toxicity of lemon eucalyptus oil constituents to acaricide–Susceptible and acaricide–Resistant *Tetranychus urticae*. *Pest Manag. Sci.* **2011**, *67*, 1583–1588. [CrossRef]

161. Choi, W.I.; Lee, S.G.; Park, H.M.; Ahn, Y.J. Toxicity of plant essential oils to *Tetranychus urticae* (Acari: Tetranychidae) and *Phytoseiulus persimilis* (Acari: Phytoseiidae). *J. Econ. Entomol.* **2004**, *97*, 553–558. [CrossRef]

162. Xu, J.; Fan, Q.J.; Yin, Z.Q.; Li, X.T.; Du, Y.H.; Jia, R.Y.; Wang, K.Y.; Lv, C.; Ye, G.; Geng, Y.; et al. The preparation of neem oil microemulsion (*Azadirachta indica*) and the comparison of acaricidal time between neem oil microemulsion and other formulations *in vitro*. *Vet. Parasitol.* **2010**, *169*, 399–403. [CrossRef]

163. Chaisri, W.; Chaiyana, W.; Pikulkaew, S.; Okonogi, S.; Suriyasathaporn, W. Enhancement of acaricide activity of citronella oil after microemulsion preparation. *Jpn. J. Vet. Res.* **2019**, *67*, 15–23. [CrossRef]

164. Pedrini, N.; Ortiz-Urquiza, A.; Zhang, S.; Keyhani, N. Targeting of insect epicuticular lipids by the entomopathogenic fungus *Beauveria Bassiana*: Hydrocarbon oxidation within the context of a host-pathogen interaction. *Front. Microbiol.* **2013**, *4*, 24. [CrossRef]

165. dos Santos, D.S.; Boito, J.P.; Santos, R.C.V.; Quatrin, P.M.; Ourique, A.F.; dos Reis, J.H.; Gebert, R.R.; Glombowsky, P.; Klauck, V.; Boligon, A.A.; et al. Nanostructured cinnamon oil has the potential to control *Rhipicephalus microplus* ticks on cattle. *Exp. Appl. Acarol.* **2017**, *73*, 129–138. [CrossRef]

166. Federal, U.; Maria, D.S.; Maria, S.; Maria, S.; Federal, U.; Maria, D.S.; Catarina, S. Archivos de Zootecnia. *Agric. Biol. Sci. Anim. Sci. Zool.* **2018**, *67*, 494–498.

167. Volpato, A.; Grosskopf, R.K.; Santos, R.C.; Vaucher, R.A.; Raffin, R.P.; Boligon, A.A.; Athayde, M.L.; Stefani, L.M.; Da Silva, A.S. Influence of rosemary, andiroba and copaiba essential oils on different stages of the biological cycle of the tick *Rhipicephalus microplus in vitro*. *J. Essent. Oil Res.* **2015**, *27*, 244–250. [CrossRef]

168. Mossa, A.T.H.; Afia, S.I.; Mohafrash, S.M.M.; Abou-Awad, B.A. Formulation and characterization of garlic (*Allium sativum* L.) essential oil nanoemulsion and its acaricidal activity on eriophyid olive mites (Acari: Eriophyidae). *Environ. Sci. Pollut. Res.* **2018**, *25*, 10526–10537. [CrossRef] [PubMed]

169. Badawy, M.E.I.; Abdelgaleil, S.A.M.; Mahmoud, N.F.; Marei, A.E.S.M. Preparation and characterizations of essential oil and monoterpene nanoemulsions and acaricidal activity against two-spotted spider mite (*Tetranychus urticae* Koch). *Int. J. Acarol.* **2018**, *44*, 330–340. [CrossRef]

170. Echeverría, J.; de Albuquerque, D.G.; Diego, R. Nanoemulsions of essential oils: New tool for control of vector–borne diseases and *in vitro* effects on some parasitic agents. *Medicines* **2019**, *6*, 42. [CrossRef] [PubMed]

171. Montoya, J.G.; Liesenfeld, O. Toxoplasmosis. *Lancet* **2004**, *363*, 1965–1976. [CrossRef]

172. Pinto, B.; Mattei, R.; Moscato, G.A.; Cristofano, M.; Giraldi, M.; Scarpato, R.; Buffolano, W.; Bruschi, F. Toxoplasma infection in individuals in central Italy: Does a gender-linked risk exist? *Eur. J. Clin. Microbiol. Infect. Dis.* **2017**, *36*, 739–746. [CrossRef] [PubMed]

173. Dunay, I.R.; Gajurel, K.; Dhakal, R.; Liesenfeld, O.; Montoya, J.G. Treatment of toxoplasmosis: Historical perspective, animal models, and current clinical practice. *Clin. Microbiol. Rev.* **2018**, *31*, e00057-17. [CrossRef] [PubMed]

174. Azami, S.J.; Amani, A.; Keshavarz, H.; Najafi-Taher, R.; Mohebali, M.; Faramarzi, M.A.; Mahmoudi, M.; Shojaee, S. Nanoemulsion of atovaquone as a promising approach for treatment of acute and chronic toxoplasmosis. *Eur. J. Pharm. Sci.* **2018**, *117*, 138–146. [CrossRef] [PubMed]

175. Bruschi, F.; Gradoni, L. *The Leishmaniases: Old Neglected Tropical Diseases*; Springer: Berlin, Germany, 2018; ISBN 3319723863.

176. da Silva Cardoso, V.; Vermelho, A.B.; Ricci Junior, E.; Almeida Rodrigues, I.; Mazotto, A.M.; Supuran, C.T. Antileishmanial activity of sulphonamide nanoemulsions targeting the β-carbonic anhydrase from *Leishmania* species. *J. Enzym. Inhib. Med. Chem.* **2018**, *33*, 850–857. [CrossRef] [PubMed]

177. Dhorm Pimentel de Moraes, A.R.; Tavares, G.D.; Soares Rocha, F.J.; de Paula, E.; Giorgio, S. Effects of nanoemulsions prepared with essential oils of copaiba and andiroba against *Leishmania infantum* and Leishmania amazonensis infections. *Exp. Parasitol.* **2018**, *187*, 12–21. [CrossRef] [PubMed]

178. de Oliveira de Siqueira, L.B.; da Silva Cardoso, V.; Rodrigues, I.A.; Vazquez-Villa, A.L.; dos Santos, E.P.; da Costa Leal Ribeiro Guimarães, B.; Dos Santos Cerqueira Coutinho, C.; Vermelho, A.B.; Junior, E.R. Development and evaluation of zinc phthalocyanine nanoemulsions for use in photodynamic therapy for *Leishmania* spp. *Nanotechnology* **2017**, *28*, 65101.

179. Shokri, A.; Saeedi, M.; Fakhar, M.; Morteza-Semnani, K.; Keighobadi, M.; Teshnizi, S.H.; Kelidari, H.R.; Sadjadi, S. Antileishmanial activity of *Lavandula angustifolia* and *Rosmarinus officinalis* essential oils and nano-emulsions on *Leishmania major* (MRHO/IR/75/ER). *Iran. J. Parasitol.* **2017**, *12*, 622.

180. Bouyahya, A.; Et-Touys, A.; Bakri, Y.; Talbaui, A.; Fellah, H.; Abrini, J.; Dakka, N. Chemical composition of *Mentha pulegium* and *Rosmarinus officinalis* essential oils and their antileishmanial, antibacterial and antioxidant activities. *Microb. Pathog.* **2017**, *111*, 41–49. [CrossRef] [PubMed]

181. Baldissera, M.D.; Da Silva, A.S.; Oliveira, C.B.; Zimmermann, C.E.P.; Vaucher, R.A.; Santos, R.C.V.; Rech, V.C.; Tonin, A.A.; Giongo, J.L.; Mattos, C.B. Trypanocidal activity of the essential oils in their conventional and nanoemulsion forms: *In vitro* tests. *Exp. Parasitol.* **2013**, *134*, 356–361. [CrossRef] [PubMed]

182. *World Malaria Report 2018*; World Health Organization: Geneva, Switzerland, 2018.

183. Dwivedi, P.; Khatik, R.; Chaturvedi, P.; Khandelwal, K.; Taneja, I.; Raju, K.S.R.; Dwivedi, H.; kumar Singh, S.; Gupta, P.K.; Shukla, P. Arteether nanoemulsion for enhanced efficacy against *Plasmodium yoelii nigeriensis* malaria: An approach by enhanced bioavailability. *Colloids Surf. B Biointerfaces* **2015**, *126*, 467–475. [CrossRef] [PubMed]

184. Torgerson, P.R. Economic effects of echinococcosis. *Acta Trop.* **2003**, *85*, 113–118. [CrossRef]

185. Budke, C.M.; Deplazes, P.; Torgerson, P.R. Global socioeconomic impact of cystic echinococcosis. *Emerg. Infect. Dis.* **2006**, *12*, 296. [CrossRef]

186. Moazeni, M.; Borji, H.; Darbandi, M.S.; Saharkhiz, M.J. *In vitro* and *in vivo* antihydatid activity of a nano emulsion of *Zataria multiflora* essential oil. *Res. Vet. Sci.* **2017**, *114*, 308–312. [CrossRef]

187. Mahmoudvand, H.; Mirbadie, S.R.; Sadooghian, S.; Harandi, M.F.; Jahanbakhsh, S.; Saedi Dezaki, E. Chemical composition and scolicidal activity of *Zataria multiflora* Boiss essential oil. *J. Essent. Oil Res.* **2017**, *29*, 42–47. [CrossRef]

188. Lymbery, A.J. Phylogenetic pattern, evolutionary processes and species delimitation in the genus *Echinococcus*. In *Advances in Parasitology*; Elsevier: Amsterdam, The Netherlands, 2017; Volume 95, pp. 111–145, ISBN 0065-308X.

189. Monteiro, D.U.; Azevedo, M.I.; Weiblen, C.; Botton, S.D.A.; Funk, N.L.; Da Silva, C.D.B.; Zanette, R.A.; Schwanz, T.G.; De La Rue, M.L. *In vitro* and *ex vivo* activity of *Melaleuca alternifolia* against protoscoleces of *Echinococcus ortleppi*. *Parasitology* **2017**, *144*, 214–219. [CrossRef] [PubMed]

190. Ntalli, N.G.; Caboni, P. Botanical nematicides in the mediterranean basin. *Phytochem. Rev.* **2012**, *11*, 351–359. [CrossRef]

191. Ntalli, N.; Caboni, P. A review of isothiocyanates biofumigation activity on plant parasitic nematodes. *Phytochem. Rev.* **2017**, *16*, 827–834. [CrossRef]

192. Ntalli, N.G.; Caboni, P. Botanical nematicides: A review. *J. Agric. Food Chem.* **2012**, *60*, 9929–9940. [CrossRef] [PubMed]

193. Caboni, P.; Ntalli, N.G. Botanical nematicides, recent findings. In *Biopesticides: State of the Art and Future Opportunities*; ACS Publications: Washington, WA, USA, 2014; pp. 145–157, ISBN 1947-5918.

194. Ntalli, N.G.; Ferrari, F.; Giannakou, I.; Menkissoglu-Spiroudi, U. Synergistic and antagonistic interactions of terpenes against *Meloidogyne incognita* and the nematicidal activity of essential oils from seven plants indigenous to Greece. *Pest Manag. Sci.* **2011**, *67*, 341–351. [CrossRef]

195. Ntalli, N.G.; Ferrari, F.; Giannakou, I.; Menkissoglu-Spiroudi, U. Phytochemistry and nematicidal activity of the essential oils from 8 Greek Lamiaceae aromatic plants and 13 terpene components. *J. Agric. Food Chem.* **2010**, *58*, 7856–7863. [CrossRef]

196. Kim, C.T.; Kim, C.J.; Cho, Y.J.; Choi, S.W.; Choi, A.J. Nanoemulsion and Nanoparticle Containing Plant Essential Oil and Method of Production Thereof. U.S. Patent US20100136207A1, 3 June 2010.

197. Magdassi, S.; Dayan, B.; Levi-Ruso, G. Pesticide Nanoparticles Obtained from Microemulsions and Nanoemulsions. U.S. Patent US9095133B2, 4 August 2015.

198. Enan, E.; Porpiglia, P.J.; Lindner, G.J. Methods for Pest Control Employing Microemulsion-Based Enhanced Pest Control Formulations. U.S. Patent US20120251641A1, 4 October 2012.

199. ECHA REACH Guidance for Nanomaterials Published. Available online: https://echa.europa.eu/it/-/reach-guidance-for-nanomaterials-published (accessed on 2 August 2019).

200. Villaverde, J.J.; Sevilla-morán, B.; López-goti, C.; Alonso-prados, J.L.; Sandín-españa, P. Considerations of nano-QSAR/QSPR models for nanopesticide risk assessment within the European legislative framework. *Sci. Total Environ.* **2018**, *634*, 1530–1539. [CrossRef]

201. Puzyn, T.; Leszczynski, J.; Leszczynska, D.; Leszczynski, J. Toward the Development of Nano-QSARs: Advances and Challenges. *Small* **2009**, *5*, 2494–2509. [CrossRef]

202. Gajewicz, A.; Rasulev, B.; Dinadayalane, T.C.; Urbaszek, P.; Puzyn, T.; Leszczynska, D.; Leszczynski, J. Advancing risk assessment of engineered nanomaterials: Application of computational approaches. *Adv. Drug Deliv. Rev.* **2012**, *64*, 1663–1693. [CrossRef]

203. Puzyn, T.; Rasulev, B.; Gajewicz, A.; Hu, X.; Dasari, T.P.; Michalkova, A.; Hwang, H.M.; Toropov, A.; Leszczynska, D.; Leszczynski, J. Using nano-QSAR to predict the cytotoxicity of metal oxide nanoparticles. *Nat. Nanotechnol.* **2011**, *6*, 175. [CrossRef]

204. Durdagi, S.; Mavromoustakos, T.; Papadopoulos, M.G. 3D QSAR CoMFA/CoMSIA, molecular docking and molecular dynamics studies of fullerene-based HIV-1 PR inhibitors. *Bioorg. Med. Chem. Lett.* **2008**, *18*, 6283–6289. [CrossRef] [PubMed]

Environmentally-Friendly Green Approach for the Production of Zinc Oxide Nanoparticles and their Anti-Fungal, Ovicidal and Larvicidal Properties

Naif Abdullah Al-Dhabi * and Mariadhas Valan Arasu

Addiriyah Chair for Environmental Studies, Department of Botany and Microbiology, College of Science, King Saud University, P. O. Box 2455, Riyadh 11451, Saudi Arabia; mvalanarasu@ksu.edu.sa
* Correspondence: naldhabi@ksu.edu.sa

Abstract: Green synthesis of nanoparticles can be an important alternative compared to conventional physio-chemical synthesis. We utilized *Scadoxus multiflorus* leaf powder aqueous extract as a capping and stabilizing agent for the synthesis of pure zinc oxide nanoparticles (ZnO NPs). Further, the synthesized ZnO NPs were subjected to various characterization techniques. Transmission electron microscope (TEM) analysis showed an irregular spherical shape, with an average particle size of 31 ± 2 nm. Furthermore, the synthesized ZnO NPs were tested against *Aedes aegypti* larvae and eggs, giving significant LC_{50} value of 34.04 ppm. Ovicidal activity resulted in a higher percentage mortality rate of 96.4 ± 0.24 at 120 ppm with LC_{50} value of 32.73 ppm. Anti-fungal studies were also conducted for ZnO NPs against *Aspergillus niger* and *Aspergillus flavus*, which demonstrated a higher inhibition rate for *Aspergillus flavus* compared to *Aspergillus niger*.

Keywords: *Scadoxus multiflorus*; leaf; ZnO NPs; larvicidal; ovicidal; anti-fungal

1. Introduction

Currently, nanotechnology is a field of intense interest. The process of nanotechnology has been generally classified into three techniques: computational, wet, and dry. While the computational process deals solely with nano-sized structures, the wet process deals with components present in the cells, tissues, and membranes of living organisms. Additionally, the dry process deals with the synthesis of inorganic materials with the help of physical chemistry techniques. The major function of nanotechnology is said to be the synthesis of nanoparticles, mainly relying on the three methodologies such as physical, chemical and biological methods. Of these methodologies, biological synthesis plays a major role when compared with the two other methodologies [1–5]. Biologically-mediated synthesis is further classified into eco-friendly synthesis, which is comprised of plants and plant sources with the corresponding advantages of simplification and lower cost [6–13]. Therefore, we decided to mainly focus on the green synthesis of nanoparticles. For this method of green synthesis of nanoparticles, our research group chose *Scadoxus multiflorus* (*S. multiflorus*) leaf powder aqueous extract (SA) as a green source. This plant is also said to be one of the ancient medicinal plants of India, and belongs to the Caesalpiniaceae family. Different sources of this plant are highly recommended for various treatment purposes, such as irregular menstruation [14]. *S. multiflorus* is a bulbous plant found in most of sub-Saharan Africa which has been used as traditional medicine.

Metal oxide nanoparticles have various significant application possibilities, such as anti-microbial, cell line studies and dye degradation properties. Zinc oxide nanoparticles (ZnO NPs) have a band gap of 3.37 eV, which is relevant for various human applications [15,16].

In this manuscript, we synthesized ZnO NPs with the help of SA. Furthermore, the synthesized ZnO NPs were used to treat one of the major diseases, dengue fever, causing death in India. Dengue is a global disease, with nearly 3 million people affected [17]. *Aedes ageypti* has been stated to be a common vector for causing dengue fever [18]. This manuscript concludes that ZnO NPs are anti-fungal agents effective against *Aspergillus flavus* (*A. flavus*) and *Aspergillus niger* (*A. niger*). Many researchers had reported on the anti-fungal activity of ZnO NPs, which proved to us that ZnO NPs could be utilized as fungicidal agents [19–23].

Overall, this manuscript describes the green synthesis of ZnO NPs using SA, and the subjection of the synthesized particles to various application studies, such as larvicidal and ovicidal activities against *Aedes ageypti* (*A. ageypti*). Furthermore, the synthesized particles, subjected to two different fungal strains, i.e., *A. flavus* and *A. niger*, were studied and are reported herein.

2. Materials and Methods

2.1. Materials and Reagents

The *S. multiflorus* leaf powder was directly procured from the local market and utilized in our research. Zinc acetate was obtained from Sigma-Aldrich (Riyadh, Saudi-Arabia). Reverse-osmosis and double-distilled water was used for the other experiments performed in this study.

2.2. Extraction of the Scadoxus multiflorus Leaf Powder Aqueous Extract Sample

The 30 g of procured powder material of the *S. multiflorus* leaf was immersed in 100 mL of distilled water and placed in a water bath at 60 °C for 1 h. Then, the solvent and powder layer were separated using a Buchner funnel and Whatmann filter paper. The filtrate solution of SA was collected and stored in a refrigerator to be utilized for the future synthesis of ZnO NPs.

2.3. Production of Zinc Oxide Nanoparticles

By using a pipette and mechanical stirrer, 20 mL of collected SA filtrate was added, drop by drop, to 80 mL of 1 mM of zinc acetate under stirring at room temperature (RT). Then, the resultant solutions were placed in a water bath at 60 °C for 3 h and monitored using UV–visible spectroscopy (Hitachi, Tokyo, Japan). Once the reaction mixtures confirmed the formation of ZnO NPs, the resultant solution was subjected to centrifugation at 3000 rpm for 20 min. The centrifugation processes were repeated three times with the help of distilled water to synthesize pure ZnO NPs. Once the centrifugation process was over, the supernatant was discarded, and the pellets were collected and placed in a furnace at 400 °C to obtain the desired product in powder form.

2.4. Analytical Techniques

After synthesis of the ZnO NPs, various analytical techniques, such as UV–visible spectrophotometry (Hitachi, Tokyo, Japan) were performed for the determination of the absorption maximum of the particles. The prepared material was mixed along with KBr to form pellets, to determine the Fourier-transform infrared (FTIR) spectroscopy, using a Shimadzu FTIR Spectrophotometer (Hitachi, Tokyo, Japan). The crystalline nature of the material was characterized by applying an X-ray diffractometer (XRD) (Model D8, Bruker, Germany). Transmission electron microscopy (TEM) (FEI company, Hillsboro, OR, USA) was performed to determine the morphology of the material. A particle-size histogram was developed using Image J software and the Zeta potential was determined by a Horiba nanoparticle analyzer (Horiba scientific, Kyoto, Japan), to identify the stability of the nanoparticles. Shimadzu atomic absorption spectrometry (Shimadzu, Kyoto, Japan) was used to determine Zn, with the help of a deuterium lamp.

2.5. Larvicidal and Ovicidal Properties of Synthesized Zinc Oxide Nanoparticles

Aedes ageypti (*A. ageypti*) larvae were cultured in the laboratory at RT. The third instar larvae were collected and utilized for larvicidal studies; the eggs were collected for ovicidal activity under various concentrations of ZnO NPs—15, 30, 60, and 120 ppm—which were studied and reported using a MANOVA; LSD-DMRT Test. LC50 and LC90 values were also calculated, and identified to be statistically significant at $p < 0.05$. In this study, Neem azal, which is a commercially available insecticide, was utilized as the standard for ovicidal activity [24,25].

2.6. Antifungal Activity of Zinc Oxide Nanoparticles

The studied fungal strains, such as *A. flavus* MTCC 873 and *A. niger* MTCC 282, were procured from IMTECH (Chandigarh, India) and were then processed by using Clinical Laboratory and Standard Institute (CLSI) methods. An amount of 100 mL of PDB (Potato Dextrose Broth) was autoclaved, and *A. flavus* and *A. niger* fungal strains were inoculated into the broth. Test samples of 1 mg/mL were placed in an incubator while being stirred at 120 rpm at RT. After two weeks, the strains were collected, and the biomass of the fungi was filtered and kept for drying. This dried biomass was utilized for further studies, with carbendazim as the standard [26,27]. The mortality percentage of the fungal biomass was calculated using the formula below.

$$\frac{\text{Weight of the control} - \text{Weight of the test}}{\text{Weight of the control}} \times 100$$

3. Results and Discussion

3.1. UV–Visible Spectroscopy

The reaction mixtures of SA and zinc acetate were monitored using UV–visible spectroscopy at the wavelengths of 200 to 800 nm. From the observed results it can be inferred that the highest absorbance of 274 nm is at 90 min, which relies on the conversion of the starting material to end product, as clearly illustrated in Figure 1.

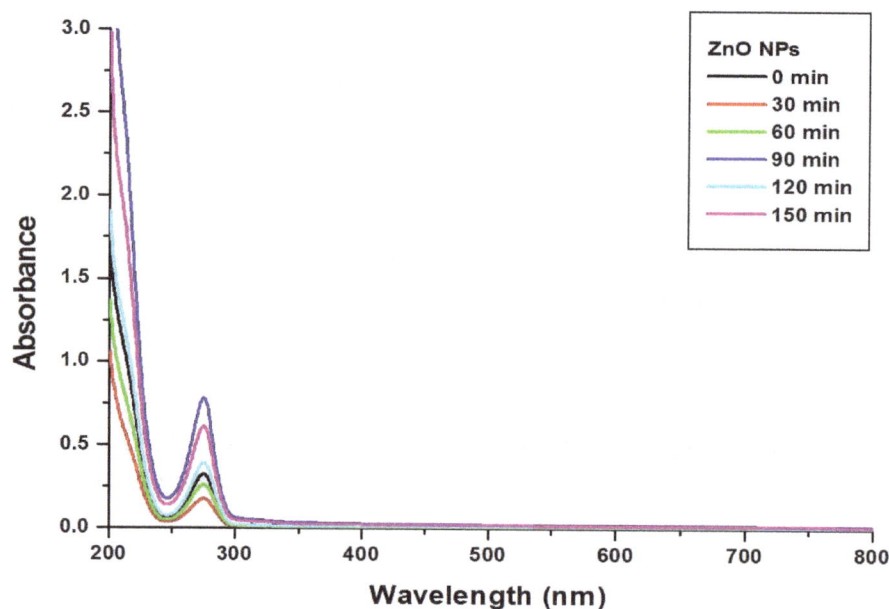

Figure 1. UV–visible spectroscopy of ZnO NPs.

3.2. FTIR Analysis of Zinc Oxide Nanoparticles

The specimens were subjected to FTIR study, as illustrated in Figure 2. Sample *S. multiflorus* leaf extract and ZnO NPs were both recorded to give the FTIR spectra. The FTIR spectrum of the *S. multiflorus* leaf extract shows peaks at 3003 and 1730 cm^{-1}, which correspond to functional groups such as –C=O and C–H (stretch), present in organic molecules. These peaks completely disappear in the ZnO NPs spectrum, which clearly illustrates that the organic molecules are acting as capping and stabilizing agents. The ZnO NPs spectrum showed a characteristic Zn–O stretching at ~417 cm^{-1}, which confirms the formation of ZnO.

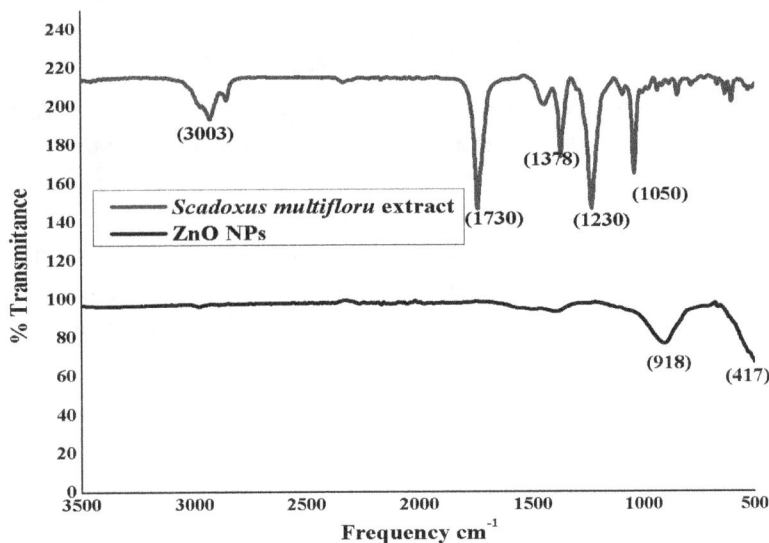

Figure 2. FTIR analysis of ZnO nanoparticles and extract.

3.3. XRD Analysis of Zinc Oxide Nanoparticles

The obtained ZnO NPs were investigated to study their crystalline nature by XRD spectroscopy. From the results it can be inferred that the synthesized ZnO NPs were synthesized in their pure phase, without any impurities. The results also confirmed the h k l values of the (100), (002), (101), (102), (110), (103), (200), (112), (201), and (004) crystalline pattern. Furthermore, the crystalline structure was matched with the JCPDS data of 36-1451, and with the help θ of full-width and half-maximum data, with $d = 1.64056$ and $2\theta = 37$. Twenty-three plane crystalline data were calculated by Scherrer's formula $D = k\lambda/\beta \cos\theta$ [28]. The synthesized crystalline particles were said to be 31.8 nm in size, as illustrated in Figure 3.

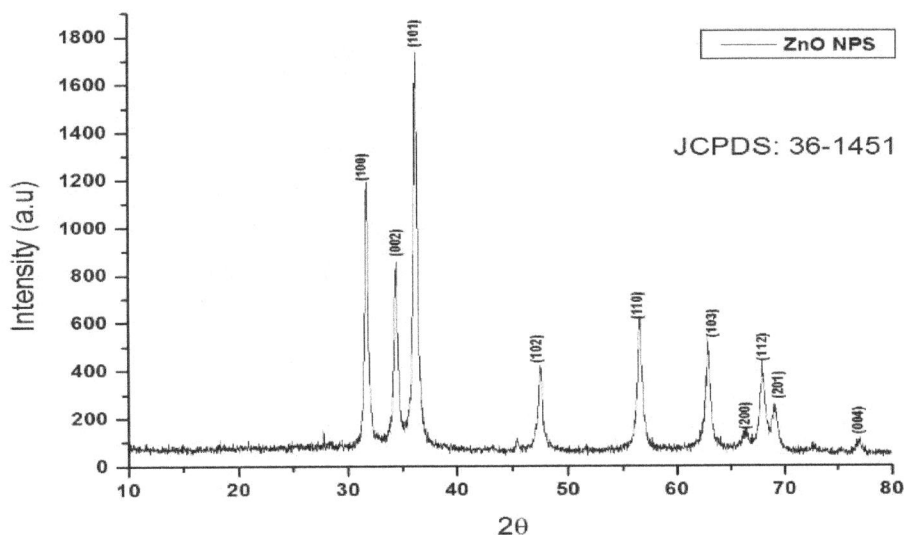

Figure 3. X-ray diffraction (XRD) analysis of the ZnO NPs.

3.4. Zinc Oxide Nanoparticles Morphological Studies

Eco-friendly synthesized ZnO NPs were identified by their morphology using transmission electron microscopy (TEM). From the observed results it can be inferred that the synthesized pure ZnO NPs show irregular, spherical-shaped particles, as illustrated in Figure 4a–c. The particles seem to be legitimately agglomerated, with sizes in the range of ~100 nm. The Selected area (electron) diffraction pattern also clearly cuts the crystalline nature of eco-friendly synthesized ZnO NPs, as shown in Figure 4d. This is a typical phenomenon, taking place due to interaction of H_2O and ZnO. Due to inter-particle interactions, such as van der Waals and electrostatic or magnetic forces, the ZnO NPs in aqueous medium have a tendency to exhibit as an aggregated particle, leading to the development of soft agglomerates. Conversely, particle agglomeration is not complex, because the application purpose (i.e., larvicidal, ovicidal, and fungicidal activity) of the ZnO NPs depends upon the particle size and not on the agglomerate size.

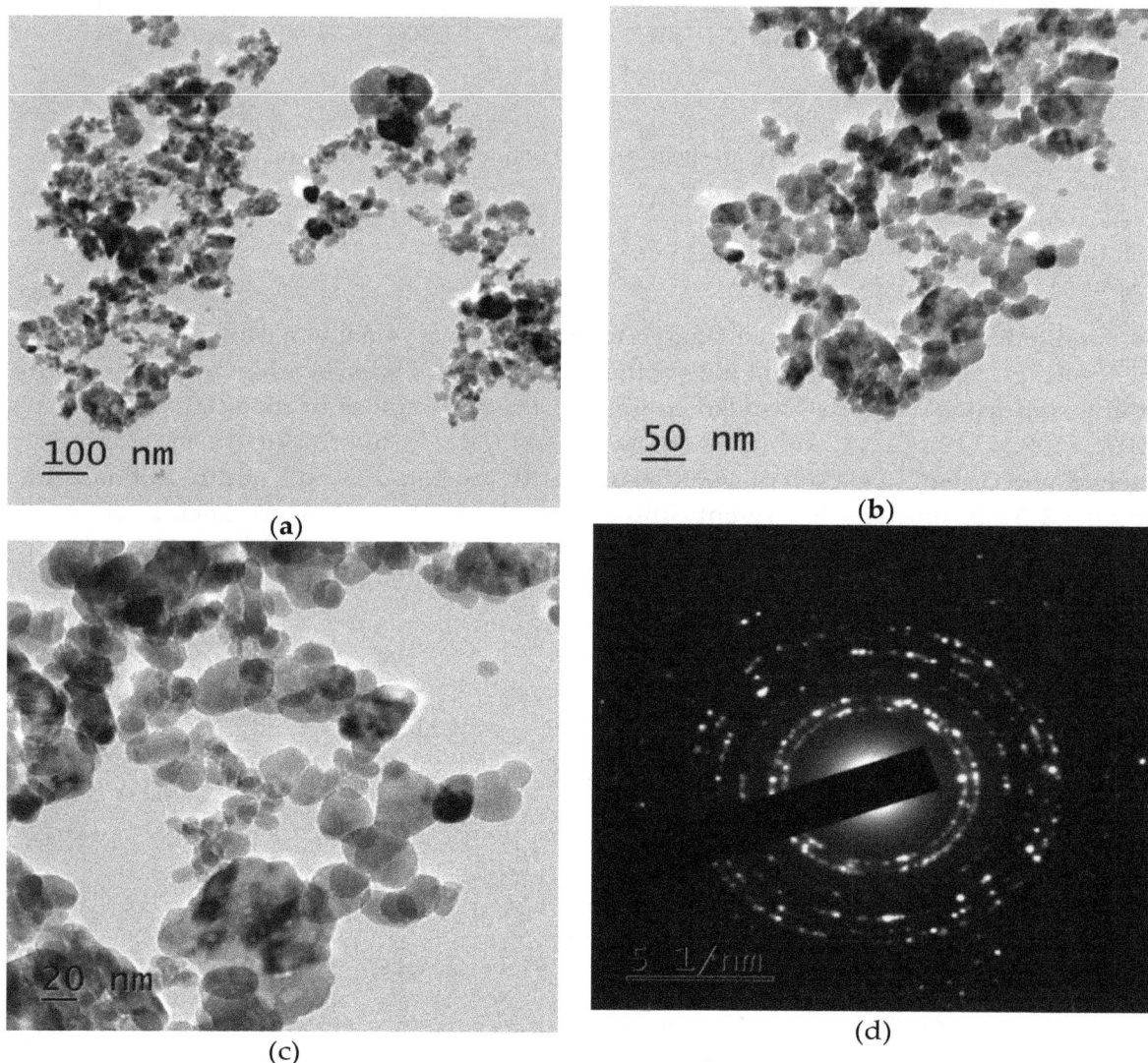

(a)

(b)

(c)

(d)

Figure 4. *Cont.*

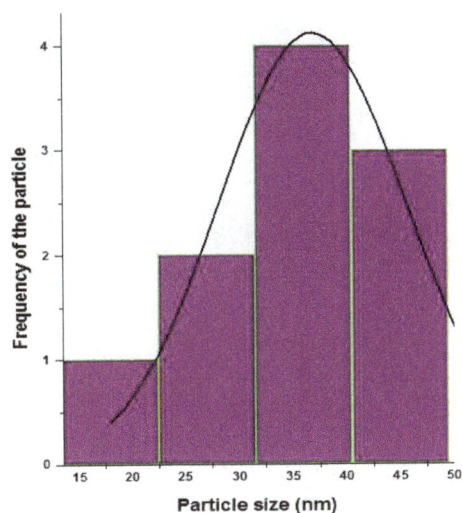

(e)

Figure 4. (**a–c**) TEM images of ZnO NPs and (**d**) SAED pattern of ZnO NPs-particle size histogram (**e**) Particle size histogram.

3.5. Particle Size Histogram Analysis of the Zinc Oxide Nanoparticles

Our research group utilized the ImageJ software for plotting the particle size histogram. The results show that the eco-friendly synthesized ZnO NPs had an average particle size of 31 ± 2 nm, as shown in Figure 4e.

3.6. Energy Dispersive X-ray Analysis (EDAX) Spectrum of Zinc Oxide Nanoparticles

The synthesized ZnO NPs were subjected to an EDAX spectrum to quantify the mixture of metal and oxides present in the sample. The results showed that 64.12% of Zn and 35.76% of O were present on the surface area, as clearly shown in Figure 5.

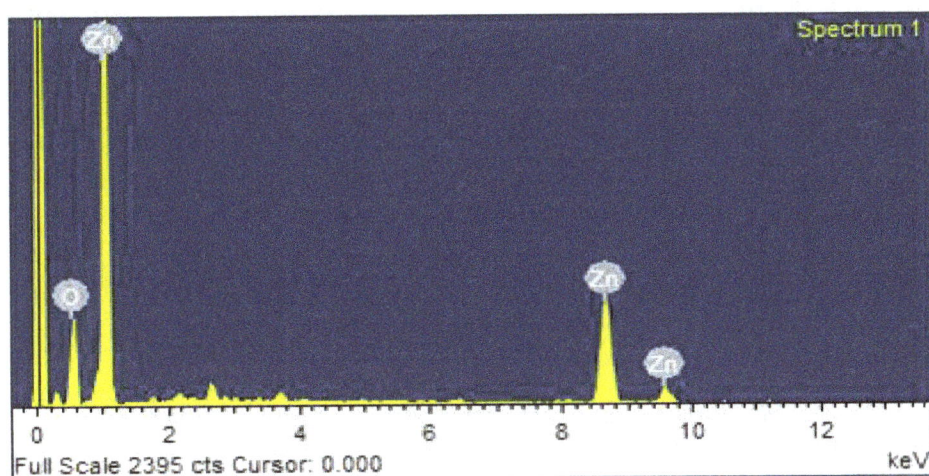

Figure 5. EDAX spectrum of ZnO NPs.

3.7. Stability of Synthesized Zinc Oxide Nanoparticles

The resultant ZnO NPs were subjected to determine the Zeta potential to test their stability, which resulted in a value of -51.8 mV, as clearly illustrated in Figure 6.

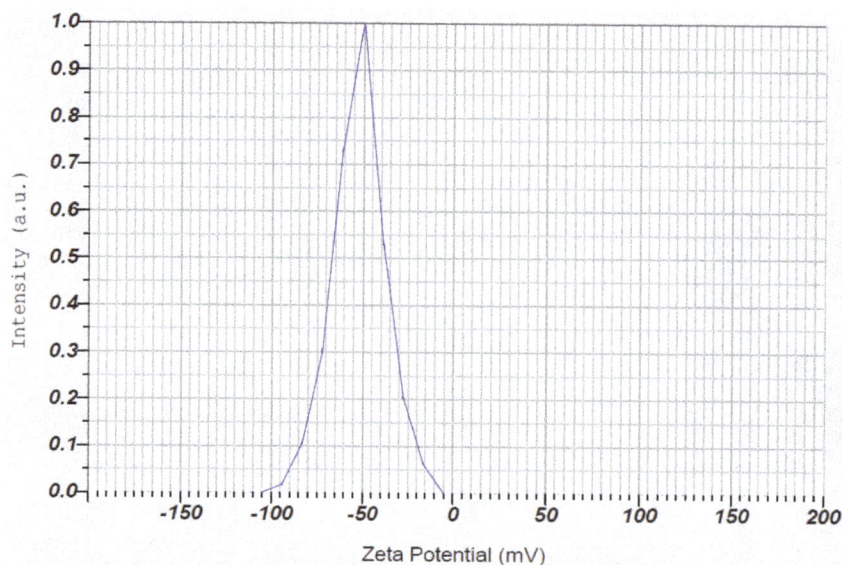

Figure 6. Zeta potential analysis of ZnO NPs.

3.8. Atomic Absorption Spectroscopy

The synthesized quantity of ZnO NPs was analyzed by atomic absorption spectroscopy (AAS) after adding the zinc acetate, with the intention of realizing the remaining concentration of zinc. AAS analysis for the nanoparticle preparing solution, performed at regular intervals of time, exhibited the formation of ZnO NPs. Initially, the standard solution of 5.02 ppm of zinc acetate was prepared and analyzed with AAS at 0 min. After adding *S. multiflorus* leaf extract and the zinc acetate, the formation of nanoparticles was observed at regular time intervals (Figure 7). The result showed a decrease in the concentration of zinc (5.02, 4.22, 3.13, 2.84, 1.87, and 0.08 ppm at 30 min intervals, respectively), indicating the conversion of zinc acetate to ZnO NPs. Additionally, in this present study, 1 gram dry weight of *S. multiflorus* leaves could synthesize 1.15 mg of ZnO NPs within 90 min. Furthermore, this is a sustainable method that does not use toxic chemicals.

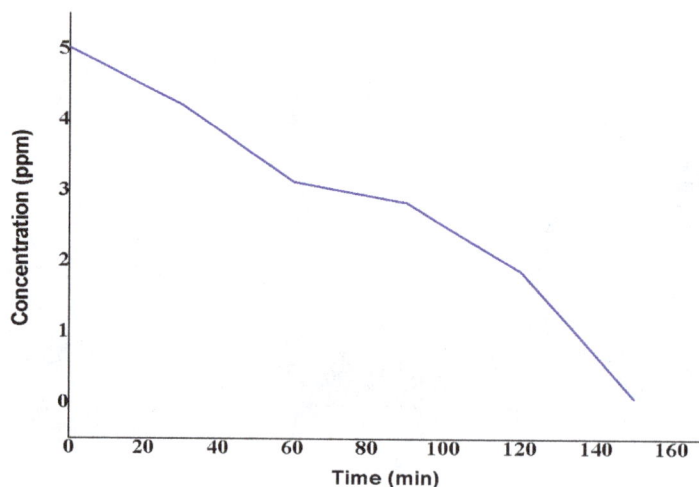

Figure 7. Atomic absorption spectroscopy analysis of zinc acetate in the nanoparticle-forming solution.

3.9. Larvicidal Activity of Zinc Oxide Nanoparticles

Dengue-causing vectors were treated with ZnO NPs at various concentrations: 15, 30, 60, and 120 ppm. The percentage mortality figures are 1.6 ± 0.4, 28.6 ± 7.5, 42.4 ± 2.5, 82.2 ± 6.4, and 98.4 ± 2.3, respectively. This mortality percentage indicates a dose-dependent reaction at higher concentrations, as well as an increasing death rate. Lastly, with the help of LSD tests, we calculated LC_{50} and LC_{90} values with upper and lower confidence limits, as clearly illustrated in Table 1 showing significant results at $p < 0.05$ [10,28]. When compared to the literature [29], our methodologically synthesized ZnO NPs had less larvicidal activity, which may be due to the absence of a bio-organic phase on the surface of the ZnO nanoparticles. *Sargassum wightii*-mediated prepared ZnO NPs have a higher LC_{50} value (49.22 ppm) compared to our result [30]. In another paper [31], *Ulva lactuca*-fabricated ZnO NPs were screened for larvicidal activity against *A. aegypti*, which showed an IC_{50} value of 22.38 ppm. Our methodology provides highly crystalline, pure, and no-bio-organic-phase ZnO NPs. For the control experiment, 1.6% mortality was recorded. The LC_{50} value for larval toxicity was 34.04 ppm.

Table 1. Larvicidal activity of synthesized ZnO NPs.

Concentration (ppm)	Mortality * (%)	LC_{50} (ppm)	95% Confidence Limits (ppm)		LC_{90} (ppm)	95% Confidence Limits (ppm)		χ^2 Value
			LCL	UCL		LCL	UCL	
Control	1.6 ± 0.4 [a]							
15	28.6 ± 7.5 [b]							
30	42..4 ± 2.5 [c]	34.04	14.82	50.32	78.06	58.75	143.75	3.189
60	82.2 ± 6.4 [d]							
120	98.4 ± 2.3 [e]							

The value represents the mean ± S.D. of five replications. * mortality of the larvae observed after 24 h of the exposure period, WHO (2005). LC_{50}: lethal concentration that causes 50% mortality; LC_{90}: lethal concentration that causes 90% mortality. LCL: lower confidence limit; UCL: upper confidence limit. Values in a column with a different superscript alphabet are significantly different at $p < 0.05$ (MANOVA; LSD-DMRT Test).

3.10. Ovicidal Activity of Zinc Oxide Nanoparticles

The eco-friendly synthesized pure form of ZnO NPs was subjected to *A. ageypti* eggs with Neem azal as a standard, with various concentrations: 15, 30, 60, and 120 ppm. The obtained results showed that the ovicidal activity relied on a dose-dependent reaction, with a higher mortality percentage of 96.4 ± 0.24 at 120 ppm. The obtained results after five replicates are depicted in Table 2 [10,28]. Our results relate to the literature [32], i.e., *Terminalia chebula* extracts against *A. ageypti*. The ovicidal activity of ZnO NPs was reported, and may be affected by diverse factors, predominantly egg age and contact period. The egg age influenced the ovicidal action of ZnO NPs. The exposure of freshly laid eggs to ZnO NPs causes higher mortality rates. Our output shows 96.4% mortality at 120 ppm, while *Terminalia chebula* (*T. chebula*) extracts exhibit only 66% mortality. The LC_{50} value for ovicidal toxicity was 32.73 ppm.

Table 2. Ovicidal activity by green synthesized ZnO NPs.

Concentrations (ppm)	% of Mortality
15	35.5 ± 0.23
30	47.2 ± 1.21
60	63.7 ± 0.38
120	96.4 ± 0.24
Neem azal (120)	100 ± 0.00

Values represent mean ± S.D. of five replications. Different alphabets in the column are statistically significant at $p < 0.05$. (MANOVA; LSD-DMRT Test). Eggs in the control groups were not sprayed with phytochemicals. LC_{50}—32.73 ppm; LCL—24.20 ppm; UCL—44.27 ppm.

ZnO NPs were screened for ovicidal activity against which showed an IC_{50} value of 32.73 ppm. Concerning the mechanisms of action of nanoparticles, Volker et al. noted that nanoparticles can affect various physiological parameters in treated organisms, both in vitro and in vivo. The results of in vitro assays showed dose-dependent cell death with oxidative stress as the main likely toxicity pathway. In addition, silver nanoparticles may affect cellular enzymes by interference with free thiol groups and mimicry of endogenous ions. The nanoparticles affect the physiological process of the target organism [33]. On the other hand, strictly limited specific studies have been carried out to elucidate the precise mechanisms of the action of metal nanoparticles on insect pests and vectors [24,25]. However, in the present study, effort has been made to find the mechanism behind the mortality of the mosquitos. The scientific findings have been claimed that, the death of the mosquito may be due the absorption of the nanoparticle into the system and might affect the epithelial cell/ midgut or cortex [25]. It has been predicted that, when the ZnO nanoparticles were absorbed they gets accumulated in the midgut which leads to the shrinkage abdomen leads to the alteration of the mosquitos system. Alternatively, ZnO may affect the functions of other parts such as thorax and midgut, as well as other effects namely lateral hair loss, deformation in gills as well as brushes. Due to these damages in the system it might be the fact the mosquitoes could not undergo respiration hence forth leads to death.

3.11. Zinc Oxide Nanoparticles as Fungicides

Green synthesized pure ZnO NPs were subjected to two fungal pathogens: A. flavus MTCC 873 and A. niger MTCC 282. The ZnO NPs played a prominent role against A. flavus, with 75% inhibition at 500 ppm and 76% inhibition at 1000 ppm, while A. niger resulted in 57% and 63% inhibition, respectively [27], as clearly illustrated in Figure 8. The results were compared with the reported work [34]. The prepared ZnO NPs are active only at higher concentrations. Therefore, there is not much activity against A. flavus and A. niger. Many reports are available on ZnO NPs and their biological activity. These reports clearly state that smaller-sized nanoparticles (NPs) will have higher activity [35]. ZnO NPs might be toxic to some strains, but they are considered essential nutrients. The second reason for the antibacterial activity is that when the Zn^{2+} released by ZnO comes into contact with the cell membranes of the microbe, the cell membranes with negative charge and Zn^{2+} with positive charge mutually attract, and the Zn^{2+} penetrates into the cell membrane and reacts with sulfhydryl groups inside the cell membrane. As a result, the activity of synthetase in the microbe becomes so damaged that the cells lose the ability of growth through cell division, which leads to the death of the microbe (Figure 9) [36,37].

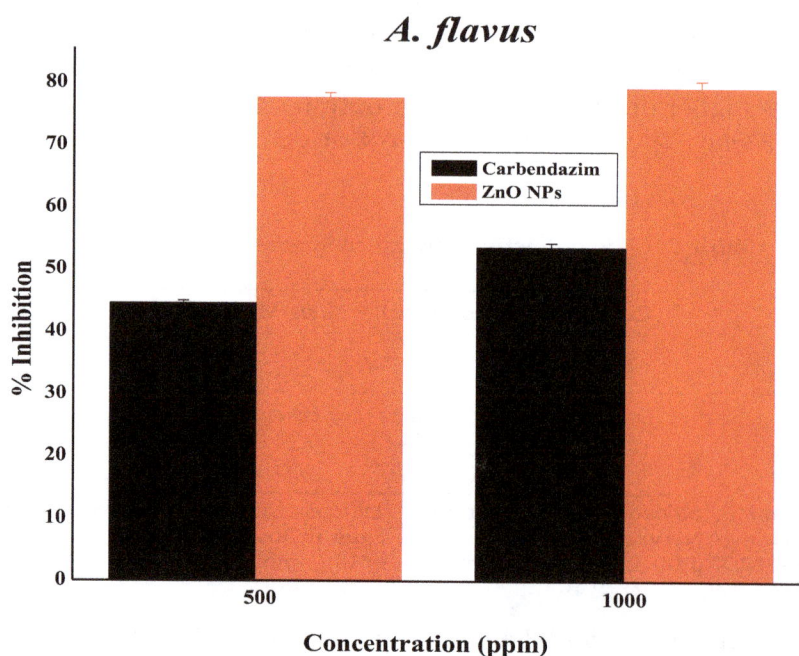

Figure 8. *Cont.*

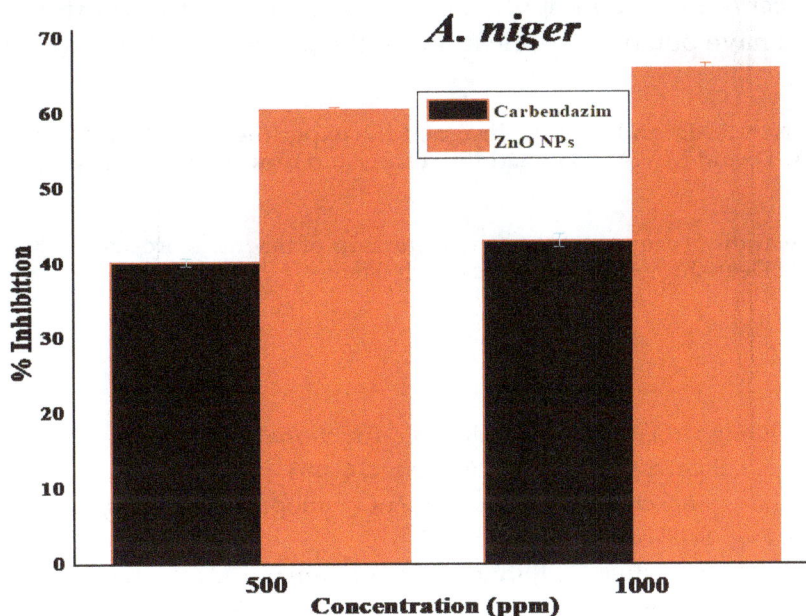

Figure 8. ZnO NPs' anti-fungal activity against *A. flavus* and *A. niger*.

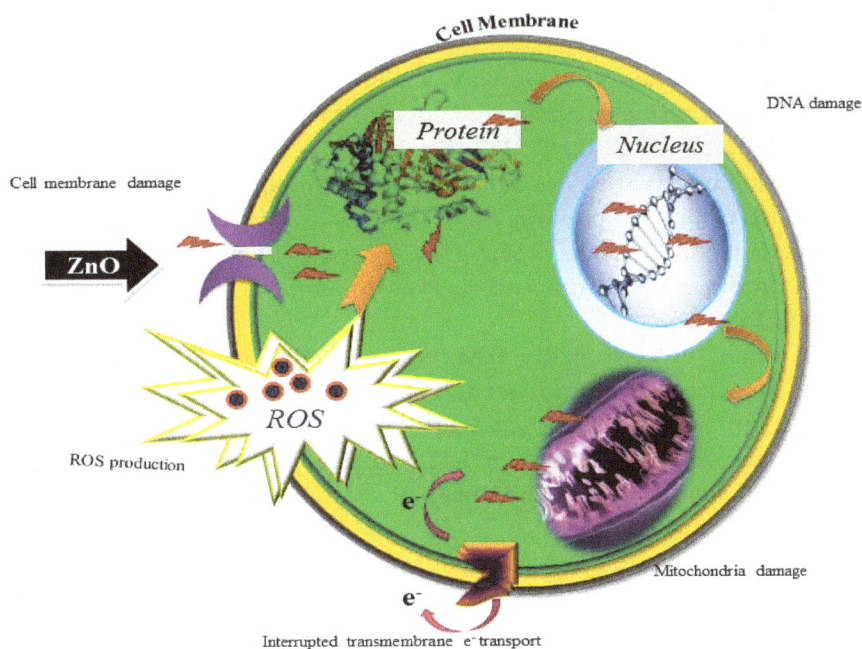

Figure 9. Mode of action of ZnO NPs on microbes.

4. Conclusions

In this manuscript, we proposed a simple process of ZnO NP synthesis by using SA. The results on the synthesized ZnO NPs showed they were irregular, spherical in shape, with an average particle size of 31 ± 2 nm. In addition, the material portrayed promising activity for dengue fever treatment by acting against *A. ageypti*, showing a significant difference at $p < 0.05$. Ovicidal activity was dose-dependent, with an increasing mortality rate at 120 ppm. The activity of the nanoparticles against two fungal pathogens resulted in a higher inhibition rate of *A. flavus* compared to *A. niger*. Moreover, they exhibited effective larvicidal properties against tested fungi and insects. Hence, this study concludes that *S. multiflorus* mediated ZnO NPs may be used as effective control tools against mosquito larval populations and have potential applications in the pharmaceutical and biomedical field.

Author Contributions: N.A.A.-D. and M.V.A. designed the experiment. N.A.A.-D. and M.V.A performed the laboratory work. N.A.A.-D. and M.V.A. analyzed the results and drafted the manuscript.

Acknowledgments: The authors are grateful to the Deanship of Scientific Research, King Saud University for funding through the Vice Deanship of Scientific Research Chairs.

References

1.   Madhumitha, G.; Elango, G.; Roopan, S.M. Bio-functionalized doped silver nanoparticles and its antimicrobial studies. *J. Sol-Gel Sci. Technol.* **2014**, *73*, 476–483. [CrossRef]

2.   Fatimah, I. Biosynthesis and characterization of ZnO nanoparticles using rice bran extract as low-cost templating agent. *J. Eng. Sci. Technol.* **2018**, *13*, 409–420.

3.   Roopan, S.M.; Rohit; Madhumitha, G.; Rahuman, A.A.; Kamaraj, C.; Bharathi, A.; Surendra, T.V. Low-cost and eco-friendly phyto-synthesis of silver nanoparticles using *Cocos nucifera* coir extract and its larvicidal activity. *Ind. Crop. Prod.* **2013**, *43*, 631–635. [CrossRef]

4.   Madhumitha, G.; Rajakumar, G.; Roopan, S.M.; Rahuman, A.A.; Priya, K.M.; Saral, A.M.; Khan, F.R.N.; Khanna, V.G.; Velayutham, K.; Jeyaseelan, C.; et al. Acaricidal, insecticidal, and larvicidal efficacy of fruit peel aqueous extract of *Annona squamosa* and its compounds against blood-feeding parasites. *Parasitol. Res.* **2012**, *111*, 2189–2199. [CrossRef] [PubMed]

5.   Roopan, S.M.; Bharathi, A.; Kumar, R.; Khanna, V.G.; Prabhakarn, A. Acaricidal, insecticidal, and larvicidal efficacy of aqueous extract of *Annona squamosa* L. peel as biomaterial for the reduction of palladium salts into nanoparticles. *Colloids Surf. B* **2012**, *92*, 209–212. [CrossRef] [PubMed]

6.   Kumar, R.; Roopan, S.M.; Prabhakarn, A.; Khanna, V.G.; Chakroborty, S. Agricultural waste *Annona squamosa* peel extract: Biosynthesis of silver nanoparticles. *Spectrochim. Acta A* **2012**, *90*, 173–176. [CrossRef] [PubMed]

7.   Kumar, D.A.; Palanichamy, V.; Roopan, S.M. Green synthesis of silver nanoparticles using *Alternanthera dentata* leaf extract at room temperature and their antimicrobial activity. *Spectrochim. Acta A* **2014**, *127*, 168–171. [CrossRef] [PubMed]

8.   Rajakumar, G.; Rahuman, A.A.; Roopan, S.M.; Khanna, V.G.; Elango, G.; Kamaraj, C.; Zahir, A.A.; Velayutham, K. Fungus-mediated biosynthesis and characterization of TiO$_2$ nanoparticles and their activity against pathogenic bacteria. *Spectrochim. Acta A* **2012**, *91*, 23–29. [CrossRef] [PubMed]

9.   Begum, S.; Ahmaruzzaman, M.; Adhikari, P.P. Ecofriendly bio-synthetic route to synthesize ZnO nanoparticles using *Eryngium foetidum* L. and their activity against pathogenic bacteria. *Mat. Lett.* **2018**, *228*, 37–41. [CrossRef]

10.   Elango, G.; Roopan, S.M.; Al-Dhabi, N.A.; Arasu, M.V.; Dhamodharan, K.I.; Elumalai, K. Coir mediated instant synthesis of Ni-Pd nanoparticles and its significance over larvicidal, pesticidal and ovicidal activities. *J. Mol. Liq.* **2016**, *223*, 1249–1255. [CrossRef]

11.   Jeyaseelan, C.; Rahuman, A.A.; Roopan, S.M.; Krithi, A.V.; Venkatesan, J.; Kim, S.K.; Iyappan, M.; Siva, C. Biological approach to synthesize TiO$_2$ nanoparticles using *Aeromonas hydrophila* and its antibacterial activity. *Spectrochim. Acta A* **2013**, *107*, 82–89. [CrossRef] [PubMed]

12.   Anzabi, Y. Biosynthesis of ZnO nanoparticles using barberry (*Berberis vulgaris*) extract and assessment of their physico-chemical properties and antibacterial activities. *Green Process. Syn.* **2018**, *7*, 114–121. [CrossRef]

13.   Velayutham, K.; Rahuman, A.A.; Rajakumar, G.; Roopan, S.M.; Elango, G.; Kamaraj, C.; Marimuthu, S.; SanthoshKumar, T.; Iyappan, M.; Siva, C. Larvicidal activity of green synthesized silver nanoparticles using bark aqueous extract of *Ficus racemosa* against *Culex quinquefasciatus* and *Culex gelidus*. *Asian Pac. J. Trop. Med.* **2013**, *6*, 95–101. [CrossRef]

14.   Singh, S.; Krishna, T.H.A.; Kamalraj, S.; Kuriakose, G.C.; Valayil, J.M.; Jayabaskaran, C. Phytomedicinal importance of Saracaasoca (Ashoka): An exciting past, an emerging present and a promising future. *Curr. Sci.* **2015**, *109*, 1790–1801. [CrossRef]

15. Raut, S.; Thorat, P.V.; Thakra, R. Green Synthesis of Zinc Oxide (ZnO) Nanoparticles Using *Ocimum tenuiflorum* Leaves. *Int. J. Sci. Res.* **2015**, *4*, 7–13.

16. Ramesh, P.; Rajendran, A.; Subramanian, A. Synthesis of zinc oxide nanoparticle from fruit of *Citrus aurantifolia* by chemical and green method. *Asian J. Phytom. Clin. Res.* **2014**, *2*, 189–195.

17. Harrington, L.C.; Scott, T.W.; Lerdthusnee, K.; Coleman, R.C.; Costero, A.; Clark, G.G.; Jones, J.J.; Kitthawee, S.; Kittayapong, P.; Sithiprasana, R.; et al. Dispersal of the dengue vector *Aedes aegypti* within and between rural communities. *Am. J. Trop. Med. Hyg.* **2005**, *72*, 209–220. [PubMed]

18. Nimmannitya, S.; Halstead, S.B.; Cohen, S.N. Dengue and Chikungunya virus infection in man in Thailand, 1962–1964. I. Observations on hospitalized patients with haemorrhagic fever. *Am. J. Trop. Med. Hyg.* **1969**, *18*, 954–971. [CrossRef] [PubMed]

19. He, L.; Liu, Y.; Mustapha, A.; Lin, M. Antifungal activity of zinc oxide nanoparticles against *Botrytis cinerea* and *Penicillium expansum*. *Microbiol. Res.* **2011**, *166*, 207–215. [CrossRef] [PubMed]

20. Baskar, G.; Chandhuru, J.; Fahad, K.S.; Praveen, A.S. Mycological Synthesis, Characterization and Antifungal Activity of Zinc Oxide Nanoparticles. *Asian J. Pharm. Technol.* **2013**, *4*, 142–146.

21. Vijayakumar, S.; Mahadevan, S.; Arulmozhi, P.; Sriram, S.; Praseetha, P.K. Green synthesis of zinc oxide nanoparticles using *Atalantia monophylla* leaf extracts: Characterization and antimicrobial analysis. *Mater. Sci. Semicond. Process.* **2018**, *82*, 39–45. [CrossRef]

22. Espinel-Ingroff, A.; Pfaller, M.; Messer, S.A.; Knapp, C.C.; Holliday, N.; Killian, S.B. Multicenter comparison of the sensititre yeast one colorimetric antifungal panel with the NCCLS M27-A2 reference methods for testing new antifungal agents against clinical isolates of *Candida* spp. *J. Clin. Microbiol.* **2004**, *42*, 718–721. [CrossRef] [PubMed]

23. Madhiyazhagan, P.; Murugan, K.; Kumar, A.K.; Nataraj, T.; Subramaniam, J.; Chandramohan, B.; Panmeerselvam, C.; Dinesh, D.; Suresh, U.; Nicoletti, M.; et al. One pot synthesis of silver nanocrystals using the seaweed *Gracilaria edulis*: Biophysical characterization and potential against the filariasis vector *Culex quinquefasciatus* and the midge *Chironomus circumdatus*. *J. Appl. Phycol.* **2016**. [CrossRef]

24. Volker, C.; Oetken, M.; Oehlmann, J. The biological effects and possible modes of action of nanosilver. *Rev. Environ. Contam. Toxicol.* **2013**, *223*, 81–106. [PubMed]

25. Benelli, G. Mode of action of nanoparticles against insects. *Environ. Sci. Pollut. Res.* **2018**, *25*, 12329–12341. [CrossRef] [PubMed]

26. Surendra, T.V.; Roopan, S.M.; Al-dhabi, N.A.; Arasu, M.V.; Sarkar, G.; Suthindhiran, K. Vegetable Peel Waste for the Production of ZnO Nanoparticles and its Toxicological Efficiency, Antifungal, Hemolytic, and Antibacterial Activities. *Nanoscale Res. Lett.* **2016**, *11*, 546–556. [CrossRef] [PubMed]

27. Basnet, P.; Inakhunbi Chanu, T.; Samanta, D.; Chatterjee, S. A review on bio-synthesized zinc oxide nanoparticles using plant extracts as reductants and stabilizing agents. *J. Photochem. Photobiol. B* **2018**, *183*, 201–221. [CrossRef] [PubMed]

28. Elango, G.; Roopan, S.M.; Dhamodaran, K.I.; Elumalai, K.; Al-dhabi, N.A.; Arasu, M.V. Spectroscopic investigation of biosynthesized nickel nanoparticles and its larvicidal, pesticidal activities. *J. Photochem. Photobiol. B* **2016**, *162*, 162–167. [CrossRef] [PubMed]

29. Ashokan, A.P.; Paulpandi, M.; Dinesh, D.; Murugan, K.; Vadivalagan, C.; Benelli, G. Toxicity on Dengue Mosquito Vectors Through *Myristica fragrans*-Synthesized Zinc Oxide Nanorods, and Their Cytotoxic Effects on Liver Cancer Cells (HepG2). *J. Clust. Sci.* **2017**, *28*, 205–226. [CrossRef]

30. Ishwarya, R.; Vaseeharan, B.; Subbaiah, S.; Nazar, A.K.; Govindarajan, M.; Alharbi, N.S.; Kadaikunnan, S.; Khaled, J.M.; Al-anbr, M.N. *Sargassum wightii*-synthesized ZnO nanoparticles—From antibacterial and insecticidal activity to immunostimulatory effects on the green tiger shrimp *Penaeus semisulcatus*. *J. Photochem. Photobiol. B* **2018**, *183*, 318–330. [CrossRef] [PubMed]

31. Ishwarya, R.; Vaseeharan, B.; Kalyani, S.; Banumathi, B.; Govindarajan, M.; Alharbi, N.S.; Kadaikunnan, S.; Al-anbr, M.N.; Khaled, J.M.; Benelli, G. Facile green synthesis of zinc oxide nanoparticles using *Ulva lactuca* seaweed extract and evaluation of their photocatalytic, antibiofilm and insecticidal activity. *J. Photochem. Photobiol. B* **2018**, *178*, 249–258. [CrossRef] [PubMed]

32. Veni, T.; Pushpanathan, P.; Mohanraj, J. Larvicidal and ovicidal activity of *Terminalia chebula* Retz. (Family: Combretaceae) medicinal plant extracts against *Anopheles stephensi*, *Aedes aegypti* and *Culex quinquefasciatus*. *J. Parasit. Dis.* **2017**, *41*, 693–702. [CrossRef] [PubMed]

33. Athanassiou, C.G.; Kavallieratos, N.G.; Benelli, G.; Losic, D.; Usha Rani, P.; Desneux, N. Nanoparticles for pest control: current status and future perspectives. *J. Pest. Sci.* **2018**, *91*, 1–15. [CrossRef]

34. Malaikozhundan, B.; Vaseeharan, B.; Vijayakumar, S.; Pandiselvi, K.; Kalanjiam, M.A.R.; Murugan, K.; Benelli, G. Biological therapeutics of *Pongamia pinnata* coated zinc oxide nanoparticles against clinically important pathogenic bacteria, fungi and MCF-7 breast cancer cells. *Microb. Pathogen.* **2017**, *104*, 268–277. [CrossRef] [PubMed]

35. Yamamoto, O. Influence of particle size on the antibacterial zinc oxide. *Int. J. Inorg. Mater.* **2013**, *3*, 643–646. [CrossRef]

36. Zhang, L.; Jiang, Y.; Ding, Y.; Povey, M.; York, D. Investigation into the antibacterial behavior of suspensions of ZnO nanoparticles (ZnO nanofluids). *J. Nanopart. Res.* **2007**, *9*, 479–489. [CrossRef]

37. Ramachandran, R.; Krishnaraj, C.; Stacey, L.H.; Soon-Il, Y.; Thangavel, P.K. Plant extract synthesized silver nanoparticles: An ongoing source of novel biocompatible materials. *Indus. Crop. Prod.* **2015**, *70*, 356–373.

Elucidating the Chemistry behind the Reduction of Graphene Oxide using a Green Approach with Polydopamine

Cláudia Silva, Frank Simon, Peter Friedel, Petra Pötschke and Cordelia Zimmerer *

Leibniz Institute of Polymer Research Dresden (IPF), 01069 Dresden, Germany; frsimon@ipfdd.de (F.S.);
friedel@ipfdd.de (P.F.); poe@ipfdd.de (P.P.)
* Correspondence: zimmerer@ipfdd.de

Abstract: A new approach using X-ray photoelectron spectroscopy (XPS) was employed to give insight into the reduction of graphene oxide (GO) using a green approach with polydopamine (PDA). In this approach, the number of carbon atoms bonded to OH and to nitrogen in PDA is considered and compared to the total intensity of the signal resulting from OH groups in polydopamine-reduced graphene oxide (PDA-GO) to show the reduction. For this purpose, GO and PDA-GO with different times of reduction were prepared and characterized by Raman Spectroscopy and XPS. The PDA layer was removed to prepare reduced graphene oxide (RGO) and the effect of all chemical treatments on the thermal and electrical properties of the materials was studied. The results show that the complete reduction of the OH groups in GO occurred after 180 min of reaction. It was also concluded that Raman spectroscopy is not well suited to determine if the reduction and restoration of the sp^2 structure occurred. Moreover, a significant change in the thermal stability was not observed with the chemical treatments. Finally, the electrical powder conductivity decreased after reduction with PDA, increasing again after its removal.

Keywords: graphene oxide; reduced graphene oxide; X-ray photoelectron spectroscopy; Raman spectroscopy; electrical conductivity; functionalization

1. Introduction

Graphene, a 2D monolayer of sp^2-hybridized carbon atoms arranged in a hexagonal lattice with a carbon–carbon bond length of 0.142 nm, has been extensively studied since it was first isolated in 2004 by Novoselov et al. [1,2]. It has a great potential for several applications due to its Young's modulus of 1 TPa, intrinsic strength of 130 GPa [3], room temperature (RT) electron mobility of 250,000 cm^2 V^{-1} s^{-1} [4], and optical transmittance of 97.7% [5]. The promising application areas for graphene are photonics, optoelectronics, energy generation and storage, sensors for gas detection, reinforcement of composite materials, and biomedical areas, particularly in biosensing, drug and gene delivery, and tissue engineering [4].

Several techniques have been reported to produce graphene such as mechanical and electrochemical exfoliation of graphite, chemical vapor deposition, plasma enhanced chemical vapor deposition, thermal decomposition on silicon carbide (SiC), and reduction of graphene oxide (GO) [6]. Mechanical exfoliation of graphite can be either performed through atomic force microscope (AFM) probe techniques or adhesive tape exfoliation and results in high quality graphene. However, the high production cost makes this technique only feasible for research purposes [6,7]. Chemical vapor deposition is the most used technique for large-scale production of single or few-layer graphene [6,8]. Chemical vapor deposition is an expensive process due to the large energy consumption and the necessity of removing the substrate. Besides, controlling the grain size and the number of graphene layers produced is still a challenge [8]. Graphene produced by this technique has high quality when the processing parameters are properly controlled. The main drawbacks are the high cost of SiC wafers and the high temperatures involved in the process (around 1200 °C) [4,8].

Reduction of GO appears to be another viable route to produce single-layer graphene [9]. GO can be prepared from graphite through various methods such as those reported by Brodie, Staudenmeier and Hummers [10], either following their original protocol or introducing some variations. The oxidation process of graphite introduces hydroxyl, epoxy, carbonyl, and carboxyl groups in the hexagonal lattice of graphene [8]. Then, the reduction of GO can be performed, for instance, through a chemical approach to remove the oxygen-containing functional groups and restore the conjugated graphene structure [11]. Compared to others, this process allows the production of large quantities of graphene at a low cost since no special equipment or high temperatures are needed and the starting materials, graphite and chemical reductants, such as hydrazine and sodium borohydride, are usually cost-effective [12].

The major drawback of the traditional routes for chemical reduction of GO is the use of toxic and hazardous chemicals both to living organisms and to the environment [12]. Thus, special care with the handling of these chemicals must be taken and, at an industrial scale, the remediation of the hazardous wastes generated might result in a substantial increase of the production costs. In addition, if toxic residues are still present in the final material, applications in the biological and biomedical fields are unsuitable [13].

Regarding the reduction of graphene oxide, an improvement in the use of green chemistry [14] has been observed and several potential environmental-friendly chemicals have been studied. Among these, vitamins [15–17], saccharides [18], amino acids [19–21], organic acids [22–25], microorganisms [26,27], proteins and peptides [28,29], hormones [30], urea [31], and plant extracts [32–35] have been tested as reducing agents for GO. Dopamine, a nature-based, commercially available, and inexpensive reagent, was first employed for this purpose in 2010 by Xu et al. [36]. Nevertheless, a substantive interpretation of data showing the successful reduction of GO at a molecular lever was not provided. In the presence of the oxygen functional groups in GO and at a weak alkaline pH, dopamine self-polymerizes to form polydopamine (PDA) with the catechol groups undergoing oxidation to form quinone groups. Thus, the polymerization of dopamine at the surface of GO is accompanied by the reduction of the last with dopamine acting both as a reducing and functionalization agent [36]. The presence of PDA on the surface of reduced graphene oxide (RGO) allows the preparation of more stable dispersions, compared with RGO prepared using other reducing agents, which can be a crucial factor for further processing through liquid assisted techniques. Furthermore, PDA strongly adheres to a wide range of substrates, which makes it a good material for applications in functional coatings [37]. Since 2010, several works have been published on the production of PDA-GO envisaging applications such as water purification [38], anion and proton exchange membrane fuel cells [39], functional coatings, and biomedical applications, for instance cancer treatment [40], drug delivery [41], antibacterial materials [42], biosensing [43], and tissue engineering [44]. However, these works focus mainly on the use of PDA-GO as a platform for anchoring of nanoparticles or covalent grafting of other molecules and on the characterization of the materials prepared envisaging the final application and lack deep insight into the mechanism of the GO reduction by PDA.

Considering the limited knowledge on this topic, in this work, we propose a new approach using X-Ray photoelectron spectroscopy (XPS) to supply evidence for the molecular reduction of GO by PDA. This reduction process is a good example of the use of green chemistry to replace traditional methods since no toxic solvents are used and PDA is a natural and renewable raw material, and only a little amount of waste is generated. Our investigation contributes to the field of conductive composite material development based on the strategy to use inexpensive and easily available graphite as basic raw material. In addition, the removal of the PDA to obtain RGO is studied. For this purpose, GO, PDA-GO, and RGO were prepared and characterized by Raman spectroscopy, XPS, thermogravimetric analysis (TGA), and electrical powder conductivity measurements. Thus, the effect of the chemical treatments in the thermal and electrical properties of GO and PDA-GO was studied and a better understanding of the chemistry behind the green reduction of GO with PDA is presented.

2. Materials and Methods

2.1. Materials

Natural graphite flakes (99%; −325 mesh), tris(hydroxymethyl)aminomethane (tris base, ≥99.8%), dopamine hydrochloride (DA, 98%), and sodium hydroxide (NaOH, >97%) were purchased from Sigma Aldrich® (Munich, Germany). Sulfuric acid (H_2SO_4, 98%), potassium permanganate ($KMnO_4$, ≥99%), and hydrochloric acid (HCl, 10% v/v) were received from VRW® (Darmstadt, Germany). Hydrogen peroxide (H_2O_2, 30% w/v) was purchased from Merck (Darmstadt, Germany). Distilled water (DW) was used in all chemical treatments.

2.2. Synthesis of Graphene Oxide, Polydopamine-Reduced Graphene Oxide, and Reduced Graphene Oxide

Graphene oxide (GO) powders were prepared through a modified Hummers' method according to our previous publication [45]. The oxidation step was performed using H_2SO_4 and $KMnO_4$ (graphite:$KMnO_4$ = 1:1). DW, H_2O_2, and HCl were used for the purification step.

For the synthesis of PDA-GO, first a tris base solution in DW (0.1 M) was prepared and degassed by N_2 bubbling during 20 min. Then, 40 mg of GO were placed in a round bottomed flask and degassed for 10 min with N_2. After that, 20 mL of tris base were added, and the suspension was magnetically stirred for 15 min at 60 °C. After stirring, DA was added to the GO suspension (GO:DA = 1 w/w) and the suspension was degassed by N_2 bubbling during 10 min. The reactions between GO and DA took place at 60 °C for the times indicated in Table 1. After the reaction, the suspension was vacuum filtered with a nylon filter membrane (0.45 µm pore size, Whatman, Kent, UK). Finally, the powders were collected and dried overnight in vacuum at 60 °C.

Table 1. Time of reaction between dopamine hydrochloride (DA) and graphene oxide (GO) and respective sample description.

Time of Reaction (min)	Sample Description (PDA-GO_reduction Time in min)
30	PDA-GO_30
60	PDA-GO_60
90	PDA-GO_90
120	PDA-GO_120
150	PDA-GO_150
180	PDA-GO_180
210	PDA-GO_210

Reduced graphene oxide (RGO) was prepared from PDA-GO. First, 20 mg of PDA-GO_30, PDA-GO_90, and PDA-GO_180 were stirred each in a 40 mL of NaOH solution (5 M), under N_2 flow for 6 h. Then, the suspension was vacuum filtered with a nylon filter membrane (0.45 µm pore size, Whatman, Kent, UK), and the powders collected and dried at 60 °C under vacuum, overnight, to obtain RGO_30, RGO_90, and RGO_180.

2.3. Modeling and Characterization of GO, PDA-GO, and RGO

2.3.1. Ab-Initio Calculations

Ab-initio calculations were performed to get information about an optimized geometry of GO by minimizing the corresponding Hartee–Fock energy applying the software package GAMESS (freeware version 2017-09-30R2, Iowa State University, Ames, IA, USA) [46] using the basis set STO-6G. For this purpose, first, a graphene-like structure consisting of four rows of heptacene was considered. Then, as a first oxidation step, the formation of external hydroxyl and carboxyl groups was simulated following the production of oxirane groups in the plane of the graphene-like molecule. Finally, the oxirane rings opening and conversion into hydroxyl groups, due to the acidic medium where the reaction takes place, was calculated.

2.3.2. Raman Spectroscopy

Raman Spectroscopy was performed on a Raman microscope alpha300R (WITec, Ulm, Germany) using a laser excitation wavelength of 532 nm, laser power of 1 mW, a spectral resolution of 6 cm^{-1}, and integration time of 0.5 s. Two hundred scans were accumulated to record each spectrum. For this purpose, GO, PDA-GO, and RGO aqueous suspensions were prepared and sprayed on a glass slide positioned on a heating plate for fast water evaporation and deposition of the graphene products to be analyzed. The I_D/I_G or I_{D+PDA}/I_{G+PDA} ratios were calculated considering the area under curve of the bands, which were estimated with a mean error of about 10%.

2.3.3. X-Ray Photoelectron Spectroscopy (XPS)

XPS studies were carried out by means of an Axis Ultra photoelectron spectrometer (Kratos Analytical, Manchester, UK). The spectrometer was equipped with a monochromatic Al Kα (hν = 1486.6 eV), X-ray source of 300 W at 15 kV. The kinetic energy of photoelectrons was determined with hemispheric analyzer set to pass energy of 160 eV for wide-scan spectra and 20 eV for high-resolution spectra. For the C 1s region, the maximum information depth of the XPS method is about 8 nm [47,48]. Employing Scotch double-sided adhesive tape (3M Company, Maplewood, MN, USA), the powdery samples were prepared as thick films on a sample holder. During all measurements, electrostatic charging of the sample was avoided by means of a low-energy electron source working in combination with a magnetic immersion lens. Later, all recorded peaks were shifted by the same value that was necessary to set the component peak Gr showing the sp^2-hybridized carbon atoms of the graphite-like lattice (–C=C– $\leftrightarrow$ =C–C=) to 283.99 eV [49]. In the case of considerable amounts of saturated hydrocarbons (PDA-GO samples), their corresponding component peaks in the C 1s spectrum was used as a reference with a binding energy of 285.00 eV [50]. Quantitative elemental compositions were determined from peak areas using experimentally determined sensitivity factors and the spectrometer transmission function. The shapes of the high-resolution element spectra were used to analyze the different binding states of the elements. For this purpose, the high-resolution element spectra were deconvoluted into component peaks (Kratos spectra deconvolution software, software version 2.2.9, Kratos Analytical Ltd., Manchester, UK), in which their binding energy values (BE), height, full width at half maximum and the Gaussian–Lorentzian ratios were free parameters.

2.3.4. Thermogravimetric Analysis (TGA)

TGA measurements were carried out on a TGA Q 5000 (TA Instruments Inc., New Castle, DE, USA) under N$_2$ atmosphere between 40 to 800 °C and heating rate of 10 K min^{-1}. Prior to the measurements, the powders, except graphite, were dried at 100 °C for 12 min using the same equipment.

2.3.5. Electrical Powder Conductivity

The electrical conductivity of the powders was measured using the equipment shown in Figure 1, which was developed and constructed at the Leibniz Institute of Polymer Research Dresden. The equipment consists of a transparent cylinder of 40 mm length with a capillary hole with diameter of 5 mm, which is mounted on a gold electrode on its bottom. After a certain amount of powder (~25 mg) was filled inside the hole, the upper movable cylindrical gold electrode with the same diameter compresses the material stepwise up to a pressure of 30 MPa using a stepper motor. The resistance was measured between the two gold electrodes using a Keithley 2001 electrometer (Tektronix, Köln, Germany) and the conductivity was calculated [51]. At least three measurements were performed to get mean values and standard deviations. In addition, based on the weighed powder mass and the volume of the sample, given by the geometrical conditions of the cylinder and the position of the stepper motor, the bulk density and its development at different pressures can be determined. This also gives a measure of the compressibility of the powder.

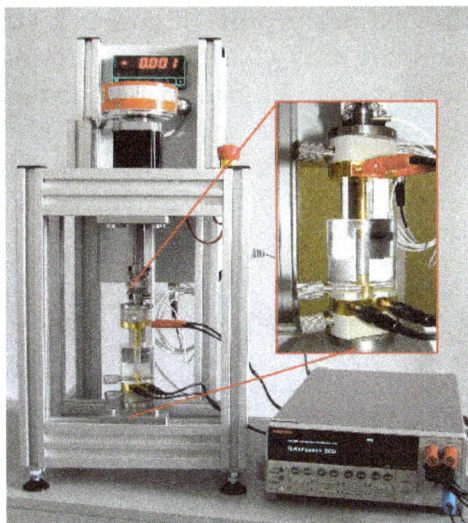

Figure 1. Equipment used for electrical powder conductivity measurements (photo by Dr. Wolfgang Jenschke, Leibniz Institute of Polymer Research Dresden, Dresden, Germany).

3. Results and Discussion

3.1. The Oxidation of Graphene Stacks

Graphene is an allotrope form of carbon with a 2D honeycomb structure. According to the sp^2 orbital hybridization of all its carbon atoms, it can be considered as an infinitely large polycyclic aromatic molecule. While the electrons in the s, p_x and p_y orbitals form σ-bonds, the p_z electronsare involved in conjugated π-bonds hybridizing to π-band and $\pi*$-bands. The half-filled π-band covering the whole molecule permits free-moving electrons, which are responsible for the graphene's electrical conductivity. In fact, the structure of a real graphene sample must be limited. Figure 2a shows a graphene-like model molecule consisting of four rows of heptacene. The degree of oxidation of all these graphene-like carbon atoms can be theoretically given to an average number of -0.307. To understand the oxidation of graphene, in a model based on the Lerf–Klinowski model [52,53], the edges of a graphene sample were decorated with hydrogen atoms to avoid a geometric disorder and keep the aromaticity and electrical conductivity of the molecule [54,55].

(a) (b) (c) (d)

Figure 2. Results of ab-initio calculations to model the geometries of graphene-like and GO-like molecules: (**a**) graphene-like structure consisting of four rows of heptacene; (**b**) oxidized graphene carrying sterically demanding carboxylic groups on the edge; (**c**) oxidative attack on the carbon atoms in the molecular plane formed oxiran groups instable in acidic media; and (**d**) GO-like molecule decorated with hydroxyl groups on the edge and in the molecular plane.

As can be seen in Figure 2a, the optimized hydrogen-decorated graphene-like model molecule is flat and planar. With increasing degree of oxidation (exchange of 7 external H atoms with 7 carboxyl groups increases the average oxidation number of carbon atoms to 0.047), the original honeycomb structure was disturbed. At first, external hydroxyl and carboxyl groups (probably also other carbonyl groups, such as quinone-like groups, which were not studied here) slightly deformed the planarity of the graphene-like molecule (Figure 2b). The oxidative attack on carbon atoms in the plane of the graphene-like molecule may produce oxirane groups (Figure 2c). However, due to the strongly acidic medium where the oxidation of graphene took place, the oxiran rings were opened immediately and converted into hydroxyl groups, as shown in Figure 2d. Regardless of whether hydroxyl or oxiran groups were formed in the molecular plane, the former flat plane bended up and needed more space. This effect is often used to separate single layers of graphene from their stacks. The combination of hydroxyl groups with the replacement of edge-standing hydrogens by carboxyl groups increases the average degree of oxidation of carbon atoms to 0.176.

3.2. Raman Spectroscopy

Raman Spectroscopy was performed to evaluate the level of "disorder" of the sp^2 hybridized structure of the materials prepared. The Raman spectra, presented in Figure 3, show the three major bands characteristic of sp^2 carbon materials. The D band, near 1350 cm^{-1}, is related to the presence of structural defects in the hexagonal sp^2 carbon lattice of graphene and to edge effects [56,57]. The G band, at approximately 1580 cm^{-1}, is related to the in-plane vibration of the sp^2 carbon atoms [56]. The band near 2700 cm^{-1}, the $2D$ band, originates on a second-order Raman scattering process and its shape, width, and position is related to the number of layers for n-layer graphene. Ferrari et al. reported that an increase in the number of layers originates a broader $2D$ band shifted to higher Raman shifts. [57,58]. The relative signal intensity of the D band to the G band (I_D/I_G) provides information about the level of "disorder" in terms of covalent modification of the graphene structure [57,59]. In pristine graphite, the $2D$ band consists of two components and appears at $\approx$2720 cm^{-1} while graphene presents a single sharp peak centered at a Raman shift lower than 2700 cm^{-1} [59].

Regarding PDA, there are two bands at about 1358 and 1588 cm^{-1} [45] that are assigned to the stretching vibration and deformation of chatecol groups [60]. Thus, the D and G typical for the graphene derivatives overlap with the PDA and a proper assignment of these bands is not possible. As long as PDA is present at the GO surface, contributions of these bands have to be considered. Thus, particularly the band at 1350–1358 cm^{-1} presents an overlap of the signals coming from the structural defects of GO, functional groups on the surface, and the amount of PDA in the material. Nevertheless, the ratios between the intensities at about 1350 and 1580 cm^{-1} were calculated and referred to as I_{D+PDA}/I_{G+PDA} for the PDA containing samples.

The Raman spectra in Figure 3a show an increase of the I_D/I_G ratio when graphite was chemically converted in GO, from 0.2 to 0.5. This is a consequence of an increase of the content of structural defects caused by the introduction of oxygen-containing functional groups during the oxidation process or by the decrease in the flake size when GO was exposed to sonication [61]. Moreover, the downshift to 2717 cm^{-1} after oxidation indicates a reduction in the number of graphene stacks.

After reduction with PDA, a further increase in I_{D+PDA}/I_{G+PDA} was observed, as shown in Table 2. Raman spectroscopy is a standard method to evaluate the molecular structure of graphene and graphene derivatives. However, considering the overlapping bands of GO and PDA, in this case, the method is not well suited to determine if the reduction and restoration of the sp^2 structure occurred since the typical I_D/I_G ratios also have a contribution of the PDA signal. Nevertheless, the position of the $2D$ band shows a further reduction on the number of graphene stacks on the PDA-GO materials.

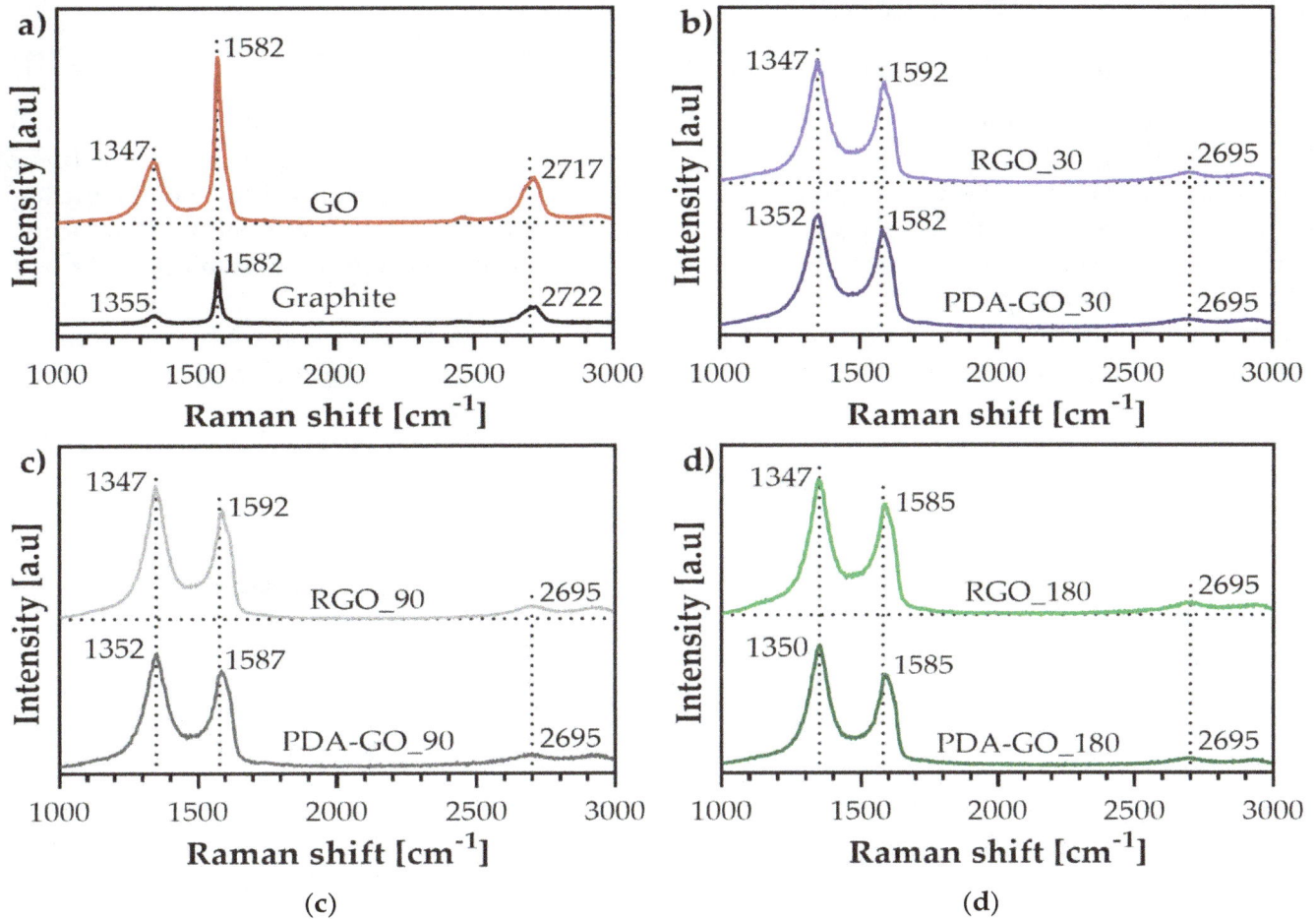

Figure 3. Raman spectra and position of the D, G, and $2D$ bands of: (**a**) graphite and GO; (**b**) PDA-GO_30 and RGO_30; (**c**) PDA-GO_90 and RGO_90; and (**d**) PDA-GO_180 and RGO_180.

Table 2. I_D/I_G or I_{D+PDA}/I_{G+PDA} of the graphene derivatives.

Material	I_D/I_G	I_{D+PDA}/I_{G+PDA}
Graphite	0.2	—
GO	0.5	—
PDA-GO_30	—	1.2
PDA-GO_90	—	1.2
PDA-GO_180	—	1.2
RGO_30	—	1.2
RGO_90	—	1.2
RGO_180	—	1.2

The evaluation of the structure of RGO should be possible since no contribution from PDA peaks is expected. However, no significant changes were found between the spectra of RGO and the corresponding PDA-GO, which might indicate that the PDA was not completely removed after the treatment with NaOH. Thus, additional experiments are required to study the molecular structure of these materials by Raman Spectroscopy and to optimize the removal of the PDA layer. In addition, different methods, such as XPS, are required to properly characterize the molecular structure of PDA-GO.

3.3. X-Ray Photoelectron Spectroscopy (XPS)

3.3.1. XPS Spectra of Graphene Stacks and GO

The XPS studies corroborate the information obtained by Raman spectroscopy. The shape of the C 1s spectrum of graphite in Figure 4a is very characteristic for carbonaceous substances consisting of sp^2 hybridized lattices. Due to the numerous excited states, the spectrum noticeably tailed on the high energy side. The main component peak *Gr* found at 283.99 eV results from photoelectrons that escaped from sp^2-hybridized carbon atoms ($-C=C- \leftrightarrow =C-C=$). Photoelectrons removed from molecules with exited electron states appear as wide shake-up peaks (gray lines in Figure 4). At 285.69 eV, a further small component peak *C* was observed. This component peak shows the presence of C̲–O bonds on the sample surface (detailed XPS data are presented in Tables S1–S18 of the Supplementary Materials).

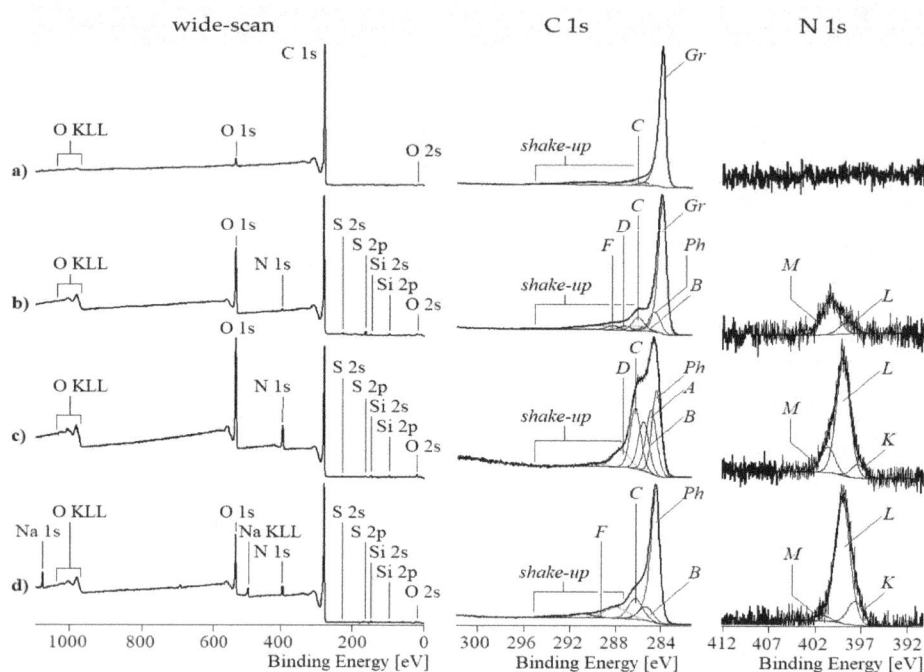

Figure 4. Wide-scan (left column), high-resolution C 1s (middle column) and N 1s (right column) XPS spectra recorded from: (**a**) graphene stacks; (**b**) GO; (**c**) PDA-GO_30; and (**d**) RGO_30.

The oxidation of graphite to GO significantly increases the relative oxygen content from 0.028 to 0.166 (Figure 4b). This increase results from the presence of numerous oxygen-containing functional groups on the sample surface. However, the C 1s spectrum (Figure 4b, middle column) shows that the oxidation reaction was gently performed. It seemed to be beneficial to minimize the destruction of the conjugated π-electron system by breaking σ-bonds during the formation of carboxylic acid groups. On the other hand, the planar graphene sheets should be sufficiently oxidized to bend them and expand the former stacks. The predominant preservation of the π-electron system becomes clear by the appearance of the intense component peak *Gr* and the shake-up peaks. It also seems necessary to introduce a component peak *Ph* (284.75 eV), indicating the presence of sp^2-hybridized carbon atoms that are not involved in the highly conjugated π-system of the original graphene lattice [50]. Oxygen-containing functional groups were analyzed as component peaks *C* (286.07 eV), *D* (287.29 eV), and *F* (288.36 eV). Component peak *C* results from photoelectrons that escaped from phenolic C̲–OH groups. The binding energy value found for component peak *C* is also characteristic for ether groups ($-C̲-O-C̲-$). However, as mentioned above, it is unlikely that oxirane groups are stable on the graphene surface. Component peak *D* shows carbonyl carbon atoms of quinone-like (C̲=O) groups. Photoelectrons from the carbonyl carbon atoms of carboxylic acid groups (O=C̲–OH) and their corresponding carboxylates (O=C̲–O$^\ominus$ $\leftrightarrow$ $^\ominus$O–C̲=O) were observed as the small component peak *F*. The presence of nitrogen ([N]:[C] = 0.009) required the introduction of an additional small component peak B (at 285.72 eV) showing $\underline{C}$–N bonds. These bonds can be constituents of amino and/or amide groups. In the case of the presence of amide groups, photoelectrons of the carbonyl carbon atoms (O=$\underline{C}$–NH–C) contributes to component peak D while component peak B presents the amine-sided carbon atoms (O=C–NH–$\underline{C}$).

3.3.2. XPS Spectra of PDA-GO

The application of dopamine and its subsequent polymerization on the GO surface significantly increased the relative nitrogen content from [N]:[C] of the samples from 0.009 to ca. 0.090. The variation of the time for the polymerization reaction from 30 to 210 min does not correlate with the relative nitrogen content on the GO surfaces and thus with the amount of PDA deposited there. As can be seen in Figure 4c, the PDA layer strongly changes the shape of the C 1s spectra. Carbon atoms bonded in the phenyl rings of the PDA's catechol groups having no heteroatoms as binding partner and sp^2-hybridized carbon atoms from the graphite-like lattice of the substrate materials were identified as component peak Ph (ca. 284.49 eV). Component peak A (285.00 eV) results from the presence of carbon atoms in the sp^3-hybrid state of saturated hydrocarbons. Some of these carbon atoms are constituents of the PDA layer, but the presence of component peak A can be also considered as a first hint of reduced carbon species on the sample surface. Carbon–nitrogen bonds were assigned as component peak B (ca. 285.72 eV). The intensities of the component peaks B equal the twice of the [N]:[C] ratios determined from the corresponding wide-scan spectra. According to the structural formula of the PDA in Figure 5, the number of carbon atoms carrying phenolic OH groups ($\underline{C}$–OH) should equal the number of carbon atoms bonded to nitrogen.

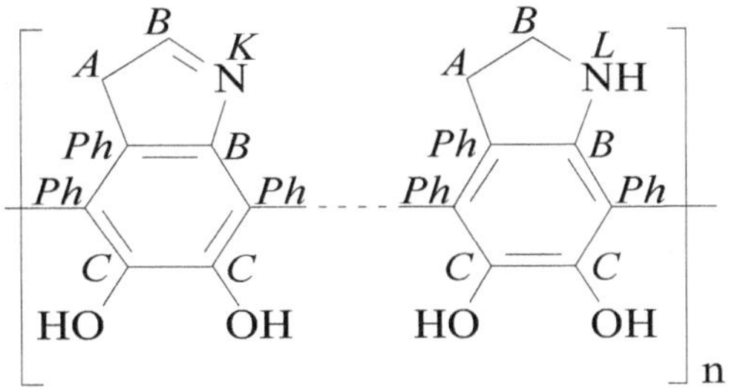

Figure 5. Characteristic cutouts from the chemical structure of PDA. Italic letters denote the assignment of the carbon and nitrogen atoms to the component peaks in the C 1s and N 1s high-resolution spectra.

However, the C 1s spectra recorded from the samples that reacted for 30 and 90 min showed component peaks C with higher intensities than their corresponding component peaks B, as shown in Figure 6a. It is assumed that the excess ([C'] = [C] − [B]) of the intensities of the component peaks C result from the contribution of $\underline{C}$–OH groups from the GO substrate material. Quinone-like groups cannot be safely detected because component peak D is overlapped by intense shake-up peaks.

After longer polymerization times, such as 180 and 210 min, the excess component peaks C' disappears completely (Figure 6b). Obviously, the reduction of the GO by the oxidative polymerization of the adsorbed dopamine molecules requires longer periods of reaction time. Figure 6b (right column) shows that the high-resolution N 1s spectra recorded from the PDA-GO samples is deconvoluted into the three component peaks K (ca. 398.69 eV), L (ca. 400.1 eV) and M (ca. 401.68 eV). Corresponding to the intensities of the component peaks B (in the C 1s spectra), which are the twice of the [N]:[C] ratios and the binding energy values found, component peak L shows cyclic secondary amino groups (C–$\underline{N}$H–C), such as the pyrrolidine structures in Figure 5 on the right side [50,62]. The binding energy values of the component peaks K are unusually small for organically bonded nitrogen. The observation of such low binding energy values indicates a high electron density at the nitrogen atom, which is characteristic for the cyclic imide nitrogen atoms (C–$\underline{N}$=C) exemplary shown in Figure 5 (left). Protonated nitrogen species (C–$\underline{N}^+$H) reflecting the protonation/deprotonation equilibrium of Brønsted basic nitrogen species are observed as component peak M.

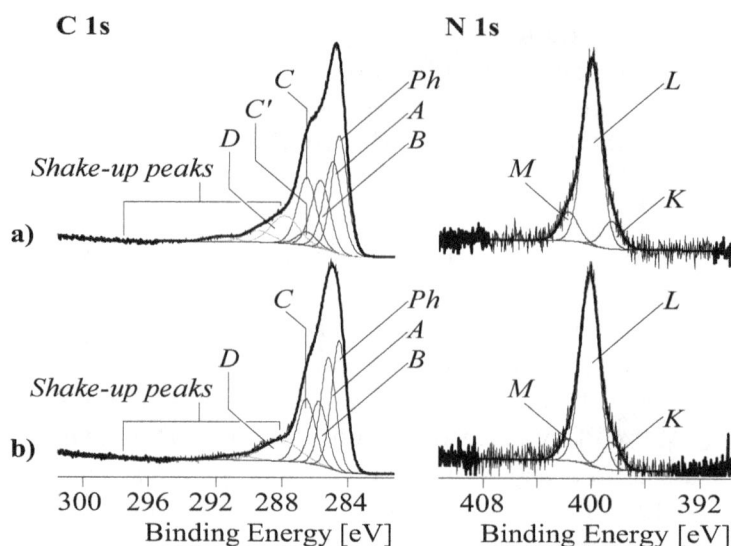

Figure 6. High-resolution C 1s (left) and N 1s (right) XPS spectra recorded from: (**a**) PDA-GO_90; and (**b**) PDA-GO_180.

3.3.3. XPS Spectra of RGO

NaOH was employed to hydrolyze and remove the PDA layer covering the carbonaceous substrate materials. However, the wide-scan spectra shows the presence of considerable amounts of nitrogen ([N]:[C] $\approx$ 0.04) after treating the samples with NaOH (Figure 4d, left column). These findings can be considered as a hint that PDA was not completely removed from the substrate materials. Nevertheless, the corresponding C 1s spectra in Figure 4d exemplary shows the C 1s spectrum of sample RGO_30 with clearly intensity-reduced shoulders in the region of 286.5 eV and tailings on the high energy sides, which were previously observed in the C 1s spectrum of the graphene reference sample (Figure 4a, middle column). Corresponding to the findings of the PDA-coated samples, component peak B (285.34 eV) appears with an intensity that is twice the [N]:[C] ratio found in the wide-scan spectrum of sample RGO_30. An equal number of photoelectrons escapes from the PDA's catecholic C–OH groups and contribute to component peak C at 286.29 eV. The excess of component peak C ([C'] = 0.061) has to be assigned to the C–O groups remaining on the surface of the substrate material. Probably, component peak D disappears completely while small traces of carboxylic acid groups (O=C–OH) and/or carboxylic ester groups (O=C–O–C) were observed as component peak F (289.36 eV). Latter groups could be formed by esterification of the carboxylic acid groups with some of the PDA's catechol groups. Photoelectrons of the corresponding alcohol-sited carbon atoms (O=C–O–C) contribute to component peak C. As discussed above, the N 1s spectrum was deconvoluted into the three component peaks K, L and M.

In summary, in the case of organic materials, it is often very difficult to determine or estimate the contribution of surface contaminations to the spectra. This is especially true for very complex-shaped high-resolution spectra, which were recorded for the samples studied here. C 1s spectra recorded from the graphene stacks, GO, and RGO_30 provide no information about the degrees of contaminations. A component peak showing the presence of surface contaminations arise at the same peak position. A separation was not possible. If they are present and can be separated [63] the photoelectrons of their carbon atoms would contribute to component peak observable at about 285 eV.

As shown in Figure 4, we found slightly different binding energy values for the component peaks C. The reason of these differences was the different chemical character of the C–OH groups. In the case of the PDA-GO samples (PDA-GO_30, PDA-GO_90, and PDA-GO_180), C–OH groups of the di-phenolic catechol units mainly contributed to the component peaks C (about 286.47 eV). After their removal with NaOH (sample RGO_30), a few residual di-phenolic catechol units and numerous OH groups from the not fully reduced substrate material contributed to component peak C. The binding energy value was slightly lowered (286.29 eV). The phenolic OH groups on the GO, which were more intensively involved in the delocalized p-electron system of the (more or less disturbed by oxidation) graphite-like lattice, showed a binding energy value of 286.07 eV. The component peak C resulting from photoelectrons of traces of oxygen-carrying functional groups found on the surfaces of the graphene stacks had a binding energy value of 285.69 eV, which is significantly lower as usually expected for C–OH bonds. However, here we have to take into account that these few phenol groups are embedded in a largely undisturbed p-system, and - in contrast to the catechol rings - their dissociations (C–OH + H_2O to $C-O^-$ + H_3O^+) increasing the electron density at the carbon atoms are not hindered by negative charges in their immediate molecular neighborhood.

In all cases, the shapes of the component peaks were the result of convolutions of a Gaussian normal distribution and Cauchy–Lorentz distribution. With exception of the distribution of the component peak *Gr* and the component peak *Ph* in the RGO_30 sample, all other component peaks had the same shapes as suggested by the reviewer. High number of excited states in the well-ordered graphite-like lattice led to a tailing of the photoelectron distribution at the high-energy side of the component peak. Hence, it seemed to be necessary to adapt the line shape of the component peaks *Gr* by an increased asymmetry. In the case of the component peaks *Ph* in sample RGO_30, we have to consider that this component peak *Ph* summarized photoelectrons from the p-conjugated carbon atoms of the remaining PDA molecules and the carbon atoms of the more or less disturbed graphite-like structures of the substrate.

Furthermore, the XPS spectra recorded from the PDA-GO samples clearly show the reduction of the GO. As can be seen in the high-resolution C 1s spectrum, the reduction is mainly due to the decrease in the C–OH groups of the former GO substrate. The majority of the C–OH groups found in that spectrum are constituents of the PDA, which remain on the sample surface after washing with NaOH. The PDA wrapping the carbonaceous material can be used as stable anchor layer for subsequent modification reactions to functionalize and compatibilize carbon-based nanomaterials.

3.4. Thermogravimetric Analysis (TGA)

TGA was performed to evaluate the thermal stability of graphite, GO, PDA-GO, and RGO. It was expected that the amount of PDA on the GO surfaces could be evaluated from the TGA results. As shown in Figure 7, PDA has the lowest thermal stability, starting thermal degradation around 180 °C due to the partial decomposition of the main chain in three different stages with a maximum at around 260 °C [64]. The weight loss observed for graphite is very low and mainly due to the release of adsorbed water and residual oxidation [61]. Regarding GO, Figure 8 shows three loss stages. The first weight loss stage of ca. 1.7% occurring between 40 and 120 °C was attributed to the evaporation of adsorbed water [65]. The second weight loss (5.7%) has its maximum at about 220 °C and might be due to the decomposition of oxygen-containing functional groups (C–OH), whereas the main weight loss (8.8%) occurs at about 400 °C and is related to the loss of more stable oxygen functionalities such as carboxyl groups. The total weight losses of graphite, GO, and PDA at 800 °C are 0.7%, 19.1%, and 86.3%, respectively.

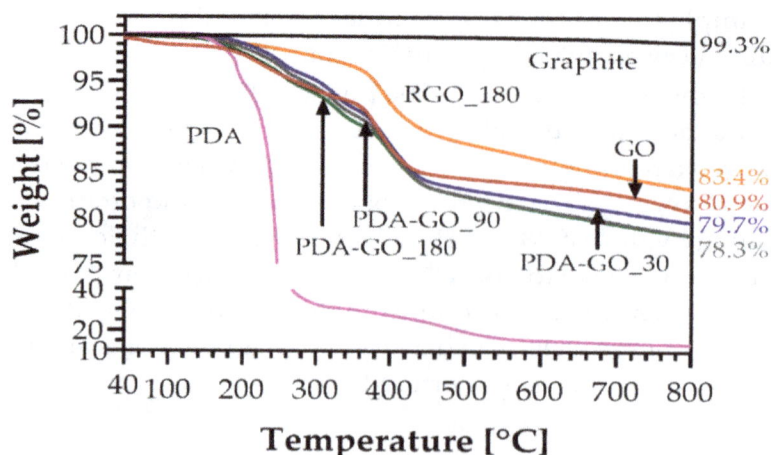

Figure 7. Thermogravimetric analysis (TGA) curves of graphite, GO, PDA, PDA-GO_30, PDA-GO_90, PDA-GO_180, and RGO_180.

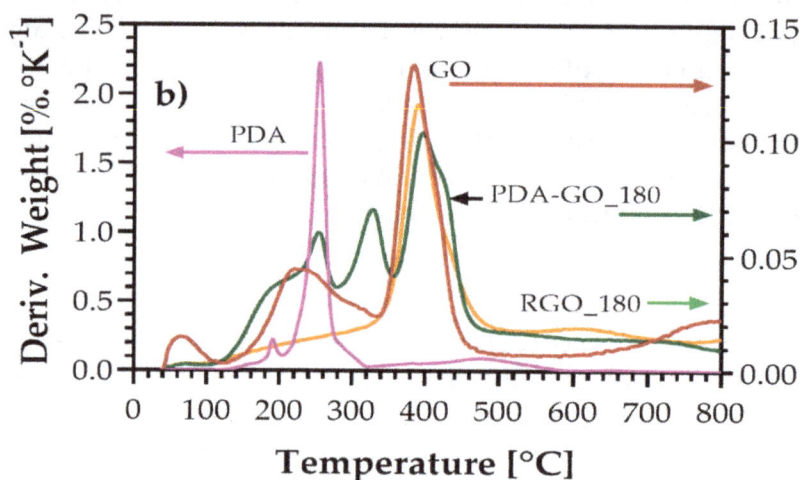

Figure 8. Derivative weight loss of GO, PDA, PDA-GO_180, and RGO_180.

The TGA curves of the PDA-GO materials show significant thermal degradation starting at around 180 °C due to the decomposition of PDA and the removal of non-reduced oxygen-containing functional groups from GO. The different polymerization times only slightly affected the thermal stability of the PDA-GO materials, which underwent a total weight loss of between 20.3% (PDA-GO_30) and 21.7% (PDA-GO_90 and PDA-GO_180) at 800 °C, as shown in Figure 7. However, a decrease in the thermal stability in comparison with GO was observed, possibly related to the increase of the oxygen content as determined by XPS due to the presence of PDA. Regarding RGO_180, the curve in Figure 8 shows only a weight loss stage with a maximum at around 400 °C, relative to the decomposition of stable oxygen-containing functional groups which were not reduced by PDA as determined by XPS. No peaks from PDA contribution were found throughout the temperature range analyzed; however, the small weight loss that occurs below 300 °C might be attributed to the decomposition of the residual PDA remaining after NaOH treatment. A total weight loss of 16.6% was observed, which reflects an improvement of the thermal stability after PDA removal. As calculated by XPS, RGO_180 possesses the lowest amount of oxygen species, which is reflected by its highest thermal stability among all prepared materials.

3.5. Electrical Powder Conductivity

The electrical powder conductivity of the graphene derivatives was measured to evaluate the effect of the reduction process on the restoration of the electrical properties. The electrical powder conductivity generally increases with the pressure applied during the measurements. As shown in

Figure 9, the starting material graphite has the highest electrical conductivity among all materials with values between 68.8 and 93.4 S/cm. As expected, after oxidation, there was a remarkable decrease in the conductivity of about 18% to 16.6 S/cm (at 30 MPa). This value of GO is higher than values reported previously when using the same measurement equipment but differently oxidized GO materials [66,67]. The reduction of GO with PDA resulted in a substantial decrease in the electrical powder conductivity, which scales with the reaction time between GO and DA and is most significant in PDA-GO_180 with a decrease to 1.9 S/cm at 30 MPa. Although the reduction process allowed the removal of oxygen-containing functionalities and potentially a restoration of the sp^2-hybridized lattice of graphene, the electrical insulating nature of the PDA layer might contribute to this effect. By this, a decrease in the contact area between conductive graphene areas and an increase in contact resistance between the partially PDA covered conductive graphene occur, which justifies the decrease in the electrical conductivity. The conductivity of PDA-GO_180 increases with the pressure until 20 MPa, remaining constant afterwards. Regarding PDA-GO_30 and PDA-GO_90, similar electrical powder conductivity values were measured for all pressures considered. For PDA-GO_30, there was an increase of 48% from 1.8 to 3.7 S/cm while for PDA-GO_90 an increase of 44% from 1.6 to 3.6 S/cm was observed when increasing the pressure from 5 to 30 MPa.

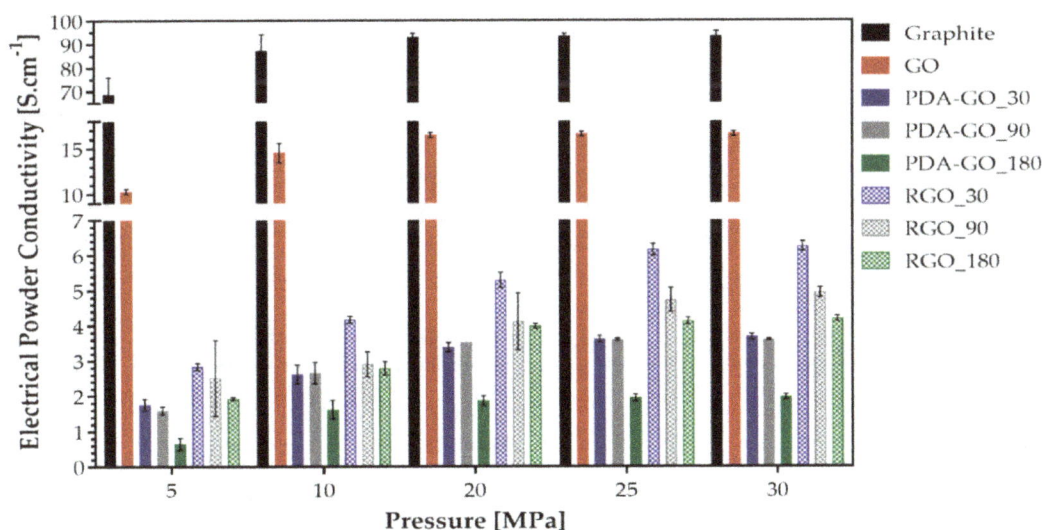

Figure 9. Electrical conductivity of the powdery materials at different pressures.

After the removal of PDA with NaOH, the electrical conductivity increases. When the PDA layer is removed, leaving only a small residue, the resultant RGO possesses a structure more suited for electrical conduction. Following the trend of decreasing conductivity with reaction time between GO and PDA, the samples after PDA removal also show this dependency. However, despite RGO_180 has the lowest conductivity among the RGO samples, the difference between PDA-GO and RGO is most pronounced in this sample which shows an increase of two times after the treatment with NaOH. The values achieved are higher than those reported for differently reduced graphite oxides measured using the same equipment in Ref. [66]. However, even after reduction and PDA removal, the desired higher electrical conductivity than that of GO was not obtained, possibly due to the presence of residual PDA, a more defected sp^2 carbon structure, or different compressibility of the powders. The representation of the electrical powder conductivity as a function of the bulk density of the various materials in Figure 10 confirms the significant differences between them. Graphite and GO have the highest bulk density values at the applied pressures, which illustrates a higher packing density and better contact between the graphene flakes inside the powder materials. Graphite has a much higher compressibility (especially at pressures between 5 and 20 MPa) than GO, which has the lowest increase in conductivity with pressure, showing the lowest compressibility of all samples. The PDA_GO and RGO powders show lower bulk densities at the initial pressure of 5 MPa and stronger density changes at increasing pressure than graphite and GO. The increase in conductivity with pressure is slightly lower for the PDA_GO than for the RGO samples, demonstrating a slightly higher compressibility of the RGO. This may be a hint that the powder particles in the RGO samples are more flexible and less stiff. The dependencies in Figure 10 illustrate that the materials are in different packing states during the measurements, which affects the contact surfaces and the contact resistance between the material powder particles and flakes in the measured sample volume. If only very small amounts of PDA remain on the surface of graphene flakes, especially at locations where the flakes are in close contact during measurement, this may result in high contact resistance and reduced conductivity through the sample. The comparison of the electrical powder conductivity of all materials at the same density, selected here at 1.85 g cm^{-3}, however, confirms the sequence described above at different measuring pressures. Graphite has the highest conductivity, followed by GO, while PDA_GO has a lower conductivity than RGO.

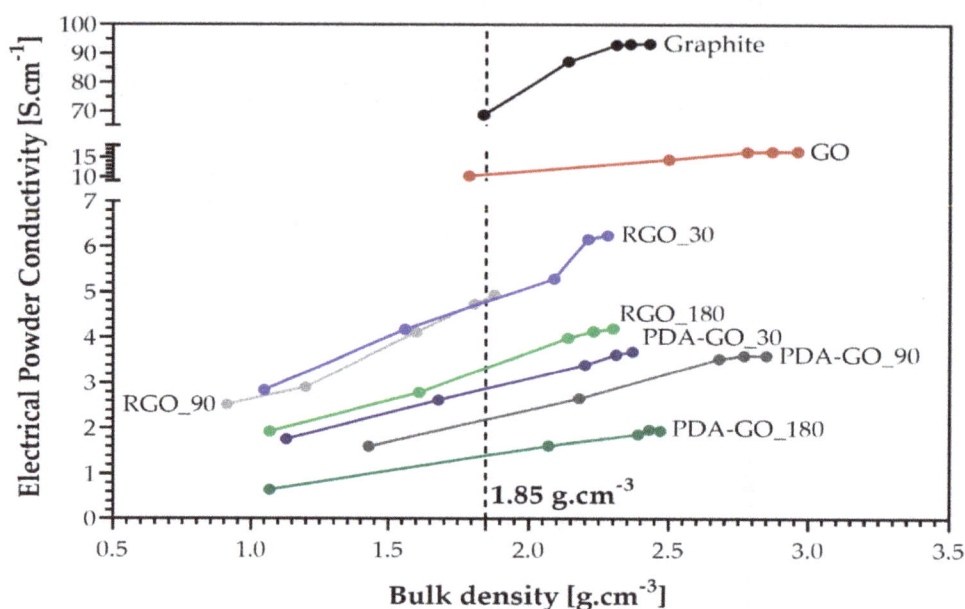

Figure 10. Dependency of electrical conductivity of the powdery materials on bulk density (density increases with pressure).

Nevertheless, despite the lower electrical powder conductivity, both PDA-GO and RGO materials possess higher conductivity values than reported for other examples in the literature [66] and show a better dispersibility in several solvents than GO, which is crucial when, for example, these materials are to be used in solvent assisted techniques to prepare conductive films.

4. Conclusions

In this work, we propose a new approach to investigate the molecular reduction state of GO by PDA and the removal of the PDA, using NaOH, to obtain RGO. It was shown that Raman spectroscopy is not well suited to determine the reduction and restoration of the sp^2 structure. However, a first hint for the presence of PDA, even after the treatment with NaOH, can be obtained using this method since no significant differences were found between the PDA-GO and RGO spectra. The reduction of GO by PDA was proven by XPS through a new approach that considers the number of carbon atoms bonded to OH and to nitrogen in PDA and compares it to the total intensity of the signal resulting from OH groups in PDA-GO to finally determine that the reduction occurs. In addition, it was shown that there was no complete removal of the PDA layer with NaOH, corroborating the Raman spectroscopy results. Regarding the thermal analysis, it was observed that the presence of PDA in PDA-GO results in a decrease in the thermal stability. However, after PDA removal, the thermal stability improved and revealed to be higher than in GO, which agrees with the XPS studies that showed RGO_180 possesses the lowest amount of unstable oxygen-containing species. In addition, the graphene derivatives prepared in the present work possess considerably higher electrical powder conductivity values than those reported in the literature, even if the desired higher electrical conductivity than that of GO was not obtained. The small proportion of PDA remaining on the material surface can be used for subsequent functionalization, which often plays the key role for the intended application.

To summarize, in the present work, deep insight into the chemistry of PDA-GO and RGO was given. The green reduction of GO by PDA proved to be a way to replace typical reduction methods that involve toxic, corrosive, and hazardous solvents and chemicals. In the future, further studies are needed to better understand the reduction of GO by such green approaches and make the processes reproducible and scalable.

Supplementary Materials:
Table S1: Elemental compositions investigated by XPS, Table S2: Wide-scan spectrum of the sample recorded from the graphene stacks, Table S3: C 1s spectrum of the sample recorded from the graphene stacks, Table S4: Wide-scan spectrum of the GO sample, Table S5: C 1s spectrum of the GO sample, Table S6: N 1s spectrum of the GO sample, Table S7: Wide-scan spectrum of the PDA-GO_30 sample, Table S8: C 1s spectrum of the PDA-GO_30 sample, Table S9: N 1s spectrum of the PDA-GO_30 sample, Table S10: Wide-scan spectrum of the RGO_30 sample, Table S11: C 1s spectrum of the RGO_30 sample, Table S12: N 1s spectrum of the RGO_30 sample, Table S13: Wide-scan spectrum of the PDA-GO_90 sample, Table S14: C 1s spectrum of the PDA-GO_90 sample, Table S15: N 1s spectrum of the PDA-GO_90 sample, Table S16: Wide-scan spectrum of the PDA-GO_180 sample, Table S17: C 1s spectrum of the PDA-GO_180 sample, Table S18: N 1s spectrum of the PDA-GO_180 sample.

Author Contributions: Sample preparation, electrical powder conductivity measurements, and data treatment and interpretation of Raman spectroscopy, TGA, and electrical powder conductivity measurements were performed by C.S. as well as writing. Ab initio calculations were performed by P.F. Conceptualization, reviewing, editing, project administration, and supervision were realized by C.Z. and P.P. who also interpreted the powder conductivity measurements. XPS was performed by F.S. including data treatment, interpretation and discussion.

Acknowledgments: We thank Julia Muche for the Raman Spectroscopy measurements, Kerstin Arnhold for TGA measurements and Beate Krause (all IPF) for introduction to the equipment used for electrical powder conductivity measurements, Matthias Holzschuh for collecting XPS data and editing.

References

1. Novoselov, K.S.; Geim, A.K.; Morozov, S.V.; Jiang, D.; Zhang, Y.; Dubonos, S.V.; Grigorieva, I.V.; Firsov, A.A. Electric field effect in atomically thin carbon films. *Science* **2004**, *306*, 666–669. [CrossRef] [PubMed]
2. Novoselov, K.S.; Jiang, D.; Schedin, F.; Booth, T.J.; Khotkevich, V.V.; Morozov, S.V.; Geim, A.K. Two-dimensional atomic crystals. *Proc. Natl. Acad. Sci. USA* **2005**, *102*, 10451–10453. [CrossRef] [PubMed]
3. Lee, C.; Wei, X.; Li, Q.; Carpick, R.; Kysar, J.W.; Hone, J. Elastic and frictional properties of graphene. *Phys. Status Solidi (b)* **2009**, *246*, 2562–2567. [CrossRef]
4. Novoselov, K.S.; Fal'ko, V.I.; Colombo, L.; Gellert, P.R.; Schwab, M.G.; Kim, K. A roadmap for graphene. *Nature* **2012**, *490*, 192–200. [CrossRef] [PubMed]
5. Zhu, Y.; Murali, S.; Cai, W.; Li, X.; Suk, J.W.; Potts, J.R.; Ruoff, R.S. Graphene and graphene oxide: Synthesis, properties, and applications. *Adv. Mater.* **2010**, *22*, 3906–3924. [CrossRef] [PubMed]
6. Liu, W.-W.; Chai, S.-P.; Mohamed, A.R.; Hashim, U. Synthesis and characterization of graphene and carbon nanotubes: A review on the past and recent developments. *J. Ind. Eng. Chem.* **2014**, *20*, 1171–1185. [CrossRef]
7. Warner, J.H.; Schäffel, F.; Bachmatiuk, A.; Rümmeli, M.H. Chapter 4—Methods for obtaining graphene. In *Graphene*; Warner, J.H., Schäffel, F., Bachmatiuk, A., Rümmeli, M.H., Eds.; Elsevier: Amsterdam, The Netherlands, 2013; pp. 129–228.
8. Singh, V.; Joung, D.; Zhai, L.; Das, S.; Khondaker, S.I.; Seal, S. Graphene based materials: Past, present and future. *Prog. Mater. Sci.* **2011**, *56*, 1178–1271. [CrossRef]
9. Liu, J.; Cui, L.; Losic, D. Graphene and graphene oxide as new nanocarriers for drug delivery applications. *Acta Biomater.* **2013**, *9*, 9243–9257. [CrossRef]
10. Compton, O.C.; Nguyen, S.T. Graphene oxide, highly reduced graphene oxide, and graphene: Versatile building blocks for carbon-based materials. *Small* **2010**, *6*, 711–723. [CrossRef]

11. Taniselass, S.; Md Arshad, M.K.; Gopinath, S.C.B. Current state of green reduction strategies: Solution-processed reduced graphene oxide for healthcare biodetection. *Mater. Sci. Eng. C* **2019**, *96*, 904–914. [CrossRef]

12. Thakur, S.; Karak, N. Alternative methods and nature-based reagents for the reduction of graphene oxide: A review. *Carbon* **2015**, *94*, 224–242. [CrossRef]

13. De Silva, K.K.H.; Huang, H.H.; Joshi, R.K.; Yoshimura, M. Chemical reduction of graphene oxide using green reductants. *Carbon* **2017**, *119*, 190–199. [CrossRef]

14. De Marco, B.A.; Rechelo, B.S.; Tótoli, E.G.; Kogawa, A.C.; Salgado, H.R.N. Evolution of green chemistry and its multidimensional impacts: A review. *Saudi Pharm. J.* **2019**, *27*, 1–8. [CrossRef] [PubMed]

15. Gao, J.; Liu, F.; Liu, Y.; Ma, N.; Wang, Z.; Zhang, X. Environment-friendly method to produce graphene that employs vitamin c and amino acid. *Chem. Mater.* **2010**, *22*, 2213–2218. [CrossRef]

16. Zhang, J.; Yang, H.; Shen, G.; Cheng, P.; Zhang, J.; Guo, S. Reduction of graphene oxide vial-ascorbic acid. *Chem. Commun.* **2010**, *46*, 1112–1114. [CrossRef] [PubMed]

17. Fernández-Merino, M.J.; Guardia, L.; Paredes, J.I.; Villar-Rodil, S.; Solís-Fernández, P.; Martínez-Alonso, A.; Tascón, J.M.D. Vitamin C is an ideal substitute for hydrazine in the reduction of graphene oxide suspensions. *J. Phys. Chem. C* **2010**, *114*, 6426–6432. [CrossRef]

18. Zhu, C.; Guo, S.; Fang, Y.; Dong, S. Reducing sugar: New functional molecules for the green synthesis of graphene nanosheets. *ACS Nano* **2010**, *4*, 2429–2437. [CrossRef] [PubMed]

19. Bose, S.; Kuila, T.; Mishra, A.K.; Kim, N.H.; Lee, J.H. Dual role of glycine as a chemical functionalizer and a reducing agent in the preparation of graphene: An environmentally friendly method. *J. Mater. Chem.* **2012**, *22*, 9696–9703. [CrossRef]

20. Chen, D.; Li, L.; Guo, L. An environment-friendly preparation of reduced graphene oxide nanosheets via amino acid. *Nanotechnology* **2011**, *22*, 325601. [CrossRef]

21. Ma, J.; Wang, X.; Liu, Y.; Wu, T.; Liu, Y.; Guo, Y.; Li, R.; Sun, X.; Wu, F.; Li, C.; et al. Reduction of graphene oxide with l-lysine to prepare reduced graphene oxide stabilized with polysaccharide polyelectrolyte. *J. Mater. Chem. A* **2013**, *1*, 2192–2201. [CrossRef]

22. Bo, Z.; Shuai, X.; Mao, S.; Yang, H.; Qian, J.; Chen, J.; Yan, J.; Cen, K. Green preparation of reduced graphene oxide for sensing and energy storage applications. *Sci. Rep.* **2014**, *4*, 4684. [CrossRef] [PubMed]

23. Wan, W.; Zhao, Z.; Hu, H.; Gogotsi, Y.; Qiu, J. Highly controllable and green reduction of graphene oxide to flexible graphene film with high strength. *Mater. Res. Bull.* **2013**, *48*, 4797–4803. [CrossRef]

24. Zhang, Z.; Chen, H.; Xing, C.; Guo, M.; Xu, F.; Wang, X.; Gruber, H.J.; Zhang, B.; Tang, J. Sodium citrate: A universal reducing agent for reduction/decoration of graphene oxide with au nanoparticles. *Nano Res.* **2011**, *4*, 599–611. [CrossRef]

25. Li, J.; Xiao, G.; Chen, C.; Li, R.; Yan, D. Superior dispersions of reduced graphene oxide synthesized by using gallic acid as a reductant and stabilizer. *J. Mater. Chem. A* **2013**, *1*, 1481–1487. [CrossRef]

26. Kuila, T.; Bose, S.; Khanra, P.; Mishra, A.K.; Kim, N.H.; Lee, J.H. A green approach for the reduction of graphene oxide by wild carrot root. *Carbon* **2012**, *50*, 914–921. [CrossRef]

27. Khanra, P.; Kuila, T.; Kim, N.H.; Bae, S.H.; Yu, D.-S.; Lee, J.H. Simultaneous bio-functionalization and reduction of graphene oxide by baker's yeast. *Chem. Eng. J.* **2012**, *183*, 526–533. [CrossRef]

28. Pham, T.A.; Kim, J.S.; Kim, J.S.; Jeong, Y.T. One-step reduction of graphene oxide with L-glutathione. *Colloids Surf. A Physicochem. Eng. Asp.* **2011**, *384*, 543–548. [CrossRef]

29. Guo, C.; Book-Newell, B.; Irudayaraj, J. Protein-directed reduction of graphene oxide and intracellular imaging. *Chem. Commun.* **2011**, *47*, 12658–12660. [CrossRef] [PubMed]

30. Esfandiar, A.; Akhavan, O.; Irajizad, A. Melatonin as a powerful bio-antioxidant for reduction of graphene oxide. *J. Mater. Chem.* **2011**, *21*, 10907–10914. [CrossRef]

31. Lei, Z.; Lu, L.; Zhao, X.S. The electrocapacitive properties of graphene oxide reduced by urea. *Energy Environ. Sci.* **2012**, *5*, 6391–6399. [CrossRef]

32. Lee, G.; Kim, B.S. Biological reduction of graphene oxide using plant leaf extracts. *Biotechnol. Prog.* **2014**, *30*, 463–469. [CrossRef] [PubMed]

33. Jana, M.; Saha, S.; Khanra, P.; Murmu, N.C.; Srivastava, S.K.; Kuila, T.; Lee, J.H. Bio-reduction of graphene oxide using drained water from soaked mung beans (*Phaseolus aureus* L.) and its application as energy storage electrode material. *Mater. Sci. Eng. B* **2014**, *186*, 33–40. [CrossRef]

34. Wang, Y.; Shi, Z.; Yin, J. Facile synthesis of soluble graphene via a green reduction of graphene oxide in tea solution and its biocomposites. *ACS Appl. Mater. Interfaces* **2011**, *3*, 1127–1133. [CrossRef] [PubMed]

35. Muthoosamy, K.; Bai, R.G.; Abubakar, I.B.; Sudheer, S.M.; Lim, H.N.; Loh, H.-S.; Huang, N.M.; Chia, C.H.; Manickam, S. Exceedingly biocompatible and thin-layered reduced graphene oxide nanosheets using an eco-friendly mushroom extract strategy. *Int. J. Nanomed.* **2015**, *10*, 1505–1519. [CrossRef]

36. Xu, L.Q.; Yang, W.J.; Neoh, K.G.; Kang, E.T.; Fu, G.D. Dopamine-induced reduction and functionalization of graphene oxide nanosheets. *Macromolecules* **2010**, *43*, 8336–8339. [CrossRef]

37. Lee, H.; Dellatore, S.M.; Miller, W.M.; Messersmith, P.B. Mussel-inspired surface chemistry for multifunctional coatings. *Science* **2007**, *318*, 426–430. [CrossRef] [PubMed]

38. Guo, L.Q.; Liu, Q.; Li, G.L.; Shi, J.B.; Liu, J.Y.; Wang, T.; Jiang, G.B. A mussel-inspired polydopamine coating as a versatile platform for the in situ synthesis of graphene-based nanocomposites. *Nanoscale* **2012**, *4*, 5864–5867. [CrossRef]

39. Cong, H.P.; Wang, P.; Gong, M.; Yu, S.H. Facile synthesis of mesoporous nitrogen-doped graphene: An efficient methanol-tolerant cathodic catalyst for oxygen reduction reaction. *Nano Energy* **2014**, *3*, 55–63. [CrossRef]

40. Wang, F.Y.; Sun, Q.Q.; Feng, B.; Xu, Z.A.; Zhang, J.Y.; Xu, J.; Lu, L.L.; Yu, H.J.; Wang, M.W.; Li, Y.P.; et al. Polydopamine-functionalized graphene oxide loaded with gold nanostars and doxorubicin for combined photothermal and chemotherapy of metastatic breast cancer. *Adv. Healthc. Mater.* **2016**, *5*, 2227–2236. [CrossRef]

41. Zhang, X.Y.; Nan, X.; Shi, W.; Sun, Y.N.; Su, H.L.; He, Y.; Liu, X.; Zhang, Z.; Ge, D.T. Polydopamine-functionalized nanographene oxide: A versatile nanocarrier for chemotherapy and photothermal therapy. *Nanotechnology* **2017**, *28*. [CrossRef]

42. Zhang, Z.; Zhang, J.; Zhang, B.L.; Tang, J.L. Mussel-inspired functionalization of graphene for synthesizing Ag-polydopamine-graphene nanosheets as antibacterial materials. *Nanoscale* **2013**, *5*, 118–123. [CrossRef] [PubMed]

43. Tian, J.; Deng, S.Y.; Li, D.L.; Shan, D.; He, W.; Zhang, X.J.; Shi, Y. Bioinspired polydopamine as the scaffold for the active AuNPs anchoring and the chemical simultaneously reduced graphene oxide: Characterization and the enhanced biosensing application. *Biosens. Bioelectron.* **2013**, *49*, 466–471. [CrossRef] [PubMed]

44. Liu, H.Y.; Xi, P.X.; Xie, G.Q.; Shi, Y.J.; Hou, F.P.; Huang, L.; Chen, F.J.; Zeng, Z.Z.; Shao, C.W.; Wang, J. Simultaneous reduction and surface functionalization of graphene oxide for hydroxyapatite mineralization. *J. Phys. Chem. C* **2012**, *116*, 3334–3341. [CrossRef]

45. Da Silva, C.A.; Pötschke, P.; Simon, F.; Holzschuh, M.; Pionteck, J.; Heinrich; Wießner, S.; Zimmerer, C. Synthesis and characterization of graphene derivatives for application in magnetic high-field induction heating. *AIP Conf. Proc.* **2019**, *2055*, 130006. [CrossRef]

46. Schmidt, M.W.; Baldridge, K.K.; Boatz, J.A.; Elbert, S.T.; Gordon, M.S.; Jensen, J.H.; Koseki, S.; Matsunaga, N.; Nguyen, K.A.; Su, S.; et al. General atomic and molecular electronic structure system. *J. Comput. Chem.* **1993**, *14*, 1347–1363. [CrossRef]

47. Briggs, D. Comprehensive Polymer Science. In *Characterization of Surfaces*; Chapter 24; Booth, C., Price, C., Briggs, D., Eds.; Pergamon: Oxford UK, 1989; Volume 1, pp. 543–559.

48. Seah, M.P.; Dench, W.A. Quantitative electron spectroscopy of surfaces: A standard data base for electron inelastic mean free paths in solids. *Surf. Interface Anal.* **1979**, *1*, 2–11. [CrossRef]

49. Paiva, M.C.; Simon, F.; Novais, R.M.; Ferreira, T.; Proença, M.F.; Xu, W.; Besenbacher, F. Controlled functionalization of carbon nanotubes by a solvent-free multicomponent approach. *ACS Nano* **2010**, *4*, 7379–7386. [CrossRef] [PubMed]

50. Watts, J.F. High resolution XPS of organic polymers: The Scienta ESCA 300 database. G. Beamson and D. Briggs. 280pp., £65. John Wiley & Sons, Chichester, ISBN 0471 935921, (1992). *Surf. Interface Anal.* **1993**, *20*, 267. [CrossRef]

51. Krause, B.; Boldt, R.; Häußler, L.; Pötschke, P. Ultralow percolation threshold in polyamide 6.6/MWCNT composites. *Compos. Sci. Technol.* **2015**, *114*, 119–125. [CrossRef]

52. Lerf, A.; He, H.; Forster, M.; Klinowski, J. Structure of Graphite Oxide Revisited. *J. Phys. Chem. B* **1998**, *102*, 4477–4482. [CrossRef]

53. He, H.; Klinowski, J.; Forster, M.; Lerf, A. A new structural model of graphite oxide. *Chem. Phys. Lett.* **1998**, *287*, 53–56. [CrossRef]

54. Bellunato, A.; Arjmandi Tash, H.; Cesa, Y.; Schneider, G.F. Chemistry at the edge of graphene. *ChemPhysChem* **2016**, *17*, 785–801. [CrossRef] [PubMed]

55. Grätz, S.; Beyer, D.; Tkachova, V.; Hellmann, S.; Berger, R.; Feng, X.; Borchardt, L. The mechanochemical scholl reaction—A solvent-free and versatile graphitization tool. *Chem. Commun.* **2018**, *54*, 5307–5310. [CrossRef] [PubMed]

56. Malard, L.M.; Pimenta, M.A.; Dresselhaus, G.; Dresselhaus, M.S. Raman spectroscopy in graphene. *Phys. Rep.* **2009**, *473*, 51–87. [CrossRef]

57. Ferrari, A.C. Raman spectroscopy of graphene and graphite: Disorder, electron–phonon coupling, doping and nonadiabatic effects. *Solid State Commun.* **2007**, *143*, 47–57. [CrossRef]

58. Ferrari, A.C.; Meyer, J.C.; Scardaci, V.; Casiraghi, C.; Lazzeri, M.; Mauri, F.; Piscanec, S.; Jiang, D.; Novoselov, K.S.; Roth, S.; et al. Raman spectrum of graphene and graphene layers. *Phys. Rev. Lett.* **2006**, *97*, 187401. [CrossRef]

59. Ferrari, A.C.; Basko, D.M. Raman spectroscopy as a versatile tool for studying the properties of graphene. *Nat. Nanotechnol.* **2013**, *8*, 235. [CrossRef]

60. Song, J.; Dai, Z.; Li, J.; Tong, X.; Zhao, H. Polydopamine-decorated boron nitride as nano-reinforcing fillers for epoxy resin with enhanced thermomechanical and tribological properties. *Mater. Res. Express* **2018**, *5*, 075029. [CrossRef]

61. Parviz, D.; Das, S.; Ahmed, H.S.T.; Irin, F.; Bhattacharia, S.; Green, M.J. Dispersions of non-covalently functionalized graphene with minimal stabilizer. *Acs Nano* **2012**, *6*, 8857–8867. [CrossRef]

62. Tripathi, B.; Das, P.; Simon, F.; Stamm, M. Ultralow fouling membranes by surface modification with functional polydopamine. *Eur. Polym. J.* **2018**, *99*, 80–89. [CrossRef]

63. Biesinger, M. Available online: http://www.xpsfitting.com/search/label/Adventitious (accessed on 11 June 2019).

64. Yang, D.; Kong, X.X.; Ni, Y.F.; Ruan, M.N.; Huang, S.; Shao, P.Z.; Guo, W.L.; Zhang, L.Q. Improved mechanical and electrochemical properties of XNBR dielectric elastomer actuator by poly(dopamine) functionalized graphene nano-sheets. *Polymers* **2019**, *11*, 218. [CrossRef] [PubMed]

65. He, Y.; Wang, J.; Zhang, H.; Zhang, T.; Zhang, B.; Cao, S.; Liu, J. Polydopamine-modified graphene oxide nanocomposite membrane for proton exchange membrane fuel cell under anhydrous conditions. *J. Mater. Chem. A* **2014**, *2*, 9548–9558. [CrossRef]

66. Santha Kumar, A.R.S.; Piana, F.; Mičušík, M.; Pionteck, J.; Banerjee, S.; Voit, B. Preparation of graphite derivatives by selective reduction of graphite oxide and isocyanate functionalization. *Mater. Chem. Phys.* **2016**, *182*, 237–245. [CrossRef]

67. Srivastava, S.K.; Pionteck, J. Recent advances in preparation, structure, properties and applications of graphite oxide. *J. Nanosci. Nanotechnol.* **2015**, *15*, 1984–2000. [CrossRef] [PubMed]

A Polyol-Mediated Fluoride Ions Slow-Releasing Strategy for the Phase-Controlled Synthesis of Photofunctional Mesocrystals

Xianghong He *, **Yaheng Zhang, Yu Fu, Ning Lian and Zhongchun Li**

School of Chemistry and Environmental Engineering, Jiangsu University of Technology, Changzhou 213001, Jiangsu, China; Zhangyaheng@jsut.edu.cn (Y.Z.); fuyu@jsut.edu.cn (Y.F.); ln@jstu.edu.cn (N.L.); lizc@jstu.edu.cn (Z.L.)

* Correspondence: hexh@jsut.edu.cn

Abstract: There are only a few inorganic compounds that have evoked as much interest as sodium yttrium fluoride ($NaYF_4$). Its extensive applications in various fields, including transparent displays, luminescence coding, data storage, as well as biological imaging, demand the precise tuning of the crystal phase. Controlling the emergence of the desired α-phase has so far remained a formidable challenge, especially via a simple procedure. Herein, we represented a polyol-assisted fluoride ions slow-release strategy for the rational control of pure cubic phase $NaYF_4$ mesocrystals. The combination of fluorine-containing ionic liquid as a fluoride source and the existence of a polyalcohol as the reactive medium ensure the formation of uniform α-phase mesocrystallines in spite of a higher temperature and/or higher doping level.

Keywords: polyol-assisted fluoride ions slow-release strategy; $NaYF_4$ mesocrystals; crystallographic phase control

1. Introduction

Since inorganic micro/nanocrystals usually exist in various forms or phases, the phase transformation from kinetically stable ones to thermally stable ones is a normal phenomenon [1–4]. The intrinsic properties of a micro/nanomaterial are largely determined by its unique crystal structure [5,6]. Hence, controlling the phase formation is essential for both scientific interests and extended applications. As a typical example, sodium yttrium fluoride ($NaYF_4$) owns two polymorphs under ambient condition, i.e., the cubic (α-) and hexagonal (β-) phase, which is a commonly used matrix lattice for up-conversion luminescence. The former is a high-temperature metastable phase, while the latter remains thermodynamically stable [7,8]. The past decades have witnessed much exploration of its controlled synthesis and up-/down-conversion luminescent properties [1,4,9–34]. Compared with considerable work on $\alpha \rightarrow \beta$ phase transformation [1,4,9–19], the fabrication of α-$NaYF_4$ as well as the investigation involving the $\beta \rightarrow \alpha$ transformation process have been neglected [26,30,33,34]. So far, some strategies have been developed to fabricate α-phase $NaYF_4$ nano-/mocro-crystal, such as a liquid–solid–solution (LSS) procedure [21], polyol method [22], two-phase interfacial route [23], microwave-assisted ionic liquid (IL)-based technique [24], modified solvothermal approach [25], and self-sacrificing template multiple-step route [26,27]. Furthermore, introducing Mn^{2+} (r = 81 pm) with a smaller size than Y^{3+} (r = 89 pm) into an $NaYF_4$ host can dominate, forming pure α-phase $NaYF_4$ nanoparticles [28]. However, α-phase $NaYF_4$ inevitably transforms into the hexagonal ones due to its thermodynamic instability. Additionally, the cubic $NaYF_4$ nanoparticles are usually formed preferentially in the solution system of non-equilibrium reactions [20]. As a consequence, rationally controlling α-NaYF$_4$ and simultaneously avoiding the generation of β-phase or a mixture of α and β phases remain formidable challenges, especially via a simple procedure [28,29,34].

On the other hand, the above-mentioned progress focused on NaYF$_4$ micro/nanocrystals instead of mesocrystals. Mesocrystals are three-dimensional (3D) order nanoparticles superstructures with unique properties and various potential applications as functional materials [35–38]. Nevertheless, the range of known mesocrystallines remains quite limited, in which few investigations to fluorine-containing compound mesocrystallines are available [39–45]. More recently, our group fabricated yttrium hydroxide fluoride mesocrystalline, as well as its Eu^{3+} doped analogue, by means of an additive-free hydrothermal procedure, which involved the reaction of Y(NO$_3$)$_3$, NaF, and NaOH aqueous solution without any organic additives [39]. Furthermore, we explored the preparation of rare-earths trifluoride mesocrystals by a solvothermal route involving IL 1-butyl-3-methylimidazolium hexafluorophosphate (BmimPF$_6$) as the fluorine source in the presence of 1,4-butanediol [40]. However, no effort has been made to reveal the phase control related to rare-earths fluoride mesocrystallines. Herein, we present a facile, one-pot route called a polyol-mediated fluoride slow-releasing strategy for the rational control of pure phase α-NaYF$_4$ mesocrystals. In spite of a higher temperature or/and higher doping level, cubic phase can be maintained.

2. Experimental Procedure

2.1. Chemicals and Materials

Analytical grade rare earth chlorides and/or nitrates (yttrium chloride hexahydrate, gadolinium chloride hexahydrate, ytterbium nitrate pentahydrate, and erbium nitrate pentahydrate, 99.9%) were provided by Aladdin Industrial Inc. Shanghai, China. NaNO$_3$ (99.0%), 2,2′-oxydiethanol (99.0%, diethylene glycol, abbreviated as DEG), 1,2-ethanediol (99.0%), and ethanol (99.8%) were obtained from Sinopharm Chemical Reagent Company, Shanghai, China. 1-Butyl-3-methylimidazolium hexafluorophosphate (BminPF$_6$, 99%) was purchased from Aldamas-beta Co., Shanghai, China. All of the reagents and solvents were directly used without further treatment.

2.2. Synthesis

NaYF$_4$:Yb^{3+},Er^{3+}(20/2 mol%) (abbreviated as NYF:Yb^{3+},Er^{3+} hereafter) and Gd^{3+} tri-doped NYF:Yb^{3+},Er^{3+}(20/2 mol%) nanocrystals (NCs) were synthesized via a polyol-mediated solvothermal procedure. Here, we took the synthesis of NYF:Yb^{3+},Er^{3+} (20/2 mol%) as an example. The starting chemicals including NaNO$_3$, yttrium chloride hexahydrate, ytterbium nitrate pentahydrate, and erbium nitrate pentahydrate in the stoichiometric ratio were well mixed with 1,2-ethanediol (or DEG) under stirring, to form solution. Thereafter, the solution was slowly added into a 25-mL polytetrafluoroethylene (PTFE) vial containing a proper amount of BminPF$_6$ under vigorous stirring. The autoclave was sealed after vigorous stirring at room temperature for around 15 min, and then heated at 120 °C for 24 h. The final products were collected by centrifugation, and then washed sequentially using ethanol and H$_2$O three times. After drying at 70 °C under dynamic vacuum for 24 h, an NYF:Yb^{3+},Er^{3+} sample was obtained. The synthetic procedure of Gd^{3+} tri-doped NYF:Yb^{3+},Er^{3+} (20/2 mol%) NCs was the same as that which was used to fabricate NYF:Yb^{3+},Er^{3+}, except that the stoichiometric amount of gadolinium chloride hexahydrate was also added to 1,2-ethanediol (or DEG).

2.3. Characterization

The crystal structure and phase analysis were determined via X-ray diffraction (XRD) using a Bruker D8 Advanced X-ray diffractometer (Ni filtered, Cu K$_\alpha$ radiation, 40 kV and 40 mA) (Bruker, Billerica, MA, USA). The morphology the products were recorded on a transmission electron microscope (TEM, JEM-2010, JOEL Ltd., Tokyo, Japan) and a Hitachi S4800 field-emission scanning electron microscope (FE-SEM) (Hitachi Ltd., Tokyo, Japan). The selected area electron diffraction (SAED) pattern were characterized by the above-mentioned TEM (JEM-2010). An up-conversion fluorescence spectrum was obtained on an Edinburgh Instrument FLS920 phosphorimeter (Edinburgh Instruments Ltd., Livingston, UK) with a 980-nm laser diode Module (K98D08M-30mW, Changchun, China) as the excitation light source. The above-mentioned measurements were performed at room temperature from powder samples.

3. Results and Discussion

NaYF$_4$-based mesocrystals were fabricated via solvothermal treatment of Na$^+$, Y^{3+}, and BminPF$_6$ in the presence of viscous polyol-like diethylcol or 1,2-ethanediol. Apart from serving as solvent and complexant (i.e., bonding with Na$^+$ and Y^{3+}), polyol also acts as a stabilizer that limits particle growth and suppresses the $\alpha \rightarrow \beta$ phase transition of NaYF$_4$ [22,46]. BmimPF$_6$ was chosen as a task-specific fluorine source (the reason why it was chosen as the fluorine source is given in the Supplementary Materials). The required fluoride anion (F$^-$) was provided by BmimPF$_6$ as a result of its slow decomposition and hydrolysis [23,45,47]. Even without additional water, BmimPF$_6$ can hydrolyze with the aid of the trace water and hydration water molecules from yttrium chloride hexahydrate [45,48]. During the treatment procedure, PF$_6{}^-$ can slowly hydrolyze and then produce F$^-$ through slowly increasing the temperature [45], as revealed in Equation (1):

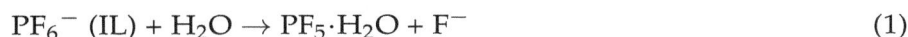

$$PF_6{}^- \ (IL) + H_2O \rightarrow PF_5 \cdot H_2O + F^- \tag{1}$$

Therefore, this procedure was defined as a fluoride slow-release strategy, which involved fluoride releasing from BmimPF$_6$ with the assistance of polyol [49].

Powder XRD patterns of a Yb^{3+}-Er^{3+} co-doped and pure NaYF$_4$ submicrocube in the case of DEG as the reaction medium are illustrated in Figures 1a and 2a, respectively. All of the diffraction peaks matched the α-phase NaYF$_4$ crystals (PDF No.77-2042), and no impurities were found. The sharp and narrow diffraction peaks revealled the highly crystallinity of these submicrocubes despite treatment at relatively low temperature (120 °C).

As exhibited in Figure 1b–d, all of the NYF:Yb^{3+},Er^{3+} submicrocrystals show cubic shapes and edge lengths of about 120 nm. Both FE-SEM and TEM photos illustrated their novel microstructure features, which are built from many nanoparticles and exhibited rough surfaces. A few nanoparticles were found attached on its surface (Figure 1e). Especially, the SAED pattern (Figure 1f) of a single NaYF$_4$ cube shows sharp and periodic spots, revealing its noticeable single crystal-like feature. According to Cölfen et al. [35–40], the regular-shaped NaYF$_4$ cubes actually belong to typical mesocrystals. The combination of a coarse surface pattern and the attachment of nanoparticles reveal that these mesocrystallines resulted from the self-assembling of nanoparticle subunits rather than the classic crystalline growth [18,35–40].

When DEG was replaced by 1,2-ethanediol, the product can also be indexed as pure-phase cubic NaYF$_4$ crystal (Figure 2b). All of these results indicated that IL BmimPF$_6$ in the presence of polyol also acts as a crystal-phase manipulator during the formation of NaYF$_4$ [18].

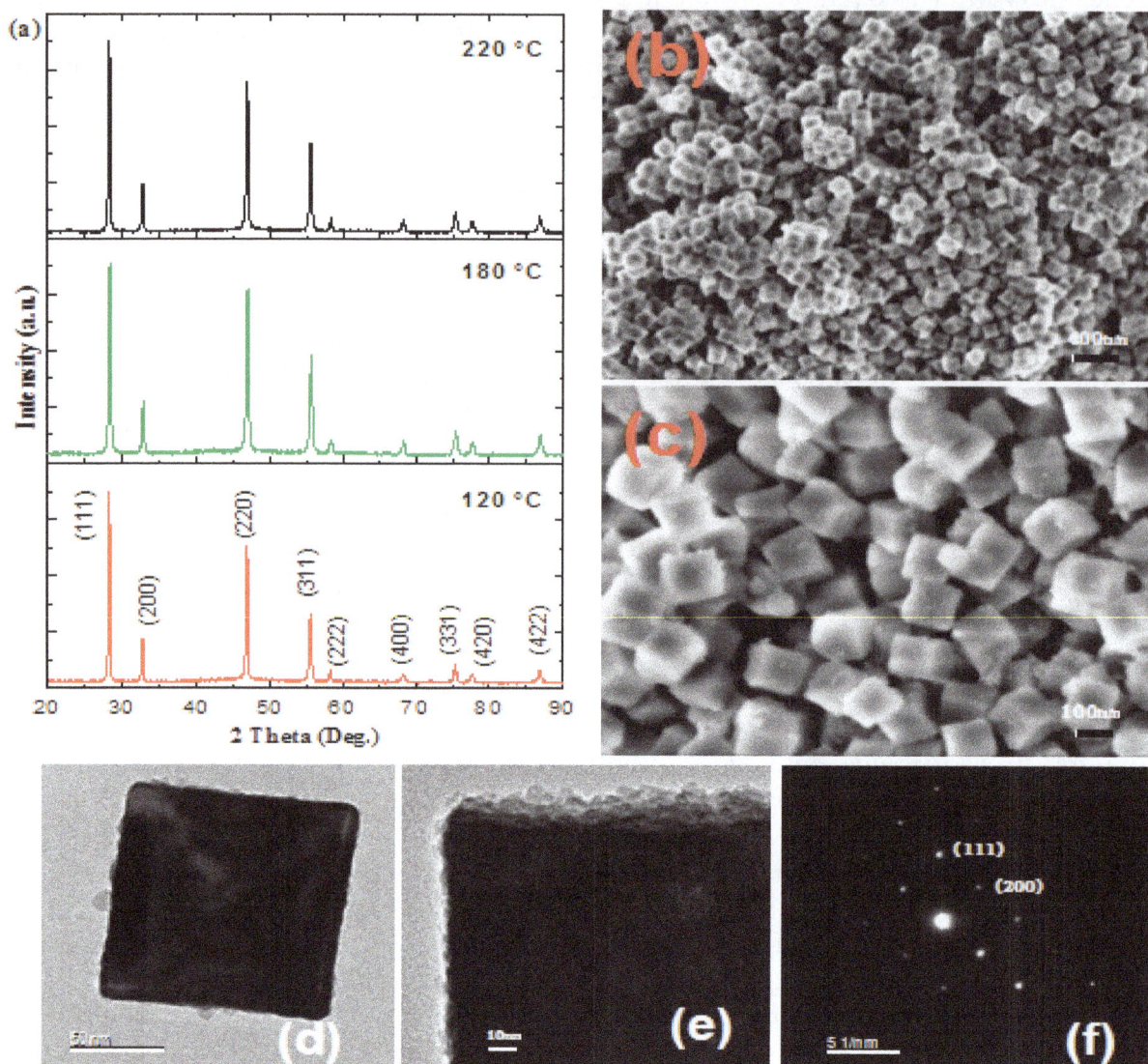

Figure 1. (**a**) X-ray diffraction (XRD) patterns of sodium yttrium fluoride (NYF):$Yb^{3+}Er^{3+}$ (20/2 mol%) sample at various solvothermal temperatures using diethylene glycol (DEG) as the reaction medium (all of the diffraction peaks are attributed to cubic-phase $NaYF_4$), field-emission scanning electron microscope (FE-SEM) images ((**b**) low-magnification; (**c**) high-magnification), (**d,e**) TEM images, and (**f**) selected area electron diffraction (SAED) pattern of as-obtained NYF:Yb^{3+},Er^{3+} (20/2 mol%) submicrocrystals at 120 °C. (Note the nanoparticles aggregated to form submicrocubes).

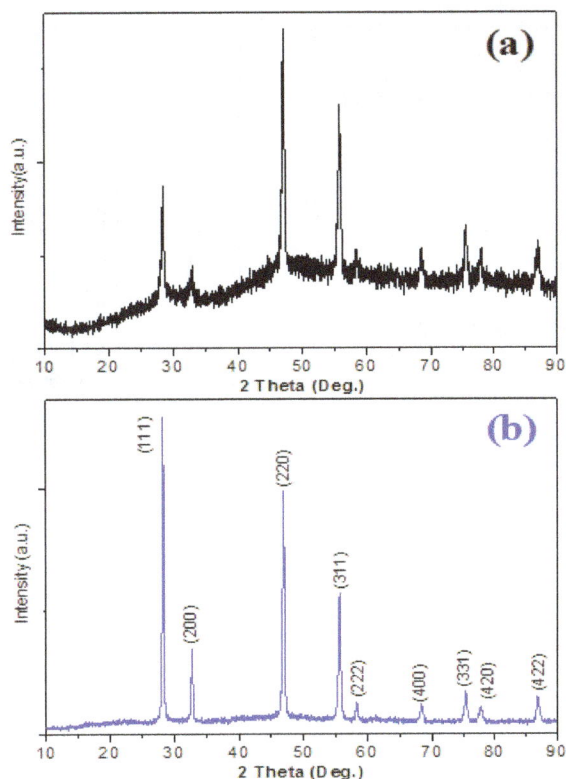

Figure 2. XRD patterns of (**a**) $NaYF_4$ host, and (**b**) $NYF:Yb^{3+},Er^{3+}$ (20/2 mol%) samples using 1,2-ethanediol as solvent.

For comparison, the preparation of $NYF:Yb^{3+},Er^{3+}$ was also conducted through an LSS procedure using NH_4F as the F^- source and the mixture of ethanol–H_2O–oleic acid as the medium at 120 °C and 220 °C [11,21]. Figure S2 (see Supplementary Materials) revealed the XRD patterns of as-obtained $NYF:Yb^{3+},Er^{3+}$ (20/2 mol%) at 120 °C and 220 °C. Obviously, the product obtained at the lower temperature can be ascribed to a pure α-phase $NaYF_4$, as expected for $NaYF_4$ synthesized under mild conditions [21,23,32]. However, in the case of higher temperature (220 °C), only β-phase $NaYF_4$ was fabricated. These results demonstrated that promoting the reaction temperature can induce the α→β phase change of $NaYF_4$, which is consistent with previous reports [9–12]. However, as shown in Figure 1a, even if the reaction temperature reached 220 °C, the as-prepared nanoparticles via the fluoride ions-slow-release procedure unambiguously remained in a pure cubic phase. In a word, regardless of the treatment temperature, an α-phase $NaYF_4$ can be obtained by this slow-release strategy.

As mentioned above, without the tri-doping of Gd^{3+}, the XRD pattern of the $NYF:Yb^{3+},Er^{3+}$ (20/2 mol%) sample matched a cubic phase of $NaYF_4$ (PDF No.77-2042). As for the $NYF:Yb^{3+},Er^{3+}$ (20/2 mol%) sample, previous works revealed that introducing lanthanide ions (such as Gd^{3+}) with a larger size than the Y^{3+} ion in the $NaYF_4$ lattice not only induced an alteration from the α phase to the β phase, it also dominated the forming of pure β-phase $NaYF_4$ NCs [1,20,50]. However, in this work, as revealed in Figure 3a, the pure cubic phase of $NaYF_4$ remained when Gd^{3+} of 15 mol% was incorporated into host lattices. With the further increasing of the Gd^{3+} ion content (Figure 3b), no impurity diffraction peaks were found, showing the forming of a homogeneous solid solution, which is due to the small structural difference between the cubic-phase $NaGdF_4$ and $NaYF_4$. Obviously, phase transformation did not occur upon a higher-level doping of the dopant.

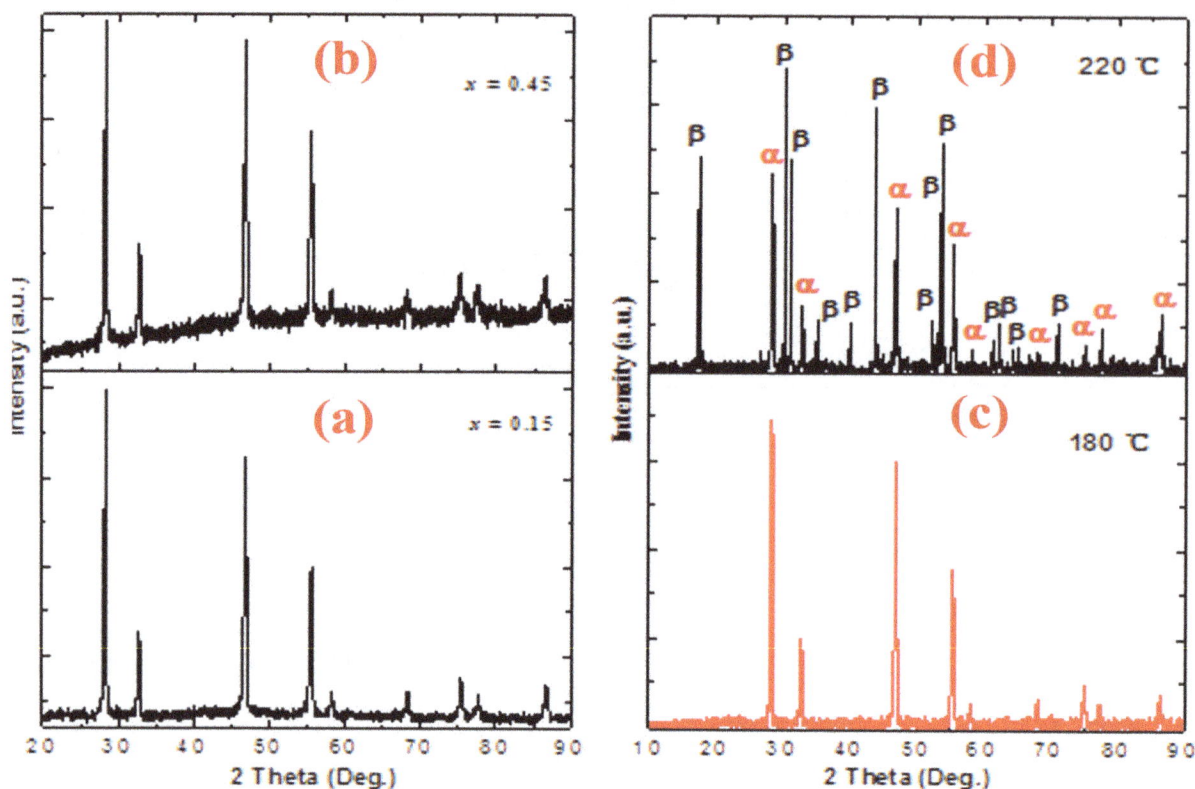

Figure 3. XRD patterns of Na $(Y_{0.78-x}Gd_x)$, F_4:Yb^{3+}, and Er^{3+}(20/2 mol%) samples with different tri-doping levels of Gd^{3+} ((**a**) $x = 0.15$, (**b**) $x = 0.45$), and Na $(Y_{0.48}Gd_{0.30})F_4$:Yb^{3+},Er^{3+}(20/2 mol%) samples obtained at higher solvothermal temperatures ((**c**) 180 °C, (**d**) 220 °C; the symbol α and β represent cubic and hexagonal phases, respectively).

High-level doping usually leads to an $\alpha \rightarrow \beta$ phase transition of $NaYF_4$ in the LSS reaction system [1,20]. However, in present work, by using $BminPF_6$ and polyol as the F^- source and reaction medium, respectively, as shown in Figure 3c, the as-synthesized submicrocubes remained in the cubic phase of $NaYF_4$ in spite of higher total doping concentrations (52 mol%) as well as a higher solvothermal treatment temperature (180 °C). Even if the total doping contents were set as high as 52 mol%, and the treatment temperature simultaneously approached 220 °C (near to the work-limited temperature of the PTFE vial), the α-phase $NaYF_4$ still existed in the products (see Figure 3d).

According to He et al. [51], the $\alpha \rightarrow \beta$ phase change of $NaYF_4$ can be attributed to the elevated content of F^- and the alteration to the reaction environment of Y^{3+} ions. In an LSS system involving oleic acid and a high active F^- source such as NH_4F and NaF, it is found that the oleate anions are more likely to be combined with Y^{3+} in comparison with Na^+ ions [4]. The interaction between oleate anions and Y^{3+} could effectively lower the energy barrier of the $\alpha \rightarrow \beta$ phase transition [4]. Moreover, effective concentration of F^- ions was elevated, resulting from the rapid supply of F^-. All of these could effectively promote the $\alpha \rightarrow \beta$ phase transformation of $NaYF_4$ [4,52,53]. Ultimately, β-phase $NaYF_4$ was formed in an LSS system [9–12,21].

However, in the case of $BminPF_6$ as the F^- source in the presence of polyol, the interaction between PF_6^- ions and Y^{3+} was quite limited compared with the case of the above-mentioned LSS system [17,18]. In addition, noting the solubility product of $NaYF_4$, $a_{Na}a_Ya_F{}^4$, the supersaturation degree ($a_{Na}a_Ya_F{}^4/K_{SP}$) drastically varies with the content of fluoride ions, with an exponential relationship. Thus, the content of fluoride ions in a reactive system is of importance to the phase control of $NaYF_4$ [15,19,33]. Herein, $BmimPF_6$ slowly decomposes and hydrolyzes to create the required F^- during the elevation of the reaction temperature [23,45,47]. Therefore, $BmimPF_6$ is a low-active F^- source relative to NH_4F and NaF. Since the equilibrium constant of the hydrolyzed reaction (Equation (1)) is extremely small, the effective concentration of F^- ions was relatively low in the reaction system, which consequently results in a very slow precipitation with Na^+ and Y^{3+} [45] (Equation (2)).

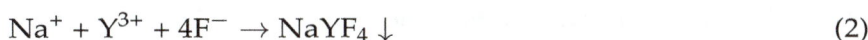

$$Na^+ + Y^{3+} + 4F^- \rightarrow NaYF_4 \downarrow \tag{2}$$

In such circumstances, the supersaturation degree of the reaction system is not adequate to form the nuclei of the hexagonal phase $NaYF_4$ [23]. Therefore, the formation of cubic phase $NaYF_4$ was favored [51].

As mentioned above, polyol can complex with Na^+ and Y^{3+}. In addition, according to Chaumont et al. [54], PF_6^- (IL) can coordinate with Y^{3+}. Consequently, when $BmimPF_6$ was uniformly dispersed in 1,2-ethanediol (or DEG) solution containing Na^+ and Y^{3+} ions, these metal ions were believed to be simultaneously bonded by PF_6^- anions as well as polyol [45]. In this case, Na^+ and Y^{3+} ions were in the same shell surrounded by the imidazolium cation of $BmimPF_6$ [23]. Upon thermal treatment, PF_6^- slowly hydrolyzed and released F^-, which was accompanied by forming $NaYF_4$ nanosized grains; this can be evidenced by the nanoparticles that were attached on the surface of the as-obtained mesocrystals (Figure 1e). Subsequently, the polyol and IL co-stabilized nanoparticles aggregated to form $NaYF_4$ mesocrystals, which possibly occurred through oriented attachment or mesoscale assembly processes due to the coexistence of a Coulombic force, van der Waals interaction, and hydrogen bonds in the system of polyol and $BminPF_6$ [55,56]. Finally, it should be pointed out that the above-proposed forming course is only one of several possible mechanisms. Further studies about this issue are underway, and will be reported in future work.

Under the excitation of a 980-nm laser, α-phase $NYF:Yb^{3+},Er^{3+}$ (20/2 mol%) mesocrystalline emitted bright yellow fluorescence, which demonstrated its photo functionality performance. The related luminescence spectrum is shown in Figure 4a. The green-emitting bands at about 521 nm and 540/552 nm are due to the $^2H_{11/2} \rightarrow {}^4I_{15/2}$ and $^4S_{3/2} \rightarrow {}^4I_{15/2}$ energy-level transitions of Er^{3+}, respectively, while the red band at around 651/669 nm is assigned to the $^4F_{9/2} \rightarrow {}^4I_{15/2}$ transition of Er^{3+}. The related Commission Internationale de l'Eclairage (CIE) coordinates are calculated as ($x = 0.3984$, $y = 0.5854$), which are situated in the region of yellowish light (point "×" in Figure 4b), revealing that it emitted yellowish light.

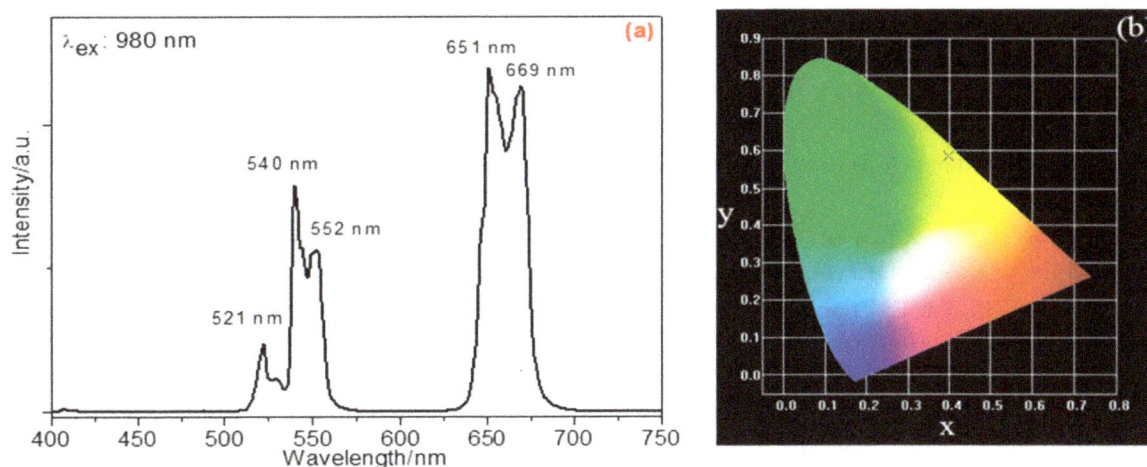

Figure 4. **(a)** Up-conversion luminescence spectrum at room temperature and **(b)** Commission Internationale de l'Eclairage (CIE) chromaticity diagram of $NYF:Yb^{3+},Er^{3+}$(20/2 mol%) sample (λ_{ex}: 980 nm).

4. Conclusions

In summary, cubic-phase well-defined $NaYF_4$ based photofunctional mesocrystallines were successfully prepared at relatively low temperature by using IL $BmimPF_6$ and viscous polyol as the fluorine source and reaction medium, respectively. Combining slow-releasing fluoride via the decomposition and hydrolysis of fluorine-containing IL and the assistance of polyol, the formation of cubic-phase NaYF$_4$ was favored, despite the higher treatment temperature or/and higher content of dopant. We believed that the key to the formation of uniform α-NaYF$_4$-based mesocrystals is the use of fluorine-containing IL as a fluorine source as well as the existence of a polyalcohol. Our contribution offers a new alternative in constructing mesocrystal and other hierarchical nanostructured materials with an object phase under mild conditions.

Author Contributions: Supervision, X.H.; data curation, X.H.; writing—original draft preparation, X.H.; writing—review and editing, Y.Z.; investigation, Y.F.; formal analysis, N.L.; methodology, Z.L.

References

1.   Wang, F.; Han, Y.; Lim, C.; Lu, Y.; Wang, J.; Xu, J.; Chen, H.; Zhang, C.; Hong, M.; Liu, X.G. Simultaneous phase and size control of upconversion nanocrystals through lanthanide doping. *Nature* **2010**, *463*, 1061–1065. [CrossRef] [PubMed]

2.   Farvid, S.S.; Radovanovic, P.V. Phase Transformation of Colloidal In$_2$O$_3$ Nanocrystals Driven by the Interface Nucleation Mechanism: A Kinetic Study. *J. Am. Chem. Soc.* **2012**, *134*, 7015–7024. [CrossRef] [PubMed]

3.   Song, S.; Kuang, Y.; Liu, J.; Yang, Q.; Luo, L.; Sun, X. Separation and phase transition investigation of Yb^{3+}/Er^{3+} co-doped NaYF$_4$ nanoparticles. *Dalton Trans.* **2013**, *42*, 13315–13318. [CrossRef] [PubMed]

4.   Sui, Y.Q.; Tao, K.; Tian, Q.; Sun, K. Interaction between Y^{3+} and Oleate Ions for the Cubic-to-Hexagonal Phase Transformation of NaYF$_4$ Nanocrystals. *J. Phys. Chem. C* **2012**, *116*, 1732–1739. [CrossRef]

5.   Zhang, F.; Li, G.; Zhang, W.; Yan, Y.L. Phase-Dependent Enhancement of the Green-Emitting Upconversion Fluorescence in LaVO$_4$:Yb^{3+},Er^{3+}. *Inorg. Chem.* **2015**, *54*, 7325–7334. [CrossRef] [PubMed]

6.   Zhua, Y.; Chen, D.; Huang, L.; Liu, Y.; Brikd, M.G.; Zhong, J.; Wang, J. Phase-transition-induced giant enhancement of red emission in Mn^{4+}-doped fluoride elpasolite phosphors. *J. Mater. Chem. C* **2018**, *6*, 3951–3960. [CrossRef]

7.   Thoma, R.E.; Hebert, G.M.; Insley, H.; Weaver, C.F. Phase Equilibria in the System Sodium Fluoride-Yttrium Fluoride. *Inorg. Chem.* **1963**, *2*, 1005–1012. [CrossRef]

8.   Arnold, A.A.; Terskikh, V.; Li, Q.Y.; Naccache, R.; Marcotte, I.; Capobianco, J. Structure of NaYF$_4$ Upconverting Nanoparticles: A Multinuclear Solid-State NMR and DFT Computational Study. *J. Phys. Chem. C* **2013**, *117*, 25733–25741. [CrossRef]

9.   Mai, H.X.; Zhang, Y.W.; Si, R.; Yan, Z.G.; Sun, L.D.; You, L.P.; Yan, C.H. High-Quality Sodium Rare-earth Fluoride Nanocrystals: Controlled Synthesis and Optical Properties. *J. Am. Chem. Soc.* **2006**, *128*, 6426–6436. [CrossRef]

10.   Wei, Y.; Lu, F.; Zhang, X.; Chen, D. Synthesis of Oil-Dispersible Hexagonal-Phase and Hexagonal-Shaped NaYF$_4$:Yb,Er Nanoplates. *Chem. Mater.* **2006**, *18*, 5733–5737. [CrossRef]

11.   Zhang, F.; Wan, Y.; Yu, T.; Zhang, F.; Shi, Y.; Xie, S.; Li, Y.; Xu, L.; Tu, B.; Zhao, D. Uniform Nanostructured Arrays of Sodium Rare-Earth Fluorides for Highly Efficient Multicolor Upconversion Luminescence. *Angew. Chem. Int. Ed.* **2007**, *46*, 7976–7979. [CrossRef] [PubMed]

12.   Wang, L.; Li, Y. Controlled Synthesis and Luminescence of Lanthanide Doped NaYF$_4$ Nanocrystals. *Chem. Mater.* **2007**, *19*, 727–734. [CrossRef]

13.   Yi, G.S.; Chow, G.M. Synthesis of Hexagonal-Phase NaYF$_4$:Yb,Er and NaYF$_4$:Yb,Tm Nanocrystals with Efficient Up-Conversion Fluorescence. *Adv. Funct. Mater.* **2006**, *16*, 2324–2329. [CrossRef]

14.   Shan, J.; Ju, Y. A Single-Step Synthesis and the Kinetic Mechanism for Monodisperse and Hexagonal-Phase NaYF$_4$:Yb,Er Upconversion Nanophosphors. *Nanotechnology* **2009**, *20*, 275603. [CrossRef] [PubMed]

15.   Li, C.; Zhang, C.; Hou, Z.; Wang, L.; Quan, Z.; Lian, H.; Lin, J. β-NaYF$_4$ and β-NaYF$_4$:Eu^{3+} Microstructures: Morphology Control and Tunable Luminescence Properties. *J. Phys. Chem. C* **2009**, *113*, 2332–2339. [CrossRef]

16.   Wang, Z.L.; Hao, J.; Chan, H.L.; Wong, W.T.; Wong, K.L. A strategy for simultaneously realizing the cubic-to-hexagonal phase transition and controlling the small size of NaYF$_4$:Yb^{3+},Er^{3+} nanocrystals for in vitro cell imaging. *Small* **2012**, *8*, 1863–1868. [CrossRef] [PubMed]

17. He, M.; Huang, P.; Zhang, C.; Chen, F.; Wang, C.; Ma, J.; He, R.; Cui, D. A general strategy for the synthesis of upconversion rare earth fluoride nanocrystals via a novel OA/ionic liquid two-phase system. *Chem. Commun.* **2011**, *47*, 9510–9512. [CrossRef]

18. He, M.; Huang, P.; Zhang, C.; Ma, J.; He, R.; Cui, D. Phase- and Size-Controllable Synthesis of Hexagonal Upconversion Rare-Earth Fluoride Nanocrystals through an Oleic Acid/Ionic Liquid Two-Phase System. *Chem. Eur. J.* **2012**, *18*, 5954–5969. [CrossRef]

19. Zhang, Q.; Yan, B. Phase control of upconversion nanocrystals and new rare earth fluorides though a diffusion-controlled strategy in a hydrothermal system. *Chem. Commun.* **2011**, *47*, 5867–5869. [CrossRef]

20. Chen, D.; Huang, P.; Yu, Y.; Huang, F.; Yang, A.; Wang, Y. Dopant-induced phase transition: A new strategy of synthesizing hexagonal upconversion NaYF$_4$ at low temperature. *Chem. Commun.* **2011**, *47*, 5801–5803. [CrossRef]

21. Wang, X.; Zhuang, J.; Peng, Q.; Li, Y. A general strategy for nanocrystal synthesis. *Nature* **2005**, *437*, 121–124. [CrossRef] [PubMed]

22. Wei, Y.; Lu, F.; Zhang, X.; Chen, D. Polyol-mediated synthesis and luminescence of lanthanide-doped NaYF$_4$ nanocrystal upconversion phosphors. *J. Alloys Compd.* **2008**, *455*, 376–384. [CrossRef]

23. Zhang, C.; Chen, J. Facile EG/ionic liquid interfacial synthesis of uniform RE^{3+} doped NaYF$_4$ nanocubes. *Chem. Commun.* **2010**, *46*, 592–594. [CrossRef] [PubMed]

24. Chen, C.; Sun, L.D.; Li, Z.X.; Li, L.L.; Zhang, J.; Zhang, Y.W.; Yan, C.H. Ionic Liquid-Based Route to Spherical NaYF$_4$ Nanoclusters with the Assistance of Microwave Radiation and Their Multicolor Upconversion Luminescence. *Langmuir* **2010**, *26*, 8797–8803. [CrossRef] [PubMed]

25. Li, H.; Wang, L. Controllable Multicolor Upconversion Luminescence by Tuning the NaF Dosage. *Chem. Asian J.* **2014**, *9*, 153–157. [CrossRef] [PubMed]

26. Han, Y.; Gai, S.; Ma, P.; Wang, L.; Zhang, M.; Huang, S.; Yang, P. Highly Uniform α-NaYF$_4$:Yb/Er Hollow Microspheres and Their Application as Drug Carrier. *Inorg. Chem.* **2013**, *52*, 9184–9191. [CrossRef] [PubMed]

27. Lv, C.; Di, W.; Liu, Z.; Zheng, K.; Qin, W. Synthesis of NaLuF$_4$-based nanocrystals and large enhancement of upconversion luminescence of NaLuF$_4$:Gd, Yb, Er by coating an active shell for bioimaging. *Dalton Trans.* **2014**, *43*, 14001–14008.

28. Tian, G.; Gu, Z.; Zhou, L.; Yin, W.; Liu, X.; Yan, L.; Zhao, Y. Mn^{2+} Dopant-Controlled Synthesis of NaYF$_4$:Yb/Er Upconversion Nanoparticles for in vivo Imaging and Drug Delivery. *Adv. Mater.* **2012**, *24*, 1226–1231. [CrossRef]

29. Heer, S.; Kömpe, K.; Güdel, H.U.; Haase, M. Highly Efficient Multicolour Upconversion Emission in Transparent Colloids of Lanthanide-Doped NaYF$_4$ Nanocrystals. *Adv. Mater.* **2004**, *16*, 2102–2105. [CrossRef]

30. Gao, L.; Ge, X.; Chai, Z.; Xu, G.; Wang, X.; Wang, C. Shape-controlled synthesis of octahedral α-NaYF$_4$ and its rare earth doped submicrometer particles in acetic acid. *Nano Res.* **2010**, *2*, 565–574. [CrossRef]

31. Liang, X.; Wang, X.; Zhuang, J.; Peng, Q.; Li, Y. Synthesis of NaYF$_4$ Nanocrystals with Predictable Phase and Shape. *Adv. Funct. Mater.* **2007**, *17*, 2757–2765. [CrossRef]

32. Zhang, F.; Li, J.; Shan, J.; Xu, L.; Zhao, D. Shape, Size, and Phase-Controlled Rare-Earth Fluoride Nanocrystals with Optical Up-Conversion Properties. *Chem. Eur. J.* **2009**, *15*, 11010–11019. [CrossRef] [PubMed]

33. Qin, R.; Song, H.; Pan, G.; Zhao, H.; Ren, X.; Liu, L.; Bai, X.; Dai, Q.; Qu, X. Polyol-mediated synthesis of well-dispersed α-NaYF$_4$ nanocubes. *J. Cryst. Growth* **2009**, *311*, 1559–1564. [CrossRef]

34. He, L.; Zou, X.; He, X.; Lei, F.; Jiang, N.; Zheng, Q.; Xu, C.; Liu, Y.; Lin, D. Reducing Grain Size and Enhancing Luminescence of NaYF$_4$:Yb^{3+}, Er^{3+} Upconversion Materials. *Cryst. Growth Des.* **2018**, *18*, 808–817. [CrossRef]

35. Cölfen, H.; Antonietti, M. Mesocrystals: Inorganic superstructures made by highly parallel crystallization and controlled alignment. *Angew. Chem. Int. Ed.* **2005**, *44*, 5576–5591. [CrossRef] [PubMed]

36. Niederberger, M.; Cölfen, H. Oriented attachment and mesocrystals: Non-classical crystallization mechanisms based on nanoparticle assembly. *Phys. Chem. Chem. Phys.* **2006**, *8*, 3271–3287. [CrossRef] [PubMed]

37. Bergström, L.; Sturm (née Rosseeva), E.V.; German, S.; Cölfen, H. Mesocrystals in Biominerals and Colloidal Arrays. *Acc. Chem. Res.* **2015**, *48*, 1391–1402. [CrossRef]

38. Zhou, L.; O'Brien, P. Mesocrystals-Properties and Applications. *J. Phys. Chem. Lett.* **2012**, *3*, 620–628. [CrossRef]

39. He, X.; Yan, B. Yttrium hydroxide fluoride based monodisperse mesocrystals: Additive-free synthesis, enhanced fluorescent properties, and potential application in temperature sensing. *CrystEngComm* **2015**, *17*, 621–627. [CrossRef]

40. He, X.; Yan, B. Double role of the hydroxy group for water dispersibility and luminescence of REF$_3$ (RE = Yb, Er, Tm) based mesocrystals. *New J. Chem.* **2015**, *39*, 6730–6733. [CrossRef]

41. Zhuang, J.; Yang, X.; Fu, J.; Liang, C.; Wu, M.; Wang, J.; Su, Q. Monodispersed β-NaYF$_4$ Mesocrystals: In Situ Ion Exchange and Multicolor Up-and Down-Conversions. *Cryst. Growth Des.* **2013**, *13*, 2292–2297. [CrossRef]

42. Zhong, S.L.; Lu, Y.; Gao, M.R.; Liu, S.J.; Peng, J.; Zhang, L.C.; Yu, S.H. Monodisperse Mesocrystals of YF$_3$ and Ce^{3+}/Ln^{3+} (Ln=Tb, Eu) Co-Activated YF$_3$: Shape Control Synthesis, Luminescent Properties, and Biocompatibility. *Chem. Eur. J.* **2012**, *18*, 5222–5231. [CrossRef] [PubMed]

43. Lausser, C.; Kumke, M.U.; Antonietti, M.; Cölfen, H. Fabrication of EuF$_3$-Mesocrystals in a Gel Matrix. *Z. Anorg. Allg. Chem.* **2010**, *636*, 1925–1930. [CrossRef]

44. Wang, J.; Liu, B.Q.; Huang, G.; Zhang, Z.J.; Zhao, J.T. Monodisperse Na$_x$Y(OH)$_y$F$_{3+x-y}$ mesocrystals with tunable morphology and chemical composition: pH-mediated ion-exchange. *Cryst. Growth Des.* **2017**, *17*, 711–718. [CrossRef]

45. Zhang, C.; Chen, J.; Zhou, Y.; Li, D. Ionic liquid-based "all-in-one" synthesis and photoluminescence properties of lanthanide fluorides. *J. Phys. Chem. C* **2008**, *112*, 10083–10088. [CrossRef]

46. Zhao, Q.; You, H.; Lü, W.; Guo, N.; Jia, Y.; Lv, W.; Jiao, M. Dendritic Y$_4$O(OH)$_9$NO$_3$:Eu^{3+}/Y$_2$O$_3$:Eu^{3+} hierarchical structures: Controlled synthesis, growth mechanism, and luminescence properties. *CrystEngComm* **2013**, *15*, 4844–4851. [CrossRef]

47. Swatloski, R.P.; Holbrey, J.D.; Rogers, R.D. Ionic liquids are not always green: Hydrolysis of 1-butyl-3-methylimidazolium hexafluorophosphate. *Green Chem.* **2003**, *5*, 361–363. [CrossRef]

48. Xie, N.; Luan, W. Ionic-liquid-induced microfluidic reaction for water-soluble Ce$_{1-x}$Tb$_x$F$_3$ nanocrystal synthesis. *Nanotechnology* **2011**, *22*, 265609. [CrossRef]

49. Lin, J.; Wang, Q. Systematic studies for the novel synthesis of nano-structured lanthanide fluorides. *Chem. Eng. J.* **2014**, *250*, 190–197. [CrossRef]

50. Klier, D.T.; Kumke, M.U. Upconversion Luminescence Properties of NaYF$_4$:Yb:Er Nanoparticles Codoped with Gd^{3+}. *J. Phys. Chem. C* **2015**, *119*, 3363–3373. [CrossRef]

51. He, M.; Huang, P.; Zhang, C.; Hu, H.; Bao, C.; Gao, G.; He, R.; Cui, D. Dual Phase-Controlled Synthesis of Uniform Lanthanide-Doped NaGdF$_4$ Upconversion Nanocrystals Via an OA/Ionic Liquid Two-Phase System for In Vivo Dual-Modality Imaging. *Adv. Funct. Mater.* **2011**, *21*, 4470–4477. [CrossRef]

52. Ghosh, P.; Patra, A. Tuning of crystal phase and luminescence properties of Eu^{3+} doped sodium yttrium fluoride nanocrystals. *J. Phys. Chem. C* **2008**, *112*, 3223–3231. [CrossRef]

53. Xue, X.; Duan, Z.; Suzuki, T.; Tiwari, R.N.; Yoshimura, M.; Ohishi, Y. Luminescence Properties of α-NaYF$_4$:Nd^{3+} Nanocrystals Dispersed in Liquid: Local Field Effect Investigation. *J. Phys. Chem. C* **2012**, *116*, 22545–22551. [CrossRef]

54. Chaumont, A.; Wipff, G. Solvation of M^{3+} lanthanide cations in room-temperature ionic liquids. A molecular dynamics investigation. *Phys. Chem. Chem. Phys.* **2003**, *5*, 3481–3488. [CrossRef]

55. Yasui, K.; Kato, K. Oriented Attachment of Cubic or Spherical BaTiO$_3$ Nanocrystals by van der Waals Torque. *J. Phys. Chem. C* **2015**, *119*, 24597–24605. [CrossRef]

56. Ye, J.; Liu, W.; Cai, J.; Chen, S.; Zhao, X.; Zhou, H.; Qi, L. Nanoporous anatase TiO$_2$ mesocrystals: Additive-free synthesis, remarkable crystalline-phase stability, and improved lithium insertion behavior. *J. Am. Chem. Soc.* **2011**, *133*, 933–940. [CrossRef] [PubMed]

Eco-Friendly Method for Tailoring Biocompatible and Antimicrobial Surfaces of Poly-L-Lactic Acid

Magdalena Aflori [1,*], **Maria Butnaru** [1], **Bianca-Maria Tihauan** [2] **and Florica Doroftei** [1]

[1] Petru Poni Institute of Macromolecular Chemistry, 41A Grigore Ghica Voda Alley, Iasi 700487, Romania; mariabutnaru@yahoo.com (M.B.); florica.doroftei@icmpp.ro (F.D.)

[2] Sanimed International IMPEX SRL, Sos. Bucuresti—Magurele, nr. 70F, Sector 5, Bucharest 051434, Romania; bianca.tihauan@sanimed.ro

* Correspondence: maflori@icmpp.ro

Abstract: In this study, a facile, eco-friendly route, in two steps, for obtaining of poly-L-lactic acid/chitosan-silver nanoparticles scaffolds under quiescent conditions was presented. The method consists of plasma treatment and then wet chemical treatment of poly-L-lactic acid (PLLA) films in a chitosan based-silver nanoparticles solution (Cs/AgNp). The changes of the physical and chemical surface proprieties were studied using scanning electron microscopy (SEM), small angle X-Ray scattering (SAXS), Fourier transform infrared spectroscopy (FTIR) and profilometry methods. A certain combination of plasma treatment and chitosan-based silver nanoparticles solution increased the biocompatibility of PLLA films in combination with cell line seeding as well as the antimicrobial activity for gram-positive and gram-negative bacteria. The sample that demonstrated from Energy Dispersive Spectroscopy (EDAX) to have the highest amount of nitrogen and the smallest amount of Ag, proved to have the highest value for cell viability, demonstrating better biocompatibility and very good antimicrobial proprieties.

Keywords: chitosan; poly-L-lactic acid; plasma; silver nanoparticles; antimicrobial

1. Introduction

Nowadays, there is a great demand for obtaining new environmentally-friendly and cost-effective materials to replace plastic products [1,2]. One of the most promising polymers is poly-L-lactic acid (PLLA); however, it has limited practical applications due to its low thermal stability and inherent brittle nature. Reinforcements with different substances prove to be a powerful tool in designing clean, eco-friendly materials for several applications [2]. PLLA is widely used in tissue engineering due to its slow degradation rate (almost 6 months to 1 year for complete degradation) [3,4] and cost-effective and effortless large scale production [5]. In this context, a fairly new area is represented by nanocomposites—the reinforcing material having the dimensions in nanometric scale [6]. Noble metal nanoparticles are of increased interest because of their potential applications in novel technologies due to their different properties compared to bulk metals [7,8]. Recently, silver and gold nanoparticles biosynthesis under eco-friendly conditions by using plant extracts, bio-organisms, proteins, and polysaccharides have gained an increased interest from material science researchers [9,10]. In this context, chitosan is one of the most used biopolymers for such approach mainly due to unique physicochemical properties in the presence of largely free amino and hydroxyl groups [11]. Chitosan is also recommended by proprieties like biocompatibility, non-toxicity, bioadhesivity, biodegradability, safety, and the promotion of drug absorption [12]. PLLA has been blended with chitosan to improve its wettability [13] or its tensile strength [14]. Some authors used chitosan as both capping and reducing agent for the incorporation of silver nanoparticles into the polymer matrices [15–17]. Due to the ion–dipole intermolecular forces, chitosan stops the aggregation of silver in clusters at the macroscopic level and has a crucial role in stabilization of the formed nanoparticles [18].

Surface treatment procedures, such as plasma discharges in different kind of gases, can modify the physico-chemical proprieties at a scale of only a few atoms layers thick, without changing the bulk material properties [19,20]. Within the last decade, plasma technique has been applied to improve PLLA surface hydrophilicity, roughness and morphology and to proliferate the selective interaction between polymer surface and proteins. Some authors observed that melt extruded PLLA sheets treated with different kind of inert gas discharges did not affect PLLA biodegradation rate in soil [21]. Other authors used plasma treatment in reactive gases for enhancing cell (human skin fibroblast) adhesion on PLLA by obtaining reactive amine groups and further to immobilize collagen through polar and hydrogen bonding interactions at the treated film surfaces [22,23]. By increasing plasma treatment time, an increase of PLLA film degradation takes place [20]. It is well known that some issues related to non-permanent surface modification can occur, making polymers films unsuitable for certain applications in medicine or package food industry [24].

In this study, a facile, eco-friendly route, in two steps, for obtaining of PLLA/Chitosan-silver nanoparticles nanocomposite scaffolds under quiescent conditions was presented. This scaffold proved to be suitable for cell (preosteoblastic cell line MC3T3-E1 established from mouse C57BL/6 calvaria) attachment and proliferation and to have very good biocide proprieties.

2. Material and Methods

2.1. Materials

Chitosan flakes (molecular weight 50,000–190,000 Da based on viscosity, 75–85% deacetylated), silver nitrate and acetic acid were purchased from Sigma-Aldrich Chemie GmbH, Steinheim, Germany. Solutions were prepared with MilliQ water. Poly-L-lactic acid films of 50 μm thick were purchased from Goodfellow Cambridge Ltd, Huntingdon, UK and were cut in square shapes of 5×5 cm size.

Culture media and solutions: Alpha Minimum Essential Medium (α MEM, with ribonucleosides, deoxyribonucleosides, 2 mM L-glutamine and 1 mM sodium pyruvate, without ascorbic acid GIBCO, Custom Product, Catalog No. A1049001); Bovine fetal serum (BFS); Penicillin/Streptomycin/Neomycin solution (P/S/N) for cell culture; Phosphate Buffered saline (PBS) for cell culture; 3-(4,5-Dimethyl-2-thiazolyl)-2,5-diphenyl-2H-tetrazolium bromide (MTT), solution in PBS (5 mg/mL) were purchased from Sigma-Aldrich Chemie GmbH, Steinheim, Germany.

Cells: Preosteoblasts of MC3T3-E1 line, subclone 4 (passage 21) (purchased from Sigma-Aldrich Chemie GmbH, Steinheim, Germany) were thawed and multiplied in culture flasks with a surface of 75 cm^2, in culture media α MEM, without ascorbic acid, supplemented with 10% BFS and 1% mixture of antibiotic. The initial density of cells culture was 2000 cells/cm^2 culture surface.

2.2. The Two Step Method

2.2.1. In Situ Formation Silver Nanoparticles on Chitosan

Chitosan in a concentration of 6.87 mg/mL was dissolved in 1% glacial acetic acid solution. To avoid nanoparticle sedimentation due to the poor solubility of chitosan, the mixture was kept two days until a clear solution was obtained (Figure 1 Step 1) [25,26]. The solution was prepared by adding 5 mL of 9 mM AgNO$_3$ to 10 mL chitosan solution under stirring for 30 min at room temperature. The mixture was transferred to glass tubes and it was kept at 90 °C for 6 h in a temperature controlled bath. After synthesis, the AgNPs colloids were cooled at room temperature for the removal of the majority of the unreduced Ag$^+$ to avoid the toxicity of the solution [27,28]. Then, the CS/AgNP solution was stored at room temperature and in dark glass tubes.

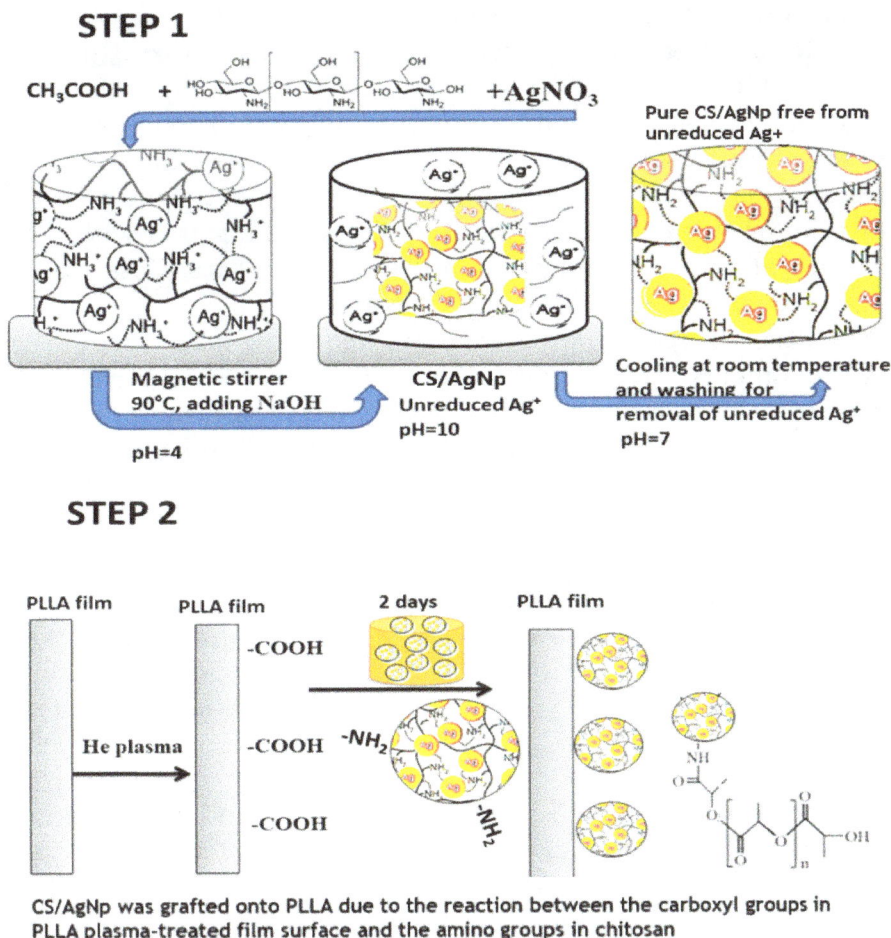

Figure 1. Scheme of two steps experimental method.

2.2.2. Combined Nonconventional and Conventional Treatments of PLLA Films

Briefly, the plasma treatments of PLLA films (Figure 1, Step 2) were performed in an EMITECH RF plasma device. Different times of treatments were performed to obtain a certain concentration of functional groups at the surface layer, without affecting the polymer bulk. The polymer was introduced in a gas vessel containing helium (He) at a pressure $p = 5 \times 10^{-2}$ mbar. Right after plasma treatments, to avoid the aging effect, the polymer was immediately immersed in the CS/AgNp solution at room temperature, in a dark place. After 2 days, the samples were rinsed with MilliQ water. Different input RF powers and times were performed for plasma treatments; however, for the present work, the following samples were selected: neat PLLA (P0) film, PLLA films treated in plasma for 4 min at 30 W, immersed in CS/AgNp (P1) and PLLA films treated in plasma for 10 min at 30 W, immersed in CS/AgNp (P2). For a lower time, no significant changes in the surfaces were observed, while for higher times the surface of PLLA film starts the degradation process, by becoming opaque and brittle.

2.3. Characterization Methods

A LUMOS Microscope Fourier Transform Infrared (FTIR) spectrophotometer (Bruker Optik GmbH, Ettlingen, Germany), equipped with an Attenuated Total Reflection (ATR) device. ATR-FTIR was used to acquire spectra in the range 600–4000 cm^{-1}.

The scanning electron micrographs of PLLA samples were registered with a Quanta 200 microscope at an accelerating voltage of 15 kV and with an Energy Dispersive Spectroscopy (EDAX) system of elemental analysis (FEI Company, Brno, Czech Republic).

Transmission electron microscopy (TEM) images of CS/Ag Np solution were obtained on a HT1700 (Hitachi High-Technologies Corporation - Hitachi High-Tech, Tokyo, Japan) microscope using an acceleration voltage of 120 kV.

A Nanostar U-system (Bruker AXS GmbH, Karlsruhe, Germany) equipped with a Vantec detector and an X-ray micro source was used to perform small-angle X-ray scattering measurements (SAXS). The sample-to-detector distance was 107 cm and the wavelength of the incident X-ray beam was $\lambda =$ 1.54 Å (Cu Kα). The solutions were loaded in capillary tubes and together with the film samples were measured under vacuum at a constant temperature, 25 °C for 10,000 s. Data analysis was performed by the model fitting approach using the DIFFRAC^{plus} NanoFit.

Particle sizes from Cs/AgNp solution were measured using a Zetasizer instrument (Zetasizer Nano ZS, Malvern Instruments Ltd, Malvern, Worcestershire, UK) using UV (ultra violet) Grade cuvette after treatment in an ultrasonic water bath (Model FB11012, Fisherbrand, Loughborough, UK) for 30 min to break up any aggregates present. All measurements were performed in triplicate.

A sample of the CS/AgNP solution was diluted with distilled water (1:10 sample:water) and analyzed by spectrophotometry Ultraviolet–visible (UV-vis) produced by Barloworld Scientific Ltd, Dunmow, Essex, UK.

Films roughness were determined using a Tencor Alpha-Step D-500 stylus profiler (KLA Tencor Corporation, Milpitas, CA, USA) both before and following the immersion in the Cs/AgNP for the samples at 1000 μm scan length, and 100 μm/s scan speed. The arithmetic average of the absolute values of the profile heights over the evaluation length Ra was measured by applying a stylus force of 2.3 mg, and a long-range cutoff filter of 25 μm.

Static contact angle measurements using the sessile-drop method were performed on a CAM-101 (KSV Instruments Ltd., Helsinki, Finland) system equipped with a video camera, liquid dispenser and drop-shape analysis software (KSV CAM Optical Contact Angle and Pendant Drop Surface Tension Software, version 3.99, KSV Instruments Ltd., Helsinki, Finland). Liquid drops (double distilled water or ethylene glycol) of ~1 μL were placed at room temperature, with a Hamilton syringe, on the polymer surface. For each drop 10 photos were recorded at an interval of 0.016 s. To obtain a statistical result, three different surface regions were selected for each liquid.

2.4. Culture Media Preparation and MTT Test

The samples were cut in fragments of 5 × 5 mm size and were decontaminated by immersion in a sterile solution of 70% ethyl alcohol for 20 min. Then the samples were rinsed three times in sterile PBS and were pre balanced in complete culture media for 24 h at 37 °C.

The MTT test was performed by a direct contact method, by using as test product the samples prepared as described above in 24-well culture plates populated with preosteoblasts MC3T3-E1 line, subclone 4. The initial cell population density was 1×10^4 cells/well, in 0.5 mL α MEM. The sample contact with the cells was made after 48 h after culture initiation, to obtain a cell monolayer semi-confluent. One piece of material with the size of 5 × 5 mm was then placed in each well over the cell culture for 72 h, at 37 °C, humidity 95% and 5% CO_2. Each sample was tested in triplicate material and the results were compared with the ones obtained for control samples, without testing material. The MTT test with 3-(4,5-dimethylthiazol-2-yl)-2,5-diphenyltetrazoliumbromid was carried out according to techniques from the literature [27,28]. The principle of the method is based on the reduction of yellow MTT compound, in a violet-colored product (formazan), as a result of mitochondrial dehydrogenase activity of viable cells. To achieve the MTT assay, the culture medium and the fragment of the material from each well was removed, the cells MTT solution was added in α MEM without BFS. After 3 h of incubation solution, the formazan absorbance was measured using a spectrophotometer plate reader (Tecan) at a wavelength of 570 nm. By reporting formazan absorbance from the wells with experimental samples to the control ones, the percent cell viability corresponding to the incubated culture with one of the tested materials was calculated.

2.5. Antimicrobial Activity

The polymers were cut into squares (1 cm side), plated in 24-well plates, sterilized with 70% ethanol (1) and washed with PBS (3 times). The test polymers were then incubated with 1 mL of bacterial suspension (*S. aureus strains*, *P. aeruginosa*, 0.5 McFarland turbidity). The plates were placed in the incubator at 37 °C for 2 h. Thereafter, the polymers were removed from the 24-well plates using a sterile forceps and were washed three times with PBS to remove the non-adherent bacteria. The films were then placed in tubes with 1 mL of PBS and vortexed for 120 s to remove all solutions from the adhering bacteria. Then, the solution was serially diluted in PBS, cultured on nutrient agar and the number of colonies forming units per ml (UFC/mL) was calculated.

3. Results and Discussions

In this work, chitosan was used as a mild reducing agent in silver nanoparticle synthesis. Protonized chitosan with NH_3^+ functional groups was obtained by reaction of chitosan with H^+ from the acetic acid solution. At the same time, the positions of Ag^+ were fixed by coordination to the functional groups of chitosan, which simultaneously act as a stabilizing agent. The biopolymer behaves as a template or matrix which prevents the nanoparticles agglomeration.

The effect of surface modification experiments can be permanent (in the case of covalent attachment of functional groups) or non-permanent (non-covalent attachment). In the case of PLLA plasma treatment, the advantage is due to the improvement of surface wettability and cell affinity, while the disadvantage is due to surface rearrangement to minimize the interfacial energy, which affects the effectiveness of the surface modification, making the effect of plasma treatment non-permanent [29,30]. Another disadvantage of plasma treatment is degradation of PLLA in certain conditions. The two-step method described in our paper overcomes those disadvantages by immersing the PLLA plasma treated film in the chitosan-based AgNp solution immediately after the treatment, without any delay. In this way, the surface rearrangement does not have time to take place and the newly-introduced functional groups are efficient in permanent attachment of AgNP to the treated polymer surface. On the other hand, the input plasma parameters (the treatment time) were tailored to obtain no degradation of the PLLA film, the samples obtained in optimal conditions being selected to be presented in this paper. At lower values of the plasma treatment, no important changes in the film surface were noticed, while for higher values of the treatment, the PLLA film becomes brittle and opaque, the bulk proprieties being seriously affected.

The synthesis of AgNp in CS was demonstrated by UV–vis spectroscopy. The concentration of the chitosan-capped silver nanoparticles was approximately 0.11 ng/mL. The presence of surface plasmon resonance in the UV–vis characteristic optical spectrum indicates the presence of silver nanoparticles of certain particle size [31]. The UV-vis results (Figure 2) show a typical silver absorption peak at 415 nm which is in the reported range of silver and silver oxide nanoparticles [32–34]. The symmetry of the nanoparticles can be determined by the number of surface plasmon resonance peaks. If only one peak is observed in the UV-vis spectra, spherical silver nanoparticles were synthesized. The absorption peak is relatively narrow; thus, this method revealed a small size distribution of the particles in solution and the absence of unreduced positively charged ions, which is consistent with the data observed in the SEM and TEM images.

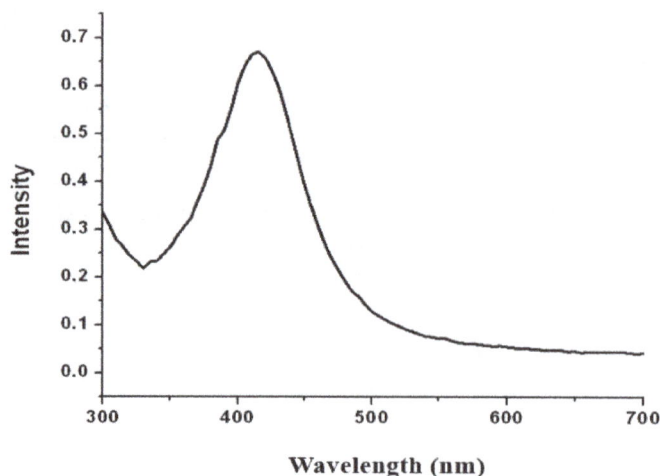

Figure 2. Ultraviolet–visible (UV-vis) spectrum of CS/AgNp solution.

3.1. TEM Images and Particle Size Results of CS/AgNp

The representative TEM images of AgNPs show that the particles were randomly distributed in the solution and due to the presence of the polymer, no agglomeration was detected (Figure 3a). The particles in Cs/Ag Np solution have a spherical shape (Figure 3b) and an average particle size of about 30 nm (Figure 3c) was obtained, in good concordance with the particle distribution obtained by dynamic light scattering method (Figure 3a). Other authors obtained similar results [35]. The contrast of TEM micrographs is correlated with the nature and size of the particles. The main discussion is between organic and inorganic particles or different organic-inorganic composites. Soft materials are predominantly composed of low number atoms, such as C, O and N. These elements, compared to heavy metals, exhibit a low level of electron-optical contrast. Thus, TEM micrographs of CS/AgNPs show AgNP formation—which appears as dark areas due to the high electron density of Ag. The contrast difference for CS/AgNP is because by drying the nanoparticle suspension, chitosan remains at the surface of the particles, and thus gives rise to areas of low contrast compared to AgNPs.

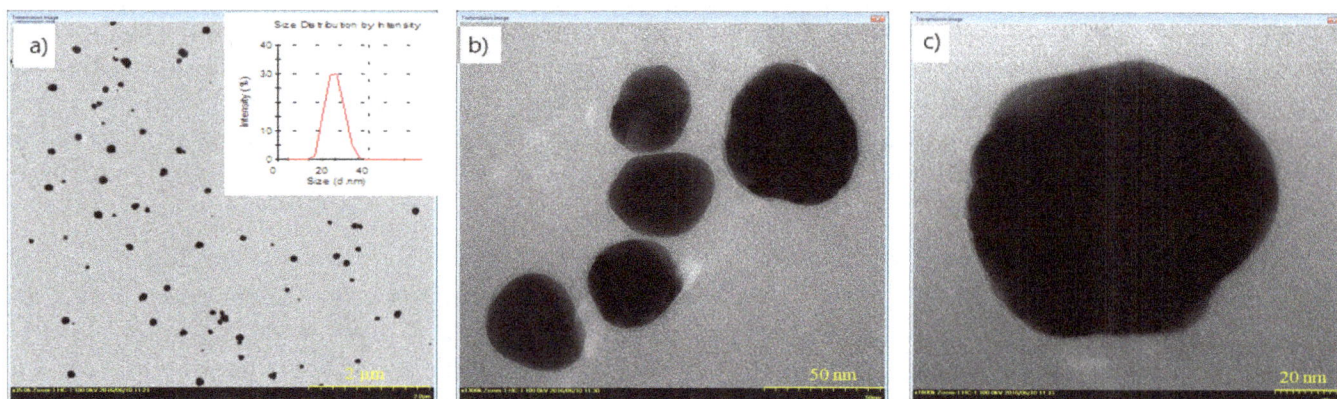

Figure 3. Transmission electron microscopy (TEM) images of CS/AgNp solution and Ag particle size distribution for scale: (**a**) 2 μm; (**b**) 50 nm; (**c**) 20 nm.

3.2. SAXS Results

The confirmation of particle size as obtained from TEM images was further authenticated by the SAXS analysis performed for all studied samples and the results are illustrated in Figure 4. The SAXS plots on a double logarithmic scale of the pristine chitosan and CS/AgNp solution presented in Figure 4a) demonstrate no important modification in chitosan morphology after the reduction of silver. The change in the slope of the CS/AgNP SAXS pattern is due to the presence of the silver nanoparticles in the system, and by applying the spherical model using DIFFRAC^{plus} NanoFit, a value of 35 nm was obtained, in good concordance with TEM results.

The structural changes in the PLLA samples after applying the two-step treatment is demonstrated by the Kratky Plot obtained from SAXS measurements (Figure 4b). It is well known that plasma treatments are accompanied by the heating of the polymer due to the interaction with high-energy plasma particles with the material surface. On the other hand, the crystal modification of PLLA is easily obtained from the melt [36]; therefore, the increase in the intensity and the shift at smaller angles of the PLLA peak in the Kratky plots is a clear indication that the process of PLLA crystallization takes place (Figure 4b). The appearance of the second SAXS peak as the treatment time in plasma increased suggests the formation of regular aligned lamellar structures in the polymer matrix due to the plasma treatment observed by other authors [37]. The average long period of a lamella L can be estimated from the maximum of the peak (q_{max}) in the Kratky plots (Figure 4b) according to Bragg's law ($L = 2\pi/q_{max}$). The thickness increased with the increase of plasma treatment time and lamellar structures grew in size from 0.25 nm for P1 to 0.27 nm to P2.

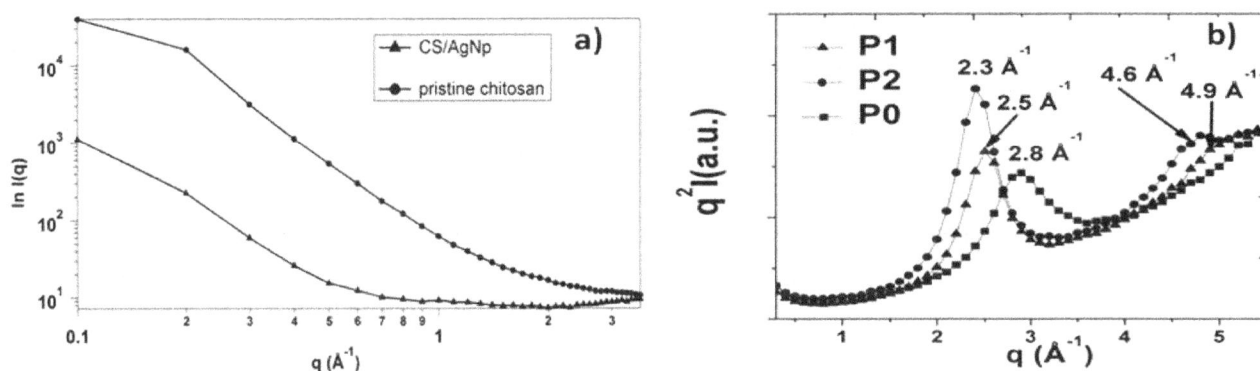

Figure 4. SAXS patterns: (a) double-logarithmic plot for pristine chitosan and CS/AgNp solution; (b) Kratky plot for pristine and treated PLLA samples.

3.3. FTIR Results

FTIR measurements were carried out to elucidate the interactions that take place in the reduction process of silver nitrate in the presence of chitosan.

In the pristine chitosan spectrum from Figure 5a, the broad absorption peak in the 2250–3800 cm^{-1} region is attributed to symmetric and asymmetric vibrations of CH$_2$ (2250–3050 cm^{-1}), and vibrations of O–H, N–H and intermolecular hydrogen bonds of polysaccharides (3050–3800 cm^{-1}). The peaks at 1588 and 1642 cm^{-1} were assigned to amino (–NH$_2$), amide I (C=O) and C=O of O–C–O–R groups in the chitosan structure, respectively. The peaks at 1060, 1075 and 1176 cm^{-1} are the characteristic absorptions due to C–O vibrations in the C–O–C band [38]. The formation of silver/chitosan nanoparticles was confirmed by Fourier transform infrared spectroscopy. As shown in Figure 5b, the spectra of the CS/AgNP exhibited a few differences from the chitosan. The peak intensities in the range 1000 cm^{-1} and 1350 cm^{-1} due to C–N stretching and bending decreased because of the reduction of silver in chitosan. The absence of 1588 cm^{-1} peak that exists in pristine chitosan and the appearance of additional peaks at 1707 and 1744 cm^{-1} (Figure 5b), corresponding to carbonyl stretch vibrations in ketones, aldehydes and carboxylic, indicate that the silver is bound to the functional groups of chitosan. The formation of chitosan- silver nanoparticles was achieved after the reduction of silver ion through the amino group. The presence of these functional groups on the surface of the synthesized silver nanoparticles and the disappearance of the NH$_2$ double spike peak indicates that the polymer successfully capped the nanoparticles and the polymer network restricts the diffusion of Ag$^+$. Moreover, the reduction of the silver ions is coupled to the oxidation of the hydroxyl groups in chitosan molecular and/or its hydrolyzates acids [39]. The band at 3434 cm^{-1} assigned to the overlap between the O–H stretching vibration and the N–H stretching vibration of the biopolymer moieties, shifted to 3421 cm^{-1} due to co-ordination bond between the silver and electron rich groups [40,41]. After silver binding during reduction of silver nitrate with chitosan, the molecule weight was heavier and the vibration intensity of the N-H bond decreased, suggesting the attachment of silver to nitrogen atoms from chitosan [42].

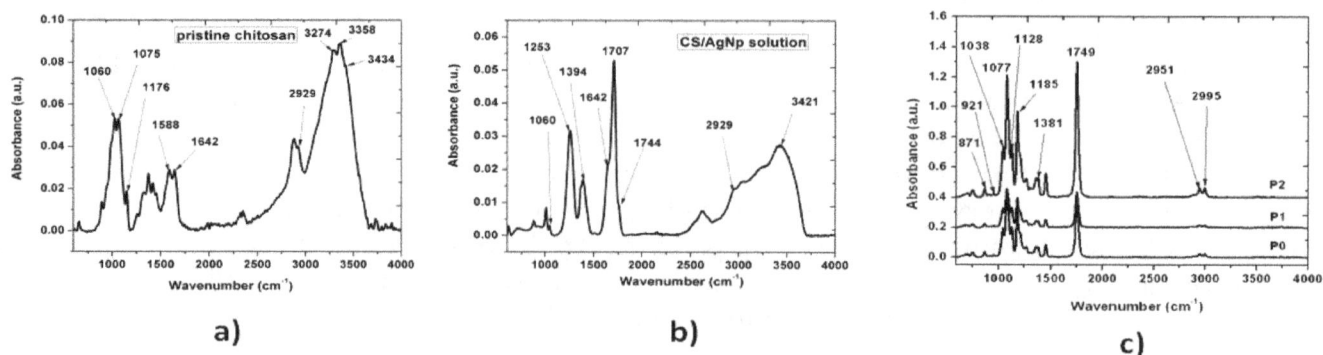

Figure 5. Fourier Transform Infrared (FTIR) results for (**a**) pristine chitosan; (**b**) CS/AgNp solution; (**c**) pristine and treated PLLA samples.

Figure 5c shows the IR spectra of the PLLA samples. The 1185 and 1077 cm^{-1} bands were assigned to C–O–C asymmetric and symmetric stretching, while the peak at 1749 cm^{-1} was attributed to the stretching of C=O. The C–CH$_3$ stretching caused the peak at 1038 cm^{-1} and the C–H (of CH$_3$ groups) rocking mode was present in the spectrum at 1128 cm^{-1}. An increase in the degree of PLLA crystallinity due to the heating of the polymer during the plasma treatments can be observed, which is in good concordance with SAXS measurements. There was an increase of the 1749 cm^{-1} and of 1381 cm^{-1} band intensities. A shift with 3 cm^{-1} to lower wavenumbers of those bands can be observed. Those bands were assigned to the carboxylic groups and demonstrated an increase of those functional groups quantity after plasma treatments and the presence of silver ions at the polymer surface [43,44]. There was an increase of the 1749 cm^{-1} and of 1381 cm^{-1} band intensities assigned to the carboxylic groups, which demonstrate an increase of those functional groups quantity after plasma treatments and the presence of silver ions at the polymer surface. A shift with 3 cm^{-1} to lower wavenumbers of those bands can be observed, which revealed the reaction between the carboxyl groups in PLLA and the amino groups in CS; PLLA was grafted onto the backbone of CS.

3.4. Surface Roughness and Wettability

The polymer surface was affected by the plasma treatment due to the breaking of chemical bonds, heating, degradation, etc. [26]. Because all of these processes may significantly change the structure and morphology of the polymer film, the surface roughness and contact angle measurements were performed.

Hydrophobicity and hydrophilicity proprieties of polymer surface are key factors in further cell adhesion. Surface functional groups and the surface roughness of the material are very important in determining surface wettability.

To determine the wettability of neat and treated PLLA surfaces, water contact angle measurement (two liquids method) was used to provide the information on the wetting properties (Table 1). The values of the static contact angle (θ_W for water and θ_{EG} for ethylene glycol) can be used to estimate the wettability and surface tension of a solid surface. Based on these measurements, some parameters such as surface free energy (γ_{SV}), solid–liquid interfacial tension (γ_{SL}), or work of adhesion (W) were calculated using Owens–Wendt–Rabel and Kaelbe methods [45–47] and the results are listed in Table 1. The polar and the dispersive components of surface free energy were also listed in Table 1, to evaluate the surface modifications after the two step treatment.

Table 1. Surface roughness and wettability results.

	Roughness	Contact Angle Measurements Parameters							
	Ra (nm)	θ_w	θ_{EG}	W_w	W_{EG}	γ_{SV}^P	γ_{SV}^d	γ_{SV}	γ_{SL}
P0	43.103	79.9	52.19	86.19	77.42	8.05	23.93	31.98	18.58
P1	86.448	51.54	34	118.07	87.79	38.93	9.61	48.55	3.27
P2	63.831	70.11	42.22	97.56	83.54	13.96	22.39	36.36	11.60

The roughness parameters measured with profiler revealed that the increase of treatment time causes an increase of the surface roughness parameters (Table 1).

Both treatments caused a significant increase in the surface roughness and a decrease in the static water contact angle compared to the untreated polymer. The lowest water contact angles and the highest value for surface roughness were obtained for P1 treatment (Figure 6).

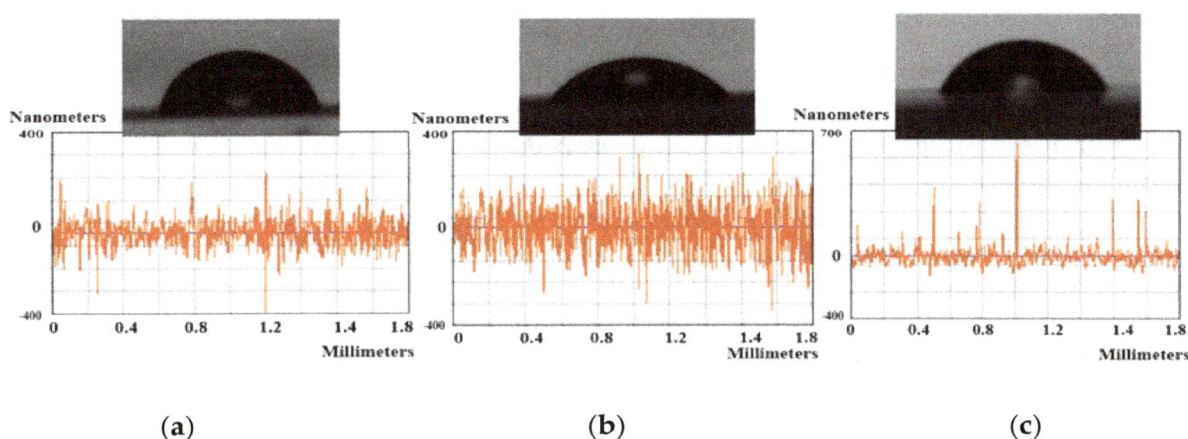

Figure 6. Surface modification results: top figures—contact angle images; bottom figures—profiler roughness measurements of samples: (a) P0; (b) P1; (c) P2.

The values listed in Table 1 revealed that both plasma treatments significantly increased the surface energy, mainly due to the increase in its polar component. Furthermore, the lowest value of the water contact angle was achieved at the largest values of the polar component of the surface energy.

3.5. SEM and EDAX Results

The morphological changes in the PLLA surface after the two-step treatments are presented in Figure 7. From Figure 7a, pristine PLLA film has a smooth surface without any irregularities. The PLLA films surfaces, after the combined plasma-wet chemical treatment, have patterns of different size and shape due to the surface interactions with different reactive species formed in plasma and due to the presence of AgNp and chitosan (Figure 7b,c, respectively). EDAX measurements (Table 2) show the presence of the AgNp on the surfaces of the P1 and P2 films after the treatments. The presence of nitrogen in the treated samples supports the presence of chitosan at the polymer surface.

It can also be observed that the highest amount of nitrogen was obtained in the P1 sample, while the high amount of Ag was obtained in the P2 sample. The carbon concentration was slightly increased with the increase of the plasma treatment time as a result of the polymer chain destruction. The low concentration of Ag nanoparticles assures low toxicity at the surface of the polymer and avoids the agglomeration of nanoparticles. Moreover, the presence of N at the treated samples surfaces demonstrates the presence of chitosan.

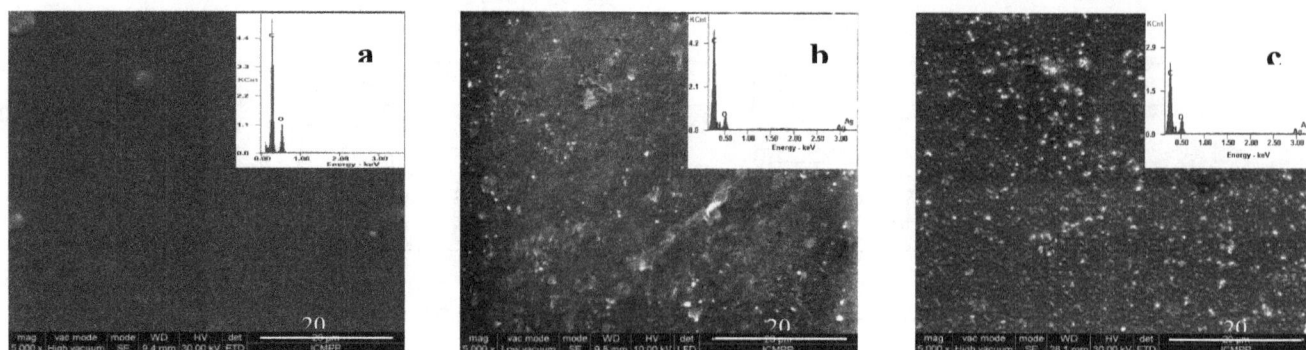

Figure 7. SEM images and Energy Dispersive Spectroscopy (EDAX) spectra of: (**a**) P0; (**b**) P1; (**c**) P2.

P1 and P2 samples were treated in plasma at different conditions and then introduced in the same solution of CS/AgNp. After plasma treatment, functional groups of different concentrations (depending on the input plasma parameters) were present at the PLLA surface. Those groups are responsible for the presence of chitosan and silver at the PLLA treated surfaces, as presented in FTIR section. From Table 2, the modified N/C ratio on the surface was 0.19 for P1 and 0.17 for P2, while Ag/C ratio was 0.006 for P1 and 0.011 for P2. A possible mechanism responsible for the higher amount of Ag and the smaller amount of N in P2 sample compared to P1 is: unreduced silver from CS/AgNp solution can easily direct bond to the PLLA surface containing functional groups, without the aid of chitosan. On the other hand, if the larger chitosan molecule does not find sufficient functional groups with which to form bonds at the PLLA surface, it is removed by washing after treatment.

Table 2. EDAX results of PLLA studied samples.

Element (%)	CK	OK	NK	AgL
P0	70.49	29.51	-	-
P1	71.73	14.36	13.46	00.45
P2	72.43	14.61	12.14	00.82

3.6. Proliferation and Morphology of on the MC3T3-E1 Cells on the PLLA Samples

Clinical biomaterial applications require good biocompatibility of the material. The biocompatibility of neat PLLA and two step treated PLLA films can be primarily evaluated by utilizing MC3T3-E1 cell lines, to be used for applications such as bone tissue engineering.

Fluorescent staining was used to study cell density and morphology after culturing MC3T3-E1 cells on P0, P1 and P2 films for 48 h and 72 h, as shown in Figure 8. For plasma—treated samples P2 and P0, spherical and round cells can be observed. The growth of cells on the surface of P1 films was better and more rapid than that on the P0 and P2 samples (Figure 8h) after culturing for 72 h. Moreover, the cells incubated on P1 show a higher degree of fibroblast cell adhesion and proliferation and a well-preserved morphology, which was flat and fully spread.

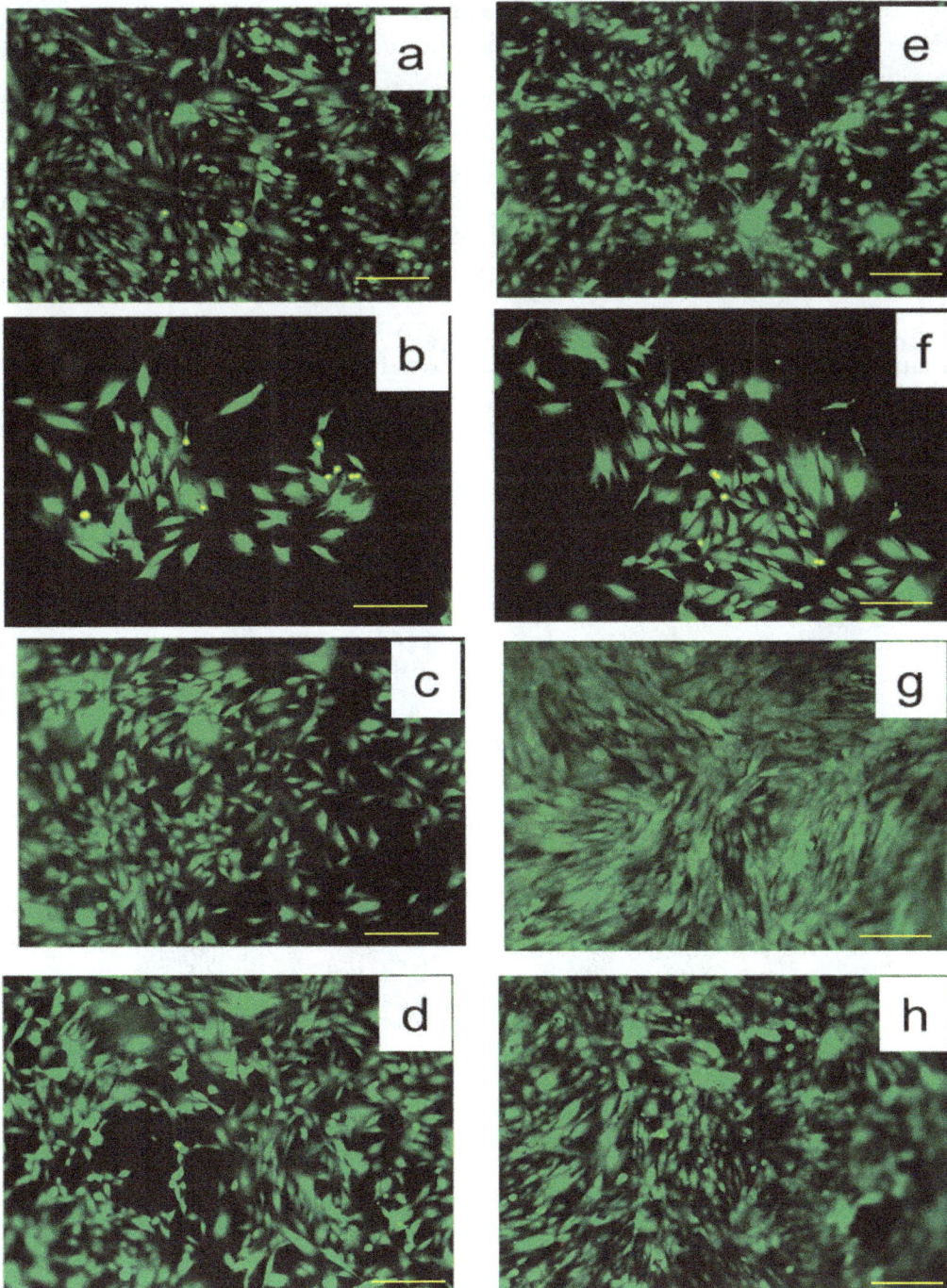

Figure 8. Cell culture (cell line MC3T3-E1) after 48 h for (**a–d**) and after 72 h (**e–h**); control (**a,e**), P0 (**b,f**), P1 (**c,g**), P2 (**d,h**). Scale: 200 μm.

As a measure of unsaturated bond energy resulting from dangling bonds of surface material [42,43] surface energy is an important chemical cue on polymer surfaces. A polymer in contact with biological fluids has a surface energy which influences cell activities, such as serum protein adsorption and cell attachment. It was found that more fibroblasts can adhere and spread widely on the more hydrophilic polymer surface. The improvement or the suppression of cell adhesion at the polymer surface is in good concordance with high or respectively low surface free energy values [44,45]. In the case of equal surface free energy, a higher value for the polar component will induce a higher degree of cell adhesion and proliferation on the surface. The surface energy of the

PLLA surface was tailored by using plasma treatment and the surface free energy and the polar component of the P1 film was higher (Table 1) than that of the P0 and P2 films.

Two different cell morphologies were found on the studied polymer surfaces (Figure 8): (1) elongated cells well spread into polymer surface and (2) rounded cells, which are attached but have not begun to spread. Both types of morphologies are seen in different proportions on studied samples. The majority (>80%) of cells grown on PLLA films are elongated. On each of the P0, P1 and P2 substrates, up to 20% of cells were rounded. The highest density of elongated cells was found at P1 treatment after 72 h.

Figure 9 shows the MTT assay results of the MC3T3-E1 cell lines on P0, P1 and P2 films on 48 h and 72 h. The number of cells in each group increased with culture time on all of the tested groups. The MC3T3-E1 cells cultured on all samples have similar proliferation on the first day compared with the control. From the first day, the viable cell numbers on P1 were higher than those on P0 and P2.

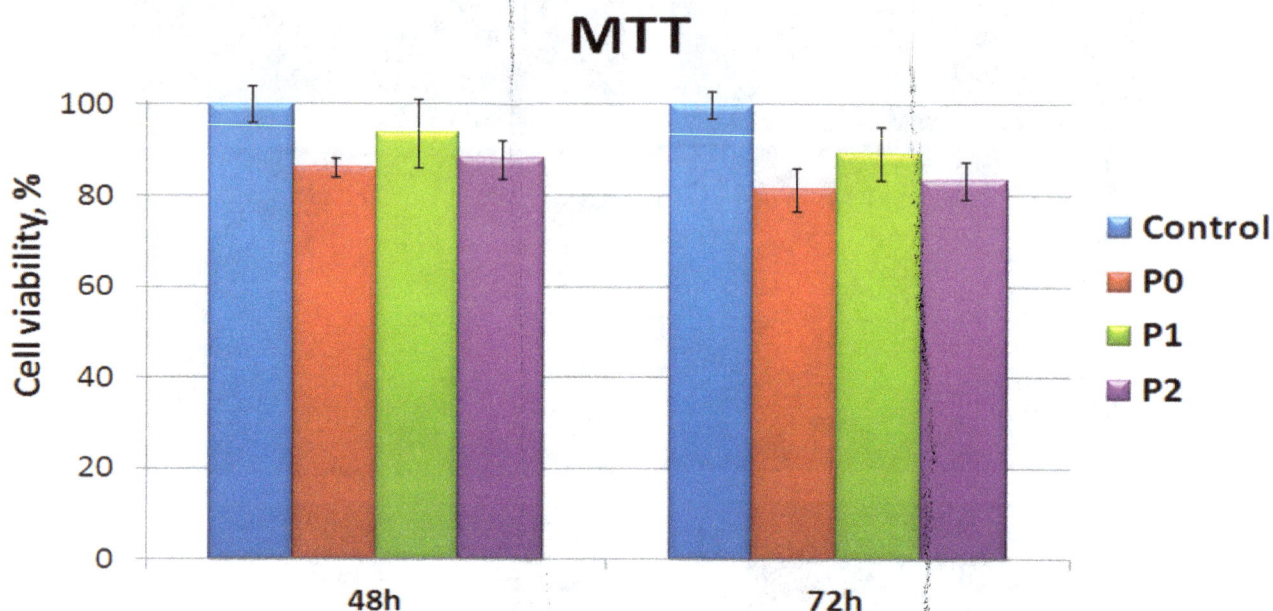

Figure 9. MTT assay results of MC3T3-E1 cells on the P0, P1 and P2.

The PLLA/Chitosan based-silver nanocomposite scaffolds appeared to be in vitro biocompatible and noncytotoxic to cells (Figure 9). The higher density of cells on P1 samples can be attributed to a higher concentration of N and a lower Ag concentration at film surfaces (Table 2 EDAX measurements). The shape, dimensions and low concentration of the silver ions prove to be nontoxic for the MC3T3-E1 cells culture. Moreover, the highest concentration of N and the highest polar component obtained for P1 samples assures better biocompatibility of P1 samples even compared to the neat PLLA.

3.7. Antimicrobial Activity

Strains adherence of *S. aureus* and *P. aeruginosa* (ATCC, clinical isolates) to polylactic acid films were studied. Figure 10 demonstrates the very good antimicrobial proprieties of P1 and P2 samples compared to the untreated P0 sample. The antimicrobial behavior of P1 sample is more pronounced in the case of *P. aeruginosa* compared to *S. aureus* and can be explained considering that both chitosan and AgNp have a bactericidal effect.

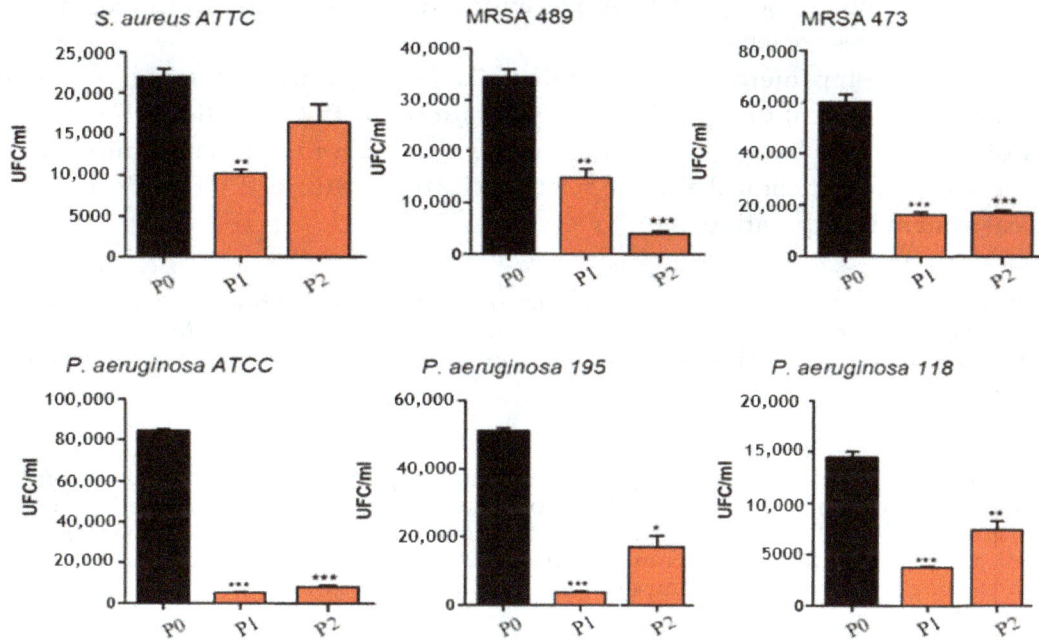

Figure 10. Strains adherence of *S. aureus* and *P. aeruginosa* (ATCC, clinical isolates) to polylactic acid films; *, ** and *** correspond to the magnitude of antimicrobial activity.

One of the proposed mechanisms of Ag action was based on AgNp capability to easily enter into the bacterial cell and form a less dense region in the center of the bacteria, causing the cell death by interacting with thiol containing enzymes [46,47]. Another proposed mechanism involves disruption of DNA/RNA caused by Ag reaction with the weak acid groups in the genetic material, such as phosphate [48].

On the other hand, chitosan has antimicrobial behavior and the main mechanism is based on the electrostatic interaction between positively charged chitosan groups and negatively charged sites on microbial cell [49]. In concordance with other authors [50], Figure 10 demonstrated that chitosan has stronger influence on Gram negative than on Gram-positive strains because the cell wall of *P. aeruginosa* (Gram-negative) has a thickness of 7–8 nm while the wall of *S. aureus* (Gram-positive) is around 20–80 nm [51].

4. Conclusions

In this paper, an environmentally-friendly synthesis of metallic nanoparticles in the presence of chitosan was performed. Silver ions underwent coordination and reduction thanks to the presence of numerous amino and hydroxyl groups in the chitosan chains. Bounding of silver nanoparticles of 30 nm average diameter to the polymer functional groups ensured a long-term stability and prevented their agglomeration. FTIR data pointed out the possible interactions of the hydroxyl or amino groups of chitosan and the carboxyl groups of PLLA. The silver nanoparticles were successfully adsorbed on PLLA films exposed to plasma treatments, by simply immersing the treated films in the chitosan solution containing silver nanoparticles. In this way, chitosan was used to fix silver nanoparticles on PLLA films surfaces. This is a time saving, inexpensive and eco-friendly synthesis that minimizes the use of toxic chemicals and does not produce toxic waste.

The biopolymer-based nanocomposite scaffolds with bioactive inorganic phases are of high interest due to their biocompatibility in combination with preosteoblastic cell line MC3T3-E1 (established from mouse C57BL/6 calvaria) seeding. The sample, which has demonstrated from EDAX to have the highest amount of nitrogen and the smallest amount of Ag, proved to have the highest value for cell viability. Moreover, it demonstrated better biocompatibility and very good antimicrobial activity against gram-negative and gram-positive bacteria. The effective component for the biocompatibility seemed to be both PLLA and chitosan, while for the antimicrobial property both chitosan and silver were responsible.

The described two-step method is a promising technology for obtaining: poly(L-lactic acid) for tissue engineering applications like bone regeneration. In this direction, the new approach of biopolymer-polysaccharides based composite enables the scaffold surface to mimic complex local biological functions. To target clinical and medical applications, the need for additional investigations in the biological system is imperative.

Author Contributions: Conceptualization, M.A.; methodology, M.A., M.B. and F.D.; investigation, M.A., M.B., B.-M.T. and F.D.; project administration, M.A.; resources, M.A. and M.B.; validation, M.A., M.B., B.-M.T. and F.D.; writing—original draft preparation, M.A., M.B.; writing—review and editing, M.A. and F.D.; supervision, M.A.; funding acquisition, M.A.

Acknowledgments: The authors acknowledge the financial support of this research through the Project "Partnerships for knowledge transfer in the field of polymer materials used in biomedical engineering" ID P_40_443, Contract no. 86/8.09.2016, SMIS 105689, co-financed by the European Regional Development Fund by the Competitiveness Operational Programme 2014–2020, Axis 1 Research, Technological Development and Innovation in support of economic competitiveness and business development, Action 1.2.3 Knowledge Transfer Partnerships.

References

1. Nofar, M.; Sacligil, D.; Carreau, P.J.; Kamal, M.R.; Heuzey, M.C. Poly (lactic acid) blends: Processing, properties and applications. *Int. J. Biol. Macromol.* **2019**, *125*, 307–360. [CrossRef]

2. Duarte, A.R.C.; Mano, J.F.; Reis, R.L. Novel 3D scaffolds of chitosan–PLLA blends for tissue engineering applications: Preparation and characterization. *J. Supercrit. Fluid* **2010**, *54*, 282–289. [CrossRef]

3. Zhou, Q.; Xie, J.; Bao, M.; Yuan, H.; Ye, Z.; Lou, X.; Zhang, Y. Engineering aligned electrospun PLLA microfibers with nano-porous surface nanotopography for modulating the responses of vascular smooth muscle cells. *J. Mater. Chem. B* **2015**, *3*, 4439–4450. [CrossRef]

4. Van Dijk, M.; Tunc, D.C.; Smit, T.H.; Higham, P.; Burger, E.H.; Wuisman, P.I. In vitro and in vivo degradation of bio absorbable PLLA spinal fusion cages. *J. Biomed. Mater. Res.* **2002**, *63*, 752–759. [CrossRef] [PubMed]

5. Pavia, F.C.; La Carrubba, V.; Mannella, G.A.; Ghersi, G.; Brucato, V. Poly (lactic acid) based scaffolds for vascular tissue engineering. *Chem. Eng. Transact.* **2012**, *27*, 1–6.

6. Evanoff, D.D.; Chumanov, G. Synthesis and optical properties of silver nanoparticles and arrays. *Chem. Phys. Chem.* **2005**, *6*, 1221–1231. [CrossRef] [PubMed]

7. He, J.H.; Kunitake, T.; Nakao, A. Facile in situ synthesis of noble metal nanoparticles in porous cellulose fibers. *Chem. Mater.* **2003**, *15*, 4401–4406. [CrossRef]

8. Wiley, B.; Sun, Y.G.; Mayers, B.; Xia, Y.N. Shape-controlled synthesis of metal nanostructures: The case of silver. *Chem. A Eur. J.* **2005**, *11*, 454–463. [CrossRef]

9. Ocwieja, M.; Barbasz, A.; Walas, S.; Romand, M.; Paluszkiewicz, C. Physicochemical properties and cytotoxicity of cysteine-functionalized silver nanoparticles. *Colloid Surf. B* **2017**, *160*, 429–437. [CrossRef]

10. Chowdhury, N.R.; Cowin, A.J.; Zilm, P.; Vasilev, K. "Chocolate" gold nanoparticles—One pot synthesis and biocompatibility. *Nanomaterials* **2018**, *8*, 496–506. [CrossRef]

11. Khalil, A.; Chaturbhuj, H.P.S.; Saurabha, K.; Adnan, A.S.; Nurul Fazita, M.R.; Syakir, M.I.; Davoudpour, Y.; Rafatullah, M.; Abdullah, C.K.; Haafiz, M.K.M.; et al. Biological synthesis of triangular gold nanoprisms. *Carbohyd. Polym.* **2016**, *150*, 216–226.

12. Kim, M.C.; Masuoka, T. Degradation properties of PLA and PHBV films treated with CO_2-plasma. *React. Funct. Polym.* **2009**, *69*, 287–292. [CrossRef]

13. Wan, Y.; Wu, H.; Yu, A.; Wen, D. Biodegradable polylactide/chitosan blend membranes. *Biomacromolecules* **2006**, *7*, 1362–1372. [CrossRef] [PubMed]

14. Suyatma, N.E.; Copinet, A.; Tighzert, L.; Coma, V. Mechanical and barrierproperties of biodegradable films made from chitosan andpoly(lactic acid) blends. *J. Polym. Environ.* **2004**, *12*, 1–6. [CrossRef]

15. Slepička, P.; Slepičková Kasálková, N.; Pinkner, A.; Sajdl, P.; Kolská, Z.; Švorčík, V. Plasma induced cytocompatibility of stabilized poly-L-lactic acid doped with graphene nanoplatelets. *React. Funct. Polym.* **2018**, *131*, 266–275. [CrossRef]

16. Murugadoss, A.; Chattopadhyay, A. A 'green' chitosan-silver nanoparticle composite as a heterogeneous as well as microheterogeneous catalyst. *Nanotechnology* **2008**, *19*, 1–9. [CrossRef]

17. Sanpui, P.; Murugadaoss, A.; Prasad, P.V.D.; Ghosh, S.S.; Chattopadhyay, A. The antibacterial properties of a novel chitosan–Ag-nanoparticle composite. *Int. J. Food Microbiol.* **2008**, *124*, 142–146. [CrossRef]

18. Shameli, K.; Ahmad, M.B.; Yunus, W.M.Z.W.; Rustaiyan, A.; Ibrahim, N.A.; Zargar, M.; Abdollahi, Y. Green synthesis of silver/montmorillonite/chitosan bionanocomposites using the UV irradiation method and evaluation of antibacterial activity. *Int. J. Nanomed.* **2010**, *5*, 875–887. [CrossRef]

19. Wan, Y.; Tu, C.; Yang, J.; Bei, J.; Wang, S. Influences of ammonia plasma treatment on modifying depth and degradation of poly(l-lactide) scaffolds. *Biomaterials* **2006**, *27*, 2699–2704. [CrossRef]

20. Surdu-Bob, C.C.; Sullivan, J.L.; Saied, S.O.; Layberry, R.; Aflori, M. Surface compositional changes in GaAs subjected to argon plasma treatment. *Appl. Surf. Sci.* **2002**, *202*, 183–198. [CrossRef]

21. Hirotsu, T.; Nakayama, K.; Tsujisaka, T.; Mas, A.; Schue, F. Plasma surface treatments of melt-extruded sheets of poly(l-lactic acid). *Polym. Eng. Sci.* **2002**, *42*, 299–306. [CrossRef]

22. Yang, J.; Shi, G.X.; Wang, S.G.; Bei, J.Z.; Cao, Y.L.; Shang, Q.; Yang, G.; Wang, W. Fabrication and surface modification of macroporous poly(l-lactic acid) and poly(l-lactic-co-glycolic acid) (70/30) cell scaffolds for human skin fibroblast cell culture. *J. Biomed. Mater, Res.* **2002**, *62*, 438–446. [CrossRef]

23. Yang, J.; Bei, J.Z.; Wang, S.G. Enhanced cell affinity of poly(d,l-lactide)by combining plasma treatment with collagen anchorage. *Biomaterials* **2002**, *23*, 2607–2614. [CrossRef]

24. Rasal, R.M.; Janorkar, A.V.; Hirt, D.E. Poly(lactic acid) modifications. *Prog. Polym. Sci.* **2010**, *35*, 338–356. [CrossRef]

25. Aflori, M.; Miron, C.; Dobromir, M.; Drobota, M. Bactericidal effect on Foley catheters obtained by plasma and silver nitrate treatments. *High Perform. Polym.* **2015**, *27*, 655–660. [CrossRef]

26. Aflori, M. Chitosan-based silver nanoparticles incorporated at the surface of plasma-treated PHB. *Chem. Lett.* **2017**, *46*, 65–67. [CrossRef]

27. Beer, C.; Foldbjerg, R.; Hayashi, Y.; Sutherland, D.S.; Autrup, H. Toxicity of silver nanoparticles—Nanoparticle or silver ion? *Toxicol. Lett.* **2012**, *208*, 286–292. [CrossRef]

28. Berridge, M.V.; Herst, P.M.; Tan, A.S. Tetrazolium dyes as tools in cell biology: New insights into their cellular reduction. *Biotechn. Annu. Rev.* **2005**, *11*, 127–152.

29. Berridge, M.V.; Tan, A.S. Subcellular localization, substrate dependence, and involvement of mitochondrial electron transport in MTT reduction. *Arch. Biochem. Biophys.* **1993**, *303*, 474–482. [CrossRef]

30. Ximing, X.; Gengenbach, T.R.; Griesser, H.J. Changes in wettability with time of plasma modified perfluorinated polymers. *J. Adhes. Sci. Tech.* **1992**, *6*, 1411–1431. [CrossRef]

31. Puiso, J.; Adliene, D.; Guobiene, A.; Prosycevas, I.; Plaipaite-Nalivaiko, R. Modification of Ag–PVP nanocomposites by gamma irradiation. *Mater. Sci. Eng. B* **2011**, *176*, 1562–1567. [CrossRef]

32. Pal, S.; Tak, Y.K.; Song, J.M. Does the antibacterial activity of silver nanoparticles depend on the shape of the nanoparticle? A study of the Gram-negative bacterium *Escherichia coli. Appl. Environ. Microbiol.* **2012**, *73*, 1712–1720. [CrossRef] [PubMed]

33. Wei, D.; Sun, W.; Qian, W.; Ye, Y.; Ma, X. The synthesis of chitosan-based silver nanoparticles and their antibacterial activity. *Carbohyd. Res.* **2009**, *344*, 2375–2382. [CrossRef] [PubMed]

34. Darroudi, M.; Khorsand Zak, A.; Muhamad, M.R.; Huang, N.M.; Hakimi, M. Green synthesis of colloidal silver nanoparticles by sonochemical method. *Mater. Lett.* **2012**, *66*, 117–120. [CrossRef]

35. Occhiello, E.; Morra, M.; Morini, G.; Garbassi, F.; Humphrey, P. Oxygen plasma-treated polypropylene interfaces with air, water, and epoxy resins. Part I. Air and water. *J. Appl. Polym. Sci.* **1991**, *42*, 551–559. [CrossRef]

36. Tverdokhlebov, S.I.; Bolbasov, E.N.; Shesterikov, E.V.; Antonova, L.V.; Golovkin, A.S.; Matveeva, V.G.; Petlin, D.G.; Anissimov, Y.G. Modification of polylactic acid surface using RF plasma discharge with sputter deposition of a hydroxyapatite target for increased biocompatibility. *Appl. Surf. Sci.* **2015**, *329*, 32–39. [CrossRef]

37. Krikorian, V.; Pochan, D.J. Crystallization behavior of poly(L-lactic acid) nanocomposites: Nucleation and growth probed by infrared spectroscopy. *Macromolecules* **2005**, *38*, 6520–6527. [CrossRef]

38. Tsai, C.C.; Wu, R.J.; Cheng, H.Y.; Li, S.C.; Siao, Y.Y.; Kong, D.C.; Jang, G.W. Crystallinity and dimensional stability of biaxial oriented poly(lactic acid) films. *Polym. Degrad. Stab.* **2010**, *95*, 1292–1298. [CrossRef]

39. Elzein, T.; Nasser-Eddine, M.; Delaite, C.; Bistac, S.; Dumas, P. FTIR study of polycaprolactone chain organization at interfaces. *J. Colloid Interf. Sci.* **2004**, *273*, 381–387. [CrossRef]

40. Chandran, S.P.; Chaudhary, M.; Pasricha, R.; Ahmad, A.; Sastry, M. Synthesis of gold nanotriangles and silver nanoparticles using Aloevera plant extract. *Biotech. Prog.* **2006**, *22*, 577–583. [CrossRef]

41. Jia, R.; Jiang, H.; Jin, M.; Wang, X.; Huang, J. Silver/chitosan-based Janus particles: Synthesis, characterization, and assessment of antimicrobial activity in vivo and vitro. *Food Res. Int.* **2015**, *78*, 433–441. [CrossRef]

42. Gartner, H.; Li, Y.; Almenar, E. Improved wettability and adhesion of polylactic acid/chitosan coating for bio-based multilayer film development. *Appl. Surf. Sci.* **2015**, *332*, 488–493. [CrossRef]

43. Jia, C.; Chen, P.; Wang, Q.; Li, B.; Chen, M. Surface wettability of atmospheric dielectric barrier discharge processed Armos fibers. *Appl. Surf. Sci.* **2011**, *258*, 388–393. [CrossRef]

44. Cory, A.H.; Owen, T.C.; Barltrop, J.A.; Cory, J.G. Use of an aqueous soluble tetrazolium/formazan assay for cell growth assays in culture. *Cancer Commun.* **1991**, *3*, 207–212. [CrossRef] [PubMed]

45. Erbil, H.Y. *Surface Chemistry of Solid and Liquid Interfaces*; Blackwell Publishing: Hoboken, NJ, USA, 2006.

46. Mittal, K.L. *Contact Angle, Wettability and Adhesion*; VSP: London, UK, 2002; Volume 2.

47. Stamm, M. Polymer surfaces and interfaces. In *Characterization, Modification and Applications*; Springer: Berlin, Germany, 2008.

48. Mamonova, I.A.; Babushkina, I.V.; Norkin, I.A.; Gladkova, E.V.; Matasov, M.D.; Puchinyan, D.M. Biological activity of metal nanoparticles and their oxides and their effect on bacterial cells. *Nanotechnol. Russ.* **2015**, *10*, 128–134. [CrossRef]

49. Rabea, E.I.; Badawy, M.E.-T.; Stevens, C.V.; Smagghe, G.; Steurbaut, W. Chitosan as antimicrobial agent: Applications and mode of action. *Biomacromolecules* **2003**, *4*, 1457–1465. [CrossRef]

50. No, H.K.; Park, N.Y.; Lee, S.H.; Meyers, S.P. Antibacterial activity of chitosans and chitosan oligomers with different molecular weights. *Int. J. Food Microbiol.* **2002**, *74*, 65–72. [CrossRef]

51. Eaton, P.; Fernandes, J.C.; Pereira, E.; Pintado, M.E.; Malcata, F.X. Atomic force microscopy study of the antibacterial effects of chitosans on *Escherichia coli* and *Staphylococcus aureus*. *Ultramicroscopy* **2008**, *108*, 1128–1134. [CrossRef]

Permissions

All chapters in this book were first published by MDPI; hereby published with permission under the Creative Commons Attribution License or equivalent. Every chapter published in this book has been scrutinized by our experts. Their significance has been extensively debated. The topics covered herein carry significant findings which will fuel the growth of the discipline. They may even be implemented as practical applications or may be referred to as a beginning point for another development.

The contributors of this book come from diverse backgrounds, making this book a truly international effort. This book will bring forth new frontiers with its revolutionizing research information and detailed analysis of the nascent developments around the world.

We would like to thank all the contributing authors for lending their expertise to make the book truly unique. They have played a crucial role in the development of this book. Without their invaluable contributions this book wouldn't have been possible. They have made vital efforts to compile up to date information on the varied aspects of this subject to make this book a valuable addition to the collection of many professionals and students.

This book was conceptualized with the vision of imparting up-to-date information and advanced data in this field. To ensure the same, a matchless editorial board was set up. Every individual on the board went through rigorous rounds of assessment to prove their worth. After which they invested a large part of their time researching and compiling the most relevant data for our readers.

The editorial board has been involved in producing this book since its inception. They have spent rigorous hours researching and exploring the diverse topics which have resulted in the successful publishing of this book. They have passed on their knowledge of decades through this book. To expedite this challenging task, the publisher supported the team at every step. A small team of assistant editors was also appointed to further simplify the editing procedure and attain best results for the readers.

Apart from the editorial board, the designing team has also invested a significant amount of their time in understanding the subject and creating the most relevant covers. They scrutinized every image to scout for the most suitable representation of the subject and create an appropriate cover for the book.

The publishing team has been an ardent support to the editorial, designing and production team. Their endless efforts to recruit the best for this project, has resulted in the accomplishment of this book. They are a veteran in the field of academics and their pool of knowledge is as vast as their experience in printing. Their expertise and guidance has proved useful at every step. Their uncompromising quality standards have made this book an exceptional effort. Their encouragement from time to time has been an inspiration for everyone.

The publisher and the editorial board hope that this book will prove to be a valuable piece of knowledge for researchers, students, practitioners and scholars across the globe.

List of Contributors

Obakeng P. Keabadile, Saheed E. Elugoke and Omolola E. Fayemi
Department of Chemistry, Faculty of Natural and Agricultural Sciences, North-West University (Mafikeng Campus), Private Bag X2046, Mmabatho 2735, South Africa
Material Science Innovation and Modelling (MaSIM) Research Focus Area, Faculty of Natural and Agricultural Sciences, North-West University (Mafikeng Campus), Mmabatho 2735, South Africa

Adeyemi O. Aremu
Indigenous Knowledge Systems Centre, Faculty of Natural and Agricultural Sciences, North-West University (Mafikeng Campus), Mmabatho 2735, South Africa

Harsh Kumar and Dinesh Kumar
School of Bioengineering & Food Technology, Shoolini University of Biotechnology and Management Sciences, Solan-173229, H. P., India

Kanchan Bhardwaj and Rachna Verma
School of Biological and Environmental Sciences, Shoolini University of Biotechnology and Management Sciences, Solan-173229, H. P., India

Kamil Kuča and Eugenie Nepovimova
Department of Chemistry, Faculty of Science, University of Hradec Kralove, Hradec Kralove 50003, Czech Republic

Anu Kalia
Electron Microscopy and Nanoscience Laboratory, Punjab Agricultural University, Ludhiana-141004, Punjab, India

Shakeel Ahmad Khan and Chun-Sing Lee
Center of Super-Diamond and Advanced Films (COSDAF) and Department of Chemistry, City University of Hong Kong, 83 Tat Chee Avenue, Kowloon 999077, Hong Kong

Sammia Shahid
Department of Chemistry, School of Science, University of Management and Technology, Lahore 54770, Pakistan

Anna Frank and Benjamin Breitbach
Max-Planck-Institut für Eisenforschung GmbH, Max-Planck-Straße 1, 40237 Düsseldorf, Germany

Jan Grunwald
Ludwig-Maximilians-Universität, Butenandtstraße 5-11, 81377 Munich, Germany

Christina Scheu
Max-Planck-Institut für Eisenforschung GmbH, Max-Planck-Straße 1, 40237 Düsseldorf, Germany
Materials Analytics, RWTH Aachen University, Kopernikusstraße 10, 52074 Aachen, Germany

Kothaplamoottil Sivan Saranya, Kunjumon Saranya and Bini George
Department of Chemistry, School of Physical Sciences, Central University of Kerala, Kerala 671316, India

Vinod Vellora Thekkae Padil, Stanisław Wacławek and Miroslav Černík
Institute for Nanomaterials, Advanced Technologies and Innovation (CXI), Technical University of Liberec (TUL), Studentská 1402/2, 46117 Liberec 1, Czech Republic

Chandra Senan
Centre forWater Soluble Polymers, Applied Science, Faculty of Arts, Science and Technology, Wrexham Glyndwr University, Wrexham LL11 2AW, Wales, UK

Rajendra Pilankatta
Department of Biochemistry and Molecular Biology, School of Biological Sciences, Central, University of Kerala, Kerala 671316, India

Renia Fotiadou, Michaela Patila and Haralambos Stamatis
Biotechnology Laboratory, Department of Biological Applications and Technologies, University of Ioannina, 45110 Ioannina, Greece

Mohamed Amen Hammami, Apostolos Enotiadis and Emmanuel P. Giannelis
Department of Materials Science and Engineering, Cornell University, Ithaca, NY 14853, USA

Dimitrios Moschovas, Kyriaki Tsirka, Konstantinos Spyrou, Apostolos Avgeropoulos, Alkiviadis Paipetis and Dimitrios Gournis
Department of Materials Science and Engineering, University of Ioannina, 45110 Ioannina, Greece

María Gabriela Villamizar-Sarmiento and Felipe A. Oyarzun-Ampuero
Advanced Center of Chronic Diseases (ACCDiS), Universidad de Chile, Santos Dumont 964, Independencia, Santiago 8380494, Chile
Departamento de Ciencias y Tecnología Farmacéuticas, Facultad de Ciencias Químicas y Farmacéuticas, Universidad de Chile, Santos Dumont 964, Independencia, Santiago 8380494, Chile

Victor Miranda
Departamento de Ciencias y Tecnología Farmacéuticas, Facultad de Ciencias Químicas y Farmacéuticas, Universidad de Chile, Santos Dumont 964, Independencia, Santiago 8380494, Chile

Ignacio Moreno-Villoslada and Sandra L. Orellana
Instituto de Ciencias Químicas, Facultad de Ciencias, Universidad Austral de Chile, Isla Teja, Casilla 567, Valdivia 5090000, Chile

Samuel Martínez and Lisette Leyton
Advanced Center of Chronic Diseases (ACCDiS), Universidad de Chile, Santos Dumont 964, Independencia, Santiago 8380494, Chile
Laboratory of Cellular Communication, Program of Cell and Molecular Biology, Institute of Biomedical Sciences (ICBM), Faculty of Medicine, University of Chile, Av. Independencia 1027, Santiago 8380453, Chile

Alejandra Vidal, Miguel Concha and Francisca Pavicic
Instituto de Anatomía, Histología y Patología, Facultad de Medicina, Universidad Austral de Chile, Valdivia 5090000, Chile

Annesi Giacaman
Instituto de Anatomía, Histología y Patología, Facultad de Medicina, Universidad Austral de Chile, Valdivia 5090000, Chile
Jeffrey Modell Center of Diagnosis and Research in Primary Immunodeficiencies. Faculty of Medicine University of La Frontera, Temuco 4780000, Chile

Judit G. Lisoni
NM MultiMat, Instituto de Ciencias Físicas y Matemáticas, Facultad de Ciencias, Universidad Austral de Chile, Valdivia 5090000, Chile

Neelika Roy Chowdhury
School of Engineering, University of South Australia, Mawson Lakes SA 5095, Australia

Krasimir Vasilev
School of Engineering, University of South Australia, Mawson Lakes SA 5095, Australia
Future Industries Institute, University of South Australia, Mawson Lakes SA 5095, Australia

Allison J. Cowin
Future Industries Institute, University of South Australia, Mawson Lakes SA 5095, Australia

Peter Zilm
Microbiology Laboratory, Adelaide Dental School, The University of Adelaide, Adelaide SA 5005, Australia

Rodrigo R. Retamal Marín, Frank Babick and Michael Stintz
Research Group Mechanical Process Engineering, Institute of Process Engineering and Environmental Technology, Technische Universität Dresden, Münchner Platz 3, D-01062 Dresden, Germany

Gottlieb-Georg Lindner
Evonik Resource Efficiency GmbH, Brühler Straße 2, 50389 Wesseling, Germany

Martin Wiemann
IBE R&D Institute for Lung Health gGmbH, Mendelstr 11, D-48149 Münster, Germany

Lucia Pavoni, Marco Cespi, Giulia Bonacucina and Filippo Maggi
School of Pharmacy, University of Camerino, via Sant'Agostino, 62032 Camerino, Italy

Roman Pavela
Crop Research Institute, Drnovska 507, 161 06 Prague 6, Ruzyne, Czech Republic

Valeria Zeni, Angelo Canale, Andrea Lucchi and Giovanni Benelli
Department of Agriculture, Food and Environment, University of Pisa, via del Borghetto 80, 56124 Pisa, Italy

Fabrizio Bruschi
Department of Translational Research, N.T.M.S., University of Pisa, 56124 Pisa, Italy

Naif Abdullah Al-Dhabi and Mariadhas Valan Arasu
Addiriyah Chair for Environmental Studies, Department of Botany and Microbiology, College of Science, King Saud University, Riyadh 11451, Saudi Arabia

Cláudia Silva, Frank Simon, Peter Friedel, Petra Pötschke and Cordelia Zimmerer
Leibniz Institute of Polymer Research Dresden (IPF), 01069 Dresden, Germany

Xianghong He, Yaheng Zhang, Yu Fu, Ning Lian and Zhongchun Li
School of Chemistry and Environmental Engineering, Jiangsu University of Technology, Changzhou 213001, Jiangsu, China

Magdalena Aflori, Maria Butnaru and Florica Doroftei
Petru Poni Institute of Macromolecular Chemistry, 41A Grigore Ghica Voda Alley, Iasi 700487, Romania

Bianca-Maria Tihauan
Sanimed International IMPEX SRL, Sos. Bucuresti—Magurele, nr. 70F, Sector 5, Bucharest 051434, Romania

Index

A

Agricultural Pests, 157, 166, 174

Antibacterial, 17-21, 24-25, 27, 29-34, 36-37, 39, 43-44, 51-53, 56, 58-60, 76-77, 96, 127, 137, 165, 180, 186, 196, 198-200, 202, 217, 243-244

Antioxidants, 20, 28, 99, 161

B

Biofilm, 36, 40, 46

Biomolecules, 16, 42-44, 48, 52-53, 56, 58, 61, 98-99, 111, 136

Biopolymer, 80, 233, 235, 241-242

C

Cavitation, 140, 142, 160

Cell Membrane, 27, 35, 48, 55-56, 164, 196

Cell Proliferation, 115, 117, 124-125

Cell Viability, 28, 41, 51, 131, 135, 173, 229, 232, 241

Chitosan, 17, 19, 58, 96, 99, 111, 113, 126-127, 229-230, 233-238, 240-244

Culture Medium, 119, 124, 232

Cyclic Voltammetry, 1, 4, 9, 11

D

De-ionized Water, 142, 144, 152

Dengue, 157, 167, 169, 176, 181, 188-189, 194, 197, 199

Dengue Fever, 188-189, 197

E

Electrical Conductivity, 19, 201, 204-205, 213-215

Electron Density, 209, 211, 234

Energy Consumption, 144, 201

Energy Density, 139-141, 144, 146-149, 152

Energy Dispersion, 98-99, 103, 140, 152

Energy Input, 61, 143-144, 146, 159-160

Energy Values, 204, 209-210, 239

F

Fabrication, 76-77, 81, 136, 181, 219, 228, 243

Fibroblast, 115, 117, 124, 127, 130-131, 135, 170, 230, 238, 243

Flower Extract, 20-21, 23, 25, 27-28, 30, 33-34, 59

Functionalization, 29, 31, 51, 201-202, 215-218

G

Graphene Stacks, 205-206, 208, 210-211, 215

Green Synthesis, 2-4, 8, 11, 17-22, 29-36, 52, 57-62, 75-77, 79, 81, 83-84, 87, 94, 96, 129, 137, 139, 181, 188-189, 198-199, 216, 243

Greenhouse, 61, 172, 185

H

Heat Production, 144-146

Hyaluronic Acid, 115-117, 120, 126-127

Hybrid Nanoflowers, 98-100, 102, 107, 109-113

Hydrophilicity, 122, 230, 236

I

Immobilization, 98-101, 104, 106-107, 110, 112-113, 127, 136

Insecticides, 28, 157, 166-167, 177, 179-180, 182

Ionic Nanocomplexes, 115-116, 121

L

Lipase, 98-102, 105-108, 110-113

Lipopolysaccharides, 27, 53-55

M

Magnetic Stirring, 117, 159

Malaria, 34, 167, 169, 173, 181, 184, 186

Medicinal Plant, 184, 199

Methylene Blue, 24, 28, 35, 79, 81-83, 90-92, 94, 96-97, 127

Morphology, 2-3, 16, 51, 62-64, 75, 77-78, 82, 86, 95, 107, 111, 118, 124, 127, 131, 134-135, 141, 146, 152, 189, 192, 220, 226, 228, 230, 234, 236, 238

N

Nanobiocatalytic System, 102

Nanoemulsions, 157, 162-163, 166, 169, 171-172, 175, 178-179, 182-186

Nanoparticles, 1-6, 8-9, 11, 16-25, 27-37, 42-43, 51-52, 57-60, 62, 67-69, 73-77, 79-100, 102, 104, 106, 110, 112-113, 120-121, 123, 152-155, 162, 170, 176, 181, 186-194, 196-200, 202, 216, 219-223, 225-230, 233-235, 237, 241-244

O

Ovicidal Activity, 188-190, 197, 199

Oxidative Stress, 27, 55, 154, 196

P

Parasites, 28, 136, 157, 167, 172-173, 176-177, 179, 181, 198

Particle Size, 11, 75, 79, 82, 86-87, 89, 94, 101, 131, 133-134, 139-141, 143-153, 155-156, 158, 167, 173, 188, 192-193, 197, 200, 233-234

Phase Transformation, 84, 219, 223-224, 226

Photocatalysis, 79, 81, 83, 90

Photoelectrons, 204, 208-211

Phytochemical, 21, 31, 51-52, 58, 135, 137

Plasma, 18, 131, 137, 161-162, 201, 229-231, 233, 235-238, 240-243

Plastic Container, 118

Polyarginine, 115, 117, 120, 126

Polymer Surface, 230, 232-233, 236-237, 239-240

Probe Sonication, 145-146, 150, 154

R

Raman Spectroscopy, 25, 79, 83, 86, 96, 98-99, 104, 201-202, 204, 206-208, 214-215, 218

Reagents, 2, 38, 94, 117, 130, 189, 216, 220

Room Temperature, 3-4, 17, 23, 28, 31, 39, 100, 116-118, 130-131, 189, 198, 201, 220-221, 225, 230-232

S

Sample Preparation, 63, 78, 139-140, 142-143, 146, 151-154, 215

Scadoxus Multiflorus, 188-189

Silver Nanoparticles, 17-20, 23, 31-36, 58-59, 130, 137-138, 181, 196, 198, 200, 229-230, 233-235, 241-244

Solvothermal Synthesis, 61-62, 67, 75, 77, 95

Sonotrode, 139-140, 142, 144-146, 149-151

Spectrophotometry, 119, 133, 189, 232

Spectroscopy, 1, 3-4, 25, 61, 63, 79, 82-83, 85-86, 96, 98-99, 103-104, 117, 129-130, 147-148, 189-191, 194, 201-202, 204, 206-208, 214-215, 217-218, 229, 231, 233, 235, 238, 243

Surface Roughness, 236-237

Suspension Volume, 122, 144

T

Terminalia Phanerophlebia, 1-3

Thermal Stability, 82, 88, 101, 106-109, 113, 201, 211-212, 214, 229

Thermogravimetric Analysis, 82, 87-88, 202, 204, 211-212

Transesterification, 101-102, 108-110

Transmission Electron Microscopy, 38, 61, 63, 79, 82, 87, 115, 118, 121, 129-130, 138, 189, 192, 232, 234

U

Ultrasonic Dispersing, 139-141, 145-146

Ultrasonication, 4, 140, 144-146, 150, 152

W

Wettability, 229, 233, 236-237, 243-244

Z

Zeta Potential, 1, 8, 11, 115, 117, 119-120, 131, 133, 155, 170, 189, 193-194